STATISTICS OF COAL.

STATISTICS OF COAL:

INCLUDING

MINERAL BITUMINOUS SUBSTANCES

EMPLOYED IN ARTS AND MANUFACTURES;

WITH THEIR GEOGRAPHICAL, GEOLOGICAL, AND COMMERCIAL DISTRIBUTION,

AND AMOUNT OF

PRODUCTION AND CONSUMPTION ON THE

AMERICAN CONTINENT.

WITH INCIDENTAL STATISTICS OF THE IRON MANUFACTURE.

BY R. C. TAYLOR,

FELLOW OF THE GEOLOGICAL SOCIETY OF LONDON; MEMBER OF THE AMERICAN PHILOSOPHICAL SOCIETY; OF THE ACADEMY OF NATURAL SCIENCES OF PHILADELPHIA; AND OF VARIOUS OTHER SOCIETIES IN EUROPE AND AMERICA.

Second Edition,

REVISED AND BROUGHT DOWN TO 1854,

BY S. S. HALDEMAN,

PROFESSOR OF NATURAL SCIENCE AND AGRICULTURE IN DELAWARE COLLEGE, MEMBER OF THE AMERICAN PHILOSOPHICAL SOCIETY; OF THE ACADEMY OF NATURAL SCIENCES OF PHILADELPHIA, ETC. ETC.

PHILADELPHIA:
PUBLISHED BY J. W. MOORE,
195 CHESTNUT STREET.
1855.

KITE & WALTON, PRINTERS.

EDITOR'S PREFACE.

The extensive practical and scientific utility of the *Statistics of Coal*, caused the first edition to be speedily exhausted; and in the year 1851, at the period of the death of the distinguished author, he was collecting materials for a second edition, which have been used in their appropriate place. A large amount of additional matter has been inserted from accessible sources upon *coal*, *iron*, and *commercial statistics*, which, it is hoped, will render the volume useful as a work of reference at the present time.

Although the first edition embraced the entire subject, in all parts of the world, the wants of the Western Continent seem to require one restricted to North and South America, and the West Indies—a scope which has accordingly been assigned to the present edition.

But whilst the details appertaining to the Eastern Continent have been omitted, other portions have been retained which include valuable material belonging to the general subject. Among these the Introduction occupies a prominent place. The omitted portions can always be consulted in the original edition, which must remain a standard work of reference in general libraries; and even had they been retained, the difficulty of collecting foreign statistics, would render it impossible to give them a proper revision.

It has been suggested to us, that the work would have been improved by having the *productive* coal measures distinguished in the shading from the extensive coal formations indicated in the maps; but this would require an amount of investigation and a knowledge of details which the best geological surveys do not give.

It was intended to distinguish the additions in this volume by placing them between brackets, which has been done in a few cases, (as in the chapter on Alabama,) but as Mr. Taylor made free use of these marks in the first edition, and as the new matter frequently entered the tables, the use of them was soon given up. In general the date of the additions will indicate the new matter.

The statistic tables have been extended from the year 1848 (the date of publication of the first edition,) to the present time, as far as the scattered nature of the materials would allow. For these we have been indebted to various state geological reports, and those of the different coal companies, as far as they have been printed; Hunt's Merchant's Magazine, De Bow's Review, American Almanac, New York Mining Magazine, London Mining Journal, Pottsville Miners' Journal, Philadelphia Commercial List; and the daily journals of Philadelphia and New York.

For the maps on the coal formations of Iowa, Wisconsin and Illinois, as also for the geological one of Alabama, we acknowledge our indebtedness to the publications of Dr. D. D. Owen and Professor Tuomey, who have permitted us to make such use of them as we might desire. Also, to Dr. Emmons, for his valuable contribution on the coal formations of North Carolina, to our "Pottsville Correspondent," and other scientific gentlemen; to the Philadelphia Franklin Institute, through Mr. Hamilton, the obliging Actuary, and to the editor of the Commercial List, who have generously given much useful statistical information for this work.

In judging of the value of a coal from its analysis, allowance must be made for contingencies which will materially modify the result. Thus, the analysis may be made from specimens better or worse than the usual average, or from a trifling vein, which does not present the characters of the entire mass. As a general rule, the chemist should visit the mines, select his specimens, and designate the strata from which his analyses have been made; or in cases where the specimens have not been carefully labelled and packed when selected, he will have no sufficient guarantee that they have come from the locality indicated. Having been formerly engaged in the State Geological Survey of Pennsylvania, with Professor Rogers, and in other geological explorations, the editor has been frequently impressed with the necessity of caution in such cases.

It will be found that the particulars of some mines exceed those of others, in many cases equally worthy of more detailed mention. This has arisen from the amount of information and material at hand being very unequal for the different localities. We have endeavoured to treat the subject as impartially as the limited time at our disposal would allow, and being entirely unconnected with any coal interest, we trust this explanation will account for any seeming neglect or injustice in our statements.

The annexed biographical sketch, from the Proceedings of the Academy of Natural Sciences of Philadelphia, for October, 1851, is used with the sanction of its author, Isaac Lea, Esq., L.L.D., &c. &c., who has kindly furnished us a revised copy for this work, and whose long intimacy with Mr. Taylor, has rendered him peculiarly competent for the task.

S. S. H.

Chickiswalungo Iron-works, Columbia,
Pa., 20th September, 1854.

BIOGRAPHICAL SKETCH

OF THE AUTHOR.

[Taken from the Proceedings of the Acadamy of Natural Sciences, of Philadelphia, vol. 5.]

MR. Lea announced the death of RICHARD COWLING TAYLOR, which took place on Sunday morning, the 26th inst., at his residence in Thirteenth street, in his 62d year, after a very short illness; and remarked, that it was very rarely that the members of the Academy had to deplore so severe a loss, as that sustained by the death of their distinguished fellow member Mr. Taylor.

In his particular branch of Geology (economic-geology) he stood pre-eminent, and as a mining engineer, no authority in this country, or, perhaps, in Europe, was superior to that of Mr. Taylor. By early education and association in England, he became versed, in the most thorough manner in these important sciences, at an age when such education usually begins. Hence, his first literary production brought him prominently before the learned world, and he was introduced into literary and scientific societies, where he took an active part. The first work of importance which he published, was one on the Monastic remains of the county in which his father lived as a country gentleman, and on whose property there was a noted Anglo-Norman ruin. It was this probably that induced Mr. Taylor first to turn his attention to this branch of knowledge, and the result was the "Index Monasticus, in the ancient kingdom of East Anglia," published in 1821, in one vol. folio,* which at once gave him a reputation for thorough investigation and exactness, which noted all his after works, and which has rarely been excelled. This work was received with so much favour, that Mr. Taylor, was induced at the request of the publishers, to undertake that thorough and learned work which he called a "General Index to Dugdale's Monasticon Anglicanum," in 1 vol. folio, with plates and maps, which was published in 1830. This took Mr. Taylor two years to complete, and was said to be so perfect as to require nothing further to be added in regard to it. In his profession, he had the great advantage of a most thorough and complete education, and he was associated in some engineering engagements

* At a public sale his private copy, with some notes, brought £30.

with the late Wm. Smith who has been considered as "the father of British Geology," on account of his having been the first geologist in England who attempted to classify the rocks of that country, by their characteristic fossils; and who was said to have preceded MM. Cuvier and Brongniart in that important step, which gave such an impulse to this great branch of human knowledge. Under such auspices, and by his assiduous application, Mr. Taylor made the most rapid advances in the art of Mining and the science of Geology. The early relations of this intercourse ripened into a friendship and sincere mutual regard, which lasted through life.

With such acquirements he was soon called into active employment, and we find him engaged, for a time, in the important ordinance survey of England; and he was also employed by the "British Iron Company," whose extensive and valuable property in South Wales, he investigated and reported upon. That portion of the Ordinance Survey which he executed, was finished in a most masterly manner, and his drafts were of the most exact and perfect kind. His report of the topography and geology of the mineral lands of the British Iron Company, were so admirably executed, that the Geological Society of London published the map and descriptive parts in its Transactions. In connection with this, he executed a model in plaster of that part of Wales,* which received so much approbation, that the Society of Arts awarded to him their *Gold Isis Medal,* which is now in possession of his family, and he had the pleasure of knowing that Sir Francis Chantry had ordered a copy to be made for Dr. Buckland. Subsequently to this, he was engaged for some years, in England, in the examination of various mining properties, after which he was induced to come to this country, and reside in Philipsburg, Pennsylvania, where he remained four years. He afterwards removed to Philadelphia, for the purpose of seeking that employment in his profession, in which he was so well qualified to excel. Previous to this, however, he was engaged in the survey of the Blossburg District, and the line of railroad which he completed, and made an extensive and able report in 1832.

It was after this time, in 1834, Mr. Lea first became personally acquainted with Mr. Taylor, which acquaintance immediately grew into a friendship, which increased through life. Shortly after this, Mr. Lea had it in his power to have Mr. Taylor placed in charge of the exploration of the extensive coal and iron property of the Dauphin and Susquehanna Coal Co., in Dauphin county, Pa., in which Mr. Lea had a large interest. Here Mr. Taylor remained about three years, and developed the mineral resources of this extensive mineral district, to the entire satisfaction of the Board of Directors. The whole of the lands embraced 42,000 acres, in a rugged, mountainous district, which required an experience and perseverance which few men had more of than Mr. Taylor. The result of this great labour was an elaborate report, of 187 pages in 8vo., together with about 150 maps, drafts, surveys and sections, which are invaluable to the Company, and in whose possession they now are. In connection with this, during a period of cessation, in this country, of activity in such works, he employed himself in the execution of a model of this part of the coal basin and its

* This was the first model of the kind executed in England.

surrounding mountains, which occupied him many months. This subsequently became the property of the Dauphin Company. It embraces in length about 45 miles, and in breadth 15 miles of the district it represents, and is about 14 feet long. It is a complete geological and topographical representation of this important district, and would be alone a monument to a man of science, if he had never executed any other labour.

As soon as the mining interests of the country had become relieved from the pressure which had prevailed for a few years, he again was called on to explore and investigate many mineral districts, connected with the working of gold, silver, lead, copper, coal, asphaltum, &c. Most of the reports of these were published, and it is believed, that in every case, they were so correctly executed as to leave no doubt in the minds of those who employed him, as to his judgment, his candor, and his scrupulous representations of that which he was employed to examine. It was one of the characteristics of Mr. Taylor, as all his intimate friends are perfectly aware, that his openness and frankness were such as to induce him never to hesitate to express a candid opinion, or to make known a fact, however much it might be against his own interests. It was this which induced the most unbounded reliance, among his friends, on his representations. Beside the numerous engagements he had in various States, he frequently had calls to examine important mines in other parts of America; the copper mines of Cuba, the gold mines of Panama, the asphaltum of New Brunswick, &c. In Cuba he was employed to examine and report on the vein of asphaltum near Havana, of which he published an account in the Transactions of the American Philosophical Society, with a plan and section. The last work of this nature he was engaged in, was the examination of the injected vein of asphaltum at Hillsborough, in the Province of New Brunswick, which is now in litigation. His testimony in this case, as taken down, and since published, is a specimen of such thorough knowledge in his profession, such clearness, exactness and completeness, as to be worthy of all praise. It should have a place in all geological libraries. He was greatly interested in this singular litigation, which seems, strange as it may appear to geologists and mineralogists to depend on the decision of a jury, whether an injected vein of asphaltum be not a seam or bed of bituminous coal, belonging to true coal measures! Nothing could be more clear, or more to the point, than Mr. Taylors evidence to the contrary. While in the examination, of various districts, in his professional employment, he carefully noted every fact connected with general geology and palæontology, and the results were generally given in the form of papers to scientific bodies, and published in their Transactions. These will be found in various learned transactions in England and this country—particularly in those of the American Philosophical Society, the Academy of Natural Sciences, and the Geological Society of Pennsylvania, on this side of the Atlantic. They all bear internal evidence of a philosophic mind, schooled in the consideration of philosophical facts.

Notwithstanding what has been said in regard to all these labours of a most industrious life, the reputation of Mr. Taylor will rest chiefly on a work which has not yet been mentioned—his "*Statistics of Coal,*" published in this city, in 8vo. pp. 754, in 1848. It included the geographical and geological distribution of mineral combustibles or fossil fuel, as well as notices of localities of the various mineral bituminous substances employed in arts and manufactures, illustrated by maps and diagrams, embracing from official reports of the great coal producing countries, the respective amounts of their production, consumption, and commercial distribution, in all parts of the world.

The execution of this work had engaged Mr. Taylor's time, not necessarily devoted to the practice of his profession, during many years of his life. His heart was set upon the completion of it, and when approaching to a conclusion it drew his mind from all other pursuits. While it was going through the press, he became so ill that, for many weeks, his physician and his family had little expectation of his living to see it completed. During this period many of the sheets had to pass through the press without his inspection, which fact naturally produced some errors. When the work reached the hands of those interested in the statistics of coal, its geology, and its geological distribution, it was received with the most entire satisfaction. His intimate friends were the first to congratulate him on the work he had achieved, and the criticisms of the press soon followed with their share of approbation.* Dr. Fitton, the distinguished geologist, reviewed it in the Edinburgh Review, and gave to its author the credit he so well deserved, of which the following is a single paragraph.

"The inquiries of the author have been extended, with marvellous industry and perseverance, to every part of the globe; but as might be expected of an engineer residing in America, the coal tracts of that country naturally occupy a large portion of the work. As these are probably less known, to most of our readers, than the coal producing states of Europe, while they are beyond all comparison the greatest depositories of coal in the world—affording to that fortunate region the prospect of almost unbounded wealth—we shall confine our attention to this part of the work. But our readers may be assured that the author's account of other countries gives equal proofs of his diligence in collecting information."

The *London, Edinburgh and Dublin Philosophical Magazine and Journal of Science*, in its notice of this work, says, "comprehensive as the title of this work appears, it does not convey a just idea of its scope, or the extent of the subject-matter. Did its title stand, '*Coal the civilizer;* its natural history, productions and applications,' it would perhaps convey to the casual reader a more just idea of the object and contents of the work." * * * "A long and intimate practical acquaintance with mines and mining operations in different parts of the world, had necessarily led him to amass a great quantity of material; the value of which, as a constant object of reference for his own use, led him to feel the utility of a digested and methodized arrangement of those materials, in a permanent shape, for the use of others. But there is found, throughout these pages, a pervading spirit beyond that merely materialistic and dry one, which the title would indicate, and which the professional engagements of the author might have led us to anticipate. We perceive, impressed on every section, the idea *not* of coal the mere *wealth producer*, the mere material instrument of the hu-

* That part relating to Europe was chiefly translated into German by Von Hauer, and published in Vienna.

man animal, but of coal as an important agent in promoting civilization. It is in the same spirit, and imbued with the same everywhere pervading high moral sentiment, that the author more than once calls attention to the vastly greater importance of iron than of gold and silver. We cannot conclude without cordially recommending this work to the attention of our readers. While it will be an invaluable book of reference to every future inquirer into the numerous economic questions connected with our most important industrial operations, and manufactures, and into the great social questions arising out of them, it will form an indispensable part of the library of every geologist."

The *London Athenæum* in its notice says, "The work of Mr. Taylor will command attention, and become standard as a reference; especially as it is the only one which endeavors to concentrate the knowledge diffused through so many channels, and often attainable only in the countries to which the statistics refer."

The press in this country did not withhold its proper appreciation of a work, so important to the great coal and iron interests this side the Atlantic.

In the notice in Silliman's Journal, it is said, "It is a sufficient guarantee for its completeness and accuracy on all points on which it touches, that it received, before publication, the highest and most unqualified praise at one of the meetings of the American Association of Geologists and Naturalists." * * * "An examination of the volume, now that it has appeared from the press, gives us a still higher opinion of the talents and industry of its author, and the great value of his labours," &c. The Journal of the Franklin Institute stated it could scarcely have been thought possible that one individual, especially in this country, could have collected together such a mass of facts, and made of them so well arranged and so delightful a book. Hunt's Merchants' Magazine says, "We venture to say that, on no kindred subject, has a more complete or perfect treatise ever been produced."

Beside the proficiency which Mr. Taylor had acquired in economic geology, he had devoted himself much to theoretic geology, and his knowledge of the various formations, which make the sum of the geological series, was rarely excelled by his colleagues. He had applied himself more particularly to the strata connected with the coal formation, and he was the first person, as Prof. Silliman stated to a meeting of the American Association of Geologists, who had referred the Old Red Sandstone, underlying the coal of this State, to its true position, corresponding with its place in the series of European rocks. He was unwilling to engage in State surveys, but his aid was sometimes required to assist in those particular branches in which he so much excelled. With this view he, for a short time, lent his services to the New York State Survey.

In the year 1832, he was elected a member of this Academy, and in 1846, a life membership was conferred upon him, "as a mark of respect and a just appreciation of its means of usefulness derived from him." His attachment to the Academy increased with his advancing years. He frequently made donations to it of specimens and books. Very recently, he presented a most elaborate geological table in manuscript, coloured to represent the different strata, and combining the analogous nomenclature of various systems. This most valuable donation was made to the Academy on condition of its never being taken from the library, and it cannot fail to be most useful for consultation and reference.

At the time of his death, Mr. Taylor was engaged in preparing a paper, for the Journal of the Academy, on the fossil plants which he had discovered in his recent visit to New Brunswick. The fossil fishes which he also discovered there, he left with his friend, Prof. Agassiz, who was to describe the new ones for him. All these Mr. Taylor intended should be deposited in the collection of this Academy, to which he had already added many valuable specimens.

Mr. Taylor was the third son of Samuel Taylor, of New Buckenham, in Norfolk, England, and a descendant of Dr. John Taylor, the author of the Hebrew Concordance. He was born at Hinton, in Suffolk, Jan. 18th, 1789. His brothers and cousins were men generally distinguished by their great literary and scientific acquirements. His younger brother, Edgar Taylor, was a distinguished member of the legal profession in London, and an accomplished scholar. He was the author of several works, and remarkable for his numerous learned reviews, published in the most prominent periodicals in Great Britain. His cousin, Richard Taylor, was the well-known and able editor of the Philosophical Magazine, which has been the leading scientific Journal of England for the last twenty-five years. John and Philip are highly distinguished as mining engineers.

The great services Mr. Taylor had rendered science, have been acknowledged, by his being made a member of the principal Societies in England and this country, which embraced those branches of knowledge which he cultivated. He was elected a member of the Geological Society of London, and of the Society of Civil Engineers, of that city. In this country he was a member of this Academy, as before mentioned; of the American Philosophical Society; of the Geological Society of Pennsylvania; of the American Association of Geologists and Naturalists, of the Franklin Institute, &c. &c.

In a rapid survey of Mr. Taylor's scientific labours, it would be difficult to give any thing more than a brief and imperfect list of his writings. In this sketch will not be introduced his professional reports, which occupied the chief part of his life, and which were generally executed in such a systematic and perfect manner, as to remain models, worthy of imitation by all engaged in such works. Whether his beautiful map of the *Ordinance Survey*, executed in 1813—14, was the first, is not certain, but it seems to bear the earliest date. He subsequently published, in the Transactions of the Geological Society of London, "Notice of two Models and Sections of about eleven square miles, forming a part of the Mineral Basin of South Wales, in the vicinity of Pontypool," (1830.) "On the Crag Strata at Bramerton, near Norwich," (1823.) "On the Alluvial Strata, and on the Chalk of Norfolk and Suffolk, and on the Fossils by which they are accompanied," (1823.) In the Magazine of Natural History he published, in 1829, a paper called the "Progress of Geology," which was followed, in 1830, by another, the "Introduction to Geology," which was succeeded by "Illustrations of Antedeluvian Zoology and Botany," and "Notes on Natural Objects observed whilst staying in Cuba." An article entitled "A description of a Fossil Marine Vegetable, of the family Fucoides, in the Transition Rocks of North America, and some considerations in Geology, connected with it, with an engraved specimen of the Fucoides Alleghanienses." "A Description of an Ice Storm at Philipsburg, 1832." "On the Geology and Natural History of the North-eastern extremity of the Alleghany Mountain Range in Pennsylvania. These papers were illustrated with many beautiful drawings and sections. In the London and Edinburgh Phil. Mag., March, 1837, occurs a notice of a vein of Asphaltum Chapapote, called in the vicinity of Havana, bituminous coal. In the Philosophical Mag., London, an article "On the Carboniferous Series of the U. S. of North America, as to the actual position of the *Old Red Sandstone* in America." His first paper published in this country, was, it is believed, in the American Monthly Magazine, in 1832, entitled "Section of the Alleghany Mountains and Moshannon Valley, in Centre County, Pennsylvania." In the Trans. of the Geological Society of Pennsylvania, followed others of great interest, "On the Geological position of certain beds, which contain numerous Fossil Marine Plants of the family Fucoides, near Lewistown, Mifflin county, Pa.," (1834.) "On the relative position of the Transition and Secondary Coal Formations in Pennsylvania, and description of some transition or Bituminous, Anthracite, and Iron ore beds, near Broad Top Mountain, in Bedford county, and of a coal vein in Perry county, Pennsylvania, with sections." "Notices of the evidences of the existence of an ancient Lake," which appears to have formerly filled the Limestone Valley of Kishacoquillas, in Mifflin county, Penna." "On the Mineral Basin or Coal Field of Blossburg, on the Tioga River, Tioga county, Penn." "Memoir of a section passing through the Bituminous Coal Field near Richmond, in Virginia." "Review of the Geological phenomena, and the deductions derivable therefrom, in 250 miles of sections, in parts of Virginia and Maryland. Also, notice of certain Fossil Acotyledonous Plants in the secondary strata of Fredericksburg," (Vir.)* In the Transactions of the American Philosophical Society he published "Memoir of the Character and Prospects of the Copper Region of Gibara, and a Sketch of the Geology of the N. E. part of the Island of Cuba." "Notice of Fossil Arborescent Ferns of the family Sigillaria and other Coal Plants, exhibited in the Roof and Floor of a Coal seam in Dauphin county, Penn." "Notice of a Vein of Bituminous Coal (Chapapote) recently explored in the vicinity of the Havana, in the Island of Cuba." (This was jointly with Mr. Clemson.) In Silliman's Journal he published, "Notes respecting certain Indian Mounds and Earthworks in the form of Animal Effigies, chiefly in the Wisconsin Territory, U. S., with Plans and Illustrations." "Notice of a Model of the Western portion of the Schuylkill, or Southern Coal Field of Penn., in illustration of an Address to the Association of American Geologists, on the most appropriate modes for representing Geological Phenomena," (with illustrative sections.) In the Journal and Proceedings of the Acad. Nat. Sciences, "Table constructed from a few Meteorological Notes, chiefly in regard to the daily

* Mr. Taylor was the first to identify the Fredericksburg sandstone with "the oolitic group of Europe," and in this memoir (Trans. Geol. Soc. of Penna., Vol. I., p. 325, 1835,) he figures the genera of fossil plants of Fredericksburg, assigning them all to that formation.

temperature of noon, on the East Coast of the Isthmus of Panama, Port Royal, in Jamaica, and on the return voyage to New York, for the month of October, 1849." "Substance of Notes made during a Geological Reconnoisance in the Auriferous Porphyry region next the Carribean sea, in the Province of Veraguas and Isthmus of Panama," 1851, with maps. Also, a pamphlet on the Anthracite and Bituminous Coal in China, and several articles in Hunt's Merchants' Magazine and in some Reviews. At the time of his sudden illness, he was engaged in a paper entitled, "On a Vein of Asphaltum of Hillsborough, in Albert county, Province of New Brunswick," which he has left in an unfinished state, but which was so far complete as to justify its publication in the Journal of the Academy of Natural Sciences, 1852.

CONTENTS OF THE INTRODUCTION.

SECTION I.

SECTION II.

SECTION III.

SECTION IV.

PLAN OF THIS WORK.

THE growing demand for the species of practical information which it has been our object in the following pages to concentrate, has often suggested itself to the author, and doubtless to numberless others. Perhaps in no country have more frequent inquiries been made in relation to COAL; to its infinite varieties, adaptations and modifications; its innumerable depositories and its geographical distribution, than in the United States of America.

This desire, probably, originates in the circumstance that in no country has such rapid progress been made in the development of mineral fuel, not only for all domestic purposes, but as a powerful agent in every department of manufacturing industry; notwithstanding that enormous and almost unbroken forests still overshadow the land. The increasing demand and corresponding supply, the rapid expansion of the field of industrial operations, have no doubt awakened this solicitude for information, local, general, statistical, commercial and scientific, on the subject of coal.

Acting under this impression the author has sought and gathered together the materials—a great number at least, to remedy the deficiency of which we speak. His design, at the outset, was limited to the collection of such coal statistics as seemed sufficient for his private guidance. As in all labours of this description, the materials, during the progress of the undertaking, accumulated to an extent far greater than was anticipated. An extended arrangement led to greatly increased labour. The sources of information as regards foreign countries, being remote, its acquisition is necessarily uncertain and tedious: in fact it has no limit, for every day furnishes new facts to be registered. The process never ends, because the elements are inexhaustible. We are reminded, however, by the bulk of the matter on hand that we have reached a point at which we may consign the work to the press.

Preparing these pages in the United States, we are not unaware of the disadvantages which result from the want of access to many official European documents, and of reference to minor authorities such as rarely find their way into American libraries. We may, in some degree, counteract these deficiencies by communicating to European inquirers a great amount of information which our position has enabled us to acquire in America. These persons cannot but contemplate with interest the enormous extent of the

North American coal-fields, whose very existence, scarce a quarter of a century back,* was unknown, even on their actual sites.

Of the surprising impulse to the interests of the New World which has been communicated by this recent knowledge, this newly acquired power; of the influence it has manifested in many of the commercial and on all of the industrial departments; of the moral consequences which are perceptible in a thousand forms, we shall hereafter submit abundant proofs. It will be much more difficult to speculate as to the position to which these combined elements of prosperity may conduct us in the next quarter or half century. We draw the most sanguine inferences with relation to the future, because the experience of the past twenty-five years fully justifies such flattering anticipations.

Something further yet remains to be said in relation to the objects contemplated in this volume, and of the several matters to which we have given a place therein. Our range would have been but narrow had we limited the investigation to mineral coal alone. It is well known that vast deposits of combustible substances have been denominated and described as coal, which the lights of science now show belong to a more recent class, and to a variety of geological ages or epochs. We refer to the brown coal or lignite class, so abundantly distributed.

In a large portion of Europe, such as the Austrian, Belgian, French and Prussian dominions, the distinction is perfectly well understood, and all official mining statistics are, in these countries, uniformly arranged under their appropriate classification. In many cases where errors have prevailed, we have been enabled to correct them by the aid of recent geological investigations. Still, modern science has not yet penetrated every where. There remains, at numerous, but rarely visited points, vast fields of so-called coal, whose true geological age we have yet to learn. For the present, therefore, we are unable to class these combustibles either with the true coals as the older series, or with the tertiary lignites as the newer, or with any intermediate deposits. This being the case, it was obviously inexpedient to exclude the LIGNITES from our pages, independently of their intrinsic value as combustibles. Brown coal is a valuable substitute for the older coal where there is a scarcity of the superior variety, as we shall have many opportunities of showing.

In like manner, while describing the lignites, PEAT seemed to demand a proportionate share of our attention, and to claim a place in our columns. The transition from one condition of these combustibles to another is oftentime so imperceptible that they seem to have almost equal claims on our notice. In its remarkable diffusion over the northern hemisphere where artificial heat is so indispensable, and where timber and other descriptions of fuel are so little abundant, turf or peat forms a substitute of inestimable worth. In its adaptation to numberless useful purposes, such as the manufacture of iron, the production of gas, &c., modern science has shown that it possesses qualities which heretofore were but little suspected. Thus, it will be seen, our list comprehends a large series of valuable products; extending upwards from carbonized peat at one extremity to hard coals and compact anthracities at the other.

* [Counting from 1823 to 1847, the date of the author's publication.—H.]

So closely do some bituminized coals approach to the mineral bitumens, some of which have even been denominated coal, as those in the West Indies and South America, that we have found it advisable to include the BITUMINOUS AND RESINOUS MINERALS.

We did not contemplate, in preparing this work, to enter extensively into the important topic of the statistics of IRON, but we have found it so interwoven with matters essential to our main subject, that a considerable mass of information has been necessarily incorporated in our pages, where will be found the latest estimates and returns of the amount of manufactured iron in all the principal producing countries, illustrated by a diagram of their respective proportions.

Explanatory tables of the current moneys, weights and measures of all the leading nations; a variety of statements of commercial facts; details of the respective tariffs, customs and international regulations, in relation to coal; the progress of railroads and canals; of steam power and navigation, and a vast series of analytical tables, besides the maps and diagrams, also occupy portions of the present volume. Among other duties, that of bringing to uniform denominations and a common standard the weights, measures and currency of so many nations, is by no means the lightest. The principal results in our tables have been calculated in the three standards of France, England, and the United States.

Where the range of inquiry is so wide, the number of documents which we have had to investigate is correspondingly large. We have endeavoured to designate our authority for every material fact which we have adopted. This recognition, we conceive, is not only in strict justice due to those authorities, but it bestows the sanction of their names, and the weight of their testimony to every page and paragraph of this volume.

Let us add further, that the practice is attended with a convenience which every inquirer can appreciate,—the enumeration of standard authors and the direct reference to their pages. The whole series thus forms, in the aggregate, a copious catalogue of statistical and scientific authorities. The Index, we cannot but think, will be found to concentrate a vast mass of information which has heretofore been dispersed through hundreds of volumes in different languages, and constitutes of itself an epitome or condensation of the entire work.

It can scarcely be expected that in so new and extensive a country as the United States of America, any organized system is in effective operation for determining the amount of coal yearly raised there. In regard to anthracite, the great avenues from the mines to tide water admit of exact returns of the quantity annually transported, and means exist, in fact, of ascertaining, through the returns of the mining establishments, the true yearly production.

Not so with the production of bituminous coal in the interior. Of this we are wholly uninformed, and the area of the coal-fields is so large, that it seems futile to hazard even the roughest calculation. In 1840, an official attempt was made to acquire that information through the instrumentality of the Census Act, but it proved, as might be expected, a decided and acknowledged failure. In 1845, the Secretary of the Treasury, in conformity with the direction of the Senate, made a report of 419 pages, 6th

January, in relation to the statistics of the United States. From no county or state in the Union was a single return obtained as to the coal mines. During the same year, the Secretary of the Treasury, pursuing the inquiry, with reference to the settlement of the proposed tariff, issued circulars throughout all the states, asking information, among other statistics, as to the mines, their produce and prices. His report thereon of 957 pages, dated 3d December, 1845, elicited no useful result on this head, nor a single return relative to coal.

The wide distribution of property in America is unfavourable to the collection of such statistics. The process must, at all times, be unpopular, and the results extremely uncertain. This species of investigation savours too much of scrutiny into the private concerns of men, and is unsuited to the spirit of republican institutions.

INTRODUCTION.

SECTION I.

INTRODUCTORY SKETCH.

We take for granted that every one who may chance to peruse the summary of *Statistics of mineral fuel* which we have embodied in the present section, will be impressed with the immense importance of those substances, particularly as developed of late years; how vastly enlarged the area and bulk of their production in all countries; how essential they now are to the comfort of the human family; how much they have done towards the extension of the useful arts; how gloriously they have aided the progress of invention and improvement; how mighty are the results which have followed their increased application! For ourselves, we may remark, that during the investigation into the geographical distribution of coal and the subordinate combustibles, nothing has struck us more forcibly than the abundant supply with which Providence has furnished the inhabitants of our globe, particularly in its northern hemisphere. We were astonished at the almost numberless positions where mineral fuel is attainable; especially in North America and Europe. With very inadequate guides at the outset, we have brought together an enormous mass of geological and statistical details, which exhibit an amount and variety of fossil combustibles which far exceeded our original expectations. We have seen how recent is the knowledge of the existence of immense regions occupied by coal, and that every year new positions, new deposits, become known to the traveller, or are demonstrated by the geologist. Through them, and the enterprise of the miner, a rich store of intelligence has been acquired, yet much remains behind. We are yet in the infancy of our knowledge as regards vast areas of country. Busy as the geologist has been during the last half century, how much is yet to be investigated; how wide the space yet untrodden; how ample the fields yet open to the scientific explorer!

The last quarter of a century has, more especially, been prolific in the discovery of the sites of useful mineral combustibles, and in the extended application of their products to the service of the community. Man has not only been taught increased facilities in adapting them to the useful arts, but practical science has apprised him of the great value of substances heretofore accounted of little worth, yet inexhaustibly abundant, and almost every where within his reach. He has acquired, for example, many new facts relative to the value of *peat*, hitherto among the humblest of the combustibles, yet the almost universal production of cold or temperate climates, and of regions which are entirely incapable of producing a growth of timber or of the larger plants. Independent of its applicability to the usual domestic and agricultural purposes, he has seen that it can be successfully applied for gas-lighting, for steam engines, for evaporation, and for every branch of the iron manufacture, commencing with smelting in the high furnaces, and ending with the most delicate manipulations practised in the working of steel. Thus, in compensation for the absence of the supposed superior descriptions of fuel, coal, for instance, nature has been bountiful of another, where most needed; and one, too, which, unlike fossil coal, is reproductive; always renewable and renewing. The fear, therefore, entertained by some theorists, that the earth will be exhausted of its mineral combustibles, may be alleviated by the contemplation of that enormous supply of vegetable fuel, which prevails where eventually it will be most needed.

It would be no difficult task to show in figures how vastly more profitable is the application of labour in the mining and working and transporting of coal, than in that of the precious metals. The annual production of all the gold and silver mines of North and South America was estimated by Baron Humboldt at £9,243,000, and at present at less than £5,000,000.* Now, the value of the coal produced annually, in Great Britain alone, is computed at near £10,000,000 at the pit's mouth, and at from £15,000,000 to £20,000,000 sterling at the places of consumption. At the same time, the value of the iron, brought into a manufactured state through the agency of this coal, is £17,000,000 more. We shall enter more particularly into this subject in a future page. We cannot but mark also the superior character and condition of the inhabitants of the coal producing and consuming countries, such as those of the northern hemisphere, especially since the introduction of steam power, to that of the people of the southern and tropical latitudes, to whom coal has either been wholly denied, or is not applied to any use. The industry, activity, moral culture and intelligence concentrated around any of the great depositories of coal and iron in the temperate regions—in the anthracite districts of Pennsylvania, for instance—have no parallel in the countries from which such treasures have been withheld.†

* [Since the above was written a vast change has taken place in the production of Gold by the discovery of California.]

† Let us be permitted to cite a very interesting illustration of the foregoing remark, from the state of Pennsylvania, just referred to. In 1825 commenced the first mining operations in Schuylkill county, and the first concentration of settlers from all countries. In 1841, the central town of Pottsville, originating at a later date than we have quoted, contained the following establishments for the education of the children of the miners and newly settled residents.

The two important mineral substances, coal and iron, have, when made available, afforded a permanent basis for commercial and manufacturing prosperity. Looking at the position of some of the great depositories of coal and iron, one perceives that upon them the most flourishing population is concentrated—the most powerful and magnificent nations of the earth are established. If these two apparently coarse and unattractive substances have not directly caused that high eminence to which some of these countries have attained, they, at least, have had a large share in contributing to it.

In preparing this volume, our investigations have in great measure been directed to one only of these simple mineral substances, coal, although the iron has not altogether been lost sight of. We will take the liberty of terminating this passage, in the words of M. Aug. Vischers.

"Coal is now the indispensable aliment of industry; it is a primary material; engendering force; giving a power superior to that which natural agents, such as water, air, &c., procure. It is to industry what oxygen is to the lungs, water to the plant, nourishment to the animal. It is to coal we owe steam and gas; it replaces, in the workshops and the domestic hearths, the charcoal which had become too costly. Under the last head, in our northerly latitudes, it is destined always to acquire increasing and more general use. The employment of coal will henceforward be no other than a question of cheapness; and, in the present age, the first interest of industry is, above all, to see ameliorated the ways of communication; to lower the tolls upon the routes and the canals. If custom-house officers still oppose shackles on manufactured products, they lower their barriers for the passage of the raw material."

PROPORTIONATE AREAS OF COAL LAND IN EUROPE AND AMERICA.

The following table* shows the relative magnitude of the principal coal producing countries, and their respective areas of coal land, together with the proportions which they severally bear to each other. Those of France and Spain are considerably less than the actual amount. Coal occurs in almost every principal subdivision of Spain, but we have only included the Asturias region.

Hence, as regards European countries, Great Britain takes the first rank: Belgium, as regards territorial proportion, occupies the second rank, although in relative coal area she is the least of the four. Pennsylvania, in respect of territorial proportion, is higher than any of these, being relatively one-third: but in absolute area of coal formation, the four eastern colonies of British America united exceeds them all, being larger than that of Great Britain, France, Belgium and Spain conjoined. This table is not strictly perfect; since we

Six private schools, numbering 479 pupils; eight public schools, numbering 472 pupils—the annual average expense for each pupil being only $5.82; eight Sunday schools, numbering 1137 pupils; teachers, 166; total, 2254, with a library of 1659 volumes.

It is but just to add, that of the Catholic school, comprising 439 of this number, every individual had taken the pledge of total abstinence from intoxicating drinks *for life*.—Notes from the Miner's Journal of Pottsville, January 1, 1842.

* See next page for table.

possess the areas of the concessions only in France; and, in Spain, only the single coal region of Asturias. We add the areas of Prussia and Austria, but cannot state the proportions of coal formations therein. The American area of coal is nearly three-fourths of the whole amount in our table.

Countries.	Entire area each country. Sq. miles. English.	Area of coal land. Square miles.	Proportions of coal to their whole areas.	Proportions, relative parts of 1000 of coal areas.
Great Britain, Ireland, Scotland and Wales,	120,290	11,859	1-10	64
Spain, (Asturias region,)	177,781	3,408	1-52	18
France, (area of fixed concessions) in 1845,	203,736	1,719	1-118	9
Belgium conceded lands,	11,372	518	1-22	3
Pennsylvania, U. States,	43,960	15,437	1-3	84
British Provinces of New Brunswick, Nova Scotia, Cape Breton, and Newfoundland,	81,113	18,000	1-4½	98
Prussian Dominions,	107,937			
Austrian Prov. containing coal or lignite,	150,000			
The United States of America,	2,280,000		1-17	
The twelve principal coal producing States,	565,283	133,132	1-4	724
		184,073		1000

PROPORTIONATE AREAS OF COAL FORMATIONS IN THE UNITED STATES.

The table we here add will be observed with great interest, on account of the enormous breadth of coal formations in the United States. There are yet several coal producing States not enumerated, of which we possess very imperfect information.

States.*	Ar. of the States. S. A. Mitchell. Sq. miles.	Coal areas. Sq. miles.	Proportion of coal.	
1. Alabama,	50,875	3,400	1-14	bituminous coal,
2. Georgia,	58,200	150	1-386	about the same amount in N. Carolina,
3. Tennessee,	44,720	4,300	1-10	
4. Kentucky,	39,015	13,500	1-3	
5. Virginia,	64,000	21,195	1-3	
6. Maryland,	10,829	550	1-20	
7. Ohio,	38,850	11,900	1-3	
8. Indiana,	34,800	7,700	1-5	
9. Illinois,	59,130	44,000	3-4	
10. Pennsylvania,	43,960	15,437	1-3	bituminous and anthracite.
11. Michigan,	60,520	5,000	1-20	
12. Missouri,	60,384	6,000	1-10	
	565,283	133,132	nearly 1-4th of 12 States.	

In the persuasion that the diagram form always conveys more accurate impressions than mere tables which embrace a number of figures, we have prepared the annexed diagram, showing the Coal Areas of various countries, and the best illustration of the preceding statement. The details of the coal areas in the United States of America, follow in the next table.

* We have adopted, in the column of square miles, the areas of Mitchell, published, 1836, rather than those of Darby or McCulloch.

Diagrams of the superficial Coal Areas of various Countries.

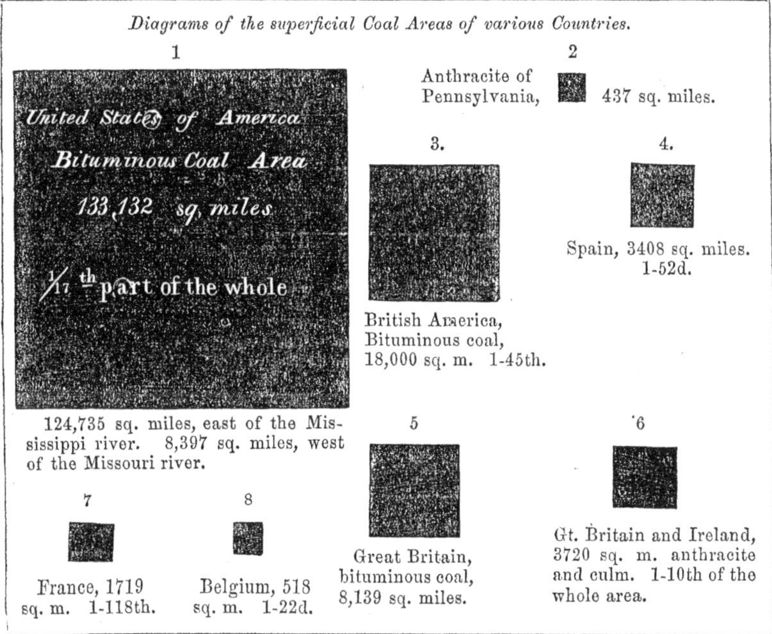

The majority of these States show a far greater proportion of coal than in those of Europe. We omit from the above table some detached coal areas in Arkansas, Missouri territory, Massachusetts, and Rhode Island, respecting which our information is incomplete.

GENERAL SUMMARY OF COAL STATISTICS.

We present, in this place, a comparative view of the coal operations of the larger coal-producing countries of the world. We should have preferred to have arranged the results simultaneously, in all of these countries; but as the dates of the latest returns are not, generally, contemporaneous, we are compelled to a slight deviation from an arrangement otherwise desirable. In general, we adopt the year 1845, for the purposes of comparison.

EUROPE.

Belgium.—In 1844–5 there were in full operation 212 mines, and not in work 97 others, making in all 309 mines, comprising 540 coal-pits in operation and construction; and employing 38,500 miners, and 500 steam-engines, of an aggregate force of 22,841 horse power. The product of their labour was 4,445,240 tons, which were returned at the value of 39,844,191 francs, at the places of extraction; equal to $7,689,900, or to £1,660,000 sterling. In 1845, the quantity raised was 4,960,077 tons.

Prussia.—In the year 1840 there were about 752 mines or pits of coal, anthracite and lignite in operation. These employed 24,024 miners, and produced 3,245,607 tons, whose value is given in the official returns at 19,687,704 francs, equivalent to $3,806,289, or to £793,860 sterling.

In 1844, the four coal provinces of Prussia produced 3,650,000 tons, of the value of 22,500,000 francs. Three German States of the Zollverein yielded 250,000 tons of the value of 2,250,000 francs.

The Prussian collieries in 1844 employed 25,000 miners:—these returns appear to be incomplete.

FRANCE.—During the year 1845, there were, according to official documents, 449 coal mines worked and unworked; employing 30,778 miners, and producing 4,141,617 tons of mineral fuel. Their value at the pit's mouth was 39,705,432 francs = $7,663,000, or to £1,603,106 sterling.

The average quantity raised in each of these three kingdoms in 1845 was remarkably similar; but there is a material difference in the value assigned to the coal at the point of production.*

GREAT BRITAIN.—As the details of the production and distribution of the coal in this country (except as regards the coasting trade and foreign exportation), are not officially registered, as in the Continental States we have previously cited, an exact comparison with them can scarcely be instituted. We may state, however, that about 1845–6, the current estimate of the total production of coal in the British mines was thirty-one and a half millions of tons; whose value at the place of extraction was considered to be £9,100,000 sterling,† = 232,000,000 francs, or $44,000,000 annually.

We have before us, however, another statement, in which the production for the year 1845, is rated as high as 34,754,750 tons; of which one third was exported or shipped coastwise, and two thirds were consumed in the interior. The value assigned is £9,450,000, equal to $45,738,000.

AUSTRIA.—The fifth European government in whose provinces coal or lignite abounds, and in which there has been a considerable increase in the extraction of those substances of late years, is the Austrian Empire. We are, however, unable to institute, with accuracy, a comparison with those countries we have just cited. The provinces which contain coal and lignite comprise an area of about 150,000 English square miles, but how much of this is covered by coal formations does not appear to be determined, except partially. In 1845, there were only 659,340 tons produced. The consumption since that period has greatly increased, owing to the extension of iron works, manufactures, railroads, and steam navigation on the Danube, on the Adriatic, &c. Bohemia alone produces the greater half of this coal.

* The Prussian official valuations and the English estimates are much lower than those of France or Belgium. While the aggregate cost of all descriptions of mineral fuel in France was fixed at more than 11⅓ francs per English ton, the coal and anthracite of Prussia were at 7 francs, and lignite only 2 francs. While the coals of France are valued at 14 francs per ton, and those of Belgium at 13.93 francs, those of Great Britain have only been nominated at 7.43 francs.

† Mr. Tennant, in 1846.

UNITED STATES OF AMERICA.

One of the most characteristic features of this immense country, is that of the enormous area of forest and mountain, which remain almost in their primitive solitude. Within these regions, vast ranges of coal formations exist; their limits, imperfectly defined by the geologist, and scarcely more productive now than at the period of the earth's first occupation by the aboriginal races. Under such a condition of things, it were scarcely just to compare them with the well worked fields of European industry. It must suffice that we exhibit the proportionate extent of surface, occupied by coal formations, as compared with the aggregate area of the whole country. This we have partly effected in a previous page, and we can, to some extent, show an approximate estimate of the annual production.

In a few of the older States which border on the Atlantic, the extraction of mineral fuel commenced, as it were, but yesterday. Yet has it advanced with a rapidity unprecedented in the world, and already has attained an importance among the industrial occupations, which it would be difficult to estimate in figures.

Proportionate Areas of Coal to the whole of the United States, and to the Coal producing States.

	Square miles.		Acres.
The United States contains, exclusive of Texas and Oregon,	2,280,000	=	1459,200,000
The exact boundaries or areas of coal and anthracite formations cannot yet be exactly defined, in each State: in a previous table we have detailed the closest approximation to those results at present attainable.	133,132	=	85,204,480
The aggregate area of the twelve coal producing States is,	565,283	=	361,781,120

The United States coal area is thus shown to be one seventeenth part of the entire area of the States (with the exceptions stated), and to be one fourth part of the aggregate area of the twelve principal coal States.

If we are to credit the Census returns in 1840, the relative proportions of capital employed in coal mining, iron making, and lead and other minerals mined, was then as follows, viz.:—

Employed in the iron trade in the United States,	$20,432,131
In lead and other minerals, - - - - -	1,820,061
In coal operations, - - - - - - - -	6,224,464
	$28,476,656

We shall demonstrate, in the progress of this volume, how rapid and enormous has been the increase in these matters, but especially as relates to coal operations in the United States.

From the final Report of the seventh Census of the United States, the total area of this country, including the territories, now called States, is set down at 2,981,123 square miles.

Mr. Kennedy estimates it at 3,230,572 square miles.

Production of coal and anthracite in the United States.—By returns to Congress, made under the census act of 1840, the following summary of the coal trade of that year was obtained.*

	Anthracite.	Bituminous coal 28 bushels to one ton.	Total.
Number of tons of 2240 lbs. each,	863,489	985,828	1,849,317
Number of workmen employed,	3,043	3,768	6,811
Capital invested,	$4,355,602	1,868,862	6,224,464
Proportion of capital to production,	$5 per ton.	$2 per ton.	

We well know that these returns are, in general, much understated, and it is also understood that no returns whatever were received from whole districts, so that, for statistical accuracy, few persons place any reliance upon the results. The difficulty arose, partly from inattention in the agents, but still more from a natural unwillingness among individuals concerned, to make known their private affairs, their capital, and the amount of their business undertakings.

Production of anthracite in 1845, 2,023,052 tons, entirely derived from Pennsylvania. In 1846, 2,343,992 tons. In 1847, 2,982,309 tons. In 1853, 5,195,000 tons.

Bituminous coal mined in Pennsylvania is supposed to have amounted to 1,300,000 tons in 1852.

Of the amount of bituminous coal annually consumed it is impossible to hazard even a guess, but it is doubtless considerably less than that of anthracite.

Were we to offer a very rough approximation to the result for the year 1847, we might say, aggregate of anthracite and bituminous coal nearly 5,000,000 tons. Value at the place of production, $7,500,000. Value at the place of consumption, $20,000,000.

In each case, being probably below the actual result.

Among all the states in the American Union we can only make selection of one, which admits in strict fairness of being compared, in its details, with the coal countries of Europe. This comparison is interesting, and gives a striking proof of the remarkable mineral importance and the flattering prospects of a country so advantageously circumstanced.

PENNSYLVANIA.—Unlike the countries of continental Europe, mineral statistics are here, owing to the free character of its political institutions, attainable with considerably difficulty. There are no records of the number of mines in operation, of the number of workmen employed, the population supported in the coal districts, the amount of production, its cost, and numerous details of interest with

* Hazard's U. S. Statistical Register of 1840, p. 359; also Hunt's Merchant's Mag. 1842, p. 287; also Baltimore Report, Nov. 16, 1843.

which the periodical returns of France, Prussia, Belgium, &c., minutely abound. The attempt to acquire this information during the process of taking the United States census of 1840, was only partially successful. The objections and difficulties attending the former inquiry, will scarcely be obviated even by a more perfectly digested plan, and a more effective organization on a future occasion.

Pennsylvania contains 43,960 square miles = 28,134,400 acres. The areas in this state which are occupied by anthracite, semi-bituminous and bituminous coals, equal to 9,862,600 acres.

Hence, it appears, that Pennsylvania has more than one-third of her whole superficies covered by productive coal formations; a proportion more than three times greater (relatively) than Great Britain, the most productive of the European countries, and almost double the proportionate coal area of the British American coal producing provinces. Our previous table has shown that there are three other states, in the Union, namely, Kentucky, Virginia and Ohio, that preserve the same ratio of one-third, as Pennsylvania. Indiana has one-fifth, and Illinois has no less than three-fourths of her entire area occupied by the carboniferous strata. These six sovereign states comprise 279,755 square miles, and average each 46,626 square miles; approaching nearly to the size of England, which has 49,643.

	Tons.	Miners.
Production in 1840.—The census return shows of anthracite, - - - - - -	859,686	2997
Of bituminous coal, returned in bushels, -	415,023	1798
In Pennsylvania, -	1,274,709	4795

The foregoing abstracts illustrate the coal statistics of the most important countries in the world. We now proceed to present the details for the purposes of comparison, in a concentrated form. The results, for the first time brought under review, are of a very interesting nature.

PRODUCTION OF COMBUSTIBLES.

Table of the Comparative Production of the Six Principal Coal Countries in the World, in the year 1845.—To enable us to exhibit the relative annual production and value of the coal, anthracite and lignite or brown coal, in the six great coal-producing countries of our globe, in the year 1845, which is the latest year in which we can now present a series of contemporaneous statistics, we have prepared the following illustrative statement. It is scarcely necessary to observe, that in the two succeeding years, up to the time of publishing this work, a regular increase has been simultaneously going on, in all the countries enumerated, and apparently at about a corresponding ratio. The present table shows the relative proportions, in each 1000 parts, yielded by each country, in 1845.

Order in 1845.	COUNTRIES.	Square miles of Coal Formations.	Tons of Fuel raised in the year 1845.	Relative parts of 1000.	Official estimated value at the places of production.	
					United States Dollars.	English Sterling.
1	Great Britain,	11,859	31,500,000	642	$45,738,000	£9,450,000
2	Belgium,	518	4,960,077	101	7,689,900	1,660,000
3	United States,	133,132	4,400,000	89	6,650,000	1,373,963
4	France,	1,719	4,141,617	84	7,663,000	1,603,106
5	Prussian States,	not defined,	3,500,000	70	4,122,945	856,370
6	Austrian States,	do.	659,340	14	800,000	165,290
	Total,		49,161,034	1000	72,663,845	15,108,729

The accompanying diagram represents these proportionate results in a simple and intelligible form.

Diagrams of the relative amounts of production of Mineral Combustibles in the Six Principal Coal-producing Countries of the World, in the year 1845.

1.

2.

Belgium
4,960,077
Tons

3.

4.

France
4,141,617
Tons

5.

6.

The returns of the productions of coal during the years 1846 and 1847 have not all reached us: those received will be found under their appropriate heads.

Comparative ratio of Increased Production in Twenty Years.—It is a matter of very great statistical interest, in illustration of the condition of a highly important branch of industrial economy, to ascertain the actual progress which has simultaneously been made, in the demand for, and consequent consumption of, the mineral combustibles, during some years past. With a view to effect such a result, and thereby to indicate this progress in a comprehensive form, we have computed the contemporaneous advance, of five principal coal producing countries, during the space of twenty years prior to 1st January, 1846—that is to say, from 1825 to 1845 inclusive. For a portion of these countries we have official returns for 1846, and even as late as 1847; but we are constrained to omit them in this table, because their insertion would prevent a just and accurate comparison with the rest; more especially as the latest years show a greatly accelerated ratio of increase. The following table exhibits the advance, *per centum*, during the twenty years aforesaid, in the mining countries named.

	Increase per cent. in twenty years.
I. Pennsylvania, production of anthracite only, that of bituminous coal being unknown,	5654
II. Great Britain, the production and general consumption are not registered.	
Exports to foreign countries and British settlements abroad,	713
General shipments at the mines, for exportation and for home consumption,	97
Brought into the port of London,	83
III. Prussia, indigenous production,	124
IV. France, do. do.	181
V. Belgium, do. do.	111

Hence it appears that as regards the highest rates of increase, Pennsylvania far outstrips all her contemporaries of the old world within the common period of twenty years. It is to be regretted that no means exist for ascertaining the advance, made in her production, by Great Britain, during the same interval. The increase, whatever it may be, is well known to be considerable, and it certainly must bear some analogous proportion to the enlargement of the manufacturing departments. The home consumption of coals in South Wales alone is not less than 5,000,000 tons per annum. In our next table we shall pursue this subject of accelerated production yet more in detail.

Onward movement of the Coal Trade.—The purport of the following comprehensive statement is to show the advance, *per cent.*, in the production, the importation, the exportation, and the consumption, of mineral fuel in the principal countries of Europe and America. We have computed it, where practicable, during three periods; that is to say, during the ten, the fifteen, and the twenty-five years prior to 1846:—the first period being from 1835 to 1845, and the second from 1830 to 1845, and the third from 1820 to 1845 inclusive. The production of Great Britain cannot be exactly known and compared; we therefore merely exhibit the increase in relation to shipments, exports, consumption in London, &c.

Proportionate increase, per centum, of the Production, Importation, Exportation, and Consumption of Mineral Combustibles, in contemporaneous periods.

Countries.		Periods of years.	Indigenous Production.	Importation.	Exportation. Increase.	Consumption. Increase.
I. Pennsylvania,		10	254	dimin'd.	increasing.	
		15	1057		to Canada.	
II. Great Britain,	General ship'ts,	15	94			
		25	134	increase.		
	Port of London,	15		66		
		25		101		
	Foreign Exports,	15			401	
		25			904	
III. Austria,		10	90			
		15	410			
IV. France,		15	125	230		154
		25	284	688		370
V. Belgium,		10	71			
		15	95		148	48
		25			514	
VI. Prussia,		10	65	328	166	
		15	112	1126	220	
		25	154		777*	

It is scarcely necessary to remark, that these proportions bear no relation to the *amount* of production, &c., in any of the countries named; but, as already announced, they simply represent the comparative periodical progress madé, per centum, in each of those coun-

* This is the increased exportation to France only. That to Holland is greater.

tries. We have in the foregoing tables, made our computations on the four epochs of ten, fifteen, twenty, and twenty-five years; because, by such subdivision, we are enabled to illustrate more faithfully the contemporaneous increase than if the comparison had been limited to a single term.

We have now placed before the reader, in the most concentrated form of which the matter is suspectible, the means of judging, with perfect accuracy, of the wonderful increase in the mining and commercial disposition of mineral fuel that has taken place in our own times. We here observe, for instance, that in the fifteen years prior to 1846, Belgium increased 95 tons on every 100, in 1830; that Prussia added 112 to every 100; that France added 125 to every 100; and Austria's rate of increase was not less than 410 on each 100. The ratio of Great Britain cannot be pointed out, except that she increased her foreign exportation five fold in the same space of time. But the most remarkable advance on record is in the case of Pennsylvania, where, on every 100 tons of coal produced in 1830, the absolute increase is represented by 1057 tons in 1845.

SUMMARY OF STATISTICS OF MINERAL FUEL.

The following pages contain a resumé of what we have elsewhere exhibited in detail; viz. of the entire range of our present knowledge, regarding the production, importation, exportation, and consumption of fossil fuel, within all the principal coal producing and coal consuming countries in the world; together with their periodical rates of increase down to the present time; derived from every official return accessible to us.

GREAT BRITAIN—*Increased General Production.*—Owing to the absence of official records, applying to the general production of the collieries throughout the United Kingdom, we are constrained to leave this as a matter of inference, from the results which we have to adduce. We know, however, that its rate of increase has been rapid, especially in all the manufacturing districts; probably even much more so than that of exportation.*

Increased Shipments for Home and Foreign Consumption, from the Ports of Production.—From 4,365,000 tons in 1819, to 11,254,750 tons in 1845; being at the rate of 158 per cent. in twenty-six years. This quantity is supposed to be about one third of the entire production of the United Kingdom. The *declared value* advanced from £145,943 in 1828, to £970,462 in 1845; or 569 per cent. in seventeen years.

Increased Exportation of Coal.—To the colonies and British possessions, from 71,000 tons in 1819, to 375,302 tons in 1845; or 428 per cent. in 26 years; to France, from 39,180 tons in 1825, to

* 1853. From a late English paper we learn the following:—
Capital invested in the British coal trade, £10,000,000.
Annual production, 37,000,000 tons.
Value at pit's mouth, £10,000,000. London alone consumes 3,600,000 tons.

647,967 tons in 1845, = 1561 per cent. in 20 years; do. from 24,800 tons in 1820, to 647,967 tons in 1845, = 2512 per cent. in 25 years; to Russia, 1450 per cent. in 25 years; do. 375 per cent. in 15 years; Denmark, 1800 per cent. in 15 years; Prussia, 1214 per cent. in 15 years; United States, 287 per cent. in 15 years; do. British and Colonial, 184 per cent. in 15 years; East Indies and Ceylon, 2025 per cent. in 15 years; British West Indies, 126 per cent. in 15 years; Germany, 417 per cent. in 14 years; Italy, 323 per cent. in 9 years. Increased number of British vessels laden with coal for foreign ports, in the six years from 1840 to 1846 inclusive, 142 per cent.

Increased Shipments from the Collieries of the North of England —viz. from the ports of Newcastle, Sunderland, and Stockton-on-Tees, for foreign and home consumption collectively.—From 820,620 tons in 1710, to 6,123,282 tons in 1842, = 646 per cent. in 132 years; from 2,985,560 tons in 1810, to 6,123,282 tons in 1842 = 151 per cent. in 32 years; from 3,160,956 tons in 1832, to 6,123,282 tons in 1842, = 93 per cent. in 10 years. For *home consumption*, 50 per. cent. in 18 years, ending 1842.

For Foreign Consumption only.—From 157,014 tons in 1820, to 1,784,988 tons in 1845, = 1036 per cent. in 25 years. In 1773, there were only 13 collieries in the Newcastle district, which number increased, in 1828, to 59, with an annual productive power of 8,123,922 tons. In 1844, this productive power was estimated at 13,000,000 tons, and the number of collieries had increased to 124, and of pits, to 192; besides 6 other collieries in other parts of the same field, and 300,000 tons which were superseded by the inland coal. The shipments of coal to foreign parts, from this district, has increased from 50,805 tons in 1810, to 1,784,988 tons in 1845; being at the rate of 3468 per cent. in 35 years.

Increased Importation of Coal into the Port of London by Sea and Land.—From 1,667,301 tons in 1822, to 3,461,199 tons in 1845, = 108 per cent. in 23 years; from 300,000 tons in 1699, to 3,461,199 in 1845, = 1057 per cent. in 146 years; from 2,079,275 tons in 1830, to 3,461,199 tons in 1845, = 66 per cent. in 15 years.

*1846,	- -	2,953,755 Tons.
1847,	- -	3,280,420 "
1848,	- -	3,418,340 "
1849,	- -	3,339,146 "
1850,	- -	3,553,304 "

Increased Foreign Shipments of Coal from Hull.—7,463 tons in 1833, to 42,789 tons in 1845, = 477 per cent. in 12 years. From Liverpool, 50,561 tons in 1833, to 123,456 tons in 1845, = 146 per cent. in 12 years.

SCOTLAND—Has a greatly increased *production* of coal, but for the same reason as in the case of England, we possess no precise returns of the aggregate. This enlarged production is, in great measure, applied to the purposes of home consumption, especially to

* From Official Reports.

the various departments of iron making, which has advanced with surprising rapidity since the discovery of the Black Band ore.

Exportation to Foreign Parts increased from 31,940 tons in 1828, to 229,513 tons in 1845; equivalent to 617 per cent. in seventeen years. But the recent excess of production is mainly reserved for home use.

SOUTH WALES.—We have no returns in relation to the advanced production. Judging from the increased number and power of the Welsh iron works, the home consumption of coal must be greatly augmented. It has been asserted, in 1844, that one third of the iron consumed in the known world, is produced in the mineral basin of South Wales, and upwards of five million tons of coal are annually consumed in its manufacture, and for other purposes, within the coal field.

Foreign Exports.—The exports of South Wales, in 1833, amounted to 24,981 tons. In 1845, four only of the principal ports exported 237,577 tons. The entire increase, probably, does not fall short of 1000 per cent. in twelve years.

Shipments to London—Increased from 34,000 tons in 1828, to 81,725 tons in 1843, or 172 per cent. in 15 years.

Increased General Shipments for Home and Foreign Consumption.—From 904,896 tons in 1828, to 1,700,000 tons in 1841, = 88 per cent. in thirteen years.

The shipments, therefore, do not keep pace with the home consumption required for the iron works, &c.

FRANCE.—*Increased Indigenous Production of Coal, Anthracite, and Lignite, in 29 years.*—From 869,410 tons in 1815, to 3,639,446 tons in 1843, the ratio is 203 per cent.

Table of the relative Increased Production of Mineral Combustibles in France; representing the production in 1787 as 1.00.

Years. Date.	Fuel, Tons.	Time, Years.	Proportion.
1787	212,910		1.00
1820	1,078,560	33	4.06
1830	1,836,950	43	7.62
1840	2,960,015	53	12.89
1843	3,639,446	56	16.09
1845	4,141,617	58	19.45

Increased value of Indigenous Production in France, in 28 and 31 years.

	In 1814.	In 1842.	In 1845.
Value in Francs,	6,082,447	33,497,779	39,705,432
" U. S. Dolls.,	1,316,950	6,476,237	7,663,000
" Engl. Sterling,	272,097	1,352,472	1,603,106

Increased ratio of value in 28 years = 397 per cent.; in 31 years, 489 per cent.

Increased Importation of Mineral Fuel into France, since 1820.

	Imported in 1820, Tons.		In 1845, Tons.		Per cent. in 25 years.
From Great Britain to France,	24,800	to	647,967	=	2512
From Belgium "	224,100	to	1,376,100	=	514
From Prussia "	27,500	to	237,200	=	762
Total imported into France,	276,400	to	2,116,272	=	665

And an increase on the total importation, since 1802, of 1756 per cent. in 43 years.

Ratio of Increased Consumption of Mineral Combustibles in France—Distinguishing the indigenous fuel from the indigenous and imported combined, in the periods of thirty and fifty-eight years, prior to 1846.

Dates.	Indigenous only.			General Consumption.	
	Tons.	Periods of years.	Increase per cent.	Tons.	Increase per cent.
1787	212,910			399,130	
1815	869,410	30	375	1,096,820	470
1845	4,141,617	58	1853	6,251,790	1467

BELGIUM.—*Increased number of Coal-pits in operation or in construction.*—From 314 pits or points of extraction, in 1830, to 660, in 1840.

Number of Miners increased from 29,253 in 1830, to 38,490 in 1844.

Production.—Within about thirty years there have been one or two periods of ebb and flow. Thus, from 1802 with 2,635,000 tons, to 1832, it was *reduced* to 2,249,000 tons, or 17 per cent. decrease in thirty years; and from 2,249,000 tons in 1832, *increased* to 4,960,077 in 1845, or 120 per cent. gain in thirteen years.

The increased *value* in the same interval being from 16,957,500 francs in 1832, to 55,400,000 francs in 1840, or 226 per cent. in eight years.

Increased annual production in the Belgian provinces,	Liege,	1830	to	1845,	537,100 tons.
	Hainault,	"		"	1,757,346 "
	Namur,	"		"	111,873 "

Total increase from 2,553,000 tons in 1830, to 4,960,000 tons in 1845, or 94 per cent. in fifteen years.

Increased Importation from Great Britain.—From 770 tons in 1831, to 36,440 in 1841, and about 20,000 tons in 1847. The import trade, being of subordinate importance, was reduced to only

11,071 tons from England, and about the same quantity from France, in the year prior to 1847.

The manufacture of iron has advanced so rapidly that there were more tons produced from the furnaces of a single province, Hainault, in 1846, than in the entire kingdom of Belgium, only five years before.

Increased Exportation of Bituminous Coal from Belgium, chiefly to France.—Advance from 621,560 tons in 1830, to 1,356,973 in 1846, = 118 per cent. in sixteen years. Upwards of 1,700,000 tons in 1847.

In the fifty-eight years from 1787 to 1845, the exportation of coa. from Belgium into France increased twenty-seven fold, or 2708 per cent. That to the Low Countries, in the seven years between 1835 and 1842, advanced nineteen fold. To other countries no advance.

Increased annual Consumption of Coal in Belgium.—Advance from 2,162,000 tons in 1830, to 2,670,000 in 1840. The domestic consumption in 1847 was probably upwards of 4,000,000 tons.

HOLLAND.—The quantity of coal received from England has increased sixty-seven per cent. in the ten years between 1831 and 1841, since which it has diminished forty-eight per cent. in 1845.

That from Belgium has augmented 1940 per cent. in the seven years from 1835 to 1842. There is a large increase from the Prussian provinces.

KINGDOM OF PRUSSIA.—*Increased Importation of Bituminous Coal.*—Advance, 154 per cent. in twenty-six years; from 1832 to 1844, 112 per cent. in twelve years.

Increased Importation of Coal from England.—From 15,956 tons in 1831, to 184,487 tons in 1845, about ten fold, or 1050 per cent., in fourteen years.

RHENISH PROVINCES OF PRUSSIA.—*Exportation to France.*—From 27,500 tons in 1820, to 237,200 in 1845; 777 per cent. increase in 25 years.

In the fifty-eight years from 1787 to 1845, the exportation into France advanced twenty-three fold, or 2322 per cent. increase.

Increased production of Bituminous Coal in the provinces of the Lower Rhine.—*Saarbrück.*—From 233,000 English tons in 1817, to 700,000 tons in 1844, = 200 per cent. in twenty-seven years.

WESTPHALIA.—*Increased production of Coal.*—From 370,268 English tons in 1819, to 1,200,000 in 1844, = 224 per cent.

PRUSSIAN SILESIA.—*Increased production of Bituminous Coal.*—From 285,000 English tons in 1817, to 850,000 tons in 1844, = 200 per cent. in twenty-seven years.

PRUSSIAN SAXONY.—Has doubled her production in twenty-one years.

HANSE TOWNS.—*Increased Importation of Coal from Great*

Britain.—From 26,500 tons in 1789, to 227,539 tons in 1845, = 759 per cent. in fifty-six years. See Zollverein.

DENMARK.—*Increased Importation from Great Britain.*—From 61,392 tons in 1828, to 168,153 tons in 1845, = 170 per cent. in seventeen years.

NORWAY.—*Increased Importation of Bituminous Coal from England.*—From 3,771 tons in 1831, to 15,894 tons in 1841.

SWEDEN.—*From Great Britain.*—Increase from 6,150 tons in 1831, to 26,941 tons in 1841.

RUSSIA.—*Increased Importation of Bituminous Coal from England.*—

From 42,061 tons in 1835, to 150,422 tons in 1845, = 257 per cent.
From 2,316 tons in 1810, to 150,422 tons in 1845, = 1150 per cent. in twenty-five years.
From 1820, to 150,422 tons in 1845, = 7400 per cent. in twenty-eight years.

HUNGARY.—*Increased production of Bituminous Coal.*—From 1823 (average of ten years) 14,500 tons, to 33,076 tons in 1845; 207 per cent. in twenty-three years.

BOHEMIA.—*Increased consumption of Coal in Prague in ten years.*—From 10,000 French tonnes in 1830, to 24,000 in 1839.

Production.—Advanced from 122,000 tons in 1832, to 340,000 tons in 1845, = 170 per cent. in thirteen years.

AUSTRIA.—Increased production of coal in the empire, from 1838 to 1846, 216 per cent.

Increased consumption of Coal in Vienna in ten years.—From 3000 French tonnes in 1830, to 10,000 in 1839.

Upper and Lower Austria increased production 47 per cent. in four years, from 1830 to 1834.

Increased *production* of combustibles in the Austrian Empire, from 209,000 tons in 1832, to 700,000 tons in 1846, = 234 per cent. in fourteen years.

GERMAN STATES.—Prussian or German Custom-house League—The *Deutsche Zollverein.*—Increased *importation* from Great Britain, from 44,033 tons in 1831, to 227,539 tons in 1845, being 417 per cent. in fourteen years.

General exportation from the Zollverein.—218,440 tons in 1834, to 349,150 tons in 1843, being 60 per cent. in nine years.

General movement of coals in the states of the German Association —from 282,760 tons in 1834, to 605,900 in 1843, = 111 per cent. of increase in nine years.

SPAIN.—Notwithstanding our inability to illustrate with precision the mining statistics of Spain, we must not overlook the fact that it seems destined to become one of the most valuable of the continental coal-producing countries. In superficial area, the Asturian coal-field is probably not exceeded by any other on the continent, and as regards the number and quality of its coal seams, it is no less distinguished, although it is one of the latest brought into operation.

Great expectations have been formed as to the national value of this district, and much enterprise has already been attracted to the development of its important mineral resources; especially those of bituminous coal and iron ore.

The coal business is comparatively in its infancy, but promises a rapid progress in future. Thus the amount shipped coastwise from the port of Gijon, in 1842, was 14,100 tons, and in 1844 was 41,400 tons.

United States of America.—*Imported Coal.*—The only countries from which coal ever finds its way into the United States, are Great Britain and British America, and the contributions from thence appear to be annually diminishing. For a time there was an increasing foreign importation; viz. from 22,123 tons in 1821, to 181,551 tons in 1839. By the operation of the American tariff, this advance was not only checked, but a retrograde movement was produced, so as in 1843 to amount to only 41,163 tons, by the United States returns. By the last annual return, that for 1847, the entry of foreign coals, whether from Europe or from British America, was 148,021 tons; of which from 12,000 to 15,000 tons were re-exported, for the service of the English steamships. 1850, 180,439 tons were imported into the United States; 1853, 231,508.

INCREASED PRODUCTION OF AMERICAN COAL AND ANTHRACITE.

Bituminous Coal.—We have already indicated that we possess no authentic data for determining the progressive production of this description of fuel in the United States. Such informal details as have reached us, will appear in the proper place; and we can only remark here that the rate of increase is evidently very rapid.

Anthracite.—Of this important combustible we shall have much to say, and we possess abundant testimony upon which to found our calculations. The production of anthracite may be said to be entirely confined to the State of Pennsylvania, which possesses a numerous and interesting group of coal basins of various sizes and characters.

Our returns show that the consumption of anthracite,—in other words, the coal trade,—commenced with 365 tons in the year 1820; that the production reached 48,047 tons in 1837; that it had increased to 881,026 tons in 1837, and advanced to three millions of tons in 1847; without including much that is consumed on the spot, in the mining districts, or in the interior of the country.

The increased production, therefore, was, in the first ten years, viz., from 1827 to 1837, 1735 per cent.; in the second ten years, viz., from 1837 to 1847, 240 per cent.; and in the twenty years previous to 1848, that is, from 1827 to 1847, 6150 per cent.

We introduce another view of this subject, which exhibits this accelerated increase in the consumption of anthracite, perhaps, with yet greater perspicuity. The amount which was periodically forwarded to market, exclusive of the consumption in and near the places of production and which has not been estimated, is as follows:

*Aggregate in the 21 years from 1820 to 1840, inclusive,	6,847,172 tons.
In the succeeding 7 years to 1847, inclusive,	12,371,961 "
Total furnished from the commencement,	19,219,133 tons.
From 1848 to 1853, inclusive,	23,841,358 tons.
Total furnished from the commencement,	43,060,491 "

With this we terminate our compendium of the coal statistics, into whose details we shall enter at large further on: our immediate object being that of showing the rapid increase in the annual production of coal, all over the globe, within the last quarter of a century.

PRODUCTION OF IRON.

We have already exhibited in the diagram form, the superficial areas of the principal coal producing countries of the world, and also the squares of the coal production of the same countries, in the year 1845. We are induced to occupy a small space here, by a similar mode of illustration in regard to the production of iron, in the same year, by the chief manufacturing countries.

In the preparation of the materials forming this volume, we never contemplated to devote any part of it to the subject of iron. The statistics of coal, which we undertook to elucidate, seemed to promise a task of quite sufficient magnitude to keep us in full occupation. Nevertheless, we have found that the rapid advancement of the coal trade was so intimately connected with the contemporaneous process of the iron manufacture, that we have, almost unconsciously, been led out of our prescribed path; and having collected some interesting results by the way, we give them insertion in their appropriate places.

We now only propose in this place, to introduce a *diagram* showing the condition, as to production, of the iron manufactory or smelting, in the year 1845, the latest year in which we could obtain a series of contemporary returns.

The respective proportions are as follows:—

1. Great Britain, - - - - -	2,200,000
2. United States, - - - - -	502,000
3. France, - - - - - -	448,000

* R. R. Report.

4.	Russia,	400,000
5.	Zollverein, or Prussian States,	300,000
6.	Austria,	190,000
7.	Belgium,	150,000
8.	Sweden,	145,000
9.	Spain, (1841)	26,000
10.	All other European countries,	50,000
	Total,	4,411,000

Diagram of the Production of Iron in 1845.

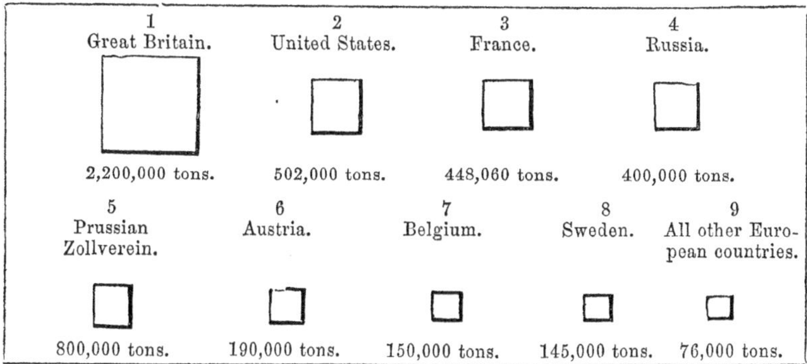

Make of Iron in England.

Years.	Furnaces in blast.	Iron produced. tons.
1830	360	678,417
1840	402	1,396,400
1843		1,215,350
1845		1,512,500
1848		1,998,568
1850	459	2,380,000
1852	655	2,701,000
1854		2,250,000
Persons employed in mining,		120,000

"During the ten months ending Nov. 5, 1853, Great Britain exported $75,000,000 worth of iron, and by far the largest portion was taken by the United States. Of pig-iron, the United States received 57,000 tons, and Holland, which comes next upon the list, took 13,000. Of bar, bolt and rod-iron, the United States took 263,530 tons, or nearly six times as much as Canada, which received the next largest amount."*

* From the London Chronicle.

Iron Trade of the United Kingdom.

Total exports from the United Kingdom in 1852:

	Tons.
Pig-iron, - - - - - -	240,491
Bars and Rails, - - - -	548,996
Rods, - - - - - - -	18,696
	808,183

Of which were shipped to the United States:

	Tons.
Pig-iron, - - - - - -	104,230
Bars and Rails, - - - -	334,224
Rods, - - - - - -	1,439
	439,893
Leaving tons,	368,290

for the requirements of the rest of the world.*

We insert the following table from page 331 of the first edition on the Annual Production of Pig and Cast-iron in Great Britain, France and other European countries, and in the United States of America:

Years.	G. Britain. Tons.	France. Tons.	Belgium. Tons.	Zollevе-rein. Tons.	U. States. Tons.	Austria. Tons.	Sweden. Tons.	Russia. Tons.
1841	1,327,612	377,142	90,000		287,000	151,000	90,000	300,000
1842	1,347,790	399,456	121,000	108,440		180,000		
1843	1,215,000	416,377		210,000				320,000
1844	1,575,260	421,388	153,791	250,000	486,000		100,000	380,000
1845	1,512,000	448,900	156,000	300,000	502,000	190,000	145,000	400,000
1848	1,998,568	348,000						
1849	2,000,000				650,000			
1850	2,380,000				400,000	160,000	133,500	189,000
1852	2,701,000							
1854	2,250,000							

RAILROADS.

After passing from coal to iron, we are almost unavoidably tempted to diverge yet further, to the subject of railroads, steam-engines, and steam vessels,—so closely do all these matters seem interwoven with each other, being at one and the same time both cause and effect, in relation to the enormous increase of coal production, in all parts of the world. Thus we are impelled to notice the astonishing extension of railroads in our day, whereby the coal, the iron, and the other minerals have become more generally accessible, and consequently more valuable, in proportion as they can be transported with cheapness and rapidity to their several markets.

* London Mining Journal, April 1st, 1854.

The following statement shows the actual number of miles of railway finished and in progress in Europe and America, in 1844.

	Miles.
In Great Britain and Ireland, - - -	2069
In Prussia and Germany, - - - -	2500
In France, 1241 *miles finished,* - - -	1750
In Russia, - - - - - -	1500
In the United States of America, - - -	3688
	11,507

In *England*, in 1845, there were obtained from Parliament new railroad acts for 3543 miles; up to 1846, the total number of miles authorized to be made in Great Britain was 7494 miles, and to January 1st, 1847, 8384 miles, besides 1862 miles already made.

In *France*, in the same year, the number of miles completed, commenced, and proposed, was 3874—whose estimated cost was $297,220,000, or £61,600,000

In *Belgium*, there were 282 miles of railroad in operation in 1842, 348 miles in 1844, and 386¼ miles in 1846,—costing £5,789,872.

In the *Zollverein*, there were completed 24 railroads in 1843, of the length of 1730¾ miles.

In all *Germany*, in 1844, 3565 English miles, in 43 railroads.

United States of America.—By an unofficial article, dated June, 1846, it appears that there were then in operation in the United States an aggregate length of 4731 miles, which was constructed at the cost of $127,417,758, equivalent to £26,325,983.

From the data furnished up to that time, we collect that the capital invested in railroads alone, independent of private and local undertakings, had augmented nearly five-fold in ten years.

During the year 1846, the total amount of completed railroads in the United States had reached the aggregate of 4864½ miles.

In the beginning of 1847, there were, according to the report of the postmaster-general, of finished railroads 4752 miles; in progress and projected, 264 miles; total, 5016 miles. Omitting the gigantic project of the Oregon railroad.

Thus, at the commencement of 1847, we find that the number of miles of completed and partly finished railroads in the principal countries of Europe and America, amounted to no less than 20,000 miles; being within a few thousand miles of the entire circumference of the globe. Those in Europe were supposed to require 6,157,000 tons of iron. Added to this, the government of British India has had surveys undertaken for 2000 miles of railroad, with a view of commencing a general system of railway in that extended empire.

1853.—The number of miles of railway now in operation upon the surface of the globe is 34,776, of which 16,180 are in the Eastern Hemisphere, and 18,590 are in the Western, and which are distributed as follows:—

	Miles.		Miles.
United States, .	17,317	Belgium, . . .	532
British Provinces, .	823	Russia, . . .	422
Island Cuba, . .	359	Sweden, . . .	75
Panama, . . .	31	Italy, . . .	170
South America, .	60	Spain, . . .	60
Great Britain, . .	6,976	Africa, . . .	25
Germany, . .	5,340	India, . . .	100*
France, . . .	2,480		

Steam Engines employed for purposes of industry, and also in mining enterprise, exclusive of that employed in navigation and locomotives.

Years.	France.		Belgium.		Years.	Great Britain.	France.		Belgium.	
	Fixed steam engines.	Horse power.	Fixed steam engines.	Horse power.		Newcastle or north of England coal field. Horse power.	Fixed steam engines.	Horse power.	Fixed steam engines.	Horse power.
1835	946	14,051			1840				1,049	26,056
1837	1,842	24,144	662	19,456	1843	19,397	2,595	35,197		
1838	2,332	27,677	1,044	25,312½	1844	Cornwall.	3,645	45,781	1,448	37,370
1839	2,459	33,308	1,044	35,512	1845	44,000	4,114	50,188		

Hence, it is shown that the amount of horse power employed in mining and manufacturing enterprise in France increased 257 per cent. in ten years; and in Belgium 94 per cent. in only five years. Our English returns are incomplete.

Steam vessels of Great Britain, the United States, France and Russia, chiefly engaged in Commerce.

Great Britain, exclusive of her Colonies.	Years.	No.	Tonnage.	Horse power.	Guns.
In 1814, only one steamboat in the British Empire.					
Merchant marine, ...	1833	386			
	1835	554			
	1837	620	69,800		
	1840	810			
	1844	900	114,000	70,000	
	1845	916	118,772		
	1846	1,000			
War steamers in commission, and building mail steamers, Indian navy, ...	1846	{ 179 134			688 310
United States.					
The first western steamboat was launched at Pittsburg, on the Ohio, by Fulton, in 1811.					
Internal navigation, ...	1840	800	155,473	57,017	
Coasting merchant marine,† sailing vessels,	1844		265,269		
" " "	1845		319,527		
Steamboats on the Western waters alone, .	1846	1,500	145,311		
" Lakes, ...	"	80	54,486		
" "	1847	86	60,825		

* Hunt's Magazine, Jan., 1854.
† Merchants' Magazine, February, 1846.

United States.	Years.	No.	Tonnage.	Horse power.	Guns.
Sailing tonnage on " . . . . . *	"		46,011		
1851, steamboats on the Lakes, U. S., . .		180	212,000		
Locomotive engines, ...	1840	350		6,980	
War steamers, ...	1846	11			
Exclusive of 13 sailing and 8 steamers, having 61 guns, and revenue vessels, . . .	1847	13			
France.					
Steamboats, ...	1833	75	3,800	2,633	
" "	1835	100	12,100	3,863	
" "	1837	124	19,900	5,408	
" "	1840	225		11,422	
" "	1844	238		12,789	
" "	1845	247		13,250	
" " †	1846	259			
War steamers in commission, and building, .	"	68			436
" " "	1847	66		14,570	
Russia.					
War steamers, ...	1847	32			

The steam marine of the United States is immense, amounting in 1852, according to Andrews's Report, to 1390 vessels, with an aggregate tonnage of 417,226 $\frac{8}{95}$, of which 96 (tonnage 91,475 $\frac{69}{95}$) were ocean steamers, 529 were coast steamers, (tonnage 121,025 $\frac{31}{95}$), 765 were interior steamers, (tonnage 204,725 $\frac{12}{95}$), of which 601 (tonnage 135,559 $\frac{15}{95}$) were on the rivers.‡

The Mercantile Marine of Great Britain and United States.

During the year 1853 there were built and registered in the United Kingdom 645 sailing vessels, of a collective burden of 154,956 tons; 153 steamers, with a tonnage of 48,215, making the total aggregate of ships building during the year of 203,171 tons, independent of the vessels built in the colonies.

A striking fact in the ship building of the United Kingdom, is the rapid increase of iron ships. Of the 153 steamers built last year, 117 were of iron. Of the sailing vessels, 10 were built of iron, averaging 857 tons each. On the Clyde, which is one of the finest ship yards in Great Britain, more than half the vessels now on the stocks are of iron. Over 250,000 tons of shipping were built in the United States in 1852, and a still greater amount in 1853. At present the aggregate amount of tonnage owned by each of the two countries does not materially vary.§

||Great Britain and Ireland have a navy
of . . . . . . 678 vessels and 18,000 guns.
France, ... 328 vessels and 8,000 guns.
Russia, ... 175 vessels and 7,000 guns.
Turkey, ... 74 vessels and 4,000 guns.

* Official Report of the Secretary of the Treasury, 1847.
† Report of the Societé Maritime, 1846.
‡ U. States Gazateer, 1854.
§ New York Tribune.
|| Map of Europe embracing latest Statistics, published by Cowperthwaite, Desilver & Butler, Phila. 1854.

Such are the results which our recent investigations have disclosed, while seeking to trace the onward march of productive industry, in opposite hemispheres. However rapid may have been that advance in the Old World, in energy and perseverance—in inventive genius and mechanical skill—in an extended application of the useful arts—in the employment of mighty agencies known to us but as yesterday—and, above all, perhaps, in the adaptation of the wonderful powers of steam—the New World has by no means suffered herself to be left behind. It is but justice to the latter to show how fully she appreciates and avails herself of this newly acquired knowledge, by her rapid advancement in the operative and industrial arts, in so wide a field for human enterprise. We cannot perform this duty so efficiently as in the language of one of her own citizens and most distinguished engineers. The admirable and truly eloquent address, from which we take the following extracts, was delivered at the successful termination of one of the most important undertakings in the system of internal improvements in Pennsylvania. It reached us soon after we had embodied in the preceding pages the statistical results which were elicited during the preparation of the present volume.*

"We have already alluded to the indications which past experience affords of the probable future consumption of coal in this country. The subject is of primary interest, and we may, therefore, venture still to add some reflections upon the causes which are now at work to extend this consumption.

"In estimating the probable growth of this trade, we must, to some extent, endeavour to free our minds from the shackles of old opinions, and the influence of ancient example. We must learn to feel the truth, that we live in an age that bears little resemblance to the past, and the progress of which cannot be safely judged by the history of the past.

"This is essentially the age of commerce and of steam—the foundations of which are our *coal mines*.

"In the machine-shop and factory—on the railroad and canal—on the rivers and the ocean—it is STEAM that is henceforth to perform the labour, overcome resistance, and vanquish space. And it is not for human intellect to assign a limit to the application of this power, in a country like that which it is our fortunate lot to inhabit—intersected by noble rivers, and penetrated by numerous bays—with an extensive sea-board, lined by flourishing cities, and possessing, along with boundless enterprise, all the elements of national wealth.

"But, look where we will, the evidence of the truth, that we live in an age of which the progress is not to be measured by examples from the history of the past, is prominent before us.

"Taking the iron trade as an example, we find that the mere increase of the production of this metal, in the valley of the Schuylkill alone, during the last eighteen months, exceed the entire production of the furnaces of Great Britain, ninety years ago. The manu-

* Report to the Stockholders of the Schuylkill Navigation Company, by Charles Ellet, Esq., President, January 4th, 1847.

facture of cotton in Great Britain which has increased about one hundred fold in the last seventy years, and of the same and many other articles, as well in Europe as in this country, exhibits results almost equally striking.

"There was, in fact, no appreciable iron trade, and, indeed, but little trade at all, in the present ordinary use of that word, anterior to the introduction of the steam engine—an instrument of power deriving its efficiency almost entirely from coal, which, through its agency, has given birth to modern commerce, to modern enterprise, and a mighty impulse, too, to modern civilization.

"A quarter of a century ago—within the memory of almost all here present—those magnificent boats which now give life to the Delaware and the Hudson—the seven or eight hundred which traverse the Mississippi—and the thousand which circulate on other waters of this country, had no existence, except, perhaps, in the imaginations of those who were then considered wild and visionary enthusiasts. Now, every year brings forth new specimens, each in its turn regarded as the noblest creation of bold invention; and each week presents some new enterprise, by which the Atlantic cities are brought into closer connection with each other, and with foreign ports.

"The use of this power on the ocean has but just commenced; yet enough has already been accomplished to point to an approaching revolution in the coasting trade and foreign commerce of all countries. The next year promises to witness new lines of ocean steamers, connecting this country with England, France, Germany, and South America, and traversing the coast from New York to New Orleans.

"A quarter of a century ago, and there were not more than a thousand tons of anthracite annually raised and exported in all this Union; now the *increase alone* is more than a thousand tons per diem, and compounding rapidly upon that.

"But still we can form no accurate estimate for the future increase from the past. New elements are daily introduced into the problem, of which no human intellect can determine the value.

"The introduction of the railway system over all Europe and even Asia—over this continent and the West India Islands—over Russia, and even into the Papal States, offer a guarantee of a future consumption of iron and coal, and all the chief mineral products of the earth, to which no bounds can be assigned.

"Each railway requires iron for its track, engines, cars, and frequently for its stations. Each new steamer requires coal to drive it—iron for its engine, and sometimes for its hull—and five tons of coal for each ton of iron it consumes.

"Every steam boat that is launched, and every road that is forced into the interior, gives birth to new enterprise, new wants and new commerce.

"The manufacture of the iron, and the propulsion of the machinery require *coal;* the quantity increases with the expansion of the railway system; the system extends the area of civilized population, and consequent agricultural wealth. This wealth needs transportation, and this transportation again needs coal and iron.

"In this country, peculiarly, the consumption of this fuel is increasing with the general increase of population where it is employed—with the wider area over which it is used—with each new purpose to which it is applied—with the growth of every description of manufacture requiring power, with every new improvement by which the cost of its conveyance is diminished, and with the extentension of inland, coast, and ocean navigation."*

Thus far has been exhibited in the foregoing pages an interesting picture of the wonderful advance made, in our day, in the application of the mineral combustibles. We have seen, and let us note the fact, that this enormous advance has not been limited to a single district, but that it has simultaneously proceeded in all the coal producing countries of the earth. Doubtless a very large portion of this is ascribable to the prodigious extension of steam power, occasioning a corresponding demand for mineral combustibles. We should exceed our prescribed limits were we to adduce the evidences of this increased application of steam, through the agency of coal. Nor, indeed, is it essential to our purpose. But we are quite sure that we cannot more appropriately terminate this introductory section, than by citing the following expressive passage, which we find in the Bulletin of the Central Commission of Statistics in the kingdom of Belgium; to the author of which we have here to acknowledge our obligations.

"Industry has undergone a complete transformation since the establishment of machinery. The development of mechanism is owing to the application of steam as the moving power. Steam has been substituted in a multitude of operations, for the natural agents. If we had to write the history of industry, we should represent man seeking at first to direct, to his advantage, the elements of nature, and subsequently creating new forces and more powerful agents. In the first period, man finds masters in every thing which surrounds him; the means at his disposal are very confined; his knowledge and his capital are limited; regulations badly conceived; the small extension of outlets; the difficulties of transportation;—all restrain his capability of production.

"In the second period, the state of affairs changes: he has subdued the natural elements; he disposes them at his will; the science of mechanics procures him the most powerful agents; natural philosophy, chemistry, discover to him a part of their treasures; capitals are no longer locked up; the slender profits of agriculture impel them back towards industrial occupations. Interior shackles have disappeared; treaties of commerce establish, between the people, fixed relations,

* The power thus convertible to the purpose of lightening the labour of man was felicitously illustrated by Sir I. F. W. Herschel, in the remark that the ascent of Mount Blanc from Chamouni, is considered, and with justice, as the most toilsome feat that a strong man can execute in two days. The combustion of two pounds of coal would place him on the summit.

which daily enlarge their social and political connections. Distances are effaced; routes are multiplied; and steam, after having ploughed the rivers and the seas, skims the earth in a rapid flight. Commerce unites together every people; the market is enlarged. Production, which outstrips all local necessities, urgently demands new outlets; embarrassment no longer attaches to production; the trouble henceforth rests in the distribution.

"The employment of the combustible mineral, COAL, in the smelting of iron, has emancipated the IRON manufactory. Henceforward the mineral comes to seek the fuel. Steam is prepared as the motive power: the forge-master, the founder, are no longer confined to the banks of rivers, or the depths of the forests, far from the inhabited places. Industry has broken her fetters; commerce is set free, at least in the interior. Gigantic high-furnaces arise; forges, bar-iron works multiply; iron receives every shape; manufactories fill the world with machines. One might even say that each operation of industry gives birth to new marvels, and that all contribute to the successive and unbounded enlagement of productive forces and of new agents.

"Thus, coal produces steam; steam fashions the metals which serve to fabricate the machines. The implements of various trades, leaving workshops, distributed through every branch of industry. Steam becomes the universal agent; if she is the producer, she is at the same time the vehicle of production.*

"The powers of man are centupled; he is no longer the serf of the creation; he is rather the king. The barons of feudality have made room, by their side, for the nobility produced by industry. The sword commands no more; it is capital which commands. To the state of strife, of warlike antagonism, succeeds a regime of industrial competition and of exchanges. Men know themselves and each other better; national characteristics are obliterated; it seems that humanity is invested with a new form; organization is established between states; between continents."

"Mineral and metallurgic industry is, with agriculture, the most vital element of our country's prosperity. Coal is the most essential agent of all industry; the foundry, the iron, constitute merely the instruments, the elements of riches."†

* "It is as yesterday only, so to say, that steam has been employed as a moving power; and yet it already furnishes the globe with a force estimated at more than ten millions of horses, or *sixty millions of men.*" M. Michel Chevalier.

We would here refer to an excellent article in Hunt's Merchants' Magazine, June, 1846, by Mr. C. Frazer, on the "Moral Influence of Steam."

† Bulletin de la Commission Centrale de Statisque, Bruxelles, 1843.

SECTION II.

MISCELLANEOUS NOTES IN RELATION TO COAL.

Terms of synonymous signification, many of which are employed in this work. Bituminous Coal, Slate Coal, Foliated Coal, Anthracite.

English, Coal, Pit Coal, Brown Coal, Sea Coal, Stone Coal; Collier, a coal miner; Colliery, a coal mine. *Saxon*, Col. *Dutch*, Koolen, Steenkull, Steenkoolen. *German*, Kohle, Steinkohle, Schwarz Kohle, Pech Kohle, Kannelkohle, Moor Kohle, Blatter Kohle, Grob Kohle. *Danish*, Kul. *Swedish*, Kol, Stenkol. *Cornish*, Kolon. *Irish*, Guel. *French*, Houille, Charbon-de-terre. *Belgic*, Houille, Houillieres, coal pits. *Italian*, Carbone Fossile. *Latin* and *Greek*, Lith-anthrax. *Portuguese*, Cârvoes de terra, ou de Pedra. *Russian*, Ugolj kamennoe. *Polish*, We̜giel ziemny. *Spanish*, Carbon de tierra, Rock coal; Carbon de Piedra, Stone coal. *Welsh*, Culm. *Swedish*, Kolm.

English, Charcoal, carbonized wood. *Italian*, Carbone di legna. Carbo ligni. *French*, Charbon de bois. *German*, Reine Kohle. *Spanish*, Carbon de lena. *Polish*, We̜giel. *Russian*, Ugolj.

English, Pitch. *German*, Pech. *French*, Poix.

English, Jet. *French*, Jayet. *German*, Gagat, Erdpech. *Russian*, *Polish*, Gagat.

English, Coke. *Swedish*, Stenkolstybb. *French*, Charbon de terre, Charbon de bois. *English*, Charcoal. *German*, Kohlenstoff.

Irish, Peat. *Scotch*, Peat. *English*, Turf. *German*, Torf. *French*, Tourbe, Tourbiere. *English*, Turbary. *New England*, *U. S.*, Tug.

GEOGRAPHICAL DISTRIBUTION OF COAL.

In his 25th chapter of the "History of Fossil Fuel," the author dilates on the influence which future discoveries of deposits of coal in foreign countries and the increased employment of the combustibles in manufactures there might have upon the industrial operations and local interests of Great Britain. Inquiries of that sort would scarcely be expedient here, inasmuch as we do not advocate the exclusive interest of any country, and acknowledge no preference for the prosperity of one section to the disadvantage of its neighbour. We espouse no cause save that of economic geology and the useful arts associated with it; contemplating these subjects with reference to their practical benefits, to their commercial and productive value, present and prospective. We estimate them in proportion as they are interesting in science, and conducive to the well-doing of the mass. With such views we seek not to define how far the possession of local advantages, the discovery of new mineral deposits or of improved appliances and facilities, may retard or accelerate rival interests. It is not our purpose to inquire into the injury which particular establishments or regions might sustain when placed in a state of competition with others which happen to enjoy a more favourable state of circumstances.

Two great facts, beyond all, stand prominent. It is certain that as manufacturing and productive industry take root and flourish almost exclusively in the cool and temperate zones, so in them do the coal formations and all the most useful mineral productions prevail in their greatest abundance. Our scientific maps and investigations confirm the one, and national statistics determine the other. Hence, the climates which are most congenial to laborious occupations, the latitudes which are best adapted to the more energetic pursuits of man, are precisely those where, fortunately, have been placed beneath his feet the raw materials most essential to his use. At the same time, the process of acquiring those materials, forms, of itself, one of the most valuable sources of his prosperity.

Between the Arctic Circle and the Tropic of Cancer repose all the principal carboniferous formations of our planet. Some detached coal deposits, it is true, exist above and below those limits, but they appear, so far as we know, to be of limited extent. Many of these southern coal-fields are of doubtful geological age. A few are supposed to approximate to the class of true coals, as they are commonly styled; others are decidedly of the brown coal and tertiary period, while the remainder belong to various intermediate ages, or possess peculiar characters which render them of doubtful geological origin.

In the high northern latitudes it has for some time been known that a species of coal exists on both sides of Greenland, and more recently it has been determined at various points of the Arctic ocean, between Baffin's Bay, and Behring's Straits. It is understood that the coal on the west coast of Greenland, and at Disco Island and Hasen Island is of the species denominated lignite, or the most recent of the mineral coals. Of the carboniferous formations discovered by the several exploring expeditions towards the North Pole, some are of the acknowledged brown coal age, others have been imperfectly examined and described, and may perhaps be of the same geological age as those enormously extended deposits which stretch through the central part of the American continent. The coals of Melville Island and Byam Martin's Island certainly appear to be of the true coal period. We know that coal exists, at numerous intermediate points, from the 75th to the 27th degree of north latitude, in America, and also that it is worked on the Salado and Rio Grande rivers in Mexico, for the use of the steamers.

Southward of the Tropic of Cancer the existence of coal, corresponding with the European and American hard coals, is somewhat uncertain. There seems to be none on the South American continent, unless it be at Cerro Pasco,—which needs confirmation,—or in the province of Santa Catherina, in Brazil. On the African continent we have had vague accounts of coal in Ethiopia and at Mozambique, also in Madagascar, and quite recently we have had intelligence of larger quantities of coal in the newly ceded territory above

Port Natal on the eastern side of Africa, but we believe no geologist has examined those sites. In the Chinese and Birmese empires only brown coal appears to approach the tropic. Southward of the Asiatic continent we are uncertain of the exact character of the coal deposits, such as occur abundantly in Sumatra, Java, and Borneo, and neighbouring islands.

In New South Wales the great coal range on the eastern margin of that continent was formerly sometimes considered to be like the Newcastle coal in England, and sometimes it was thought to be only brown coal; and indeed it is very certain that lignite does exist there; but the recent investigations of Count Strzelecki suggest that the epoch of the principal coal formations of Australia and Van Dieman's Land approaches somewhat to the oolite period. This coal differs essentially from that of any known European formation, but bears a strong resemblance to the Burdwan coal of India.

We may mention here, incidentally, that good coal is not essentially limited to the carboniferous period of the European geologists, but may and does exist, of excellent quality, in formations both of older and later origin. The Richmond coal field of America is now shown to be of an age not much, if any anterior to the older oolite series. Mr. Lyell has observed that no estimate of the probable value of the coal of India can be formed by comparing it with coal of the same age in Europe. Sir Henry de la Beche, has also remarked, that it was incorrect to suppose that in all other countries the most valuable coals would be found in rocks agreeing in age with the English coal measures. Those of Australia and Northern India, for instance, resemble each other in quality and in their fossil flora, yet both are dissimilar from those of the English coal-fields, and are evidently, like the Virginia coal alluded to, of an entirely different origin.

The evidence as to the facts contained in the foregoing sketch, will be found in detail in succeeding pages. From what has already been stated, it will be seen that it is impracticable, in numerous instances, to announce the true place in the geological scale, of formations which pass under the common denomination of coal. In some of these cases they have received no scientific investigation, and in others the results, if ascertained at all, have not reached us.

Of course, we have not yet arrived at the period when we could pronounce with any approach to certainty, on the actual number of coal basins in the world. Were we to venture an opinion, we should rate the number at from two hundred and fifty to three hundred principal coal-fields; and many of these are subdivided, by the disturbed position of the strata, into subordinate basins.

COMPARATIVE VALUE OF GOLD AND SILVER, AND OF COAL AND IRON.

A Spanish writer, not long since, instituted a comparison between the productive value of the silver and gold mines of America and that of the coal mines of England. The author exhibits a balance in favour of the latter of nearly two hundred and thirty millions of francs*=£9,286,000 sterling, annually.

Baron Humboldt, at the commencement of the nineteenth century, estimated the produce of the gold and silver in North and South America at	$43,500,000	
Which sum at the rate of 4*s.* 3*d.* a dollar amount† to		£9,233,750
Mr. Jacob estimated the annual value of precious metals from the American mines between the years 1800 and 1810, at	$47,061,000	£10,000,000
But from 1810 to 1829, the average annual production was only		£4,036,000
From thence to the present time the produce is certainly under	$24,000,000	£5,000,000
Exports of gold and silver from Mexico in 1842,‡	$18,500,000	£3,850,000
An estimate has been recently made with regard to the production of the precious metals to the following effect: In 40 years, from 1790 to 1830, the production of Mexico, Chili, Buenos Ayres, and Russia, in gold and silver, £188,000,000 sterling, equivalent to an annual average of		£4,700,000
Sir H. T. De la Beche estimates the value of the coal at the pit's mouth in Great Britain,		£9,000,000
Others estimate it at		£9,450,000
Another estimate extends the value to§		£10,000,000
The produce of the British coal mines is variously calculated at from 31½ millions to 34 millions of tons. At the respective places of consumption, in manufactures, in domestic use and that exported,	$96,800,000	
The value is probably from £17½ millions to		£20,000,000
The capital employed in the coal trade is computed at 8 or 10 millions more,‖		£10,000,000
The value of the iron produced through the agency of this coal in Great Britain at the furnace,		£8,000,000
Value of the iron when manufactured, in its various branches, which of itself		

* History of Fossil Fuel, p. 474.

† McCulloch's Geographical Gazateer, p. 80.

‡ Commerce and Resources of Mexico.—Hunt's Mag., vol. x., 1844, p. 121.

§ McCulloch.

‖ Mr. Buddle, in 1829.

greatly exceeds the value of all the gold and silver of the new world, in the most productive times,	$82,280,000	£17,000,000
Or nearly five times that of the gold and silver of Mexico, in 1842.		
The yearly value of the coal in five principal coal countries of the world, viz: Great Britain, Belgium, France, Prussia and Pennsylvania at their respective places of consumption, we have computed to be,	$145,200,000	£30,000,000
Which is nearly nine times the annual value of the gold and silver exported from Mexico, or six times that of the gross produce of the precious metals in North and South America and Russia.		
In 1847, a statement had obtained extensive circulation, which rates the value of the gold and silver produced in the world at 339,334,000 francs,	$65,489,000	£13,710,407
The value of the coal produced in the same year, upwards of		£17,000,000

EMPLOYMENT OF MINERAL COMBUSTIBLES.

In Great Britain, coal, according to some authorities, was mentioned as occurring in *England* as early as the ninth century, A. D., 853. It was certainly known and applied to various economical purposes in the middle of the twelfth century. In 1239, King Henry III. granted the privilege of digging coals to the good men of Newcastle. But it is little more than two hundred and fifty years since coal came to be in general use, as fuel, in London. Upon its first introduction there, one or two ships were sufficient for the whole trade.† At the present day there are several thousand ships constantly engaged in the transportation of that combustible.

It appears from a charter of Edward the Second, A. D. 1315, that the coal of Derbyshire was at that time known and in use. The introduction of coal for domestic purposes was retarded by the difficulty of employing it conveniently, and by the natural prejudice against such a description of fuel, as a substitute for wood, in cities.

By a proclamation of Edward the First, and again in the reign of Queen Elizabeth, we find that stone coal was prohibited in London during the sitting of Parliament, lest the health of the Knights of the Shire should suffer during their residence in the metropolis.

Blythe, an old agricultural author, writing in 1649, has the following passage:—" It was not many years since the famous city of London petitioned the Parliament of England against two anusances or offensive commodities, which were likely to come into great use and

* Williams's Mineral Kingdom.

esteem: and that was, Newcastle coals, in regard of their stench, &c., and hops, in regard that they would spoyle the taste of drinck, and endanger the people."

In France, the precise period of its adoption as a substitute for wood, is not ascertained: its introduction was probably very gradual. The commencement of its use in the city of Paris was in 1520, the coal being drawn not from the mines of France, but from the collieries of Newcastle. It would seem, however, that at the outset it met with little favour in Paris, as for some time was the case in London, doubtless owing to the difficulties attending its application. It was submitted to the decision of the faculty of medicine, in the former city, how far this new description of fuel was prejudicial to the public health. It was not probably before the middle of the sixteenth century that coal mining in France had commenced to be of any importance.*

In Scotland, mineral coal was known, probably, much earlier than in France. The privilege of digging coal is mentioned in a grant to a religious house, A. D. 1291.†

In Belgium, the earliest reference to mineral coal was in 1198 or 1200, in the country of Liege, where tradition gives the credit of the application, as a fuel, to a blacksmith. From this time there seems to be evidence of its being in ordinary use, and that the business of its extraction had, from a remote period to the fifteenth century, been subject to the supervision of an especial court or jury.‡

In these and some other countries, we have already shown the extraordinary accelerated demand for coal since the application of steam power; more especially within the last quarter of a century. We have also pointed out the vast capital which this substance keeps in motion; the numerous population which it employs and sustains.

Great as has been the rate of advance in England, that of France and of Prussia, within the same time, has somewhat exceeded hers, while that of Pennsylvania, in the United States, has far surpassed them all.

The Tyne and Wear districts, in Northumberland, are the most remarkable instances of coal production in the world. They supply above six millions of tons annually; employ about 23,000 miners; support 140,000 persons in manual labour; and, with their families and dependents, sustain 700,000 individuals.

From *South Wales* we have received no recent returns of the annual quantity of bituminous coal and anthracite, or of the number of persons engaged in their production. The bulk of the former has always been consumed in iron making in the interior, beides a vast amount exported coastwise. Since the uses of anthracite have been made apparent, the consumption of that mineral has greatly increased. As far back as 1835, the making of bar iron in that region employed 28,000 persons.

* Résumé des travaux statistiques, Paris, 1839.

† See many historical notes in the "History of Fossil Fuel."

‡ Bulletin de la Commission centrale de Statisque, 1843, Brussels.

The total number employed in England on this branch of manufactory was, at that time, near 70,000 persons; while the aggregate of persons dependent on these was upwards of 250,000. Proceeding to a more advanced stage in iron manufactures, it was announced that the value of the hardware and cutlery annually made, was above $82,280,000, giving employment to 325,000 persons. Hence, it appears, that the number of persons directly or indirectly drawing support from the production and employment of the two substances, iron and coal, amonnt, on a rough estimate, to a million and half of persons.

"It is hardly possible," says Mr. McCulloch, "to exaggerate the advantages England derives from her vast beds of coal. In this climate fuel ranks among the necessaries of life; and it is to our coal mines that we owe abundant and cheap supplies of so indispensable an article. Our coal mines are the principal source and foundation of our manufacturing and commercial prosperity. Since the invention of the steam engine, coal has become of the highest importance as a moving power; and no nation, however favourably situated in other respects, not plentifully supplied with this mineral, need hope to rival those that are, in most branches of manufacturing industry. To what is the astonishing increase of Glasgow, Manchester, Birmingham, Leeds, Sheffield, &c., and the comparatively stationary, or declining state of Canterbury, Winchester, Salisbury, and other towns in the south of England, to be ascribed? The abundance of coal in the north, and its scarcity and consequent high price in the south, is the real cause of this striking discrepancy.

"Our coal mines have conferred a thousand times more real advantage on us than we have derived from the conquest of the Mogul Empire, or than we should have reaped from the dominions of Mexico and Peru. They have supplied our manfacturers and artisans with a power of unbounded energy, and easy control; and they have enabled them to overcome difficulties insurmountable by those to whom nature has been less liberal of her choicest gifts."*

Mineral Coal applied to Iron Making.—The earliest employment of this fuel in England, in the manufacture of iron, was in 1713, at Colebrookdale. In Scotland it was introduced about the middle of the eighteenth century, and in France in 1782; in the coal field of Creusot. Numberless notes will be found in the pages of this volume, in illustration of this interesting subject.

GEOLOGICAL POSITION OF COAL BEDS.

"Coal is found in beds, and its presence characterizes, in an especial manner, the carboniferous formation. We have to seek it then, above the transition series and below the secondary deposits; —above the schistose beds, the insoluble clays and trilobite limestones; below the arenaceous deposits which contain the debris of porphyries, the limestones with ammonites, gryphites, belemnites, &c.

* Statistics of the British Empire, vol. ii. p. 2.

The coal formation is remarkable for the peculiar appearance (*facies*,) of its micaceous sandstones and its argillaceous shales. In the coal sandstones, the elements of feldspar and quartz, in very nearly equal proportions, spangled with mica in little scales, passing in the lower portions, into breccias and conglomerates with large fragments, are evidently the result of the action of the waters upon pre-existing transition rocks. The granites and gneiss have furnished the principal amount of these elements; and we can often determine the points from whence they have been drifted. The argillaceous schists, rarely soluble, but always falling to pieces in the air, form the passage of the transition argillaceous schists into the true clays of the posterior strata. They are evidently decomposed parts of the rocks which constitute the sandstones. An impure melange of kaolin, of silex and of mica, of which the elements, fine enough to have been held in suspension, were only deposited when the stagnation of the waters permitted. These beds alternate with a great predominance of the sandstones; all are frequently colored by the disseminated carbon, which gives to the ensemble a grey tint and a characteristic duskiness. The presence of the carbon manifests itself also by that of the carbonate of iron—*fer carbonaté lithoïde*—which is found, either in subordinate beds, or in disseminated nodules—*rognons*—in certain beds of clay. Finally, it manifests itself by numerous vegetable impressions, and by the frequent, but not essential, presence of seams of coal, sometimes fat and sometimes dry.

The influences which have determined the characters of the rocks that are associated with the coal beds, have been so constant, that not only are they identical all over the globe, but in the cases where coal beds are found in other formations than the coal formation, the rocks of those formations abandon their special characters to borrow those which we have described.

Thus, in the anthraxiferous formation which immediately precedes the coal period, the lean coals which are worked in the west of France, are accompanied by feldspathic, micaceous sandstones, and carbonaceous schists, with impressions of calamites, ferns, and sigillariæ. Black argillaceous schists, with nodules of carbonate of iron, accompany equally the beds of secondary coal which are found in certain points of the lias near Milhau, (Aveyron) and in Yorkshire.

To sum up the various geognostic positions of coal: they are met with, 1st, In the anthraxiferous formation; that is to say, in the upper part of the transition series, even above the silurian beds. 2d. In the coal formation, properly speaking. 3d. In the marnes irisées, where are found the coals of Noroy and Gemonval. 4th. In the lias formation. [Environs of Milhau.]

Above this last position, the vegetable debris is found most generally in the state of lignites. We find, but rarely, in the lignites of the cretaceous and tertiary formations, portions from which the ligneous texture has disappeared, and which present the appearance of coal; but this case is exceptional. Thus certain lignites in the environs of Marseilles, and others which exist in the tertiary beds of Italy, present the tissue and the characters of coal, but these accidental facts, which establish between the coals and the lignites mineralogical transitions that exist even between rocks the most distinct, strikes no blow at the rules of position, established undeniably by geological observations. It is the same with that other geognostic law which assigns peat solely to the alluvial epoch, or the actual existing epoch.

The meagre coals and anthracite appear in general to be of a more ancient age than the fat or flaming coals. This classification is sufficiently indicated by the general dry nature of the combustibles mined in the anthraxiferous formation of the west. In the north, the lean coals of Fresne, Vieux-Condé, Vicoigne, are evidently inferior to the fat beds of Anzin and Denain. The beds found in the carboniferous limestone at Château l'Abbaye are true anthracites. The anthracites of the environs of Roanne, and those of the United States,* belong to the upper formation of the transition series. But it is necessary again, more than in the preceding cases, to abstain from taking this rule in an absolute manner; for the anthracite state is very often the metamorphic state of the coal, and even of the lignites. The interesting researches of M. Elie de Beaumont upon the anthracites of the Alpine regions have demonstrated this fact, otherwise easy to conceive."

From each of the four classes or epochs of combustibles, M. Régnault has selected the most characteristic, and after having submitted them to analysis, he has acknowledged that this general succession of characters in the fossil combustibles is in accordance with a successive approach towards the composition of the vegetation; in such manner that, from the anthracites of the transition series, even to the lignites and to the peat of the existing epoch, the fossil combustibles form a series of which almost pure carbon forms the base, and which is gradually charged with four, five, and six per cent. of hydrogen, and with four, eight, twelve, sixteen, and thirty per cent. of oxygen.

We may lay down this principle, abstraction being made for the anomalies of metamorphism, that the more of gas that a combustible contains, and the higher the amount of oxygen and hydrogen, so much the more modern is the combustible.

LOCAL POSITION AND ARRANGEMENT OF BEDS OF COAL.

Coal, whatever may be the formation in which it is found, affects the form of *beds*, of very variable thickness and continuity, but whose

* Respecting the geological age of the anthracite of the United States, we think that there is good ground for dissenting from the views of M. Burat, in placing this carboniferous formation in the superior part of the transition series. It is true, the present writer formerly held and advocated precisely the same opinions, but subsequent investigations have clearly established the geological fact, that the Pennsylvania anthracites are simply in the metamorphic state; that they are based upon the old red sandstone, and that the numerous basins in which they are deposited are but isolated, or out-lying portions of the great bituminous coal field of the Alleghany Mountains.

constant character is that of conforming to all the courses [*allures*] of the beds of schist and carboniferous sandstones between which they are included. This stratification is not only indicated by the limits of the roof and the wall or floor, but also by natural variations in purity, the positions of which generally pursue or occupy lines parallel to those of the roof and floor; by the bands of intercalated slate, and by the continuous *barres* which divide the beds into several courses. Finally, the coals themselves often present a great number of interruptions, and of veins which render its structure striped, lamellar, and following the direction of the stratification.

The stratification of coal ought not, however, to be considered as absolute, and to be compared to that of the calcareous or argillaceous beds of the sedimentary formations, nor even to that of the sandstones and shales which alternate with them. Certain beds present massive undulated forms, yet without these undulations having been occasioned by the course or strike of the formation. This shows that the origin of the coal permits, at one and the same time, thin layers, continuous and of the greatest regularity, and thick beds, so limited and irregular, that they may be assimilated to masses. The ulterior details, respecting the forms and courses of the principal coal deposits, will decide perfectly our ideas in this respect. The thin and regular beds, although very much disturbed, of the basins of the north of France and Belgium, and the thick and limited beds of Montchanin represent the two extremes of position.

The number of coal beds in the same formation, as well as their thickness and continuity, appear to be subject to very great variations. Nevertheless, there is a certain connection between these different conditions. The thin and regular beds are commonly continuous and multiplied; the thick and unequal masses are, on the contrary, limited in their extent, and there are rarely more than two or three superposed in the formation which encloses them. Thus, in the basin of Mons, in Belgium, more than one hundred distinct seams of coal are counted, whose ordinary thickness varies from eight inches to near five feet. In the collieries of the department of the North, in France, there are few centres of exploitation which do not count six, eight, twelve, or more beds of coal; but their maximum thickness does not exceed three feet, and the greater part of those which are worked have only about twenty inches.

Although coal beds have frequently been traced along a distance of many miles, yet we ought not, even in the case of very great regularity, to suppose that certain beds of coal are absolutely coextensive with the whole formation. For example, there is an interruption between Valenciennes and the Belgian frontier, to such an extent that the beds of Anzin are not those of Mons, and these again have no relation of continuity with the beds of Liege or Charleroy. The same remarks can be applied to other coal basins. Thus, in the basin of the Loire, the beds of Rive-de-Gier are not the same as those of Saint-Etienne. We may then, in a basin of some extent, consider the coal as forming, within the beds of sandstones and shales, special districts; often isolated, the one from the other, by sterile portions, and of which the coal beds, differing in number and power, have no relation of continuity.

Although, then, even if we have discovered the sandstones and the shales of the coal formation, it does not follow that we have also found the coal, though we were on the prolongation, in direction or inclination, of known beds. To form a probable hypothesis on this subject, it would be necessary first to study the peculiar conditions of the formation on which to operate, and calculate, from the known portions, the chances that we may have.

The strata of the basin of the Saône-et-Loire appear to form basins subordinate to the principal basin, which is filled up with coal, sandstones and schists. These subordinate basins are bounded like the basin which contains them, and have, besides, nearly similar proportions between the axes. Further, the coal appears there to diminish in length in proportion as it acquires thickness. In the valley of Creuzot, the great bed which is worked has twelves metres, or forty feet, of mean thickness. In the enlarged portions it has 40 metres, or 130 feet from wall to roof. In direction it is not prolonged above 2000 yards; and, with regard to its limits, its divided extremities, there impoverished, present all the symptoms of a total suppression.

The bed to Montchanin, greatly inclined, whose thickness attains even to 70 metres, or nearly 230 feet, measured from roof to floor, representing, consequently, the thickest known coal bed, is equally one of the most limited in extension. In fact, in the upper part of the exploitation, this direction is about 650 yards, at the end of which the bed terminates abruptly, and is confused or entangled in the rocks of the roof and floor. At a lower stage, about thirty yards below the first, the length is reduced to 450 yards, and it is probable that at the depth of about 150 yards the bed will terminate.

The basin of the Loire contains, in the region of Rive-de-Gier, but three beds, of which the average united thickness does not exceed 32 feet; but, in the district of Saint-Etienne, the sum of the regular beds amount to 114 feet, in fifteen to eighteen beds. At Brassac they amount to from 27 to 40 feet; 45 feet at Comentry and Doyet; and 48 to 65 feet in the basin of Aubin. It is remarkable that, in all these basins, the coal beds, of 15 to 30 feet, are occasionally reduced, by contractions, to 6 or 10 feet, and at times are swelled out to the thickness of 60 to 90 feet, an ordinary and normal fact.

In the department of the North, on the contrary, 30 feet of total thickness are divided into 14 beds, worked at Fresne and Vieux-Condé. The 12 beds of Aniche only form 22 feet; four successive beds at Douchy have only 11 feet aggregate; at Denin 7½ feet only; and 38 feet are occupied by not less than eighteen beds at Anzin. There are also still more veins, which are unworked, and whose thickness is below one foot each. But these beds are regular and prolonged, and are not disturbed by those enlargements and entanglements so frequent in the beds of the southern basins.

This difference of power and continuity in the coal beds agrees also with some very important differences indicated by geological observation.

The southern basins of France appear to have been deposited, during the coal period, in isolated lakes of fresh water; encircled, and entirely commanded by the neighbouring summits, from whence the materials have often been drifted with violence: forming breccias and conglomerates. In studying this debris, especially in the lower parts of the deposit, we can recognize the transition rocks of the surrounding countries. The northern basin of Belgium and France, containing at its base the carboniferous limestone, is, on the contrary, only composed of sandstones and fine schists. It appears, from the character of the fossils, to have been formed in marine waters, and thus represents, with the coal basins of England, the pelagic accumulations of an epoch, of which the basins of the south are but the lacustrine terminations. It is, then, natural to discover, in these northern deposits, a regular and continuous disposition which comports not with the deposits of the south.

The southern basins, desposited in isolated lakes of fresh water, form the principal riches of France. The aggregate thickness of the coal is, besides, nearly as great as in the basins of the north. To sum up all, we can lay down no absolute rule for the number and power of the beds of coal, any more than for their continuity. The indices, which result from the direction of the stratification, have, nevertheless, a real value, even in the countries where the continuity presents the most frequent exceptions; because they always conduct to the possibility of finding, if not the prolongation of the beds, at least to formations analogous to such as have been already discovered.

ACCIDENTS, FAULTS, AND IRREGULARITIES, OF COAL BEDS.

The beds of coal are rarely in the position where they have been produced, for that position would approach sensibly to the horizontal; a condition compelled, if not by the mode of production of the coal itself, at least by that of the beds of sandstone and slates between which it is stratified. Most frequently, the uniformity of the formation is disturbed, not only by inclinations, more or less great, but also by the folds which change these inclinations, and distort the beds to such an extent that a vertical shaft might cut them several times. There often, also, prevails one or several systems of *faults* [*failles*] which change the levels and isolate, one from the other, the divers parts of a bed.

This *accidentation*, subsequent to the production of the beds, and resulting from dynamic perturbations, commonly influenced by determinable conditions of direction, must be distinguished from the contemporaneous accidents inherent even to the production of the coal; such as the undulations of the roof and floor, which swell or contract a bed, and the intercalations of layers or of amygdaloid rocks, which interrupt the regular order of the stratification. Nevertheless, there is an evident connection between these two causes of irregularities;

because the dynamic perturbations appear to have sometimes operated upon the coal beds before they were solidified; or, at least, when they were in such a state, that they were capable of being compressed, squeezed, and even completely suppressed, by a compression between the rocks of the walls, and, consequently, enlarged at other points.

The distorted structure, often smooth and polished, of the shales which accompany the coal, thus troubled; the state of the coal itself, which is there not only more crushed than at other places, but sometimes twisted or contorted and in a manner kneaded, seem to conform the existence of these almost contemporaneous perturbations.

We may, besides, by attentive observations, frequently distinguish the dynamic and violent perturbations of such as result even from the circumstances of the deposit. The regular seams [*nerfs*] of shale and their beds or bands [*barres*] of clay, which are almost always interposed in beds of coal, following the direction of the stratification, can furnish many indices in this respect. Thus, in a natural expansion, not only the seams (nerfs) and bands which exist, do not experience perturbations, but they add other parallels to the increasing thickness of the coal. A natural contraction is often produced by the dilatation of the "barres," and, at other times, the barres submit gradually, like the coal itself, to the influence of diminution. In the dynamic accidents on the contrary, the "nerfs" and barres are broken suddenly, and their fragments, blended with the coal, announce, in advance, to the miner the accident which comes to modify the "allure" of the bed.

The accidents to which coal beds are subjected, are those of the *inclination*, and *folds*, the *faults*, the displacements and the disturbances.

Inclination is the most general casualty: it is rarely, in fact, that the seams are presented in a horizontal position. These inclinations are evidently the result of perturbations, of upheavings or of sinkings of the earth, subsequently to the deposit of the formation.

The direction of all the beds is commonly the same in a coal basin, but their inclination varies. Thus, it is remarked that upon the opposite borders of a basin, the slopes were most frequently directed towards each other; and it has been proved that sometimes there was a junction of the two slopes in the middle of the basin by a plane or curved portion which has been called the *bottom of the boat*—["*fond de batteau,*"] because, in fact, the section of the two slopes, thus united, bears resemblance to the section of a boat. This disposition, which has been too generalized for the ensemble of the formation, it having been very frequently deranged by accidents of another nature, and as regards the coal beds, when the continuity is not always established between the beds of which the slopes tend theoretically towards each other, is nevertheless, with these exceptions, a very common occurrence.

This fact indicates that coal basins have generally been compressed by lateral upheavings or pressure.

The change of inclination often involves the existence of curves of adjustment, which are only the folds [*plis*] of the beds. In the greater part of the circumscribed basins, these folds are of great radius: but, in the beds of the great northern basin of France, the folds are sometimes so sudden and decided that they change the inclination of from 10° or 15°, to that of 75° or 80° at any given point.

The section of the basin of Mons, in Belgium, is a good example of the plication or doubling back of the coal beds of the northern basin, even to such an extent as to permit the verticle shafts to pass two or three times through the same beds. Most frequently there is an enlargement of the thickness at the angle or bend ["*crochon*,"] of a ply, and the thickness of a bed of one yard is increased to one and a half or two yards. Those beds whose inclination is below twenty degrees, permit the establishment of working galleries according to the method which bears the name of flats—"*plats*," and they call uprights—"*droits*"—those which possess a high inclination. The same beds, therefore, that occur in our section of Mons, assume alternately the disposition of flats and uprights.

These folds have, at the same time, both a direction and an inclination, and form a sort of sloping gutter which is called by the Belgian and French miners "*ennoyage*."

The plications are evidently the effect of dynamic causes, which have produced these inclinations; they result from upheavings which have undulated the surface of the basin, and from lateral pressure, which has forced the groups thus undulated, to occupy a greatly contracted space.

Contractions and enlargements are frequent accidents in coal beds; they both exist, generally, in the same vein, and, in some basins, the miners are in the habit of saying, when the roof and floor widen suddenly, "the bed enlarges, we are going to lose it." A gradual and prolonged contraction, a division of the bed, whose planes of stratification become entangled in the rocks of the roof and floor; in fine, the alteration of the coal which becomes more and more mixed with argillaceous slate, constitute an impoverishment which is the ordinary precursor of a total suppression.

When the two walls of the vein, approaching each other, come to unite, and for a while to suppress the coal bed, the accident takes the name of *fault*, [*crain* or *coufflée*.]

Faults are more frequent accidents in the thick beds than in those which do not exceed a yard. By following the carbonaceous thread which almost always exists as a trace left by the coal itself, or, in default of that, by following the rocks of the roof and floor, of which nature furnishes indices sufficient to preserve the plane of stratification, the bed will be recovered, after an interruption of greater or less extent. In the mines of Rive-de-Gier, the lines of white slate serve as a guide, in default of carbon, by which to pass over or go by the faults. In the mines of the environs of Nantes, where the suppression is equally complete, it is necessary to consult attentively the rocks of the walls, in order to pass by, without deviation, the very considerable spaces which often separate the prolongation of the same bed.

Miners are sometimes very much embarrassed, when they have pursued the trace of a fault during a long space; for, not finding again any indication of resumption, they do not know whether the interruption should be attributed to the presence of a fault or to the definitive cessation of the coal. No rule can be laid down in this case; the observation of the structure of the whole can only furnish the data which we have already indicated above.

Faults are very common accidents. They are fractures which affect the entire character of the formation, and cause greater or less disturbances. These faults have a determinate direction, and frequently a basin is affected by a system of faults parallel with each other. At other times, there are several systems which follow different directions, but are each composed of faults, connected with each other, by a parallelism of direction.

The intensity of the faults is very variable; sometimes they scarcely interrupt the formation, and appear as fissures which have changed the level of the two ruptured parts, but not enough to constitute the total interruption of the coal, which is always easy to follow, when the offset is not greater than the thickness of the coal. The bed of Lucy, in the basin of Saône-et-Loire, frequently presents such faults.

The section of the coal bed of Monceau furnishes an example of interruptions of continuity caused by movements subsequent to the formation of the coal deposit, and of a confusion [*brouillage*] which totally interrupts the coal.

The *brouillages* are nothing but the intervals comprised between the planes of fracture; in these intervals all the beds are broken and reduced to angular blocks mingled together.

When these faults form part of those which have determined the outline of the surface of the ground, the offsets or upthrows are, in some measure, proportionate to the inequalities which it presents. Thus, there are offsets and upthrows of several hundred feet in the basins of England, of Wales, and of Rive-de-Gier, whose surface is highly disturbed.

In the basins of the Saône-et-Loire, they are rarely more than from thirty to ninety feet, and are, notwithstanding, in agreement with the undulations of the surface.

Endeavours have been made to establish, from the most detailed and available information, agreements between the arrangement of the beds of coal and the superficial accidents of the soil. Thus, in a great number of basins, the direction of their beds coincide with that of the great axis of the coal formation, and this great axis is itself directed in the same course as the existing valleys, in such a manner that the direction of the beds is confounded with that of the dividing lines of the waters and the principal valleys. In some other basins, the planes of stratification of the formation have not only the same direction, but also the same inclination as that of the surface declivities.

The basin of Bert offers a striking example of this agreement. The inclinations of its coal beds change as often as five times whilst conforming to the inclinations of the surface, and the range of the valleys corresponds with the direction of the beds. The greater part of the basins which affect the boat form, present, at their surfaces, an analogous disposition; that is to say, the waters follow the direction of the strata, and the lateral margins of the basins are generally more elevated than the axes.*

English Coal-Fields.—Faults and interruptions prevail, more or less, as might be expected, in most coal-fields, but they possess different characters in different regions. The Newcastle coal-field is remarkable for its number of faults; from the dimensions of a few inches to a hundred fathoms. But in the southern coal basins, particularly those of the Forest of Dean and South Wales, there are frequently found remarkable irregularities, called "*horses*." Where these horses occur, the coal disappears all at once; but yet without any fault at all. They have to be cut through, and, after a time, the coal reappears.† These horses appear to be ascribable to interruptions in the original deposition of the vegetable matter of the coal seams.

Chemical Geology, as applied to Coal.—At the tenth annual meeting of the British Association for the Advancement of Science, as reported in the Athenæum, Professor Johnston brought forward the result of his investigations on the most important of mineral productions, coal.

Although some geologists may entertain a different opinion, he assumes for granted the vegetable origin of coal. Although it may be classified in various ways, for economic or geologic convenience, as into caking or not caking, bituminous or non-bituminous, the true basis of the classification must depend on the chemical composition. Carbon, oxygen, and hydrogen, are the component parts of living vegetables, and the same elements compose coal, but in different proportions.

In the decomposition of vegetable matter, there are two agents always at work—viz. atmospheric air and water, which resolve it into carbon, oxygen, and hydrogen; forming, with one another, those combinations—carburetted hydrogen, carbonic acid, and water. Vegetable matter, consequently, in different states, showed different proportions of these elements.

The quantity of carbon in all the different varieties of coal, in Mr. Johnston's table, was taken as a constant quantity; and from lignite, downwards, we see a progressive loss of hydrogen and oxygen; until, in anthracite, the carbon is the chief component.

This is borne out by experience. In the change from lignite to

* Géologie appliquée, par M. Amédée Burat, 1846.

† Professor Ansted's Geological Lectures, 1847–8.

fossil wood we find that carbonic acid only is parted with; and this continues, without variation, in all the kinds, down to cannel coal.

In mines of lignite and cannel coal, we find only *carbonic acid*, or *choke damp;* while in mines of coal lower in the scale, we find, in addition, *carburetted hydrogen* or *fire damp.* This also appears in the table referred to; the hydrogen diminishing in each variety as we approach anthracite.

In some mines we find a perfect confirmation of this theory. In certain Yorkshire mines, coal of different kinds, cannel coal being at the top, evidently prove that those below, having been longer subjected to chemical action, had parted with more of their hydrogen. The same occurs in mines in Lancashire.

In conclusion, Professor Johnston asserted, that bituminous matter must be of vegetable origin—in fact, chemistry proved it. Distillation of vegetable matter in a gas work, or in the laboratory of a volcano, was the same process.

In further support of this conclusion, we cite the following high authority:

Table of Analysis of Coal and certain allied Combustibles, by Berthier.

Composition in 100 parts.	Peat or Turf.	Lignite or Brown Coal.	Bituminous Coal—rich.	Pennsylvania Anthracite.	Graphite or Plumbago.
Carbon,	38	54	73	94	95
Hydrogen,		04	05	2.55	
Oxygen,		26	20	2.56	
Ashes,	17.4	14	02		
Volatile matter,	28				
Iron,					5

These different varieties of brown coal, peat, bituminous coal, anthracite, and graphite, correspond so exactly, that this alone would show the vegetable origin of them all; from the peat up to the graphite, if no other proofs were at hand. [See Table next page.]

In the belief that every species of information which makes the adaptation of the various mineral combustibles to the manufacture of iron better understood, must be useful and in strict conformity with the plan of the present work, we have arranged the following practical details. Great changes have taken place, within a few years, in the management of fuel, and in the degree of estimation in which each species is held by operative and scientific men. It is proper to know the conclusions to which those persons have arrived. We cannot here give all those results in detail; and, moreover, this is not a treatise on iron making. But we have sought to concentrate certain material facts on the nature and capabilities of the principal varieties. It will then be easy to compare them with others of corresponding character. We have therefore given, in the following page, a comparative table and characteristic analysis of the principal descriptions of coal employed in the iron works of Europe and the United States.

Varieties of Coal, with reference to their Adaptation to the making of Iron.

Countries and classification.		Localities of Coals.	By whom analysed.	Carbon. Per cent.	Volatile matter.	Ashes.
I. Fat, bituminous, adhesive coals; the greater part *close burning* or strong burning blazing coals.	America,	W. Penn'a, Ohio, Virg. Illinois,	Various persons,	52.0	44.0	4.0
	England, A.	Newcastle-upon-Tyne, Birtley,	Berthier,	60.5	35.5	4.0
		Northumberland, Tyne works,	"	67.5	30.0	2.5
		Staffordshire, Apdale works,	"	62.4	34.1	3.5
	do. B,	" Wednesbury,	"	67.5	30.0	2.5
		Derbyshire, Butterley, Cherry,	"	57.0	40.0	3.0
		" Codnor Park, soft coal,	"	51.5	45.5	3.0
	do. C,	Lancashire, cannel coal,	Karsten,	56.0	38.5	5.5
		Scotch, " Lismahago,	Mushet,	39.4	56.6	4.0
		Derbyshire, " Morely P'k,	"	45.0	45.0	10.0
	France, D.	Anzin,	Berthier,	70.5	25.0	3.5
		Rive de Gier,	"	66.5	31.5	2.0
		Saint Etienne,	Gruner,	74.3	24.2	1.5
II. Dry coals, not very adhesive: can be used crude in the furnace with heated air. *Open burning coals.*	Scotland. E.	Clyde, splint coal,	Mushet,	59.0	36.8	4.2
		" clod coal, richest,	"	70.0	26.5	4.5
		" near Glasgow,	Berthier,	64.4	31.0	4.6
		Calder, near "	"	51.0	45.0	4.0
		Monkland, "	"	56.2	42.4	1.4
III. Less adhesive or caking.	U. States, F.	Pennsylvania, Philipsburg,	Johnson,	68.0	22.0	10.0
		" Karthaus,	"	68.1	26.8	5.1
		Virginia, Richmond,	Clemson,	64.2	26.0	9.8
		Illinois, Ottawa,	Fraser,	62.6	35.5	1.9
IV. Steam coals, very dry coals, with excess of carbon. *Open burning.* Intermediate class, semi-bituminous.	S. Wales,	Dowlais, iron works,	"	79.5	17.5	3.0
		Merthyr Tydvil "	"	78.4	18.8	2.8
		Pen-y-Daran "	Mushet,	86.0	12.0	2.0
		Aberdare, "	Unknown,	87.0	11.5	1.5
		Rhymney & Tredegar, works,	Mushet,	81.0	15.0	4.0
		Steam c'l, Pembrey & Llanelly,	" [mean]	80.0	17.0	3.0
	Belgium,	Mons, Dour,	Berthier,	85.0	12.7	2.3
	France,	Auvergne, Saint Etienne,	Gruner,	74.8	21.7	3.5
	America, U. S.	Dauphin Co. Pa. Rattling Run,	Lea,	76.1	16.9	7.0
		Maryland, Savage River,	Jackson,	77.0	16.0	7.0
		Pennsylvania, Blossburg,	Clemson,	75.4	16.4	8.2
		" Broad-top,	"	70.1	16.7	13.2
V. Anthracite.	S. Wales,	South Wales, Neath Valley,	Mushet,	91.0	8.0	1.0
		" Ystal-y-ferra,	"	92.5	6.0	1.5
		" Cwm Neath,	"	95.7	2.8	1.5
	U. States.	Pennsylvania, Pottsville,	Rogers,	94.1	1.4	4.5
		" Black Sp. Gap,	Lea,	88.6	7.1	4.3
		" Mauch Chunk,	Rogers,	88.5	7.5	4.0
		" Sugar-loaf,	Johnson,	90.7	7.0	2.3
		Rhode Island, Portsmouth,	Jackson,	85.0	10.0	5.0
		Massachusetts, Mansfield,	"	92.0	6.0	2.0
	Russia,	Territory of the Don Cossacks,	Voskressensky,	94.2		

A, Coals which cannot be employed in iron works, in the crude state.

B, Coals which cement less in the fire, and which it is practicable to use raw in furnaces worked with heated air.

C, Chiefly for illuminating gas.

CLASSIFICATION OF MINERAL COALS.

In the foregoing table of analysis of coals and anthracites we have so arranged them as to exhibit their varieties or gradations, and their distinguishing properties, in different countries. Hence, the European coals can readily be compared with those of America, and the adaptations of either may be assigned with some degree of confidence. We proceed to note these characteristic differences and agreements more in detail.

I. *Fat Bituminous, blazing, coking.*—In the first class, series A, of the table, by way of illustration, the English coals of the north, and some of the coals of Silesia, of Hesse, of France, and of America, in the Ohio Valley, are chiefly fat and very adhesive or caking; swelling much in the fire. The hot air blast is successfully applied with these in the high furnaces. But, as their tendency to cement together in a solid mass, when in the fire, is such as to prevent a free draft or passage of the air through the furnace, it has been found indispensable to submit the coals to a preliminary process, and to reduce them to coke. Thus, the difficulty is wholly removed; and a light, cellular, and purely carbonaceous substance, easily ignited, is substituted for the unmanageable coal in its crude state. The average quantity of carbon which the English coals possess, is stated to be sixty-five per cent.

Series B, more southward, in Staffordshire and Derbyshire: these coals, although containing as much, and even more bitumen, do not melt together like those of Northumberland. They scarcely change their form even in the state of coke. The varieties, having this property, admit of their being used in the *raw state*, but require the introduction of hot air into the furnaces. Some of the American coals west of the Alleghany mountain have also these characters.

In regard to the manufacture of illuminating *gas*, the type of perfection, in the series C, is the Scotch cannel; then comes after it the Lancashire cannel, and, in the third order, the Yorkshire and Derbyshire cannel. With this class we would place the cannel coals of Kentucky, Indiana, Illinois and Missouri. This series can be assimilated, in many respects, to the coal of the basin of Mons in Belgium. The splint coal of Scotland is only a coarse variety of cannel, as are the greatest part of all the Scotch coals.

The Newcastle coals have a resemblance to those of Anzin, of Saint-Etienne, and of Rivè-de-Gier, the analysis of which we have placed in the series D.

II. Series E.—In the second class, the Scotch coals, although containing as much bitumen as those of the north of Scotland, are of the kind denominated *dry coals*. They cement together, but without change of form, and are not so adhesive as the fat English coals. These were heretofore coked before being put into the furnaces; but recent improvements have shown that, with the application of heated air, they can be employed without being previously carbonized. Their average proportion of carbon is about sixty per cent., and of bitumen 36 per cent. Some of the Alleghany coals will probably be found to assimilate with these. Approximating to the same class, to a certain point, will be found the coals of Auvergne and of a part of the south of France.

III. Series F.—We have assigned an intermediate space for a series of coals in the American coal basins which differ little from E, except that they contain somewhat less of bitumen and more of carbon, viz. about 66 per cent. of carbon and 27 per cent. of bitumen and volatile matter, and are less adhesive and caking. The Heraclea coal, in Anatolia, appears to belong to this series, and those of the Cantal and Puy de Dôme in France. These are convertible into coke.

IV. *Intermediate Series, very dry coals,—Semi-bituminous coals,—Steam coals.*—In the fourth class, of which the Welsh coal of the southern and eastern districts is the type, and which possess only from twelve to twenty per cent. of volatile matter and bitumen, may be arranged those denominated "very dry coals, with excess of carbon." These do not cake or cement together in a mass, although each individual fragment is susceptible of conversion separately into coke, and consequently do not offer a similar obstruction to the current of air in the furnace, like those of the first class. It has, therefore, been found that they may be employed in a crude state in the cold air furnaces of South Wales. This class contains a larger proportion of carbon than the two others, being eighty-one per cent.

There exist both in France, Saxony and Belgium, coals which bear some resemblance to these. In the United States of America, particularly in Maryland, Virginia, and Pennsylvania, are some species which closely assimilate with the foregoing, usually denominated "open-burning," and sometimes "semi-bituminous," and are not surpassed by any known species for certain valuable properties.

Under this head may be arranged the culm of Kilkenny and of Glamorganshire, and the quality which prevails in some of the southern seams of the Welsh coal-field, and now universally known by the name of "steam coal," being supplied to the British marine steamers, and even to those of France and Egypt. The Welsh culm is a very light coal, of loose texture, very glossy, and composed of capillary fibres arranged in divergent rays. It burns easily, and without smoke, makes a lively fire, and is in great request in Swansea and Cornwall for the smelting of copper. Depots of steam coals are formed in the East and West Indies, and in various parts of the world, for the service of the English steamers.

There is in England another variety of coal, but not abundant, called *flint coal*, because it is almost as hard as flint, and has a shining fracture approaching to anthracite. The *flew coal* of the mines of Wedgebury in Staffordshire, belongs to this series. In Cumberland, at Alston Moor, a variety of coal is found, almost without bitumen, called *crow coal*, which approaches to the French coal of Fresnes.

V. In the fifth class are comprised the anthracites, or the non-bituminous coals. Our tables of analysis exhibit the component parts of this mineral from all the principal known deposits. In Pennsylvania it contains from 85 to 92 per cent. of carbon; in South Wales from 88 to 95; in France 80 to 83; in Saxony 81, and in Russia reaches 94 per cent.

After many years of unsuccessful trial in endeavouring to adapt this valuable mineral combustible to the manufacture of iron, the difficulties which at one time seemed insurmountable, were overcome both in Wales and in Pennsylvania, where many furnances, using the hot blast, are now in full activity. The domestic use of anthracite in the United States is very extensive, and annually increasing; all the original objections to its use having vanished.

In the United States of America the investigation of coal is of so recent a date that we have scarcely had time to institute comparisons with the corresponding combustibles in Europe. Nor have we acquired more than a meagre amount of information in relation to the economic value of similar substances in other countries.

While in the new world, remarkable as it may appear, the most simple properties of mineral fuel have scarcely been known half a century; while the first anthracite found its way from Pottsville to Philadelphia in the year 1812; from the Lehigh region in 1814, and from Wilkesbarre in 1820;—while the first bituminous coal reached tide water down the Susquehanna only in 1804, the coals of England had been employed for fuel and manufactures from the beginning of the thirteenth century; those of Scotland towards the close of the same century; of France at the beginning of the fifteenth century, and in Belgium the coal mines had been in operation at least as early as the year 1198.

The amount of current information as to what has been effected, and as to what is the existing condition in other parts of the world in relation to coal mining industry and the enormous developments of this mineral in various countries, even during our own time, forms a department in industrial statistics which greatly needs elucidation, for the details which it embraces are by no means of easy access to the inquirer, either in the new or the old world. It is the growing necessity for such information, the demand for a multitude of essential data for which we have so often to seek in vain, that has led to the preparation of the present volume, and has encouraged its author to persevere. We feel assured, moreover, that in the concentration of such a multitude of useful facts which time has developed, but which are now, in great measure, for the first time brought together, we are conferring no slight accession to the generally prevailing knowledge, on a subject which is annually acquiring importance, and becoming more intimately connected with the advancement of the human race.

It may be useful to pursue these preliminary notes on the classification of mineral combustibles somewhat further; and we, fortunately, are not without ample scientific authority for extending this section as far as our space will permit.

It has been perceived that similarity of results in analysis, is not of itself an entire and decisive guide to the ascertainment of all the properties of coal. Even as regards chemical results, apparently parallel, discrepancies are discoverable, when the investigation is carried further, which show the absence or presence of principles that materially influence operative results. Thus, in coals containing similar quantities of carbon, those of the north of England and in Scotland, for instance, the analytical results, acquired by Dr. Thomson, prove that the relative quantities of hydrogen, carbon, and azote, materially differ. Again, external properties and characters must likewise be consulted. The structure and texture of the coal, the density, the mode in which it burns in the fire, swells or decrepitates, and other phenomena must be attended to. We have seen, for example, that some of the English coals possess so strong a tendency to melt, cement, and coke, as to form a hollow fire, and cannot be used in iron works without previous coking; while other coals, even such as possess ten or twenty per cent. more of bitumen, swell but little; and although their fragments cohere in the fire, they do not change their form and bulk, even in the process of coking.

There is yet another mode which has been employed to compare, with still greater delicacy, the respective qualities and composition of these combustibles. This is by means of the relative proportions of carbon and gaseous matters, ascertained more completely than is exhibited in the usual form of analysis. For a knowledge of these results, and some others that we propose to introduce, we are indebted in great measure, to the work of M. Pelouze on gas.*

"All the *compact coals*, even the fattest, the most coking, the most inflammable—in a word, those which the English designate by the name of "*close-burning coal*," and which yield to distillation a coke, always more or less abundant, dense, and of better quality than those of the light coals—ought to be avoided for the manufacture of gas.

But among the eligible coals how many distinctions still remain to be made. We are often astonished to find that the lightest coal—that which leaves the least residuum after its combustion—above all, that which possesses characters entirely bituminous; which kindles rapidly and gives out a fine and elongated flame, yields much less gas to distillation than some other variety of the light coals which possess the same apparent characteristics, or which were even far from promising as much.

We are acquainted with a great number of analyses of coal, made at various periods; but all at a time when the science of the analysis of organic bodies had made little advances. Besides, the only object of those analyses was that of stating the respective proportions of coke, or de-bituminized coal, and the incombustible residuum which the coke yielded by a complete incineration. Little attention was given to determining the component parts of the bituminous portion.

Mr. Richardson has devoted himself to researches in the laboratory of Professor Liebig, at Giesen. He has examined the English coals. We give the results of his analyses, the more willingly that his examinations have been directed to the produce in coke, and to the elements of the bituminous portion.

British Bituminous Coals.

With the certain means that chemists possess, now-a-days, for analysing organic substances, such a work, published by a person so competent in these matters ought to inspire confidence. Now we see that in the bituminous portion of the coals assayed by Mr. Richardson, the proportion of oxygen varies from 14.54 for 6.33 of hydrogen, to 5.50 for 5.31 of hydrogen. There is, therefore, reason to think,

* Traité de l'éclairage au Gaz, avec 24 planches, par Pelouze Pere, Paris, 1839.

that in the distillation of the first variety, in consequence of the formation of water, there would remain very little hydrogen for the production of gas for illumination; while the second variety would have yielded a much more abundant result of carbonated hydrogen gas.

Species of Combustible.	Locality.	Composition.				Composition after deducting the ashes.		
		Carbon.	Hydrogen.	Oxygen.	Ashes or Cinder.	Carbon.	Hydrogen.	Oxygen.
Splint Coal,	Wylam,	74.823	6.180	5.085	13.912	86.91	7.18	5.91
	Glasgow,	82.924	5.491	10.457	1.128	83.87	5.55	10.58
Cannel Coal,	Lancashire,	83.753	5.660	8.039	2.548	85.94	5.81	8.25
	Edinburgh,	67.597	5.405	12.432	14.566	79.13	6.33	14.54
Cherry Coal,	Newcastle,	84.846	5.048	8.430	1.676	86.29	5.14	8.57
	Glasgow,	81.204	5.452	11.923	1.421	82.38	5.53	12.09
Caking Coal,	Newcastle,	87.952	5.239	5.416	1.393	89.19	5.31	5.50
	Glasgow,	83.274	5.171	9.036	2.519	85.43	5.30	9.27

At the same time that Mr. Richardson was operating at Giesen, M. Reynault, an aspiring mining engineer, was devoting himself at Paris to similar researches, with much assiduity. We give below the principal results which he has obtained.

European Bituminous Coals.

Analysis per cent., the earthy residuum being previously abstracted. The results show the mean of three different assays for each species of coal.

Localities.		Carbon.	Hydrogen.	Oxygen.
1. Coal of Alais, basin, No. 23, mine of Rochebelle,	France,	90.55	4.92	4.53
2. Coal of Lavaysse, Dep. of the Aveyron,	"	82.12	5.27	7.48
3. Coal of Mons. 1st variety of *Flenu,*	Belgium,	86.49	5.40	8.11
4. " 2d "	"	87.07	5.63	7.30
5. Coal of Epinac, basin No. 11,	France,	83.22	5.23	11.55
6. Coal of Blanzy, No. 10,	"	78.26	5.35	16.39
7. Cannel Coal of Lancashire,	England,	85.81	5.85	8.34
8. Coal of Commentry, basin No. 13,	France,	82.92	5.30	11.78
9. Coal of Rive-de-Gier, " No. 20, Grande-Croix,	"	89.04	5.23	5.73
10. " " " " Rassaud,	"	89.07	4.93	6.00
11. " " " " Corbeyre,	"	90.53	5.05	4.42
12. " " " " Cimetiere,	"	85.08	5.46	9.46
13. " " " " "	"	87.45	5.77	6.78
14. " " " " Couzon,	"	84.49	5.75	9.36
15. " " " " "	"	86.30	5.27	8.43
16. Coal of Noroy, basin No. 3, Vosges,	"	78.32	5.38	16.30
17. Coal of Oberkirchen,	Westphalia,	90.40	4.88	4.72
18. Coal of St. Girons, Dep. of L'Arriege,	France,	76.05	5.69	18.26

We remark how much the proportions of oxygen in relation to hydrogen, vary even in their most extended limits: and if we admit that the abundance of the first is injurious to the production of gas for lighting, all the uncertainty which is generally observed in the result of the manufacture is explained.

Many of the coals comprised in the preceding table are defective by an absolute want of hydrogen; but several others, even those rich in hydrogen, by the association of that with a too strong proportion of oxygen, which in the distillation of coal disengages itself with the hydrogen, both being in a nascent state, are found in conditions favourable to combination; that is to say, to the production of water, to the detriment of the quantity of illuminating gas.

Anthracites.

Localities.		Carbon.	Hydrogen.	Oxygen.
1. Anthracite of Pennsylvania,	United States,	94.89	2.55	2.56
2. Coaly Anthracite of Rolduc, near Aix-la-Chapelle,	Belgium,	92.85	3.96	3.19
3. Anthracite of Mayenne,	France,	93.56	4.28	2.16
4. " of South Wales,	Wales,	91.05	3.38	2.57
5. " of La Mure, Dep. of Isere,	France,	94.07	1.75	4.18
6. " of Macot,	La Tarantaise,	97.23	1.25	1.52

We here perceive that the anthracites are absolutely wanting in hydrogen, independently of the consideration of oxygen. It explains then very well why this species of combustible is the least convenient substance for the manufacture of illuminating gas.*

The results, according to the English engineer, Luke Herbert, obtained from a series of experiments made upon each of the three classes of English bituminous coals, and in each case by the distillation of one ton of the coal, are as follows:

1. Cannel coal of Lancaster produced 11,600 English cubic feet of gas.
2. Coal of Newcastle, (Hartley mine,) 9,600 do.
3. Coal of Staffordshire, best quality, 6,400 do.

By experiments on a similar scale to the last, were obtained the following results:

1. Wallsend coal, - - - 10.300 cubic feet of gas.
2. Temple Main, - - - - 8.100 do.
3. Primrose Main, - - - 6.200 do.
4. Pembrey - - - - - 4,200 do.

The gas obtained possessed an illuminating power much inferior to that from the coal of the first class; but there was much coke of good quality.

In this class the series terminates with the drier and less adhesive coals, called "open burning coals." Those of this kind are preferred by blacksmiths because they better bear the blast of the bellows.

ADAPTATION OF DIFFERENT VARIETIES OF COAL TO THE PURPOSES OF STEAM NAVIGATION.

There has been recently published a very elaborate report, of 607 pages, "to the Navy Department of the United States, on American

*Pelouze on Gas.

coals applicable to steam navigation and to other purposes, by Prof. Walter R. Johnson." It includes two hundred and one tables, prepared by the author with unusual care and under peculiar advantages, and furnishes the results of a long series of experimental investigations conducted at Washington.

Our space precludes our quoting extensively from this voluminous document: but we cannot refrain from selecting the following table

Classification of American Coals, in the order of evaporative power under equal *bulks*, to which is added the relative numerical rank of the same coals under equal *weights*, also in the order of their specific gravities, and of their marketable weight.

No.	Names and Localities.	State or County.	Quality.	Pounds of steam from 212°, produced by one cubic foot of each coal.	Relative evaporative power, for equal *bulks* of coal.	The same Coals. Evaporative power under equal *weights*.	In the order of their specific gravities.	In the order of their weight in the marketable state.
						No.	No.	No.
1	Atkinsons, Cumberland coal,	Maryland,	Dry bituminous c'l,	566.2	1.000	1	29	21
2	Beaver Meadow, Slope V.,	Pennsylvania,	Ant'cite, white ash,	556.1	.982	7	3	1
3	Peach Mountain,	Schuylkill Co., Pa.,	" red ash,	545.7	.964	3	6	11
4	Forest Improvement,	" "	" white ash,	540.8	.955	4	5	13
5	Easby's, Cumberland coal,	Maryland,	Dry bituminous c'l,	535.6	.946	6	21	23
6	N. Y. and Maryland comp.,	Cumberland coal,	" free burning,	524.8	.927	9	9	12
7	Queen's Run coal,	Clinton co., Penn.,	Moderately bitu's,	517.0	.913	2	22	28
8	Blossburg,	Tioga co., Penn.,	"	515.9	.911	10	25	20
9	Neff's Cumberl'd c'l,	Maryland,	Free burning bitu's,	512.7	.906	12	20	9
10	Easby's "coal in store,"	Cumberland, Md.,	"	511.1	.903	5	30	16
11	Beaver Meadow, No. 3,	Pennsylvania,	Ant'cite, white ash,	505.5	.893	15	1	5
12	" navy yard,	"	"	500.0	.883	18		4
13	Mixture 1-5th Cumberland and 4-5ths Beaver Meadow,		Mixed,	498.5	.880	16		6
14	Lehigh coal,	Pennsylvania,	Ant'cite, white ash,	494.0	.872	23	2	3
15	Ralston,	Lycoming cr'k, Pa.,	Moderately bitu's,	493.3	.871	24	16	2
16	Summit Portage coal,	Cambria co., Pa.,	Bituminous,	486.9	.860	14	12	17
17	Mixture 1-5th Mid-Lothian and 4-5th Beaver Meadow,		Mixed,	481.1	.850	25		8
18	Barr's deep run,	ne'r Richmond, Va.,	Bituminous,	478.7	.845	19	17	19
19	Lackawanna,	Pennsylvania,	Ant'cite, white ash,	477.7	.844	8	10	30
20	Karthaus,		Moderately bitu's,	477.4	.843	17	34	22
21	Stony creek, Perseverance seam,	Dauphin co., Pa.,	Semi-bituminous,	472.8	.835	13	8	26
22	Lykens valley,	" "	Anthracite,	459.7	.812	11	15	31
23	Pictou,	Nova Scotia,	Bituminous,	450.6	.796	33	28	15
24	Mid-Lothian, av'ge,	Richmond, Va.,	"	448.5	.792	35	31	10
25	Crouche's pits,	"	"	445.0	.785	34	7	14
26	Newcastle,	England,	Fat bitu's coal,	439.6	.776	27	38	25
27	Mid-Lothian, 900 ft. shaft,	Virginia,	Bituminous,	433.7	.766	29	13	27
28	" new shaft,	"	"	418.6	.739	26	24	32
29	Pictou, Cunards,	Nova Scotia,	"	417.9	.738	30	23	29
30	Chesterfield comp.,	Richmond, Va.,	"	410.9	.726	20	32	40
31	Mid-Lothian, scr'nd,	"	"	408.7	.722	22	35	39
32	Natural coke,	"	"	395.3	.698	31	26	37
33	Creek company,	Chesterfield co., Va.,	"	391.8	.692	32	27	38
34	Pittsburg,	Pennsylvania,	Fat bitu's coal,	384.1	.678	36	39	36
35	Sydney coal,	Cape Breton,	Bituminous,	378.9	.669	37	19	35
36	Liverpool,	England,	Fat bitu's coal,	375.4	.663	38	37	33
37	Scotch,	Scotland,	Bituminous,	353.8	.625	42	4	24
38	Tippecanoe,	nr. Petersburg, Va.,	"	350.2	.618	39	18	42
39	Cannelton,	Indiana,	Cannel coal,	348.8	.616	41	36	34
40	Clover Hill,	Richmond, Va.,	Bituminous,	347.4	.614	40	33	41
41	Coke of Cumberland coal,	Maryland,	Coke,	284.0	.502	21		44
42	Coke of Richm'd c l,	Virginia,	"	282.6	.499	28		43
43	Dry pine wood,		Pine wood,	98.6	.175	43	40	45

of the relative degree of evaporative power of different coals under similar or uniform *bulks.* We select this table, at the suggestion of the author, in preference to that which exhibits "the order of evaporative power under equal *weights.*" He remarks that coal, "when sold by weight and used on shore, the weight per cubic foot is a point of little moment. Space for stowage is easily obtained. But in steam navigation, bulk, as well as weight, demand attention; and a difference of *twenty per cent.*, which experiment shows to exist between the highest and the lowest average weight of a cubic foot of different coals assumes a value of no little magnitude. This is obviously true, since, if other things be equal, the length of a voyage must depend on the amount of evaporative power afforded by the fuel which can be stowed in the bunkers of a steamer, always of limited capacity."

ADAPTATION OF COAL TO STEAM POWER.

We learn, through various channels, that the Lords of the Admiralty, in England, have taken up the subject of coal, not solely as relates to its economic working and consumption, but with reference to the probable quantities absolutely workable in Great Britain, the most economic methods of combustion, and the chemical properties and combinations of coal. An inquiry was announced as in progress, in the close of 1846, in reference to the value of coals for the use of the British steam navy. It is designed not merely to ascertain, by chemical analysis, the constituents of different sorts of coal, but, by an extensive series of comparative experiments, to determine their practical applicability. With this object in view, it is announced that steam boilers and furnaces have been erected at the Engineering College at Putney, and the examination is intrusted to Sir Henry de la Beche and Dr. Lyon Playfair, and those associated with them.

The editor of the Mining Journal remarks,* "our beds of coal have been the undoubted production of ages; and, vast as they are, it appears the height of the ridiculous to assert, that they are inexhaustible. Every succeeding year brings its increasing consumption, not simply of tons, but of millions; and perhaps there is no other question in the range of political economy that deserves so much patient investigation, and no body of men so highly competent to the task as the gentlemen alluded to."

In order the better to follow up this interesting subject, illustrated in the last table, we proceed to append the results of a series of experiments to determine the evaporative values of several varieties of American coals—chiefly those of a semi-bituminous character. These investigations were undertaken by Messrs. Thompson, of New York, on account of the New York and Liverpool Steamship Company. As some of these results differ in certain respects from those of the last table, it has been judged expedient to precede them with the introductory notices of the experimenters.

* Mining Journal, August 1st, 1846.

"E. K. Collins, Esq.

"Having received orders from you to test, by experiment, the evaporative values of several kinds of coal upon your 'Vertical Tubular Boiler,' and having discharged such duties, we hereby enclose you a copy of the report of said experiments with our views annexed.

"Our first experiment was with the Dauphin Rattling Run. It possesses high evaporative power, with considerable cohesion of its particles, so that it may not be broken into too small fragments by the constant attrition which it may experience in the vessel. It ignites very quick with natural draught, burns with a clear bright flame, leaving but a slight soot deposit upon the tubes, owing to the free circulation of the air through the grate bars, rendering the combustion more perfect. In short, it requires little if any attention from the fireman until it is necessary to charge the furnaces with a fresh supply of coal.

"Our second experiment was with the Baltimore Cumberland. It ignites rapidly, producing a strong heat; shortly after which time, it runs together, often adhering to the grate bars, requiring much time and labour to separate it. Without the strictest attention, in a short time, it would entirely exclude the air from penetrating through the interstices of the coal, thereby producing a great loss of steam. It has very small cohesion of its particles, requiring very careful handling to prevent it from crumbling into very small pieces.

"Our third and fourth experiments were with 'Young's Mining Company' and 'Maryland Mining Company.' They produce the same effect as the second experiment, the results being all forced.

"Our fifth experiment was with the Erie—possessing rapidity of ignition, makes an intense fire, throwing off an immense volume of carbonaceous matter, causing the tubes to fur up in a very short time, also having an undue quantity of sulphur in it.

"Our sixth experiment was with the Dauphin Rattling Run; varies nothing from the first experiment.

"Our seventh experiment was with the 'Maryland Mining Company;' results as before.

"Eighth experiment was with the Dauphin Backbone, slightly differing from the Rattling Run in evaporation and rapidity of ignition.

"Ninth experiment was with the Dauphin Rattling Run; still retaining the pre-eminence for marine purposes over any of the coals submitted for trial, requiring less labour by 75 per cent.—producing more steam in less time than any of the coals experimented with.

"Our tenth experiment was with the 'Maryland Mining Company.' Experienced the same difficulty as before, viz.:—running together, forming a heavy compact crust over the whole area of the grate, requiring frequent raking, that the air might circulate through and ignite the surface or top of the coals, thereby occasioning the furnace doors to remain open much longer than the circumstances of the case ought to justify.

"Having given a brief review of the coals submitted to us for trial, it may be well to add, that the Dauphin Rattling Run (for marine purposes) meets our decided approval, from the experiments we have had of it."

New York, Dec. 24th, 1850.

Table showing the economic values of the Coals consumed in a "Collins' Vertical Tubular Boiler," at the Novelty Works, New York, December, 1849.

Names of Coals employed in the experiments.	Evaporating power, or number of pounds of water evaporated from 40° by one pound of Coal.	Evaporating power, or number of pounds of water evaporated from 212° (latent heat 1000°).	Rate of evaporation or number of pounds of water evaporated per hour.	Quantity of water evaporated in all, pounds of.	Number of pounds of Coal consumed per hour.	Number of pounds of Coal consumed per square foot of grate surface per hour.	Total pounds of Coal consumed in all.	Ashes, per cent. of.	Mean pressure of steam in lbs.
Dec. 5. Dauphin, Rattling Run.	7.510	8.802	523.875	2095.5	69.75	16.41	279	12.37	24.
Dec. 6. Baltimore, Cumberland.	7.202	8.452	468.187	1872.75	65.	12.70	260	11.74	22.
Dec. 7. Young's Mining Co. do.	6.618	7.762	443.437	1773.75	67.	13.41	268	12.97	22.
Dec. 8. Maryland Mining Co. do.	6.788	7.962	459.937	1839.75	67.75	15.94	271	12.	22.25
Dec. 9. Erie, New York.	5.871	6.880	402.187	1608.75	68.5	16.11	274	6.42	19.75
Dec. 10. Dauphin, Rattling Run.	7.425	8.702	495.	2970.	66.66	15.68	400	12.54	24.
Dec. 11. Maryland Mining Co.	7.520	8.814	425.428	2978.25	56.57	13.31	396	11.62	22.61
Dec. 12. Dauphin, Backbone.	7.389	8.652	519.75	3118.5	70.33	16.43	422	12.77	23.5
Dec. 16. Dauphin, Rattling Run.	7.858	9.209	552.708	3316.25	70.33	16.43	422	12.51	24.
Dec. 17. Maryland Mining Co.	7.304	8.572	400.714	2805.	57.85	13.61	384	11.86	22.
Maryland Mining Company. Mean of three trials.	7.204	8.443	423.5	7623.	58.388	14.28	1051	11.67	22.95
Dauphin and Susqueh. Coal Co. Rattling Run. Mean of three trials.	7.597	8.903	523.859	8381.75	68.812	16.17	1101	12.40	24.

In 1852 was published the Report of the Engineer-in-Chief of the Navy, on the comparative value of anthracite and bituminous coal, from which we extract the following:

"The coals used for the experiments were the kinds furnished by the agents of the government for the use of the United States Navy Yard and steamers, were taken indiscriminately from a pile in the yard without assorting. The bituminous coal was from the Cumberland Mines. The anthracite was the kind known as "White Ash Schuylkill."

From the experiments it appears, that in regard to "getting up steam," the anthracite exceeds the bituminous 36 per cent.

Again he remarks, "that from the experiments, without allowing for the difference of weight of coal that can be stowed in the same bulk, the engine using anthracite could steam about two-thirds longer than with the bituminous.

"These are important considerations in favour of anthracite coal for the uses of the navy; without taking into account the additional amount of anthracite more than bituminous that can be placed on board a vessel in the same bunkers, or the advantages of being free from smoke, which in a war steamer may at times be of the utmost importance in concealing the movements of the vessel, and also the almost, if not altogether, entire freedom from spontaneous combustion."

"The following tabular statement shows the actual evaporation of water effected by bituminous and anthracite coals in the boilers of several naval steamers, and in those of some trans-atlantic and river steamers plying to and from New York the past few years:—"

Table of Practical Tests of Different Varieties of Coal.

NAME OF VESSEL.	TRADE.	Sea water evaporated from temperature of condenser (100 deg. Fahrenheit) by 1 lb. of bituminous coal.	Sea water evaporated from temperature of condenser (100 deg. Fahrenheit) by 1 lb. of anthracite coal.	REMARKS.
		POUNDS.	POUNDS.	
Michigan	United States Navy	*5.000		Fresh water.
Mississippi	do do	†4.780		Sea water.
Spitfire	do do	†4.870		do
Engineer	do do	‡4.531		do
Alleghany	do do	†5.600		do
Iris	do do	†5.180		Sea water & old flue boil.
Princeton	do do	†6.666		Sea water and new boil.
Princeton	do do	§5.372		do do
Princeton	do do		7.554	do do
Princeton	do do		6.639	Sea water and old boilers
McLane	United States Treas'y		7.030	Sea water.
Bibb	do do		6.030	do
United States	Transatlantic packet		7.480	do
Herman	do do	†4.487		Sea water and old boilers.
Baltic	do do		8.555	Sea water.
City of Pittsburg	do do	†4.930		do
New World	Hudson River		8.022	do
Commodore	Long Island Sound		7.262	do
Roanoke	N. York and Norfolk		6.554	do
		10) 51.416	9) 65.116	
	Averages	5.142	7.235	

"From the averages of the above table it will be seen that the economical evaporation by the anthracite exceeded that by the bituminous in the proportion of 7.235 to 5.142, or about forty-one per centum of the latter."

"In the experiments made on coals by Playfair and De la Beche, by order of the British government, in 1848, were found eleven varieties of Welch coals having a constitution almost identical with the nine specimens of Pennsylvania anthracite, experimented on by Prof. W. R. Johnson, viz:"

* Pittsburg coal. † Cumberland coal. ‡ Virginia coal. § Scotch coal.

	Welch anthracite.	Pennsylvania anthracite.
Fixed carbon	87.54	88.54
Sulphur	0.79	0.05
Other volatile matter	5.50	5.17
Earthy matter, &c.	6.48	6.51
	100.31	100.27

"The average evaporation of water by the Welch anthracite and by the Pennsylvania anthracite was as follows:

Fresh water evaporated from the temperature of 212 deg. F., by one pound of coal.

By Welch anthracite, ... pounds 9.263
By Pennsylvania anthracite,. . . . . do. 9.590

"Thus far there is a very close agreement between the results obtained by the different experimenters from substantially the same coal—that coal being anthracite.

"In the experiments of Playfair and De la Beche, above cited, we find three varieties of Welch bituminous, three varieties of Scotch bituminous, and one variety of English bituminous, having a constitution almost identical with the five specimens of Maryland (Cumberland) bituminous coal experimented on by Prof. W. R. Johnson."

	Welch, Scotch, and English bituminous.	Maryland (Cumberland) bituminous.
Fixed carbon	75.00	75.05
Sulphur	1.47	
Other volatile matter	14.55	15.45
Earthy matters, &c.	8.97	9.49
	99.99	99.99

"The average evaporation by the Welch, Scotch and English bituminous and by the Cumberland bituminous was as follows, viz:

Fresh water evaporated from a temperature of 212 deg. F., by one pound of coal

By Welch, Scotch and English bituminous. . . . . pounds. 8.02
By Maryland (Cumberland) bituminous . . . . .do. 9.93

"Here is a great discrepancy between the results obtained by the two experimenters on substantially the same coals; Prof. W. R. Johnson making the Cumberland bituminous better than the British bituminous in the proportion of no less than 24½ per centum of the latter. Had a similar difference been found in the case of anthracite between the results of the two experiments it might have been accounted for by a difference of boiler or method of conducting the experiments." We are sorry our limits will not allow of further extracts from the reports."

Dr. Higgins, in his Report for 1854, p. 72, expresses himself strongly in favour of the superiority of the Cumberland coal, especially for steam navigation, for which he deems three qualities essential in a coal, "*quickness* of combustion, *continuance* of combustion, and *steady* combustion. Fuel should take fire *rapidly*, it should burn for a *long time*, and its intensity should not be *diminished* by fresh additions of material." Excepting the first, anthracite has these qualities, and a steam-ship is not a factory where the fires are extinguished every night—and even if they were, proper coal and other free-burning fuel could be supplied for kindling, and such emergencies as may arise.

In the region supplied by the Susquehanna coals, the hard anthracites are extensively used for stationary steam-boilers, where the various grades from Karthaus bituminous and Dauphin semi-bituminous, to the Lyken's valley or Bear-gap free-burning anthracite, are accessible—in some cases where a fire has to be renewed every morning, and in others where it is kept up without intermission for several months.

Dr. Higgins says—"The policy of the world at present is for steam navigation, not only for commercial, but also for warlike purposes. A steam vessel of war requires, above all others, to have a fuel which can speedily generate and keep up a steady head of steam, whether in pursuit of, fleeing from, or in actual combat with an enemy. A minute's delay may prove disastrous; the increased revolution of the paddle-wheels for a few times will frequently insure success. Our national flag may float gloriously over the sea, or be stricken from the mast, as the ship which bears it is well or ill supplied with fuel, and these ships should always use the Cumberland coal. Our navy should learn from the experience of our commercial marine, and this teaches it that the best coal for steam navigation is the Cumberland coal." The following tables are then given of the time required to cross the Atlantic by the Collins and Cunard steamers.

Western Passage.	Days.	Hours.	Min.
Average time of the Collins line,	11	18	33
" " Cunard line,	12	16	11
In favour of Collins, each passage,	00	21	38

Eastern Passage.	Days.	Hours.	Min.
Average time of the Collins line,	11	00	29
" " Cunard line,	10	22	46
In favour of Cunard, each passage,	00	1	43

	West. Pass.			East. Pass.			Yearly Total.		
	Days.	Hrs.	Min.	Days.	Hrs.	Min.	Days.	Hrs.	Min.
Cunard,	342	04	54	284	16	06	626	21	39
Collins,	317	20	55	286	12	40	604	09	35
In favour of Collins for the year (27 voyages),							22	12	35

"From this there was the novel fact of the Cunard line, steamers of British build, being swifter on the Eastern passage than the Collins line, steamers of American build. On their Western passage quite the reverse took place. . . . . .

"Many explained the difference between these vessels as to their respective superiority on their eastern and western passage, by their different powers in going with or against the wind. The result, however, was to be explained by their different means of propulsion as to the generation of steam. I therefore addressed a letter to the proprietors of each of these lines, and received the following replies:

"'*New York, January* 19*th*, 1854.

"'DEAR SIR,—Messrs. Brown, Bro. & Co. have handed me yours of the 17th inst., making inquiries relative to the coal used by us. Our first trial was with the Cumberland; but finding so much slate and earthy matter in it, we were compelled to try the Anthracite, which we have been using for the past three years; but from what I learn of the Cumberland mines, I think we will soon give that coal another trial. From Liverpool we use the Welch. The Cunard Co. use the Cumberland, not being able to burn the anthracite in their furnaces. "'Yours respectfully,

"'E. K. COLLINS.'

"'*New York, January* 19*th*, 1854.

"'SIR:—In reply to the inquiry made in your letter of the 17th instant, I beg to inform you that on the voyage from England to America, we use Welch coal for fuel on board our steam-ships, and Cumberland coal on the voyage from America to England.

"'I remain, &c. "'E. CUNARD.'

"The reason of the difference was now satisfactorily explained. It was in the superiority of the Cumberland over the anthracite coal. When these steamers used the same coal (the Welch), American ship-building proved its superiority, for the American ships (the Collins Line) proved themselves the faster; but even superior American naval architecture could not compete with English ships, when those ships used Maryland coal, for then they surpassed in speed the American ships using the anthracite coal."

Prof. H. D. Rogers, State Geologist for Pennsylvania, has permitted us to give his authority as to the value of anthracite for steam navigation. He states that it is less adapted for steam navigation, in proportion to the quantity of volatile matter.

Mr. S. V. Merrick, of Philadelphia, who enjoys the highest reputation in all that relates to the manufacture and employment of steam machinery, assigns the *first* place to anthracite, the *second* to Welch, and the *third* to Cumberland coal. In a recent letter (March 2d), he expresses doubts about the accuracy of Dr. Higgins, in attributing the more rapid outward passage of the Cunard ships to the coal used, and were this the case, "it would prove nothing, as these boilers are adapted specially to the bituminous, and will not work economically with the anthracite."

The results of the laboratory experiments of Professor W. R. Johnson, must be adopted with great caution in deciding upon cases where the circumstances are different, and the quality of the coal possibly different from the samples used by him. Such experiments afford valuable collateral aid as indications of the direction which practice should take for their verification, but their evidence is far from being sufficient to settle the question. This must be accomplished by steady and varied practice, under different circumstances.

We conclude these various experiments and remarks on the adaptation of coal to steam power, by appending the following account of the semi-bituminous coals of the Welch Basin, from the foreign portion of the first edition of this work,* p. 362.

Semi-bituminous Coals of the Welch Basin.—Steam coals or intermediate of the southern side of the basin.

Towards the close of 1840, an association was formed in London for the encouragement and protection of the Welch coal trade. They remark that "the durability of the ordinary bituminous coal, the very peculiar qualities of the anthracite or stone coal, and the great superiority of the intermediate or steam-packet coal of South Wales, are now so well ascertained, that it would appear as if nothing more were required to insure a preference at all the places of import which can be reached at a moderate rate of freight. It has only been, however, by very small degrees, by very great individual exertions, and by very considerable private loss, that the Welch coal has just begun to obtain a reputation in the port of London."†

The semi-bituminous coal of the south part of this basin, possesses many characters in common with certain coals in Pennsylvania; both of them being admirably adapted for steam engines; so much so as to have received the specific title of "*steam coal.*"‡

The Craigola coal has been recommended for similar qualities; and the Llangennech has established for itself a higher reputation as a steam coal, and has been used on board steam ships in various parts

* Since the above publication, we understand a new geological survey of England and Wales is progressing—the maps are on a scale of 1 inch to a mile and of great accuracy and excellence. "The principal coal strata are traced nearly throughout the complex involutions of the surface with surprising accuracy and effect; and even the *Faults* are laid down with distinctness,—ranging with an approach to parallelism, from about S. E. to N. W., but in some places singularly complicated; and their connection with the features of the surface is clearly shown. The horizontal sections with the vertical leave nothing to be desired, and the whole work forms a subject of most instructive study to geologists."—Edinburgh Review, October, 1849; Review of Statistics of Coal.

† Mining Journal, Vol. X. p. 359, 1840; and Cambrian Newspaper.

‡ Monmouthshire Merlin, 1840.

of the world. In 1842, an Egyptian war corvette, belonging to Mohammet Ali, arrived from Egypt, at Port Talbot, in Glamorganshire, and was there loaded with the steam coal of that vicinity. In the same year the proprietors of the Risca colliery contracted with the Royal India Steam Navigation Company, to supply 72,000 tons of steam coals.

Analysis and Quality of the Welch Coal at different points and in different beds.—The variations are as numerous as there are coal seams and square miles in the whole area. To particularize them, would be a herculean labour; yet that task, great as it appears, has been triumphantly accomplished by Mr. Mushet; to the results of whose labours we shall take the liberty of referring in detail, in another page.

Previous to this, we shall introduce a résumé of a series of interesting observations on the coals of this basin, by Mr. T. Forster.*

Table of Welch Coals.

Coal Seams in a line of Section from South to North.	Per 100 parts.			Description of earthy residuum left after combustion.
	Carbon.	Bitumen and volatile matter.	Ashes.	
Anthracite.				
Seam on Mynydd bach, Llaneidi,	89.85	8.65	1.50	Pale yellow ashes.
Free-burning Coal.				
Clyn-wernon seam,	79.00	14.00	7.00	Heavy reddish ashes.
Pembrey seam,	82.00	14.50	3.50	White ashes.
Bituminous Coal.				
Gelly Gile seam,	80.60	16.80	2.60	Red ashes.
Lwchor colliery, five feet coal,	78.50	19.00	2.50	
Globraise seam, Adair colliery,	72.20	27.50	2.30	Yellow ashes.
Seams in different parts of the South Wales Coal Basin.				
Coxe's stone coal, Cwm. Tw. Anthracite,	91.50	7.50	1.00	do.
"Pool coal," near Llanelly and Pembrey. Bituminous seam of Killymaen Ilwyd.	77.80	19.80	2.40	Reddish ashes.
Bushy seam, Llanelly. Semi-bituminous coal,	81.60	15.90	2.50	do.
Great seam at Myrthyr. Semi-bituminous coal, of which the coke for the blast furnaces is made,	85.60	13.40	1.00	White ashes.

In further illustration of the various qualities of coal within the South Wales basin, we add a few other analyses.

* Transactions of the Natural History Society of Northumberland, part 1.

	Carbon.	Vol. matter.	Ashes.	
Welch anthracite, 1st,	92.42	5.97	1.61	Analysis of Dr. Schafhaeutl.
Welch anthracite, 2d,	94.10	4.90	93	
Ynis-ced-win,	89.00	7.00	4.00	* Corresponding with M. Mushet's analysis.
Hirwain,	90.00	8.00	2.00	
Dawlais,	81.00	15.50	3.00	
Pontypool,	69.00	28.50	2.50	
Abersychan,	66.00	29.50	4.50	
Mynydd yslwyn,	68.50	27.00	4.50	

Steam coal.—We have alluded, as far as our space permits, to this valuable product of the South Wales mineral basin. The peculiar qualities of this fuel, and the experiments and results therefrom, have been made public in various ways. Among the latest of these may be mentioned the experiments on Cameron's Coalbrook steam coal, detailed in the Mining Journal.† We cannot enter into these elaborate particulars, and our notice must be brief.

This coal is obtained within six miles of the port of Swansea, in which range a large amount of a similar species is known to prevail.

By experiments made at H. M. dockyard, Woolwich, with the steam coal and several other varieties, the results were the following.

Description of coal.	Weight of water evaporated by one pound of coal. lbs.	oz.
Merthyr, South Wales, bituminous coal,	8	0
Cameron's steam coal, S. W.,	9	7¾
Llangennech steam coal, S. W.,	8	14½
Parson's Abbey, Craigola, S. W.,	8	6⅔
Hasting's Hartley Main, bituminous,	6	14½
Carr's West Hartley, do	7	5

The proportionate weights of the clinkers, the ashes, and the time in getting up the steam, were also taken into account. As relates to the economy in fuel, its first cost, the saving effected in stowage, and the absence of smoke in the steam coals, all these results, without reference to the interests of individuals or associations, form very important matters in the economy of steam navigation and manufactures.

Sulphur and Smoke, their absence, or inconsiderable amount, in the Welch Anthracite and Steam Coal.—It has been urged, by every one having any experience in anthracite, that one of its properties, by no means unimportant, is its non-liability to spontaneous combustion; which, it is well known, occasionally takes place with bituminous coals; whereby vessels have been lost at sea, and valuable property destroyed on land. A steamer belonging to the British government was destroyed in the Mediterranean a few years ago, from the spontaneous combustion of her stock of bituminous coal. By evidence before a parliamentary committee on the coal trade of Lon-

* Mining Journal, Vol. XI. pp. 118, 133, 149, 173; articles by Llewellyn, Brough, and others.

† March 14th, 1846.

don, in 1838, it was shown that the bituminous coals are liable to ignition in ware-houses when they have been put in wet; and that they have ignited on board ships going to the East Indies.* Several instances of similar accidents have occurred in the United States. The absence of smoke, in war steamers using anthracite, is no small desideratum, to those who have witnessed the dense columns of black smoke, proceeding from steamers even at very great distances, at sea, employing smoke-producing coals.

So compact and dense a fuel also has great advantages in point of stowage space, over the ordinary weak bituminous coals. For long voyages, this concentration of power and economy of space, may easily be appreciated. It renders wholly unnecessary the adoption of patent contrivances for smoke prevention and consumption.

The value of the Welch steam or slightly bituminous coal is enhanced by this quality of burning almost wholly without smoke:—a property hitherto slightly appreciated, but which will, one day—and perhaps not far distant—be considered a prime requisite in fuel for steamers; or at least for those employed on naval service. By the ascent of the columns of smoke above the horizon, the motions of the steamers in Calais harbour are at all times observable at Ramsgate; from the first lighting the fires to the putting out to sea."†

Steamers burning the fat bituminous coal can be "tracked" at sea, at least *seventy miles*, before their hulls become visible, by the dense columns of black smoke pouring out of their pipes or chimneys, and trailing along the horizon. It is a complete tell-tale of their whereabouts; which is not the case with those burning anthracite; as the latter kind sends forth no perceptible smoke.‡

ANTHRACITE AND ITS USES.

Evaporative Power.—It will not be possible, in this work, to investigate the comparative merits of anthracite and bituminous coals, or of their intermediate varieties. There assuredly are highly appropriate and valuable properties in each. The experience of the last ten or twenty years, in the principal countries where they have been carefully experimented upon, and practically employed on the largest scale, has developed the relative advantages of each. More especially—both in Europe and in the United States—has it done "*justice to anthracite*," in pointing out the incalculable value of a species of fuel, previously rejected and despised, as amongst the most inferior and most impracticable of all the combustibles.

On this matter, the reader will find many instructive papers in Silliman's American Journal of Science—in the abundant correspondence scattered throughout the Mining Journal of London—in the Journal of the Franklin Institute of Philadelphia—and in other works appropriated to practical science on both sides the Atlantic; not forgetting

* Evidence, pp. 121–2.

† Extract from the Sun paper, February, 1841. Also Mining Journal, same date.

‡ New York Herald, October, 1841.

the *Annales des Mines*, the "*Bulletin de la Société Geologique de France*," &c., the *Archiv für Mineralogie, Geognosie, Bergbau, und Hüttenkunde*, &c., the experiments of Dr. Fyfe, of Edinburgh, and some other authorities, occasionally quoted in this volume.

Dr. Fyfe's experiments show the evaporative power of pure anthracite over all other descriptions of fuel. The analyses of these combustibles were as follows:

	Fixed Carbon.	Vol. Mat.	Ashes.	Weight of water evaporated by one pound of coal.
Middling Welch Anthracite mean of several kinds, *earthy and containing volatile matter*,	71.40	17.80	10.80	7.94lbs.
Common Scotch Bituminous, not caking, Middlerig, near Edinburgh,	50.50	42.00	7.50	5.88 "
Do. English caking coal,	67.00	31.00	2.00	7.84 "
Pure Welch Anthracite, according to the experiment of De Schafhaeutl,	92.42	5.97	1.61	10.56 "

Thus establishing the fact, that not only is the evaporative power in the ratio of the fixed carbon, but that there is a very remarkable approximation in the evaporative power to the proportion of this ingredient in each.*

Other experiments, by the same gentlemen, show that the evaporative power of anthracite is 25.41 per cent. greater than the Craigola coal of Swansea; and 33 per cent. over that of the Scotch bituminous coals. Hence, he contends, as the practical evaporative power in fuel is according to the per centage of fixed carbon, it is important, for the use of steamers in long voyages, to select that in which it is most concentrated; namely, the purest description of anthracite.

The results of another series of experiments on combustibles have been more recently made known. From these it appears as follows:

Number of lbs. of water to which 1 lb. of fuel will impart one degree of heat.

Wall's End, or Newcastle coal,	2,000 lbs.
Llangennech, South Wales, semi-bituminous,	9,000 "
Charcoal,	10,000 "
Anthracite,	12,000 "

From these, and from subsequent experiments by other parties, it appears that the Welch anthracite exceeds the medium species called in Wales "steam coal," in evaporative power, more than twenty per cent.

* Dr. Fyfe on the evaporative power of different kinds of coal. Edinburgh Philosophical Journal, April, 1841.

Anthracite and its application to Iron Making.—Respecting the adaptation of the Welch and American anthracites for the manufacture of iron, there are many valuable communications in the Mining Journal of London, Vol. X. It has been treated on in various scientific works devoted to metallurgy and the useful arts. Among others, it has formed the subject of an elaborate treatise, published in Paris, by M. Michael Chevalier, in 1840. In America, a treatise on anthracite has been published, in Boston, by Mr. Johnson. The process of iron making with this description of fuel, has become common in Pennsylvania, and the difficulties, which for so many years seemed to be insurmountable, now appear to be entirely overcome. At the present day, the anthracite iron of South Wales enjoys a high reputation. We can adduce no better testimony than that of Mr. Mushet, in favour of cold-blast anthracite pig iron. After concluding a series of most elaborate experiments, he remarks, "it is hence abundantly evident that the pig iron now making, with cold-blast and anthracite, at the Ystalyfera iron-works, greatly exceeds in strength, in deflective powers, and capacity to resist impact, any iron at this time manufactured in the United Kingdom."

The anthracite district of South Wales is rapidly rising into importance in the production of iron. In 1847, there were no fewer than nine establishments, possessing thirty-two furnaces, in the Swansea valley. Ten years previously, there was only one, of three or four furnaces, in operation at Yniscedwyn. Twenty-three of these furnaces were in blast in September, 1847, making eleven hundred and fifty tons per week, or 59,800 tons per annum—a quantity by no means insignificant.*

The following extract is from the London Mining Journal, March 4, 1854. "The question as to the capability and value of anthracite as a furnace fuel, and particularly for steam navigation purposes, may be considered as decided, the *Great Britain* having taken 1000 tons in her last voyage to Australia, which, from the report of Captain Matthews, appears to have answered the most sanguine expectations, quickly getting up steam, burning clearly, and promoting no injurious action on the fire-bars.

"The great deposits of anthracite in Pembrokeshire and Carmarthenshire, will, we have no doubt, very shortly prove of great national importance, and so convincing has been recent experiments as to its value, that the West India Royal Mail Steam Packet Company have been induced to take a colliery in Pembrokeshire, for the purpose of supplying their large steamers with anthracite coal, and thus avoid any delays which might occur should they be dependent on others. Even at the port of Llanally large steamers are now continually taking in cargoes of anthracite coal."

* London Mining Journal, September 4, 1847.

General View or Table of the relative density of different species of Coal and Anthracite, at various parts of the World.

In a previous page we took occasion, by means of the requisite tables, to exhibit the relative amounts of carbon and volatile matters which exist in the principal bituminous coals of England, Scotland, Wales, Belgium, France, and the United States of America; and hence to show their comparative adaptation to the manufacture of iron. The following table offers another method of making useful comparisons of the bituminous and non-bituminous combustibles in various parts of the world, by means of their respective specific gravities and weights per solid cubic yard, in pounds, avoirdupois.

Localities of Bituminous Coal.	Specific gravity.	Weight of one cubic yard, in lbs.	Localities of Anthracites and Anthracitous Coals.	Specific gravity.	Weight of one cubic yard, in lbs.
UNITED STATES.			UNITED STATES.		
Pennsylvania.			*Pennsylvania,—Semi-bituminous, intermediate coals, dry, blazing.*		
Pittsburg,	1.265	2134			
Mercer county,	1.275	2151			
Karthaus,	1.263	2131	Somerset co., 18 per ct. bitu'n,	1.382	2332
Farrandsville,	1.339	2257	Blossburg, 15 to 18 per ct. bit.,	1.400	2362
Philipsburg,	1.358	2292	Dauphin co., Rattling run,	1.391	2347
Blossburg,	1.371	2313	Lebanon co., Yellow Springs,	1.395	2351
			Broad Top Mn., Bedford co.,	1.700	2868
Virginia.					
Wheeling,	1.230	2075	*Maryland.*		
Kanawha, salines,	1.250	2109	Frostburg, 12 to 20 p. ct. bit.,	1.552	2619
Richmond,	1.246	2102			
States.			*Tennessee.*		
Ohio,	1.270	2140	Cumberland Mountains,	1.450	2447
Kentucky,	1.250	2106	Mean weight in the U. S.		2475
Indiana,	1.260	2126			
Illinois,	1.273	2146	*Pennsylvania,—Anthracites.*		
Average in U. States.		2160	Lykens valley,	1.327	2240
			Lebanon co., grey vein,	1.379	2327
			Schuylkill co., Lorberry creek,	1.472	2484
EUROPE.			Pottsville, Sharp Mountain,	1.412	2382
			" Peach "	1.446	2440
England.			" Salem vein,	1.574	2649
Newcastle, cherry coal,	1.266	2136	Tamaqua, vein N.,	1.600	2700
" caking,	1.270	2143	Mauch Chunk,	1.550	2615
Wigan, "	1.275	2151	Nesquehoning,	1.558	2646
Durham, "	1.274	2150	Wilkesbarre, best,	1.472	2484
Lancashire, cannel,	1.199	2023	West Mahonoy,	1.371	2313
Derbyshire, furnace coal,	1.264	2133	Beaver Meadow,	1.600	2700
Shropshire, "	1.268	2140	Girardville,	1.600	2700
			Hazelton,	1.550	2615
Scotland.			Broad Mountain,	1.700	2869
Glasgow, cherry coal,	1.268	2140	Lackawanna,	1.609	2715
" splint,	1.307	2205			

Localities of Coal.	Specific gravity.	Weight of one cubic yard, in lbs.	Localities of Anthracites, &c.	Specific gravity.	Weight of one cubic yard, in lbs.
EUROPE.			UNITED STATES.		
Scotland.			*Massachusetts.*		
Wylam, splint,	1.302	2197	Mansfield,	1.710	2885
Edinburgh, cannel,	1.318	2224	*Rhode Island.*		
France.			Portsmouth,	1.810	3054
Puy-de-Dôme,	1.310	2212	Average in U. S.		2601
Cantal,	1.300	2193			
Auvergne,	1.320	2222	EUROPE.		
Allier,	1.330	2244	*South Wales.*		
Belgium.			Swansea,	1.263	2131
Hainaut,	1.270	2140	Cyfarthfa,	1.337	2256
Liege,	1.300	2193	average,	1.350	2278
Silesia.			Ynis-cedwin,	1.354	2284
Frederich,	1.263	2131	*Ireland,* mean,	1.445	2376
Gustaw grobe,	1.270	2140			
Average in Europe,		2164	*France.*		
			Allier,	1.380	2207
			Cantal,	1.390	2283
			Brassac,	1.430	2413
ASIA.			*Belgium,* anthrac's coal of		
Bengal, Hurdwar,	1.368	2308	Mons,	1.307	2105
" Chirra Punjee,	1.447	2441	Westphalia,	1.350	2278
Assam, Kosya hills,	1.275	2151	Prussian Saxony,	1.466	2474
Aracan, Birmese,	1.308	2207	Saxony,	1.300	2193
Average in Asia,		2277	Average of Europe,		2281

From the foregoing table several useful facts are made apparent. The first is, the greater weight of the American anthracites than those of Europe; second, that the bituminous coals very closely coincide in both quarters of the globe.

	Average weight of a cubic yard.		
	Bituminous coals.	Anthracites.	Intermediate species.
Mean weight of the American,	2560 lbs.	2601 lbs.	2475 lbs.
" " European,	2164 "	2281 "	
" " Asiatic,	2277 "		

We have neither added the weight of the intermediate species of coal to the bituminous column nor to the anthracites; because, in either case, it would have unduly affected the true character of the averages. Third, as regards the table of American anthracites—and it may be correct also to include that of the bituminous coals—it will be seen, with the assistance of a map, that their specific gravity increases as we advance from west to east: confirming also the fact, noted elsewhere, that the weight of the combustible decreases in proportion to the amount of bitumen with which it may be charged.

The Anthracites of Pennsylvania, commonly distinguished as White Ash, or Red Ash Coals, and selected according to their respective qualities.—Whilst treating on the comparative value of the varieties of Pennsylvania anthracite, as applied to iron making, we have said but little in relation to their relative values for domestic use. It seems established that, for closed furnaces, for warming houses, the white ash variety, being the most compact, dense, and slow burning, is more durable, and consequently, more preferable than the softer red ash coal. In open grates, for warming apartments, the latter is decidedly preferred. We have observed a recent statement of the result of an experiment, in relation to this point, which, as regards the warming of apartments, seems tolerably decisive.

A very important and interesting experiment was recently made for the purpose of testing the comparative value of the red and white ash coals for *domestic* purposes. Two rooms of nearly the same size, and having the same temperature, were selected to ascertain how many pounds of each kind would be required to heat them to a temperature of 65 degrees, during a period of 15 hours, when the temperature out of doors at 9 A. M. was at ten degrees below the freezing point. Two days were occupied in the trial, so that the red and white ash coals might be used in *alternate* rooms. Fires were made at 9 A.M. and continued until 12 P.M. Two thermometers (one in each room) were suspended at the greatest distance from the grates, and the temperature was carefully registered every hour. The result was as follows:—

Thirty-one pounds each day of the Schuylkill *red* ash coal gave a mean temperature of 64 degrees; and *thirty-seven* pounds each day of the *white* ash, taken from a vein of high repute in the Lehigh region, gave a mean temperature of 63 degrees. Making 2000 pounds of the red ash to be equal to 2,387 pounds of the white; or, red ash coal at $5.50 per ton, to be equal to white ash at $4.61. This settles the question on the score of ECONOMY.

DEPTHS OF COAL MINES.

The following statement has been prepared from a much more extensive sèries, in order to exhibit the minimum, the maximum, and the average depths beneath the surface at which beds of coal are at this time productively worked, in the principal mining regions of the world.

Number of Coal-Fields.	Coal Fields.	Depths of Coal Mines.		
		Minimum feet.	Average feet.	Maximum feet.
	Great Britain.			
X.	Ashby de la Zouch, depth reached in the works,	"	"	1167
XI.	South Staffordshire, Christchurch,	120	498	870
XII.	Coalbrook Dale,			729
XIV.	North Staffordshire, or Pottery,	100	450	725
XVIII.	Lancashire, or Manchester coal-field, Pendleton,		750	1521
	Shaft at Sankey Brook, near St. Helens,			1377
	Victoria pit, Dukinfield, east of Manchester,			1000
XX.	Yorkshire, near Wakefield,			870
	Derbyshire, near Chesterfield,		300	500
	The Swan Banks colliery, near Halifax,			812
XXIII.	Whitehaven,		600	990
XXV.	Newcastle, Tyne district,	126	510	1020
	" Wear district, Monkwearmouth,	180	450	1794
	" Do. Murton colliery,			1488
	" Tees district,	180	233	480
XXVI.	Berwick-upon-Tweed,			360
XXIX.	Victoria colliery, Nitshill, Glasgow, Scotland, deepest mine,			1038
XXXIX.	North Wales, or Flintshire,			450
XL.	South Wales, worked by adit levels, chiefly,			480
	Duffryn colliery, near Aberdare, shafts 94 yards,			282
	Ireland.			
XLV.	Kilkenny, anthracite district,	66	102	180
LI.	Limerick, culm beds,	70		240
	France.			
I.	Valenciennes, coal pits, upwards of			1500
	Maximum of the collieries of France, 503 metres,			1635
II.	Basin of Hardinghen,			221
LVIII.	Alais, in Gard,			235
XVIII.	Decize, in Nievre,			845
	Belgium.			
I.	Mons district, Hainault province,		810	1140
II.	Charleroy, " "		482	
III.	Liege province, L'Esperance mine, at Liege, 450 metres,			1476
	North America, United States.			
	In this country, the short period in which the coal beds have been worked, has not occasioned the sinking of vertical shafts to any considerable depth.			
	The deepest anthracite mines of *Pennsylvania* are commonly worked by sloping shafts, which follow the inclination of the seams.			
	In numerous positions the coal can be mined by adit level as in Wales, several hundred feet below the mountain summits.			
V.	Virginia, Richmond coal-field, the deepest mines in America.			
	Midlothian shaft,			775
	Heth's pits,			700
	Wills,			700
	Anderson's shaft,			450
	Gowrie pits,			460
	British America.			
	Nova Scotia, Pictou mines,	60		240

SYSTEMS FOR WORKING COAL MINES.

We have devoted but small space to this subject; not that we are insensible of its extreme importance, but because it was somewhat out of the scope we had assigned to the present volume, and also because this knowledge may be separately obtained through the medium of numerous publications by experienced persons; conveying that precise description of information, for the benefit of those who are practically engaged in this service, or are interested in this description of property. It would be invidious, perhaps, to make mention of some of these, without including all.

The Parliamentary Reports embody a great amount of practical information on the methods employed in excavating coal mines. The pages of the Mining Journal, during many years, have been rich in valuable details of the same kind: and among the most recent of its articles is one from Mr. Dunn, "on the various systems practised in the conducting of coal mines, and of the methods employed in counteracting the effects of inflammable air."* This article has elicited criticism and additional facts from others, equally practical, through the same useful channel. This subject is also treated on at some length, in Dr. Ure's Dictionary of Mines, &c. The coal measures of Anzin in the coal basin of Valenciennes, as at Mons, in Belgium, are covered by an enormous thickness of horizontal cretaceous and tertiary strata, through which it is necessary to penetrate. These overlying beds are called by the French miners "*morts-terrains*," or dead lands, and being highly charged with springs of water, require great skill and enormous expense in sinking the shafts through, until they reach the inclined coal seams, at the depth of from two hundred and twenty to eight hundred feet beneath the surface.

The annexed figure affords a remarkably instructive view of these circumstances, both in a mining and geological sense; showing the *revêtement* or impervious lining of the shaft, through the "dead formations;" the mode of ascent and descent provided for the miners, and the position of the ventilating fire, near the bottom of the vertical shaft.

* Mining Journal, March 21 and 28, 1846. The reader will derive much interesting information, respecting coal mining operations, from the lectures of Professor Anstead, as reported in the London Mining Journal, 1847–8.

Shaft of the Coal Mine of Anzin, in France.

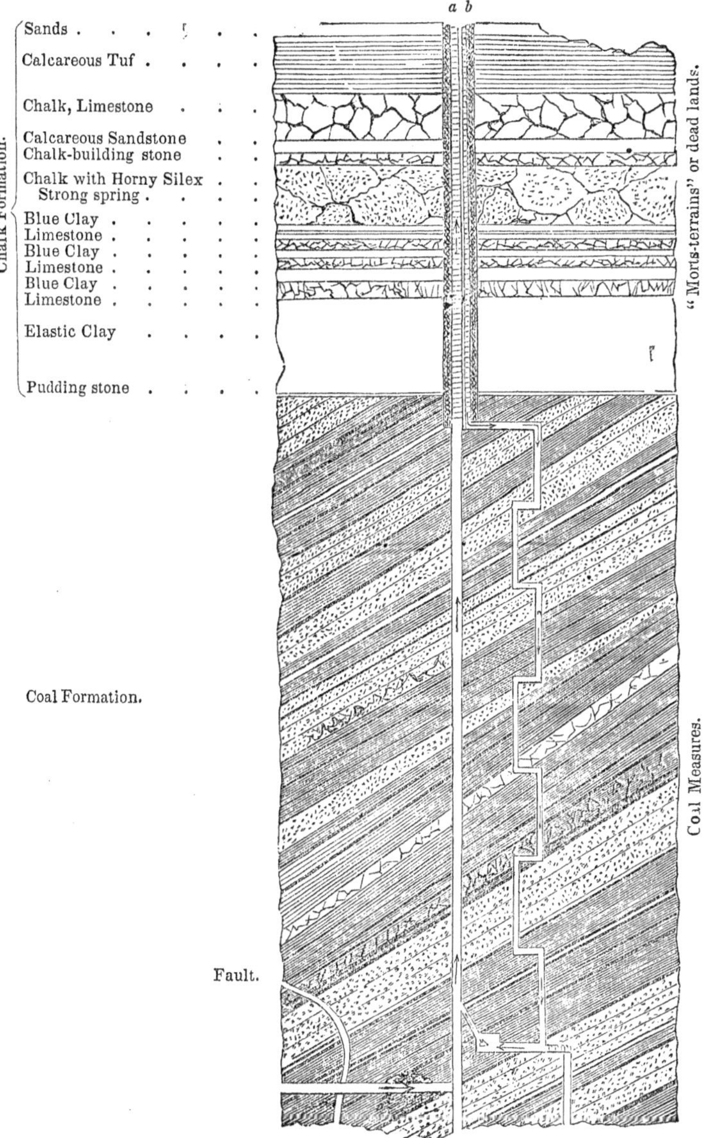

In order to show more distinctly the position and arrangement for the ventilating fires at the bottom of the shafts, we introduce the following enlarged figure of those employed in the mines of Anzin.

Diagram showing the arrangement of a ventilating furnace "foyer d'aerage."

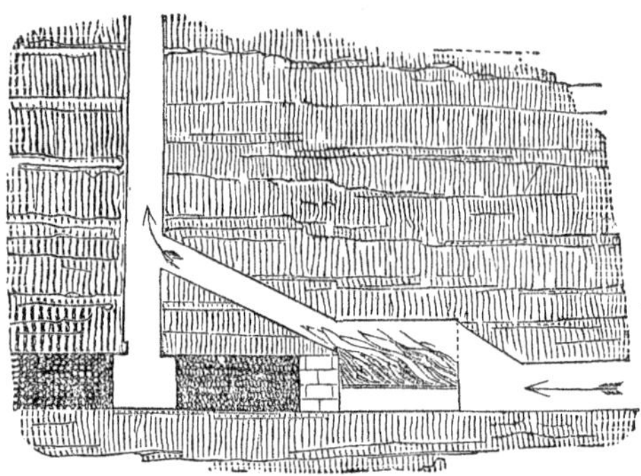

It is observed by M. Burat, that, as a general fact, applicable to all the methods of *exploitation*, it is necessary to be watchful that the pillars be not suffered to remain isolated, and for a long time exposed to the action of the air, before pulling them down. Coal alters in the mines almost as much as at the surface; the pyrites, contained therein, decompose, and the hydroxide of iron, which is the result, gives to the coal a rusty stain which depreciates its value. Finally, the schists disintegrate, effloresce, and the selection of the coal becomes much more difficult. It is necessary, therefore, to proportion the excavation, by preparatory works, to the extraction which may be required during the year.

Coal is, without contradiction, among the useful minerals, the one whose exploitation presents the greatest difficulties. In fact, it requires to be extracted in very large masses; its primitive value, scarcely more considerable than the stones of the mines, is, nevertheless, sufficiently important that we should not abandon the smallest possible amount of it. Left in the old workings, it is lost forever; besides, the interior sources of water and of deleterious gases incroach sometimes upon the immense subterranean surfaces that are exposed. It needs all the resources of science and industry to render possible the working of certain basins, which would have remained in abandonment without the modern means of safety and ventilation and the progress of the steam engine.

Obstructions generally develope themselves in connection with the surfaces placed under investigation. The engineer can then unfold, progressively, his means of action in such a manner as to remain always master of the exploitation. But it is the sudden accidents which defy all human prudence, which endanger the safety of the miners, and which, in a few hours, destroy the fruits of long labour and of powerful capital. The most terrible of these accidents arise from the collection of water and of gas, which in nearly all of these basins, are found accumulated in ancient workings, of which tradition has scarcely preserved the remembrance. When a cutting approaches the vicinity of one of these accumulations, a blow of the pick or a blast of the mine suffices to put the works in communication with the danger; and when it manifests itself, there is neither time to fly nor to resist it.

To avoid these sad rencounters, the miners are preceded in the drifts where danger is to be feared, by horizontal borings, *sondages*, some of them straight, others divergent. These soundings, to furnish sufficient security, ought to be about thirty feet around; if one of them reach a chasm, all the work of excavation ought to cease as soon as it is practicable to ascertain its nature.*

VENTILATION OF COAL MINES.

On the causes which vitiate the air in mines.—"The means of maintaining in the mines an atmosphere constantly respirable, and of preserving the workmen from the accidents which result from deleterious gases, constitutes one of the capital parts of the art of working, *exploiter*.

The causes which most frequently vitiate the air, are these: the respiration of the workmen; the combustion of the lamps; the explosions of powder; the spontaneous decomposition of certain mineral substances, such as the sulphurets which change into sulphates; the coal which heats and burns spontaneously; the corruption of the wood; the striking of the tools against rocks which contain ores of arsenic or mercury; in addition to which is the natural disengagement of deleterious gases which penetrate the rocks, or are accumulated in the crevices and natural cavities, and sometimes in old workings.

The gas thus produced or disengaged disposes itself in the drifts or galleries according to the order of density, as follows:

	Specific gravity.
Carbonated hydrogen, fire-damp, or inflammable gas,	0.558
Azote or nitrogen gas,	0.976
Atmospheric air,	1.000
Sulphuretted hydrogen,	1.191
Carbonic acid, or *choke-damp*,	1.524
Arsenical and mercurial vapours.	

The general precautions employed to get rid of these gases as soon as they are formed, in creating currents sufficiently active to effect their diffusion with the atmospheric air, and to draw the mixture out of the works before it is prejudicial, constitute the art of ventilation —*aérage*. But these general means do not always suffice, and it is necessary to add special means to avoid, or at least to restrain the sudden disengagements, until the common methods shall have restored

* Burat Géologie appliquée, p. 416.

the equilibrium. It is necessary, then, to be able to recognize the presence of each of these gases, in order to destroy them in time, and even, if possible, to diminish the causes of their production.

When the working of a mine, pit or gallery is commenced, if no particular phenomenon facilitate the renewal of air, the respiration alone of the workmen and the combustion of their lamps, are not slow to modify it sensibly. In fact, a workman respires an average of 800 litres = 210 gallons of air per hour, from which he absorbs, in part, oxygen, and substitutes for this oxygen, in the same space of time, 24 to 25 litres, = 6½ gallons of carbonic acid; his lamp, operating nearly with the same intensity as his respiration produces as much carbonic acid, and augments besides the proportion of unconnected azote.

The *carbonic acid*, or choke-damp, which is thus the most immediate and most general product of the workings in the mine, is recognized by its weight; it always occupies the lowest parts of the excavations; its intermixture with air manifests itself by the difficulty of combustion in the lamps, whose flame diminishes in brilliancy in proportion as the acid increases, and ends by extinction, when the mixture attains to one-tenth.

Upon the miners, the carbonic acid manifests itself by an oppression which overwhelms them; nevertheless, temperament and habit will greatly vary the proportions of the mixture which some men are able to breathe. Certain miners can yet work when the lights have ceased to burn; there are even some whose acquired habit is such that they pass through, we are assured, galleries where there is more than twenty per cent. of carbonic acid. Nevertheless, we should watch, on pain of the greatest dangers, that the lamps can everywhere burn with facility, and that the proportions never exceed five per cent.; for this gas, which the French miners commonly call *mofette*, has the greatest tendency to isolate after generation, and will then cause an instantaneous asphyxia.

A single example will demonstrate this energetic action. The workmen of the Creuzot mine descended one morning, the one following the other, in rotation, into a shaft below, in which carbonic acid had accumulated during the night. Arrived at the level of the "*bain*," at a few yards from the bottom of the pit, the first fell, struck with asphyxia, without having time to utter a cry; the second followed immediately; the third saw his comrades prostrated on the ground, almost within reach of his arm; he stooped to seize them, and fell himself; another quickly shared the same fate, in his desire to save the others, and the catastrophe would not have been arrested had not the fifth been an experienced master miner, who obliged those who followed him to reascend.

These accidents are often to be dreaded in coal mines, where spontaneous disengagements are capable of producing in a little time large quantities of carbonic acid. In this case, it is necessary to have within reach ammonia, caustic potash, or lime, of which a solution must rapidly be made, to be thrown into the invaded workings, either by letting it fall from a watering pot, if it be in a shaft, or projected from a pump, if it is in a slope or a gallery. It is also necessary to fight incessantly against the production of the carbonic acid, and to prevent its accumulation by leaving no wood in a state of decomposition, and proscribing all combustion beyond that of the lamps necessary for lighting. Finally, it is essential to prevent the spontaneous heating and firing which is so frequent in coal mines. When a fire is ascertained, it should immediately be circumscribed by impermeable walls, called *corrois;* walls constructed of rubbish with a mortar of clay.

The gases which result from the subterraneous decomposition of the coal, have, besides carbonic acid, carbonic oxide, azote, sulphurous acid, and the carburets of hydrogen, which have a special odour. Before the coal takes fire, the interior air is already heavy and heated by the gaseous disengagements which are the precursors of ignition. As quickly as these symptoms are remarked, the coals already mined should be raised, and we should isolate from the surrounding air the region or the crevices which enclose the fire: employing at this work the labourers whose organization is known to be the best adapted to support the deleterious influence of these gases.

Azote, or nitrogen gas, is much less to be dreaded than the carbonic acid; because its action upon the animal economy is less energetic; besides, its production can only take place by the absorption of oxygen from the air, and it does not naturally exist in the fissures or cavities of the rocks. It has, then, no spontaneous disengagement; but if we penetrate into the works which have been a long time abandoned, and where there has been combustion, the azote will occupy, in consequence of its lightness, the higher parts of the excavations, while the carbonic acid will occupy the lower parts; the respirable air forming the intermediate zone. Azote is found isolated in certain mines, where there exist pyrites in a state of decomposition; the sulphurets changing into sulphates, absorb the oxygen and isolate the azote; the sulphuret of iron is, in this respect, the most active agent.

Azote manifests itself by the red colour of the flame of the lamps, which ends by extinction; it renders respiration difficult, produces a heaviness of the head, and a hissing or singing in the ears, which seems to indicate a mode of action different from that of carbonic acid.

The ordinary lamp of the miner is extinguished when the air contains no more than 15 per cent. of oxygen: [the atmospheric air is composed of 21 per cent. of oxygen and 79 per cent. of azote,] it is also at this proportion of 85 per cent. of azote that asphyxia or suffocation is caused.

Proto-carbonated hydrogen, or inflammable air, designated by the French and Belgian miners under the name of *grisou*, is of all the gases the most dangerous; that which occasions the greatest number of accidents, not by asphyxia, which it can nevertheless produce when it is not mixed with at least twice its volume of air, but for its property of igniting when in contact with lighted flames, and of exploding when it is mixed, in certain proportions, with atmospheric air.

This gas is frequent in nature, and often designated under the denomination of *marsh* gas, because it disengages from the stagnant waters which retain vegetable matters in decomposition. Some muddy volcanoes called salses, emit it in large quantities; it also penetrates certain rocks, such as the coal series and the saliferous strata, where it is accumulated and condensed in caverns and natural vacancies; so that by soundings or borings its true sources can often be determined. There even exist natural or artificial sources, which can be lighted and which have persistance enough to be brought into useful service.

The *grisou* is more abundant in the fat and friable coals, than in the dry and meagre coals; it particularly disengages itself in the crushed places, *éboulements*, in the recent stalls whose surfaces are laid bare, and that so vigorously as often to decrepitate small scales of coal and produce a slight rustling noise. The fissures or fractures of the coal, and even the clefts of the roof or the floor, give sometimes outlets to *soufflards* or jets of gas.

The action of this gas upon the flame of the lamps is the most certain guide in ascertaining its presence and proportion. The flame dilates, elongates, and takes a bluish tint, which can readily be distinguished by placing the hand between the eye and the flame, so that only the top of it can be seen. As soon as the proportion is equal to a twelfth part of the ambient air, the mixture is explosive, and if a lamp be carried, it will produce a detonation proportionate to the volume of the mixture. When, therefore, a miner perceives at the top of the flame of his lamp the bluish nimbus which decides the presence of the fire-damp, he ought to retire, either holding his light very low or even to extinguish it.

The experiments conducted by Sir H. Davy show that the most violent explosions take place when a volume of proto-carbonated hydrogen gas is mixed with seven or eight volumes of atmospheric air.

The chemical effects of an explosion are, the direct production of the vapours of water and carbonic acid and the separation of azote. The physical effects are, a violent dilatation of gas and of the surrounding air, followed by a reaction through contraction. The workmen who are exposed to this explosive atmosphere are burned, and the fire is even capable of communicating to the wood work or to the coal; the wind produced by the expansion is so great that, even at considerable distances from the site of explosion, the labourers are thrown down, or projected against the sides of the excavations; the walls, the timbering, are shaken and broken; and crushing, or falling down, is produced. These destructive effects can be propagated even at the mouths of the pits, from which are projected fragments of wood and rocks accompanied by a thick tempest of coal in the form of dust.

The evil rests not there; considerable quantities of carbonic acid and azote, produced by the combustion of the gas, become stationary in the works, and cause those who have escaped the immediate action of the explosion to perish by suffocation. The ventilating currents, suddenly arrested by this perturbation, are now much more difficult to re-establish, because the doors which served to regulate them are partly destroyed; the fires are extinguished, and often, even the machines fixed at the mouths of the shafts, to regulate the currents, are damaged and displaced, to such an extent that it becomes impossible to convey any help to the bottom of the works.

Some examples will give a just conception of the intensity of these explosions and of their effects.

In a gallery of a coal mine of Saarbruck, in Rhenish Prussia, the explosive air took fire on the arrival of a miner carrying a common lamp. Seven dykes or walls of bricks built in the lateral works and at twenty feet from the gallery, forming with them sharp angles, in such a manner that they could not be struck by the dilatation of the air in the direction of the explosion, but only by contraction, were nevertheless thrown down. We infer from the description that these walls fell inwards; that is, towards the point of explosion. At nine hundred feet from the explosion timber of eight inches diameter were broken; a door for ventilation was torn up, and violent effects of the same nature were manifested even at near 2000 feet distance.

In a mine of Schaumburg, the fire-damp, which filled a gallery and a shaft of 1000 cubic yards in capacity, took fire in 1839. Stones which weighed more than a ton, serving as the foundation of a hydraulic machine of the weight of twelve tons, were displaced, notwithstanding the strong wooden props which consolidated them against the direction of the explosion and which were themselves broken. In another mine of the same principality, the coal was set on fire, and this coal was coked, by that cause, even to the depth of more than a yard.

The explosion, *coup de feu*, of the mine of Espérance, which occurred at Liege, in June, 1838, does not seem to have produced such fatal effects as analogous phenomena have elsewhere done. The fire being propagated without explosion, to the right and left of a working, by the effect of a blast in the mine, produced an explosion in a distant working. Sixty-nine miners were killed. In the place of explosion they were burned and broken; in that, where the fire had first taken place, all the corpses were arranged with their heads directed towards the very point where the combustion originated; these unfortunate people having evidently sought to protect themselves thus against the gas which burned behind them.* In the other workings, the miners had only perished by asphyxia.

The relation of these accidents suffices to indicate the general precautions which should be taken. Thus, it is essential to place the lamps only near the lowest parts of the excavations; to avoid all

* Is it not more probable that these bodies were instantaneously thrown into this position, by the great reaction, the collapsing of the air towards the focus or vacuum caused by the explosion?—T.

methods of working which ascend without outlets; to work, if possible, by descending, rather than by ascending, and to redouble the usual precautions on entering into excavations after an interruption of the work. A great number of accidents have taken place, for example, on Monday mornings, when the miners descend after having quitted the mine on Saturday.

M. Bischof reports that having visited a gallery which had been abandoned for several days, he found the gases liquated to such an extent that they were inflammable in every part of the area; detonating in the middle portion, while the almost pure atmospheric air filled the lower part.

It is very dangerous to allow these liquations to be produced; it is necessary that the current of air should be sufficiently active to produce immediately the diffusion of the gas in the air and its withdrawal out of the mine before the mixture has become explosive. But, notwithstanding the precautions of ventilation—*aérage*—many mines would be completely unworkable if there had not been found the special means of guarding them from the fire damp—*grisou*. The coal beds, most dangerous, are those which are the most valuable for their good qualities; science and industry have therefore been called on to seek the means of combatting the effects of the grisou, and we proceed to expose those which have been successively employed.

MEANS TO DESTROY OR CHECK THE FIRE-DAMP OR GRISOU, IN SUBTERRANEAN WORKS.

The first idea which presented itself to the explorers was to disembarrass themselves of the gas by allowing the liquation to establish itself and by setting it on fire, so as to burn it, in the absence of the miners. For this purpose, a workman, clothed in vestments of moistened leather, his visage protected by a mask with spectacles of glass, advanced, crawling on his belly, in the galleries where the fire damp was known to exist, and holding forward a long pole, at the end of which was a lighted torch; he sounded thus the irregularities of the roof, the front of the excavations, and set fire to the grisous. This method, which has been employed, within twenty years, in the basin of the Loire, and even occasionally at the present day, in some of the English fiery collieries, has numerous inconveniences. The workmen, whom they called *pénitents*, were exposed to dangers to such an extent, that a great number perished. When the gas, instead of being simply inflammable, was detonating, the solidity of the mine was constantly compromised by the explosions; the fire attacked the coal and the timbers; the gases, which resulted from the combustion, became stationary in the works, and menaced the workmen with asphyxia; at length it became necessary, in certain mines, to repeat, even three or four times a day, this perilous operation, and yet it in no respect obviated the rapid disengagements which caused these numerous accidents. This method was equally in use in the English collieries; only the penitent or *fireman*, instead of carrying the fire himself, caused it to be moved by means of a slider placed over a line of poles connected together, and directed by a system of pullies and cords. The danger was thus diminished for the fireman, who retired into a niche formed in a neighbouring gallery; but in the meanwhile many were still overtaken, and, besides, all the other inconveniences remained.

The method called the *eternal lamps* was evidently better. It consisted in placing towards the top of the excavation, and in all the points where the fire-damp collected, lamps constantly lighted, which burned the grisou as fast as it was produced; the danger was diminished in a considerable degree, because there could not be formed such large accumulations of inflammable or detonating gas. This mode of proceeding was, however, renounced in a great number of mines, on account of the production of carbonic acid and of azote; a production the more sensitive, since, to facilitate the liquation of the gases, the air ought not to be very strongly agitated.

At length it was devised to profit by the property possessed by platina in sponge to facilitate the combustion of the hydrogen with which it was brought in contact, and pellets, composed of one part of platina, and two parts of clay, were made, and were placed near the points at which the grisou or fire-damp concentrated. But all these efforts, based upon the incited combustion of the inflammable gas, proved to be only dangerous and incomplete palliatives, which substituted for a great peril a series of other dangers, less imminent, doubtless, but equally distressing.

From that time all the well disposed continued to search for processes based upon another principle. Two only could conduct to a good result: 1. The withdrawal of the gases out of the mine; 2. A mode of lighting different from that which was in use, and which would suffice for the purposes of the miner without compromising his safety.

The principle of withdrawing—*entrainement*—of the gases by a rapid ventilation is, without contradiction, that which was the most natural to conceive; because it was already applied to all the other deleterious gases. Dr. Véhrle proposed at first to effect the decanting of the gases by making the excavations (stalls?) communicate by ascending passages with a gallery embracing all the works, and uniting with an ascending shaft. But this project, otherwise impracticable, offered a remedy for only a part of these accidents; the execution alone of the necessary works could not have been made without the greatest danger, if these works had been undertaken in the coal; while, in the rocks of the roof, the expenses would have rendered them impracticable. But a good ventilation alone could not suffice to place the miners in security; it was an excellent auxiliary means, but it always left unsolved this important problem: *the prevention of the inflammation of the gases which disengage themselves from the surfaces of the stalls.*

The lighting alone could conduct to the solution of this problem, and numerous attempts had been made, under this head, when Davy discovered the safety-lamp. Before him, they had operated with a small number of lights, placed in the lowest positions, and at a distance from the stalls; the workmen kept these lamps in view, and when the blue nimbus, the indication of hydrogen, began to show itself, they extinguished them or withdrew, covering them with their hats. They made use of, also, in the most infected mines, various phosphorescent matters, and particularly a mixture of flour and lime formed from oyster shells, called Canton phosphorus, although the uncertain and ephemeral light which these materials produced, was but a very feeble resource. At length it was observed that the proto-carbonated hydrogen was somewhat difficult of ignition, and that the red heat was insufficient to accomplish it; thus it was practicable to carry a red coal, or a red hot iron into the fire-damp without inflaming it, the white heat alone having the necessary temperature. They profited by this discovery by lighting the stalls by means of a wheel of steel, which was made to turn against a fragment of flint: a workman was detailed to this service, and the sparks, which were thus produced in a continuous manner, sufficed to light the miners. It happened, occasionally, that these sparks set fire to the grisou; but this discovery, imperfect as it was, was not the less a real benefit.

Such was the state of the question, when Davy commenced the series of experiments which conducted him to the object in view. Many mines had been abandoned notwithstanding the palliatives in use, and a number of those which were maintained in activity, only produced coal at the price of the lives of a great number of men. Davy discovered that the gas, contained in a vase, which only communicated with the exterior by long and straight tubes could not be set on fire; that the flame was difficult of transmission in proportion as the tubes were reduced, and that, consequently, the more their diameters were reduced, the more their lengths might be shortened. He thus arrived at the proof that a plate of thin metal, pierced with holes of about one hundred in an inch, did not communicate fire to the exterior gas, although the interior was charged with lighted gas; the cooling produced by the gas in this small passage sufficed to reduce the temperature of the white heat of the interior down to the red heat of the exterior, and the inflammation could not be communicated. Such was the series of ideas which conducted Davy to surround the flame of the lamps with an envelope of metallic gauze, and thus to construct the *safety-lamp.**

LOCAL VENTILATION.

In the sinking of a shaft the work would soon be stopped by the want of air, were it not for the plan of dividing its total section into two unequal parts, by means of a partition of planks, the joints of which are hermetically closed with moss, &c. The smallest compartment is reserved for the ladders; and a current of air is established

* Burat Géologie appliquée, 1846, p. 472.

between these two compartments in the manner of an excavation having two orifices. This movement is sometimes facilitated by carrying up the level of the orifice of the small compartment, by means of planks, built up as a chimney.

The excavation of a long gallery or tunnel would become impossible, through the want of air, if a spontaneous ventilation were not produced by similar means. Thus, there are directed from the surface of the ground, towards the gallery, troughs which are arranged similarly to the works which have two orifices of different levels and unequal sections. At other times they establish upon the gangways a floor for carriages [*roulage*]; reserving the lower part of the gallery beneath for draining and for a current of air, which enters by the lower part and returns by the principal or upper section.

If this precaution do not suffice, the current may be rendered more active by means of a small shaft disposed in such a way as to accelerate the circulation during all the working time. For this purpose, two doors are placed at the entrance of the gallery, so that one of the two shall always remain closed during the work, and the air is forced to leave by the shafts. This disposition is equivalent to the case of an excavation having two orifices of different levels.

We have translated freely from the excellent work of M. Burat, already frequently quoted, most of the matter which is comprised in the preceding pages in relation to the deleterious gases which are constantly generated in coal mines, and on the means resorted to for ventilating them. Our work would be incomplete, without adverting to a subject so immediately connected with the mining of coal, and with the safety of the operators, whose lives are hourly perilled, who are exposed to accidents inseparable from its extraction, and peculiarly attendant on this branch of mining economy.

VENTILATION OF FIERY COLLIERIES.

There have been numerous suggestions on the ventilation of those coal mines which are subject to explosions; among others we may mention the published views of Mr. Dunn, of Newcastle. We hope to be pardoned here for remarking that it is disadvantageous to an important practical science, and especially embarrassing to inquirers into these subjects, that the valuable information conveyed by the most capable English authorities, is so compounded of absolute technicalities, for the most part also entirely local and unscientific, as to be almost untranslateable. This is, perhaps, the main cause why the English mining processes, from Cornwall to Scotland, varying in their progress through every district, continue to be so little comprehended elsewhere.

Mr. Dunn, at page 341 of Volume X., of the London Mining Journal, furnishes a sketch of the general principles of ventilating the collieries in Northumberland, which are moderately troubled with inflammable air. At page 405 of the same volume, he follows up the first communication by a description of the more elaborate process, as practised in Northumberland, of ventilating fiery collieries. We can do little more here than refer the reader to those articles, and to the instructive diagrams which accompany them. Our description would be unintelligible without the elaborate drafts which are necessary to elucidate the whole process.

The object aimed at by this system is by one series of channels to ramify, through every part of a mine, currents of respirable air, while by another series to withdraw from it and to discharge at the surface the impure air and inflammable or deleterious gases.

Mr. Dunn observes, that until late years the said current of impure air was kept as much as possible together or united; and it was no uncommon circumstance to have it travel twenty or thirty miles before reaching the upcast pit, and then loaded with gases, which steamed from the candle, and even the furnace fire, in thick vapours and flakes of blue flame, alarmingly visible to the naked eye. But, in modern practice, these currents of air are divided and subdivided in countless branches, so as to prevent the air which becomes adulterated in one quarter, from spreading the contagion among the workmen of another. They are conducted through passages either adopted for the purpose, or provided by anticipation in the laying out of the works. To guard against the dangerous influx of gas during the working of pillars, and in order to obviate the danger at the furnace, a "dumb drift" is provided, sloping up into the shaft some fathoms above the furnace, at which point the inflammable air effects a junction with the general air of the pit, and is carried upwards in safety, so that the furnace may be blazing below with good and secure air, and a perfectly inflammable portion may be coming in from above. (See the diagram, page 98.)

The fall of the barometer is a sure presage of increasing discharge of inflammable gas; for when the barometer stands steadily,—say at 29°—and the pressure is uniform, nothing exudes but the ordinary "makings" of the mine:—but when a sudden fall of the barometer portends a lightening of atmosphere, and consequently a change in the counterpoising pressure upon the orifices whence the gas escapes, or upon the main body accumulated in the wastes, then it is that extraordinary eruptions take place,—enough to overpower and adulterate even the main current of air, and consequently to subject the mine to explosion.

"Blowers" are sometimes met with in the coal, but more generally in the stone, and contiguous to the fissures of dykes. They originate in the chinks or crevices and other receptacles, which, being filled with inflammable air under high compression, are discharged momentarily, and without previous warning. They are often known to endure for many years; although, generally, they decline as the supply of pent-up gas is exhausted. Many very calamitous events have arisen from the miners unexpectedly coming in contact with these blowers; for the discharge is so sudden that the general air-course, although previously safe and satisfactory, becomes, in the course of a few minutes a mass of inflammability, and that without time being allowed to notify the workmen or to prevent an explosion.

In many collieries, but especially those of the north of England, an action is going on whereby cavities are formed in the roofs of old worked seams, by a portion of the roof giving way, and forming a dome or inverted cauldron. These cavities form dangerous receptacles or reservoirs in which the carburetted hydrogen collects in large quantities, and renders the mine liable to frequent explosion, especially during periods of the change of density in the atmosphere. These cavities are locally termed *goaves.**

In 1839 was appointed a committee, called "The South Shields Committee," immediately after the explosion of the Hilda pit, South Shields, in which fifty-two lives were lost. The report of this committee was first published in 1843, and it has been affirmed that it may with every propriety, be extolled as a monument of ability and perseverance.† We make a few extracts.

Safety Lamps.—Such is the immense abundance of *fire-damp* in most of the northern mines, that to obtain a natural light, nothing more is necessary than to bore a small hole in the coal seam, insert a tube, and a perpetual flame may be obtained. From the report, it appears that in twenty years, upwards of 680 miners were destroyed by this dreadful means, in the district of the Tyne and Wear alone. The various kinds of safety lamps are investigated, and the reporters give, as their opinion, that the lamps most to be relied upon in mines charged with the destructive gas, are those on the principle of Clanny in England, and Mueseler in Belgium. Even with these the utmost attention must be always paid to their condition ; the gauze examined daily, and every part of their construction ascertained to be perfect ;

* *Glossary* of local mining phrases, employed in the Northumberland collieries, used by Mr. M. Dunn, in the Mining Journal, 1840.

Boards—principal working places, five yards wide.

Drifts—leading places, (galleries) in the direction of the boards.

Headways—the course of passages at right angles to the boards.

Winning headways—preparatory leading places, two yards wide.

Stentings—the holings between the winning headways.

Rolley-ways or *roll-ways*—main horse-roads to the distant workings, six feet high.

Pillar—the whole coal left during the first working.

Stoppings—brick wallings to force the air to the parts required.

Waste—the mine once worked over, and pillars standing.

Hydrogen or *inflammable gas*—lighter than common air.

Choke damp or *carbonic acid gas*—heavier than common air.

Brattice—temporary partitions of slit deal, to ventilate the leading places.

Blowers—orifices in coal or stone, leading a constant stream of inflammable gas, attended with a great noise—hence the derivation of the name.

Goaf—the cavities which result from the falling down of portions of the roof, and in which fire damp accumulates.

Creep—where the pillars or *sill* give way, under the superincumbent strata.

Air courses—principal passages, [thirty feet area] along which the air is conveyed, and in which the standard quantity is taken to be upwards of two thousand cubic feet per minute.

Brattices—wooden partitions to direct the currents of air in the system of ventilation.

Intake drift—the passage through which the current of pure air circulates.

Return drift—that by which the impure air is withdrawn.

Dumb drift—for the passage of the deleterious gases.

Upcast pit—the shaft by which the foul air ascends, and is occupied by the ventilating furnace.

Downcast pit—that by which the atmospheric air descends into the workings.

† Mining Journal, February, 1843.

the workmen to be warned never to continue working in an inflammable atmosphere, with an overheated lamp; and that, instead of impressing them with the idea that they are perfectly safe instruments, they should be convinced that a reliance on lamps is fatal error, and that no mere safety lamp, however ingenious, is, of itself, sufficient to secure a fiery mine from explosion.

We learn from the London Mining Journal, May 13th, 1854, "of a method by which Mr. Septimus Piesse proposes to illuminate coal mines by means of coal gas, thus rendering useful that which has often caused the death of the miner. Mr. Piesse has suggested, that the gas might be made on the surface of the mine, and carried down by fixed piping, there to be kept burning in lamps, with gauze wire round the flame. But a practical miner, who has addressed the editor of the Mining Journal on the subject, is of opinion, that it would be better to collect the gas generated in the mine, which might be done by making large cavities above the level of the roof, and openings in the mine, to form a reservoir for the retention of the gas till it is quite full. He says, he has seen large cavities where acres of coal have been excavated, and the entrances walled up with only a ten-inch wall, and this reservoir has been completely filled with gas. Now, if the smallest hole were made in this wall, the gas would instantly escape, and could be ignited; thus forming a capital gasometer. Where, then, he asks, is the necessity of making the gas on the surface of the mine, and conveying it down in pipes, when the process can be safely carried on in the mine itself? Besides, there would, he thinks, be great risk of the pipes breaking by the falling in of the roof; whereas, by the plan he suggests, no danger need be apprehended, especially if gutta percha tubing was used.

Ventilation.—When it is considered that the explosions are always from a very limited portion of the mine, and that the air has commonly not a motion of more than three-fourths of a mile per hour, in the greatest part of the mine, it is matter of surprise that these lamentable occurrences, instead of being occasional, are not incessant and overwhelming. Living thus always on the verge of destruction, it has excited, among the officers and men in the mines, a continual watchfulness and knowledge of dangerous symptoms that alone enable them to proceed with any degree of safety in such a situation, but in which, on the smallest error, or a contingency unforseen, as a boy at sleep or at play, a heated lamp, a broken wire, a sudden eruption of gas, or change in the wind, or a sudden pressure of the atmosphere, whether from the falling of parts of the roof or otherwise, the bounds of safety can no longer be preserved; but tremblingly alive to their danger, they are plunged, unresisting victims into the abyss.

In regard to indications by means of instruments, the report states that "the combined indications of the barometer, thermometer and wind, tell the state of a mine with the greatest nicety. When the barometer indicates a fall, the thermometer a rise, and the wind blows from the E. S. E. or south, an ordinary fiery colliery will be certain to pass rapidly into a state of great danger.

VENTILATION OF COLLIERIES, IN SCOTLAND AND THE NORTH OF ENGLAND.

A report was issued, in 1847, by Mr. Tremenheere, the commissioner appointed under the provisions of the act of the 5 & 6 of Victoria, in relation to the mining population of Scotland and the north of England. This and other previous reports have formed the subject of an able article in the North British Review, for November, 1847. We take the liberty of quoting, somewhat irregularly, the substance of a portion of that article.

"Although some of the principal collieries in Scotland are pretty well ventilated, yet it must be admitted that, taken as a whole, the arrangements connected with ventilation in Scotland are, as compared with England, in a very imperfect state. Happily, our mines are almost entirely free from that dangerous element which so frequently produces such awful havoc and devastation to our neighbours in the south. In this respect, therefore, our necessities have not required us to be so particular in carrying fresh air to the mines. Hitherto, most of the mining operations in Scotland have been situated within a reasonable distance from the surface, and the ease with which one pit could be sunk, to relieve the workings of another, superseded the necessity for great outlay in connection with ventilation, and to some extent caused it to be overlooked, and a matter of indifference. In some of the old mining districts, the workings are now extending to a great depth, and the method of ventilation is assuming the most important aspect, and is conducted on the most improved principles; but at the small country collieries, when sinking a new pit, little preparation is made, even now, to have it properly ventilated.

"Accordingly, for several weeks—some seasons, even months, during a continuance of warm weather in summer, the colliers at such places are either partially or altogether idle; the extent of their work being regulated by the state of the atmosphere. The irregularities occasioned by this imperfect ventilation tell very materially both on the profits of the coal master and the incomes of the men, and ultimately on the price of coal in the market, besides doing terrible injury to the health of the people employed, by causing them to breathe in an impure atmosphere. The persons in charge of such works have generally not only a limited education, but possess very limited means of observation; and to them the advice of a properly qualified inspector will be an incalculable boon.

"In respect of Scotland, therefore, the way for an inspector is perfectly clear, and his appointment may be the means of doing much good in other departments as well as that of ventilation. But what shall we say of England, the scene of so many terrible calamities? That which has baffled the ingenuity and skill of the most talented and accomplished coal-viewers that the world ever saw, such as the late Mr. Buddle, and has proved the ingenious scientific theory of

Messrs. Faraday and Lyell to be impracticable, is not likely to be controlled by a government inspector of mines. We have often felt oppressed and overpowered at the thought, that the mightiest efforts of man could not prevent these awful explosions, which cause such a sacrifice of human life. If it were possible to get at the immediate cause, some hope might be entertained of at least mitigating the evil; but from the scene of those accidents no one has ever returned to tell the truth.

"The system of ventilation pursued at the collieries in Northumberland and Durham, where most of these explosions occur, is of the most perfect and complete kind, and entirely in accordance with the principles of scientific truth. But, however sound the principles on which the ventilation is conducted, practice declares that there is a limit to the distance to which atmospheric air can be conveyed with safety underground, from the impurities it mixes with on its way; and however much the question may be avoided, by those who have capital invested in the deep collieries, to this it must come at last,—more openings must be made from the surface; more pits must be sunk. The question must be brought to this practical issue,—whether is capital or human life to be sacrificed? and when it does appear in this shape before the British Parliament, we do not fear the result.

"It is stated by Messrs. Faraday and Lyell, in their report on the explosion which occurred at the Haswell collieries in 1844, that 'when attending the late inquest, we were much struck with the fact, that more than half of the pitmen who gave evidence, were unable to write, or even to sign their names as witnesses.'

"It is a well ascertained fact, that accidents from fire damp have generally occurred with a low barometer; and when we consider that a fall to a very small extent will render a place, which it was safe to work in at night, entirely unsafe and dangerous in the morning, we cannot help feeling that there is something grievously wrong in allowing men, who cannot write their names, to have any thing to do with ventilation at all."

The subject of Ventilation of Collieries and the nature of fire-damp being of such great importance, it is hoped the following extracts on the subject from the London Mining Journal, Nov. 1853, may not be uninteresting.

On Ventilation of Collieries.

"The following epitome of that portion of the evidence given before the Parliamentary Committee of England by the government inspector, Mr. Dickinson, contains some valuable information in relation to colliery operations. Ventilation formed of course a most important element in the inquiry of the committee, and in its consideration are necessarily comprised the natural differences in roofs, the attendant dangers, and the most improved and secure methods of underground working, in its several branches and details. In his testimony, Mr. Dickinson, the government inspector, recommends, in fire-damp mines to drive on the galleries to the extremity of the mine, and to work the coal backwards, as this insures a permanent air-way at all times; and he would prefer this system, whether the seam was perpendicular or horizontal; conceiving that there are no greater difficulties in ventilating a mine with air-ways in solid coal, than in air-ways maintained by gobbing—that is, in the refuse or rubbish thrown back into the excavations remaining after the removal of the coal. There is, besides, in his view, no liability to leakage, if the air-ways are in the solid coal; and liability to leakage is avoided by driving out the level to the extremities, and working the coal backward. Mr. Dickinson is very decided in his approval of the long-work in collieries; and while he condemns the old system, as creating a series of unsightly caverns, he says, 'It is worthy of remark, that there is no instance that I have heard of, where long-work has been introduced, that that system has been abandoned and the old system again resorted to.' He admits, that, in the beginning of working long-work, the first weight of the superincumbent strata on the face of the work, makes it dangerous for the workmen, and may crush the coal; in the course of a few days, however, after the first subsidence has taken place, the roof subsides regularly behind, and there is no more difficulty. He states that the workmen have sometimes abandoned the work until the first subsidence has taken place, and that where long-work has been attempted to be introduced into new collieries, by persons who did not understand it, when this first weight has been coming on, which is the critical weight, the system of long-work has been frequently abandoned; had they however waited until the subsidence had taken place, their efforts might have been successful.

"He observes that the great difficulty is with the workmen. He considers the long-work much better for ventilation, as in working pillar and stall-work, a number of galleries are left open, and all those galleries require to be ventilated; while, with the long-work you have simply the working face open, you leave an intake gallery for the air to go into, it then passes up the face of the work, and returns by the return drift;—while with pillar and stall-work you have a number of galleries which require to be ventilated; and he further states, as a reason for preferring the long-work as to ventilation, that the distance which air travels in long-work is shorter than that which it travels with pillar and stall-work, because you have only the intake gallery, the face of the work, and the return air course to ventilate, and as there is no pillar and stall-work the air is not to be coursed through. The evidence explains that in long-work there is simply a current of air going from the down-cast shaft to the face of the work, running along the face of the work, which is only a channel 6 or 7 feet wide, and returning by a drift to the upcast; while with the pillar and stall-work, if you have 20 stalls, you may split the air into 4 parts, each current of air, or each fourth part, has to course through 4 stalls, with a door between each stall; and therefore, each current of air has to pass up one half of the stall and down the other. In the old system there is also a risk of doors being left open, while in the long-work there are no doors except there be main doors; and where the face of the coal gives vent to an enormous quantity of gas, contingencies arising out of doors are very much fewer than in pillar and stall-work, as the air is all passing up the particular places where it is required. Mr. Dickinson clearly intimates his opinion that the long-work system is in all cases more economical and safe than the pillar and stall system, and that many persons work long-work both with good and bad roofs. He further remarks, that in working the thick coal of Staffordshire, where the top part of the seam is worked first, and there is nothing but the old gob for the roof of the second working, Mr. Gibbons, whom Mr. Dickinson considers authority in these matters, having worked the coal for several years under the long system, says he prefers a bad roof to a good one. This is explained thus: that a hard solid rock for a roof which will not break, is apt to crush your coal, and is attended with more danger than would otherwise be met with if you had a tender roof. He then explained that he had seen a modified system of long-work which, although not the ordinary system of long-work, is called long-work, in South Wales, practised very successfully under a quor roof. It is by driving a stall 8 yards wide, and bringing back the same width of pillars. All the coal is obtained in that working; and perhaps it is the only successful working of coal that there is in South Wales, for all the rest, under the bad roofs, is attended with a very considerable sacrifice of pillars; and he observed that his remarks applied to the cleanness of working, but generally equally to the ventilation."

Mr. Dickinson further stated, "that it was a general rule that a plate roof, which usually bends rather than breaks at first, is one of the best roofs for working long-work, and that long-wall work is quite applicable to it; he does not, however, seem to approve of the usual way in Staffordshire—that is, of working the upper portion first—but thinks that the best way is to work the lower part first, taking care to pack the gob very tight with rubbish. He then proceeded to detail the plan on which he would commence the long-work system. He would keep the lower levels in advance of the upper, for it was generally found that, in attempting to keep the upper levels in advance of the lower, there is a tendency to throw the weight of the roofs on the face of the work, which makes it more dangerous for the men, and also tends to crush the coal; even where the pit is sunk to the bottom of the seam, as the weight always tends to the dip, he would start the drifts, so as to keep the lower drifts in advance of the upper, and throw the weight of the work on the gob and not on the face of the work.

"After opening the pit, and getting the ventilation connected between the down cast and the up cast, if it were not a fiery mine, he would breast all the coal forward, carrying the airing along the deepest level, and bringing it back along the upper level, working straight before him. If it were a very fiery vein, he should recommend driving out the galleries to the extremity, and sinking backward instead of forward, so that the gas would be left behind, and the ventilation maintained by having the galleries in solid coal, and not subject to leakage through the gob. He would take the breast of coal forward and leave the gob behind. Mr. Dickinson then explained the mode of working which he would recommend; it is not new to experienced coal mining engineers, nor indeed to many working miners, but we have been thus minute as coming from a government inspector, to whose care a very extensive mining district is confided, and as it may be supposed to bear with it the stamp and weight of official authority."

On the Nature of Fire-Damp. From London Mining Journal, Dec. 1853.

"Mr. Dickinson entered on another branch of the subject, and observed in answer to a question, No. 98: 'An imperfectly ventilated goaf is about the most dangerous thing you can have in a colliery—that is where the fire-damp is mixed with fresh air enough to bring it to the explosive point. When the goaves are not at all ventilated, the fire-damp in them is generally too pure to be explosive; and I have known cases where a goaf which has been full of fire-damp, has fired along the edge where it has been mixed with fresh air, but there not being sufficient air with the fire-damps in the goaf, it has merely been an explosion for the width of a yard or two along the edge.' Mr. Dickinson stated that the air may be so foul as not to be inflammable, but then when there is a strong admixture of fire-damp it is not respirable. He had in his district two men who were suffocated by inhaling a strong admixture of fire-damp. The effect is to quicken the pulse; he stated that he had tried his own pulse before going in—it was 78 at entering this admixture of fire-damp, and after being in for a few minutes, it ran up to 84; he tried a manager's pulse and it ran up from 80 to 84; he also tried a fireman's pulse, it was at the very unusual height of 120, and it ran up to 126. A person is only enabled to remain in this gas a few minutes, otherwise he would soon fall down and expire, and this gas was explosive at the edges. Mr. Dickinson then stated 'that in order to be explosive there must be an admixture; an explosive mixture is, *he should think*, 1 part of gas and 7 of air—and when asked 'When does it cease to be explosive?' his answer was, I *think* at about 15—that is, 15 parts of atmospheric air to 1 of fire-damp.' We regret that the information which Mr. Dickinson supplied to the Committee was so defective, throwing no new light whatever on the subject—as after an interval of 40 years, which have witnessed a marvellous advance in philosophic knowledge, much might have been anticipated from more accurate analysis and more perfect apparatus. Specimens of fire-damp had been sent from various collieries to Sir Humphrey Davy at that period, for examination and experiment, and he found that the pure sub-carburetted hydrogen, commonly called 'fire-damp,' requires twice its bulk of pure oxygen gas to consume it completely, and that it would, for the same effect, require about 10 times its bulk of atmospheric air, as this volume of air contains about two volumes of oxygen."

"Ten volumes of atmospheric air, therefore, mixed with one volume of sub-carburetted hydrogen gas or fire damp, form the most powerful explosive mixture. If either more or less air be intermixed, the explosive power will be impaired, until 3 volumes below, and 3 above that ratio constitute non-explosive mixtures, that is, 1 of pure fire-damp mixed with either 7 or 13, or any quantity below the first or above the second number, will produce an unexplosive mixture. Davy drew a conclusion that fire-damp would not explode when mixed with less than 6 times, or with more than 14 times its volume of atmospheric air. Scientific men have for a long time been acquainted with these results, but as the experiments which led to them would appear to have been made with gas, brought in some instances from distant collieries, subject to the effects of time and carriage, and as Mr. Dickinson had ample means of procuring the fresh gas on the spot, and has, at his command, the improved appliances of modern science, we confess we would have been better pleased if he had enlightened the committee by evidence of his own philosophical skill. According to Davy, it is the carbon the fire-damp contains, which enables it to emit more light during combustion than pure hydrogen. 100 cubic inches of fire-damp weigh about 17.2 grains; its specific gravity compared with atmospheric air is 0.554. It consists of 4-3 grains of hydrogen gas, combined with 12-9 grains of carbon. A most important fact seems, however, to have been almost wholly overlooked. Davy also ascertained that 1 volume of carbonic gas to 7 of an explosive mixture, composed of fire-damp and atmospheric air deprived it of its power of exploding altogether."

"We are aware that in some collieries, the fissures in the coal resemble natural gasometers, and that even if a mine be cleared one day of inflammable gas, it often fills like a well the next. When, however, we know that carbonic acid gas is, in artificial formation, the easiest procured of all gaseous product, and when we find that the injection of it in large quantities has been recently applied, with complete success, to extinguish fire in mines, it may not perhaps be a very extravagant speculation to suggest that it might be hereafter successfully combined by a simple process with the admixture of fire-damp and atmospheric air, so as to render that admixture unexplosive, and consequently harmless, provided the combination would not itself fatally affect the respiratory organs of human life."

"When we reflect upon the extraordinary and almost marvellous achievements which human perseverance and scientific skill have accomplished in our times, we see nothing to discourage the experiment as visionary or hopeless. The safety lamp was itself the result of clear conception, with careful and cautious investigation; it owes its present perfection to the improvement of time and experience, but still it has not proved a certain or complete protection. The Middle Dyffryn Colliery is a remarkable instance. That colliery had a furnace and a steam jet to produce the ventilation, but the explosions occurred from the gas igniting the furnace. The most judicious and circumspect use of the safety lamp could not therefore have prevented the catastrophe, and some other mode must consesequently be devised of averting the recurrence of so frightful a calamity in those cases of apprehended danger, to which the safety lamp is inapplicable."

Mr. Herbert Francis Mackworth, another Government Inspector of Mines, was also examined before the committee on the two points to which we have principally referred. There is something consolatory in his statements that, taking the number of lives lost in the coal mines of his district in 1851 and 1852, he found a considerable diminution in accidents in shafts, and in the number of explosions. In accidents from explosions there was, however, a considerable increase in the number of deaths, owing to sixty-five lives having been lost by one explosion at Middle Dyffryn. He further explained that the considerable increase in the number of deaths from "miscellaneous" accidents, was owing to an irruption of water at the Gwendreath Colliery, by which twenty-six lives were lost, and this accident occurred on the same day as the great explosion at Dyffryn, 10th May, 1852.

We read in a recent number of London Mining Journal* of a frightful colliery explosion at the Arley Mine of the Ince Hall Coal and Cannel Company, near Wigan, Lancashire, in which 89 unfortunate miners perished. "During the last five years, it appears that 164 lives have been lost in this colliery by explosions, viz. in March, 1849,—12; in Feb. 1850,—4; in March, 1853,—59; and in Feb. 1854,—89."

"We trust, for the cause of humanity, the government of Great Britain will take up the subject, and that some measures will be acted on to secure a better supervision as to the state of the mines. As from the examination it is clear that the government plan of inspection is quite inefficient, as it imposes duties on a public officer which it is impossible for one human being to perform."

We also refer the reader to the valuable report of Mr. Kenyon Blackwell, Government Inspector, in the London Mining Journal.

We must again take the liberty of quoting from the London Mining Journal, May 13, 1854, an account of a meeting held in London of coal proprietors, to inquire into the lamentable subject of accidents in coal mines—a subject which is equally interesting to all connected with collieries in this country as in England.

"The document presented to the conference which led to most discussion, was the printed 'Rules and Regulations for the Safety of Coal Mines and of the Workmen employed therein.' They embrace a code of coal mining laws under the following heads: 1. Responsible charge of the mine; 2. Working-places; 3. Waste; 4. Goaves; 5. Wagon-ways and tram-ways; 6. Timbering and props; 7. Machinery; 8. Shafts; 9. Ventilation; 10. Stoppings; 11. Brattice; 12. Doors;

* March 11, 1854.

13. Furnaces; 14. Fire-damp; 15. Safety lamps in fire-damp mines; 16. General instructions; 17. Penalties. It will at once strike our readers that the foregoing topics are quite sufficient to include every kind of management, and every possible contingency; and it is but justice to those from whom these resolutions have emanated, to acknowledge that they have been framed with perfect fairness, and that they do not exhibit the slightest disposition to screen either proprietors, managers or workmen from responsibility. They are particularly stringent in their directions as to the use and management of the safety lamp, and in prohibiting that of tobacco in any part of the colliery in which safety lamps are employed; and they expressly forbid that lucifer matches, or other self-igniting apparatus, under any pretext whatever, be taken down the pit by workmen and boys. They also propose to restrict the use of ale or any other intoxicating liquors in mines.

"The last fatal explosion at the Ince Hall Collieries, at Wigan, has necessarily furnished matter for further inquiry, but apparently without throwing any further light on that distressing event.

"The evidence of Mr. James Darlington, who is himself not only a coal proprietor, but also manager of those extensive coal and cannel works, occupies a large space in the publication. He considers the practice, which prevails in many districts, of leaving pits open as waste pits, very dangerous, and recommends that it should be made compulsory by law on all colliery proprietors, that they should be either arched over or fenced round with rails or walls. With regard to the taking off the tops of safety lamps, he tells us the magistrates at Wigan will invariably convict, where the rules are clearly proved to have been read over to the men; but in other districts the magistrates say they have no power, and seem surprised that the magistrates at Wigan should commit the men under the circumstances; and he added that he had heard it stated that there was no law to authorize a conviction. Mr. Darlington, although a coal proprietor, admitted that he was favourable to a system of inspection for the satisfaction of the public, and to prevent, if possible, the recurrence of accidents; but he emphatically declared that he considered *the present to be no inspection at all*, because the extent of the districts is such, that the present inspectors are unable to visit the mines, which are in consequence only inspected at intervals of two, three, four, or five years, according to the number of pits in the district."

One of the witnesses examined, Mr. George Elliott, agent to the extensive collieries of the Marquis of Londonderry, stated, that he thought it would be wise in the present Committee to recommend the furnace system of ventilation to be made general, except under very special circumstances; and he further mentioned, that he had tried experiments with the view of ascertaining whether coal could be brought down by any process of explosion, without generating fire. We give the result in his own words; "A very painful accident happened at Usworth Colliery, of which I was the owner and manager, some time since, and which arose from the fire of shot; and since that Mr. Lee Pattinson, a practical chemist, and myself, have been endeavoring to ascertain whether we could invent some power to bring down the coal without an explosive mixture, such as gunpowder; and I am sorry to say that no very successful result has yet been arrived at; but up to the present time, I am afraid, notwithstanding gunpowder is a very old chemical invention, that very little progress has been made since it was first used, and that we are in ignorance of any substitute for it of equal power." He subsequently added that "if public attention were called to it, perhaps science might discover some substitute. And as this journal had previously solicited public notice to the subject, we again repeat that the researches of chemical and electrical discovery, which have in our times produced such marvellous results, could not be devoted to a nobler object of scientific ambition."

MEDICAL TREATMENT AFTER EXPLOSION.

The report of the South Shields committee, previously referred to, goes at considerable length into an explanation of the condition in which miners meet their death by explosions; the proportionate quantities of the gases, which create them, and the nature of the *after-damp.*

This after-damp is formed of

8 parts of nitrogen having a specific gravity of 0.9722.
2 parts of aqueous vapour.
1 part of carbonic acid gas, specific gravity, 1.5277.

The latter takes its place towards the bottom of the passages, and, probably, extends little more than six inches high. Hence it is inferred that when the men, after explosion, if not struck down at once by it, attempt to leave the mine through an atmosphere of after-damp, they are at first rendered partially insensible by the nitrogen, which has been substituted for atmospheric air, and then, falling, they come in contact with a still more deleterious gas, a positive poison, [the carbonic acid gas], which having inhaled to a small extent, they pass rapidly into a state of asphyxia, owing to the state to which their systems have been previously reduced.

Two practical inferences are thus deduced—

1. Where carbonic acid gas is abundant, the lights are instantaneously extinguished, and burn with a dull red flame as they approach it; on these indications the miner is warned to retire, as here *flame is extinguished before life;* but when there is a large admixture of nitrogen, the lamp continues to burn, as in sulphuretted hydrogen, even when the miner has been struck down—life in this case *being extinguished before flame.*

2d. That asphyxia, arising from nitrogen, and *completed* by carbonic acid gas, might probably indicate a different system of medical treatment from that hitherto pursued. The symptoms of asphyxia,—always easy to be known, are the sudden cessation of respiration; of the pulsation of the heart, and of the action of all sensitive functions; the countenance is swollen and marked with reddish spots; the eyes become protruded, the features discomposed, and the face often livid.

It is necessary to succour an asphyxed person with the utmost promptitude, and to continue the remedies with perseverance until it is certain that life is completely extinguished. The following general remedies should be adopted: immediate removal into fresh air; undress and dash the body with cold water; endeavour to make the patient swallow water slightly acidulated with vinegar; clysters of two-thirds water and one-third vinegar, to be followed by others of a strong solution of common salt, or of senna and epsom salts; introduce air into the lungs by blowing with a nozzle of a bellows, into one of the nostrils and compressing the other with the finger. Should these means not produce the desired effect, and the body still retain its natural warmth, recourse must be had to blood-letting, the necessity of which will be clearly indicated by the red face, swollen lips, and eyes protruding. If blood fails to flow from the jugular vein, an attempt should be made on the foot; the last effort which can be made is to make an opening in the trachea, and introduce air to the lungs by means of a small pipe and a pair of bellows. These various remedies should be applied with the greatest promptness. The absence of the beating of the pulse, and the want of respiration are not certain signs of death, nor should all be regarded as dead whose breath or pulmonary transpiration does not bedim the brightness of glass; nor those whose members appear stiff and insensible. In giving these brief instructions, the committee hope that some of them may be judiciously practised, instead of the injurious plans sometimes adopted, until the arrival of a medical practitioner, who will thus find the patient prepared, uninjured for his professional skill, and his object facilitated, not obstructed, by the previous treatment he has received.

DRAINAGE OF COAL MINES.

In the mines which are situated in hilly or mountainous countries, it is generally easy to intersect the beds by galleries which commence at the lower part of some valleys; the galleries furnish a natural outlet for the waters of all the works which are above their level, and on this account are called in France *galeries d'écoulement*, or drainage galleries; in England, Wales and the United States, *adits*, or *adit levels*, and occasionally *drifts* and tunnels: but the word adit is the most distinctive of its object and uses.

The advantages of these adits are numerous, and have often decided the undertaking of long and expensive works.

In fact they are not only preferable to mechanical means of drainage, because when once made they require very little management or attention, but in giving issue to the upper waters, they also create a moving power which can be employed in the service of extraction, or in the draining of the lower works; finally, they furnish the most economical means for the other services of the mine, such as the forced ventilation, or the extraction of the substances mined.

An adit level can often be so arranged as to serve, at the same time, the working of several veins. On account of all these united advantages, there has been undertaken in the district of Schemnitz, in Hungary, a gallery of 20,000 yards in length, or about 11½ miles, designed for the use of the principal mines of the district, under the double service of draining the waters, of the carriage or gangway, and the creation of mechanical powers; it has been, besides, directed with a view to explore the ground for the discovery of new veins. At the Hartz, the great "galerie d'écoulement" of the mines of Clausthal, which is 13,000 yards in length, serves equally for a great number of mines, in different branches of the service. The use of these galleries is common in the countries where the mines are numerous and near together.

In relation to pumps, and the varieties of hydraulic machines employed in mines, for the purpose of drainage, we must refer to the various authorities who have written either in England or on the continent of Europe on this important subject, and which, moreover, would require the aid of numerous illustrations to render any description intelligible.

In the third volume of the Mining Review, p. 302, our readers will find a description of the pumps used in the deep mines of Cornwall, by Mr. John Taylor. The machinery brought to such perfection, and operating with so much economy and simplicity, is celebrated throughout the world. At the period of this communication, the steam engines of the district performed the work of 44,000 horses.

TRACING OF COAL BEDS IN THE ANTHRACITE DISTRICTS OF PENNSYLVANIA.

In these basins, where the outcrops of the coal seams almost always present themselves at a very high angle, they are in general readily traced, by the subordinate depressions which may be observed ranging longitudinally along the sloping side of the mountain ridges; pursuing, of course, the direction or strike of the strata. These depressions are obviously formed by the removal of the decomposable and soft materials of the coal seams; that is to say, the shales, the under-clay, and the coal itself, and they are conspicuously in contrast with the rocky siliceous beds which flank them, and which being composed of less destructible materials, have longer resisted the atmospheric agencies. Thus, in numberless instances, these longitudinal grooves afford an unerring clue to the subjacent beds of anthracite. In the great bituminous coal region of the Alleghany mountains, where the strata closely approximate to a horizontal state, such guides as those we have mentioned can, of course, have no existence, and we have there to seek for other phenomena which may indicate the presence of coal. Happily these are so abundant, that no coal region in the world, probably, presents more ready facilities for the ascertainment of what lies at so insignificant a depth beneath the surface.

To return to the more disturbed region which is occupied by anthracite in Pennsylvania. The disposition of the outcrops, to which we have alluded, materially influences the physical features of these coal districts, and modifies the contour of the surface by a numerous succession of terraces, steps, or benches, on the inner slopes of the mountains, facing the centres of the basins. Those who have ascended, from either side, the long parallel mountains which border the southern coal region of Schuylkill county, to the height of 1350 feet above the Susquehanna river, and more than 1650 feet above tide-water level, need not to be reminded of these characteristic details.

Between the external margins of the principal coal basins of Pennsylvania, subordinate axes of elevation are of frequent occurrence. Even the undulations of the surface between these limits are all attributable to these minor axes, and correspond, in great measure, with the local inclination of the upheaved stratification beneath. These undulations of groups of coal seams, so important to the proprietor, yet whose existence, until of late years, was scarcely suspected, are daily becoming more familiar to us, as the progress of development and practical investigation gradually advances.

The long narrow troughs, of which there are so many in central Pennsylvania, owe their contour to parallel synclinal axes, which present highly inclined or vertical coal beds; and occasionally even exhibit the strata of one of their sides tilted or leaning over so much that their inclination becomes almost parallel to, or conformable with, those of the opposite side of the basin.

In proportion as the anthracite basins become wider, their interior is the more disturbed or broken by undulations, consisting of one, two or three subordinate axes, each maintaining itself for a space as a parallel inferior ridge, and thus interrupting the general trough-like arrangement of the stratification. It is to be expected that the carboniferous beds in the vicinity of the centres of these synclinal axes are liable to be too much crushed to permit an advantageous working of their contents.

The southern anthracite region, in particular, furnishes numerous instances of the modified arrangement of which we speak, and we might introduce several illustrations from our own observation, which would exemplify the extent of the forces to which the anthracite country has been subjected, in the area between the Lehigh and the Susquehanna.

Beginning near the eastern extremity, at Nesquehoning, we see the ordinary basin-form arrangement modified by an upheaving or saddle in its centre, it being here scarcely one mile in width.

Next westward is a section in the meridian of the Mauch Chunk summit mines, where the basin has now expanded to almost double the breadth that it occupied at Nesquehoning. The structure of the interior is now considerably complicated, and the enlarged breadth allows of a triplication at least of the coal series. So confused is its aspect at this point that we are by no means certain that our section embraces all the details.

Further westward we have the very interesting and magnificent transverse section formed by the stream of the Little Schuylkill, at Tamaqua, where the basin has again contracted to the simple synclinal axis, of scarcely one mile in breadth.

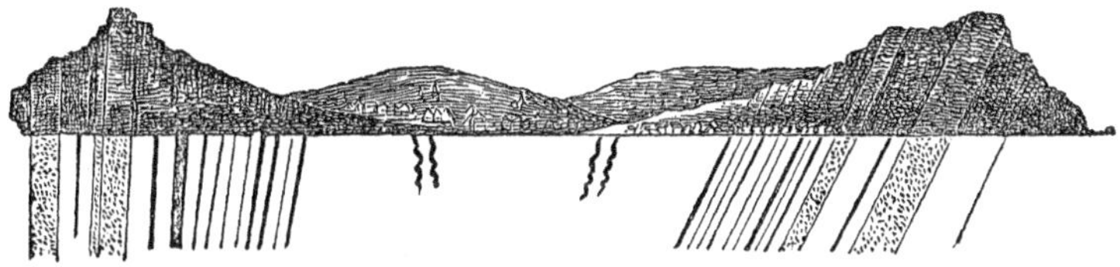

Our figure exhibits this section with the accuracy resulting from an original survey, and it is the more memorable from the presence of a particular seam in the Sharp Mountain, which is worked to the breadth or thickness of no less than seventy feet.

At Pottsville, the same region has widened to the extent of about five miles, affording, by the repetition of the coal beds, a vast industrial area; and at the head waters of the Swatara river there is now a breadth of no less than six miles. In the Pinegrove coal district we have at least three miles of breadth. Thus we perceive that in proportion to the space or breadth between the geological margins of the Schuylkill coal-field, so is the frequency of the undulations, the number of anticlinal elevations or axes, and the consequent repetitions of the same series of coal seams.

Westward of the Swatara or Pinegrove coal region, it bifurcates and stretches, with diminished breadth, for many miles towards the Susquehanna. The geological structure of these two forks is illustrated by the above diagram.

Section showing the North and South Forks of the Southern Coal Region, Pa.

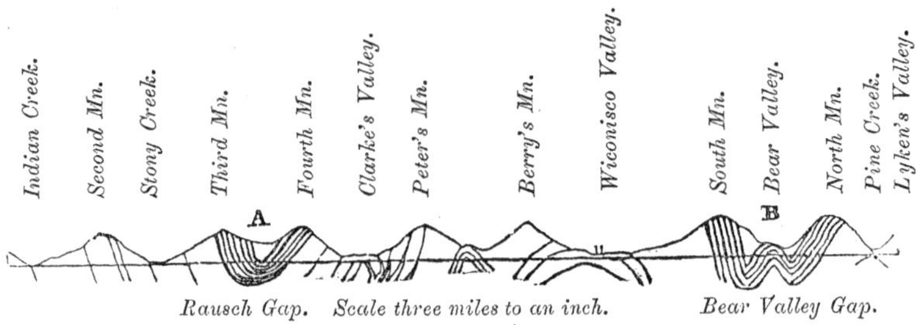

The figure represents a cross section, in a north and south direction, of a part of the coal region near the western boundary of Schuylkill county, crossing both the forks of that basin. It shows, in the first place, at A, the simple synclinal axis which forms the south-western fork of the region, and its nearly vertical strata on the southern margin of Sharp Mountain. On the same meridian, crossing the north fork at B, is a specimen of more complicated structure; not a simple anticlinal axis, but a trough which exhibits a subordinate anticlinal ridge, or central saddle at B.

The enlarged details of the portion A, are shown in the sketch below, a few miles to the eastward, at Black Spring Gap.

We have been led somewhat astray from our purpose of devoting this section to the consideration of the usual means of tracing the coal seams along their outcrops in the anthracite region of Pennsylvania. We have previously remarked that in the horizontal beds of the bituminous coal-fields of North America, their position was very readily ascertained. We showed also that in the highly inclined anthracite areas, the range of the outcrops was ordinarily distinguishable by parallel depressions along the mountain flanks of the basins.

During our own investigations we have remarked that the true positions of those veins which had their bassets on the slopes of the mountains were, in most cases, rendered obscure by the curvatures of the crops, almost at right angles to the true inclination of the veins. We ascribe this to the atmospheric agency, operating to a given depth below the surface, and to the mechanical influence of surface waters, decomposition, the sliding down of the higher masses, &c. In every instance which has come under our observation, in relation to the outcropping of coal seams on these slopes, we have perceived the manifestation of the like influences, which have deflected the "wash" or decomposed materials of the coal veins from their true courses and thrown them over among the alluvial detritus, generally in a curve, as shown by the next figure, which is merely the representative of numerous corresponding cases.

Coal Crops on the Fourth Mountain.

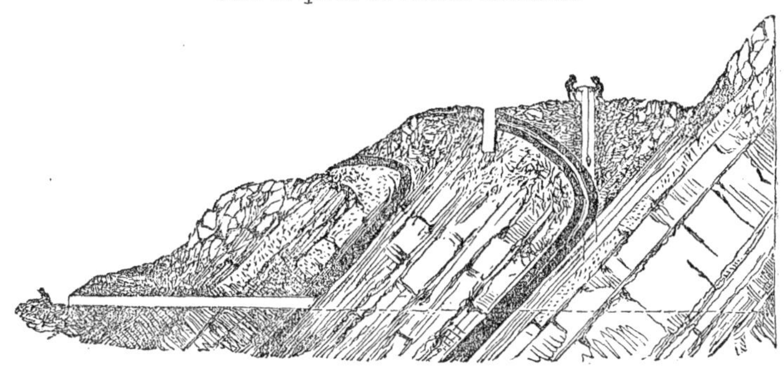

ON THE MAPS OR PLANS OF MINES.

In every working a good plan of the mine is of great utility; it is above all indispensable when the subterranean works are considerably developed. In fact, it is necessary to maintain the works in the limits of the property, in order to avoid contests with neighbouring owners, and there always exist some points from which it is necessary to keep removed, under risk of the greatest dangers. Finally, when it is suggested to effect a junction with a point, fixed beforehand by a pit or gallery; if there was not a plan constructed with precision, we should run the risk of missing the object, and of making costly works to no purpose.

The drawing of the plans of mines presents great difficulties. The mines being composed of crooked passages, isolated one from the other, how should we determine, singly, the form and the position of each of them, and render them conformable to the plans of the whole? These difficulties are still increasing from the necessity of working in obscure galleries, often low and difficult of access. To construct correctly a map of the works which only communicate with the surface, by sinuous galleries or by shafts, it is absolutely necessary to have recourse to the needle. The dial, or mining compass, is composed of a magnetic needle tinted with blue steel at the north point, and balanced on a cap of agate.

The mining compass is suspended from the middle of two axes or spindles, *tourillons,* upon the support of the brackets, *crochets;* the line N and S corresponding with the axis of the crochet. If then, after having strongly held a cord or copper wire, following the axis of the gallery of which the direction is required, the compass be suspended at this cord, the deviation of the needle from the north and south line, will give the angle of the direction. In order to facilitate the reading of this course, the letters E. and W. are commonly transposed, so that the true point of the course may be read in degrees and minutes, by means of the figure which approaches the nearest to the blue point. This method is also applicable to the compass which is carried in the hand, and appears to be generally in use in Germany and France. The difficulty of reading off, with sufficient exactness with an uncertain light, and in positions often incommodious, the angles marked by the needle which oscillates during a long time, is one of the many obstacles to the perfect accuracy of the observations, and it is admitted that by this method an observation cannot be taken with greater nicety than a quarter of a degree, or fifteen minutes.

In mines of the magnetic oxide of iron, where the action of the needle is deranged by its proximity to the mineral, the compass cannot be employed, and the graphometer is used. A theodolite, for subterranean service, is also adopted in France, with which plans and surveys of mines can be constructed with equal celerity as with the compass, and in a more exact manner.

MINE SURVEYING.

In England the usual surveys in mines of all descriptions were made with the dial. The most useful treatise on the art and on the practice of this instrument, is that of Mr. Budge, of Cornwall,* an eminent mine surveyor. Although constantly employing the dial in his business, he, from the first, by no means viewed it as the most accurate that can be employed, and remarks, "There doubtless are instruments much better adapted to the work, both for speed and accuracy, than the dial; and it is matter of surprise that they have not been more generally introduced in our mines: of these instruments the theodolite certainly stands unrivalled, for taking both horizontal and vertical angles."

In the second edition of this work, after a lapse of twenty years, the author devotes a section to the subject of "surveying without the magnetic needle." This is a valuable modern discovery; and as the general introduction of iron railways and tram roads in mines drove the surveyor to seek some substitute for the needle, which the attraction of iron rendered useless, he has happily succeeded. The best circumferentors are now made with an external graduation and vernier scale, on the theodolite principle, on purpose for the performance of this work.

The author enters into all the necessary details for proceeding with the observations ascertained by this improved instrument, and for protracting and calculating the work thus performed.

IRON ORE OF THE COAL FORMATION.

In the coal formation, iron only exists in the state of carbonate: it is generally concentrated in particular beds of a basin, and upon

* The Practical Miner's Guide, by J. Budge, second edition, 1845.

a much more limited superficies than that of the beds themselves. The usual form of the carbonate of iron of the coal series is that of oval or kidney shaped balls,—*rognons*, having a brown or greyish fracture; it is a mixture, more or less rich, of clay, and carbonate of iron. These balls occur stratified in the argillaceous beds. They appear to be assembled and precipitated during periods of repose, when the waters deposited at the same time the argillaceous particles with which they were charged. These spheroidal concretions or rognons, in consequence of their mixture with clay, are often arranged in concentric laminæ, and frequently also present in their centre, a nodule of clay or of pyrites; sometimes even a fossil substance,—nuclei, which appear to have attracted around them the chemical precipitation.

The position of the nodules of lithoide carbonate of iron is in strata parallel to the coal seams. This is the case at the mine of Treuil, in France, and almost every coal-field in England presents a similar deposition. In those of North America, we find a smaller amount of argillaceous carbonate of iron so interstratified with the coal beds, than in Europe, and the instances where iron works are supplied from these sources in America, at the present day, are but rare.

In the coal basins of England, the carbonate of iron is almost always found in the same beds, extending over very large districts. There are two large beds of ferriferous clay in the Dudley basin,* and sixteen or more in the great anthracite district of South Wales.

According to M. Burat, the numerous coal basins of France are far from containing the carbonate of iron in the same abundance as those of England. Few of the argillaceous beds contain these balls, and still more seldom do they contain those which are concentrically formed. At Saint Etienne, for instance, there exist two which furnish in the concession of Treuil, flattened rognons of fair quality, and contribute to supply the furnaces of Janon; but in the other concessions, the balls are not recognized in the equivalent beds, or they are so small as to be neglected. At some other points, the carbonate of iron appears in great abundance, but with very different characters: at the mine of Cros, it penetrates the entire beds of clay, of one or two yards thickness, and gives to that rock a remarkable solidity and density. But these massive beds are much more impure than the beds with disseminated nodules: besides a large proportion of clay, they also contain pyrites and precipitations of dark silex, to such an extent that the working which ought to have been developed, remains almost unproductive.

The coal basin of Aubin, in France, contains the iron ore in the

* The Dudley coal-field is remarkable as being one of the earliest positions where the argillaceous iron was smelted by means of pit-coal. The experiment was made by the founder of the noble house of Dudley and Ward, who published an account of it in the time of Charles II. He states that in a large stone furnace, twenty-seven feet square, he made seven tons of iron per week, "near which furnace the author discovered many new coal mines, ten yards thick, and iron mines underneath, which coal-works having brought into perfection, the author was by force thrown out of them, and the bellows of his new furnace and invention by riotous persons cut in pieces, to his no small prejudice, and loss of his invention of making iron with pit-coal."

most abundance. It exists, in the first instance, in balls in the beds of clay which accompany the coal, and, as at Saint Etienne, under that form it is the purest mineral. In the other cases, it constitutes a somewhat schistose bed of from three to fifteen feet thickness, which appears to extend under the greater part of the coal area. This bed is remarkable, inasmuch as it presents at several points a series of contradictions and enlargements, which constitute the arrangement called *en chapelet;* like a string or chaplet of beads,—an arrangement very frequent in all the substances which result from chemical precipitations in sedimentary waters, such, for instance, as is often presented by the flints in chalk. This structure is, in other respects, independent of the other accidents, faults or disturbances, which equally affect this bed as those of the coal series. The kidney ores and the bed of stony carbonate of iron are worked at several places in the basin, and supply the high furnaces of Decazeville.

There are but very few basins which do not possess beds analogous to the argillaceous carbonate of iron; but they are in such slight amount, that there are no other workings than those of the two basins of Saint Etienne and Aubin, which we have just mentioned.

On the whole, if we compare the beds of lithoide carbonate of iron with the mass of coal formations, we see that their existence, but little developed, although frequent, must only be considered an accidental circumstance. It is equally worthy of remark, that, in every case where there was a formation of coal in the series subsequent to the true coal, viz., the ferruginous elements have anew resumed the composition and the characteristic aspect of this epoch.

Thus in the coals and the shales with vegetable impressions of the epoch of the lias of Yorkshire, we find the carbonate of iron stratified in balls; whilst, in the same formations, when their appearance is in the normal state, the ferruginous infiltrations appear only in the state of oxides.

These variations of composition in the ferriferous minerals, establish no real difference in the origin to which they may be attributed. They tend merely to demonstrate that the iron, collected at certain intermittent epochs and at isolated points, most commonly in the state of oxides, has undergone through the influences of the carboniferous epoch, a mineralogical transformation. The coal period appears generally to have been a period of tranquil deposits. It is, then, natural to find that in it the ferruginous infiltrations are more concentrated than in the periods of the old red sandstone, of the new red sandstone, and even of the trias, where the products of these infiltrations are blended with the general materials of the deposit.

We have already remarked that these infiltrations only become valuable according to their concentration: the formation before mentioned, so highly coloured by the per-oxide of iron, contain, perhaps, altogether much more iron than the coal formation; but, in the latter, it is collected together, and often possesses a concentration of 30 and 40 per cent., constituting serviceable beds. In the red or mottled sandstone formations, we find iron everywhere; but the concentrations even amounting to ten per cent. are but rare exceptions. The presence of the iron would then be scarcely remarked, if the glaring colour of these red and variegated formations did not contrast with the gray and dark rocks of the coal deposits which they cover, and with the white and greenish colours of the thick limestones and clays which are about them.*

SECTION III.

FOSSIL BOTANY AND GEOLOGICAL DISTRIBUTION OF VEGETABLE REMAINS.

In intimate connection with the matter of the present volume, a knowledge of the forms, the botanical classification, the geological arrangement of the vegetable remains of an ancient world, seems to be almost indispensable. It embraces facts, at least, sufficiently valuable, to ensure for it, as a collateral branch of natural science, a conspicuous section of this book. Independently of its usefulness, there is a never failing interest attached to such an investigation, which enables us to trace the history, as it were, the past condition, the present adaptation of the primeval flora;—that magnificent vegetation which, amidst the mutations of our planet, yet survives for our use; its characters changed, it is true, but only to become more serviceable to man.

A happy provision was it that secured for the ultimate advantage of the human race, ages before its appearance upon the globe, the trees of gigantic size, the densely growing shrubs, the most delicate even of the lesser plants—that flora which covered in such profusion the islands and plateaux, and filled the humid valleys of the early world. A happy provision was it, that amidst the early catastrophes of the earth,—those convulsions which modified its entire surface, overwhelmed its primeval forests, and buried them beneath enormous accumulations of earthy debris, of sediment and of rocky debacle—still perpetuated and matured during the lapse of countless ages, that primitive vegetation, which, finally, in the form of mineral combustibles, we are now busy in exploring and mining, and appropriating in a thousand ways, and for a thousand purposes. A happy provision was that—a beneficent one, surely—which at the moment when man is compelled to level the existing forests, to make room for the progress of agriculture, and the cultivation of the present surface, he finds nigh at hand, yet buried beneath that surface, within the shallow basins and woody islands of the antediluvian world, those inex-

* Burat.—Géologie appliquée, p. 108.

haustible stores of a combustible now rendered infinitely more precious and effective than that existing vegetable fuel, whose destruction is the inevitable consequence of advancing civilization.

Respecting the wondrous influence which the employment of mineral combustibles has had, even in our own days, upon the whole world, by the acquisition of new forces; by the extension of mechanical powers, of manufacturing capabilities; by the impulse given to the industrial arts, and the creation of new sources of wealth; by rapid and cheap modes of transportation, and enlarged commercial facilities; above all, by the improved condition of the people, we will not here dilate. Abundant evidence of all these will be found in this volume.

FOSSIL BOTANY.

Classification of Plants; their families, classes, and orders.—We shall occasionally have to make mention of the varieties of plants which occur in a fossil state, and which, in common with all other organic remains, are characteristic of, or distinguish with remarkable precision, every geological epoch. It may save the reader some trouble in referring to elementary books, if we briefly explain here the mode observed in the classification of this fossil vegetation; of which the true coal formation alone contains about four hundred known species.

The system generally adopted by botanists is, that of Jussieu, which is termed the "natural system," in contradistinction to that of Linnæus, which is denominated the "artificial system." Mr. Loudon states that the former method has for its object the arrangement of plants according to their greater or lesser degree of resemblance, both externally and internally.

The *seed* is considered the most important part of the plant; as being destined for its re-production and continuance in the world. The fundamental divisions of this arrangement are, therefore, founded on the characters of seeds.

The first grand division is derived from the presence or absence of seed-lobes; the next on the union or division of the seed-lobes in such as have them. Thus we have the three primitive divisions of *Cotyledoneæ*, *Monocotyledoneæ*, and *Acotyledoneæ*.*

Every one allows, M. Decandolle observes, that plants which resemble each other by their exterior forms, resemble each other also in their internal structure; their mode of vegetation and their properties. The three primitive divisions are divided by this botanist into eleven classes; and, according to the Jussieuan method, all vegetables are furnished with seeds which arrange themselves under one or other of the following heads.

COTYLEDONEÆ.

Exogenous stems.—Furnished with two or more cotyledons, or

* Loudon's Encyclopedia of Gardening, p. 113.

seed-lobes; as the bean or the acorn; having a central column or pith, and an external band called the bark, the two being connected by medullary rays; this division being thus subdivided into I. Dicotyledons; II. Monocotyledons; III. Acotyledons.

I. DICOTYLEDONEÆ.

Having the calyx and corolla distinct.		Having the calyx and corolla, forming only a single envelope.	
Six classes and eighty-three orders.		One class,—*Monochlamydeæ.*	
Ranunculaceæ,	Samaroubeæ,		
Magnoliaceæ,	Ochnaceæ,	Seventeen orders.	
Papaveraceæ,	Terrebinthaceæ,		
Cruciferæ,	Leguminosæ,	Plumbagineæ,	Laurinæ,
Caryophylleæ,	Oleineæ,	Plantagineæ,	Santalaceæ,
Lineæ, &c.	Jasmineæ, &c.	Euphorbiacæ,	Urticeæ,
	Cacti, Ericæ, &c.	Amentaceæ, &c.	Coniferæ,
embracing 1255 genera, and 8612 species.		comprising 172 genera, and 1249 species.	
Besides 53 genera and 71 species whose orders are not fully determined.			

Fossil dicotyledonous plants of the coal formation—Until recently the fossilized dicotyledones were supposed to occur not lower than the Tilgate or Upper Oolite beds. The coniferæ also were considered as not older than the oolite series. But recent investigations, by distinguished naturalists, have shown that these groups formed the greater portion of the coal vegetation. Thus, for instance, some fossil trees, which were discovered rooted in a coal bed in the Lancashire coal-field, were identified by Mr. Bowman as sigillariæ,* while at the same time he showed that medullary rays and coniferous structure existed; a fact which M. A. Brongniart, Lindley and Hutton, Humboldt and others have fully corroborated. Hence, it seems that botanists are inclined to withdraw the Sigillaria altogether from the family of tree ferns, with which they have been heretofore classed, and even from the Endogenous class, or Monocotyledones. We are therefore to understand that the Sigillaria is a dicotyledonous and coniferous plant, and that the arborescent ferns, Caulopteræ, belong to the monocotyledonous group.

Among the dicotyledonous plants of the coal formation are now arranged

Sigillaria, 59 species,
Stigmaria, 30 "
Calamites, 18 "
Cycadea,
Lepidofloyas,
Asterophillites,
Annularia,
Sphenophyllum,
Coniferæ.

II. MONOCOTYLEDONEÆ.

Endogenous Stems—furnished with only one cotyledon or seed-lobe, [as the lily,] and having neither pith, concentric circles of

* Proceedings Geological Society, London, vol. iii. p. 270; also Mantell's Medals of Creation, p. 132.

woody fibre, nor true bark: distinguishable as follows, in the existing series:

Those in which the fructification is visible.	Those in which the fructification is concealed, unknown, or irregular.
One class, *Phanerogameæ.*	One class, *Cryptogameæ.*
25 orders. Cycadeæ, Bromeleæ, Orchideæ, Scitamineæ, &c. Irideæ, Liliaceæ, Junceæ, Palmæ, Canneæ, Gramineæ, &c.	5 orders. Naiadeæ, Equisetaceæ, Marsiliaceæ, Lycopodiaceæ, Filices, Including the arborescent and herbaceous ferns.
338 genera, and 1945 species.	99 genera, and 261 species.
Of doubtful genera, 53 genera and 71 species.	

Distribution of Fossil Vegetation.—In a memoir "on the Ancient Flora of the Earth," written some years ago by a contributor to the Edinburgh Philosophical Journal, the author concludes with the following summary:

1st. That among the universally distributed rock formations, [groups] since the first appearance of organic beings, there is not one of them in which the remains of a contemporary land vegetation are not to be observed.

2d. That the different periods of the vegetation of a former age are gradually characterized by the continual entrance of new and always more perfectly organized [?] families of plants; but there is not a complete disappearance of all the species of the preceding periods.

3d. That species of the most perfectly developed class, the dicotyledonous, are first traced in the oldest strata of the *secondary* formations, while they uninterruptedly increase in the successive formations. To similar views Humboldt opposes some objections, particularly in relation to the theory of the supposed* simplicity of the first forms of organic life, and especially the assumption that vegetable life was awakened sooner than animal life upon the face of the old earth.†

With respect to the vegetation of the true coal formation, Sir Alexander Crichton observed, that every coal country in every part of the world, which has been hitherto examined, abounds in the fossil remains of similar or corresponding vegetables. There is no material variety, let the latitude or longitude or elevation be what they may. Recent examinations of the fossil flora of remote coal beds, such as those of Australia, Van Dieman's Land, and Northern India, would seem to point out some exceptions to the rule heretofore adopted, but the evidence is by no means complete that these fields were really of the true coal period. "Every plant in the present

* Edinburgh Philosophical Journal, January, 1830.
† Cosmos, A. Von Humboldt.

condition of the globe, independently of its natural dwelling-place, has, as it were, a central spot in which it flourishes best; and considering this spot as the centre of a circle, or rather as a zone, the plant degenerates in proportion as it approaches the limits of this district." The writer goes on to point out a very important circumstance, namely, that there is a difference of mean temperature, at present, of forty-one degrees of heat between the parallels in which coal has been discovered.* Between these, as regards the existing vegetation, the diversity in the genera and species of plants, at present, is very great; so much so, indeed, that there is no resemblance between the floras of the two extreme points. At the time, however, of the true coal formation, it is now admitted that the flora of these two remote parallels was nearly the same, both as to genera and species, and in this respect strongly contrasted with the present condition of things.

Fossil Plants of the class Phanerogameæ.—The monocotyledonous family of this class, in the fossil state, commence in the London clay tertiary formation, and, until lately, were thought not to descend lower in the geological series than the oolites, or the Waelden beds, the Portland oolite and the Lias.

The cycadeæ [Cycas Zamia] form the connecting link between the ferns and the palms, while, according to the authorities last cited, the sigillaires differ not more from the aborescent ferns [Caulopteris,] yet existing, than the stems of the calamites, the bactris, and other arundinaceous palms,† which order contains, in the recent state, eighteen genera and twenty-nine species. Zamiæ were very abundant in the oolite period. Eleven species occur in the coal of the Yorkshire oolite alone.

Distribution of the Vestiges of Palms in the Geological Formations.‡—Prof. Unger states, *first*, That no vestiges of palms have been detected in the earliest rocks which contain the organic remains of maritime and terrestrial plants.

Second, That palms bore some small part in the vegetation at the period of the *coal formation*. He names four species or forms, two of which occur in the coal schist of Swina, Bohemia, one in sandstone of the Ural Mountains, and one from Rajemahl, North India; also two undescribed species from the coal formation of Silesia.

Third, The flora of the *red sandstone*, above the coal series, although it has been very imperfectly preserved, and its scanty remains but little studied, Unger thinks was not materially different from that of the coal formation. But the fossils of this era, which have been referred to palms, he thinks are very doubtful. In the Quadersandstein, Gœppert found some vestiges in Silesia. From the next series, the oolites, the four species of *Carpolythes*, described by Lindley and Hutton, may be mentioned.

Fourth, and finally. In the tertiary, palms reappear, and the

* This approaches closely to the range we have assigned to the coal formations.
† Histoire des Vegetaux Fossiles.
‡ American Journal of Science, July, 1846.

number of species far surpass that of all the other formations together.

Subdivision of tertiary positions,—

In the chalk and eocene,	4 species, also fruits.
" miocene	26 species on the European continent.
" pliocene,	4 species, island of Antigua.

Fossil Cryptogameæ.—Many years ago Count Sternberg noticed that out of one hundred and fifty species of plants belonging to the old coal formation, one hundred and thirty-eight were vascular cryptogamea: soon afterwards M. A. Brongniart stated that the vascular cryptogamous plants had a vast numerical proportion in our great coal-fields; and, in fact, even at that early period, he had ascertained that out of two hundred and sixty species, discovered in that formation, two hundred and twenty belonged to this class.

This arrangement has of late received very considerable modifications; chiefly through the aid of a microscopic elucidation of their structure as we shall proceed to show. Messrs. Lindley and Hutton, A. Brongniart and others, now withdraw the sigillariæ, the stigmariæ and the calamites, from this numerous group; separating them from the associated filices or herbaceous ferns, and the caulopteræ, which only comprise the true arborescent ferns.

The fossil cryptogamous series embraces the following:

	Species.	
Sphenopteris,	36	146 species belonging to the filices or herbaceous ferns, chiefly of the coal beds.
Cyclopteris,	6	
Nevropteris,	28	
Pecopteris,	76	
Caulopteris,	"	The true aborescent ferns.
Equisetaceæ,	"	Some species.
Lycopodiaceæ,	8	48, belonging to lycopodites and club mosses, of the coal formation.
Lepidodendrons,	40	

III. ACOTYLEDONEÆ, OR IMPERFECT.

Vegetable beings composed of a cellular tissue unprovided with vessels, and of which the embryo is without cotyledons. The divisions of this family are as follows:

With leafy expansions and known sexes.	Without leafy expansions, and not of known sexes.
1 Class, Foliaceæ, 2 Orders, { Musci or mosses. / Hepatica—liverworts. 545 species are natives of Great Britain.	1 Class, Aphylleæ. 5 Orders or subsections, { Lichenea-fuci, conferoæ, &c. / Hyponyloneæ. / Agariceæ, fungi. / Algæ, flags. / Fungi. About 1350 species natives of Great Britain.

Fucoides, of many species, are exceedingly abundant in the silurian or transition formations, from the coal series down to the primitive rocks. In certain portions of the silurian series of North American rocks, this class of plants is surprisingly prevalent, and characteristic. The oldest of these formations present us with nothing but cellular-leaved marine plants. Many species of fucoides in the copper slate of Mansfeld.

The prevailing vegetable forms of the chalk formation are those of marine and freshwater plants—fuci and najades.

Of confervæ are three fossil species; of Algæ, nine species; and of Naiades, four species, in the cretaceous group.

Distribution of Fossil Vegetation.—M. Alex. Von Humboldt has stated in a recent work, that it is in the Devonian strata that a few cryptogamic forms of vascular vegetables, equisetaceæ and lycopodiaceæ, are first encountered. After these strata, we arrive at the coal formation, the botanical anatomy of which has made such brilliant progress in recent times. These comprise nearly four hundred species, including in their number not only fern-like cryptogamic plants, and phanerogamous monocotyledons, grasses, yucca-like lilaceous vegetables, and palms, but also gymnospermic dicotyledons, coniferæ and cycadeæ. Fossil coniferæ have been found in the old coal formation of England and Upper Silesia; while cycadeæ are contained in that of Radnitz, in Bohemia, and Königshütte, in Upper Silesia. The cycadeæ attain their maximum in the Keupfer strata and the lias, where about twenty different forms make their appearance.

The lignitic or brown coal strata, which are at present in every one of the divisions of the tertiary period, amongst the earliest forms of cryptogamic land plants, exhibit a few palms, many coniferæ with distinct annual rings, and frondiferous trees, of more or less decided tropical character. In the middle tertiary period we observe the complete recurrence of the palms and cycadeans; and in the last members of this epoch, at length, strong resemblances to our present flora. We come suddenly upon our pines and firs; our cupuliferous tribes; our planes, and our poplars. The dicotyledonous stems of the lignites are frequently distinguished by gigantic thickness and vast age. A trunk was found near Bonn, in which Noggerath counted 792 annual rings.

With relation to coal vegetation, M. Humboldt remarked that where several series of coal strata lie over one another, the genera and species are not always mixed; they are rather, and for the major part, generically arranged, so that only lycopodites and certain ferns occur in one series of beds, and stigmariæ and sigillariæ in another.*

In elucidation of the progress made in fossil botanical discovery, Mr. Adolphe Brongniart† has lately observed that the further we proceed in the series of ages towards the earliest geological periods,

* Cosmos.

† Comptes Rendu, Dec. 29th, 1845—and Annual and Mag. Nat. Hist., February, 1846.

the further are we removed from the actual creation, and the greater do the differences between the living and fossil beings become.

Thus, most of the fossil plants of the tertiary strata belong to *genera* in actual existence, and merely present specific differences.

Those of the secondary strata may, undoubtedly, almost always be referred to known *families*, but appear in most cases, to require the formation of new genera.

Lastly, in the older strata, particularly in the coal formations, many of the fossil plants cannot be classed in families at present existing, and ought to constitute new *groups* of equal importance.

He adds that new and hitherto very rare specimens, which have been collected and carefully studied in England, Germany, and France, have caused important changes relative to the plants which he had previously considered as vascular cryptogamia. This advance is owing to the discovery of portions of stems of these plants having the internal structure in a state of preservation. They have shown that the *sigillariæ*, stigmariæ, and probably most of the calamites, are not plants nearly related to the *ferns*, *lycopodia*, and *equiseta*, but to distinct families of the dicotyledonous gymnospermous group, more nearly approaching the *coniferæ* and *cycadeæ*.

Hence, at the period of the coal formation, vegetation would have consisted entirely, or nearly so, of two of the great divisions of the vegetable kingdom: the ACROGENOUS CRYPTOGAMIA, represented by the herbaceous and arborescent ferns, [the latter reduced to the true *caulopteris*,] the *lepidodendreæ*, a family nearly related to the *lycopodiaceæ*, and some *equisetaceæ;* and the GYMNOSPERMOUS DICOTYLEDONS, comprising the *sigillariæ*, [*sigillaria*, *stigmaria*, *lepidofloyos*,] the *calamitaceæ*, the *coniferæ*, and probably the *asterophylleæ*, [*asterophyllites*, *annularia*, and *sphenophyllum*.]

Mr. Brongniart proceeds to describe a plant which closely approaches a family of the gymnospermous dicotyledons still in existence,—the *cycadeæ*, and of the genus *noggerathia*. This plant, at first known to M. de Sternberg, by the impression of a single leaf, from the coal formation of Bohemia, has since been observed in the coal shales of Newcastle, in those of Silesia, in the Permian sandstones of Russia, and many new species of the same genus are in the schists and coal sandstones of France.

He considers, with M. Humboldt, that each stratum of coal is the product of a peculiar vegetation, frequently different from that which precedes and that which follows it,—vegetations which have given rise to the superior and inferior layers of coal; each stratum resulting, in this manner, from a distinct vegetation, is frequently characterized by the predominance of certain impressions of plants, and the miners, in numerous cases, distinguish the different strata, which they remove, by the practical knowledge they possess of the accompanying fossils. Any seam of coal and its overlying rock or slate, should consequently contain the various parts of the living plants at the period of its formation; and by carefully studying the association of these various fossils, which form so many special floras, containing generally but few species, we may hope to be able to reconstruct these anomalous forms of the ancient world.

Distribution of fossil plants.—Notes from the Quarterly Journal of the Geological Society of London, Vol. I., 1845, p. 566, and Vol. II., 1846, p. 83.

The following extract, [with some modifications derived from Mr. Murchisson's paper on the Permian system,] from a memoir by M. Gœppert, of Breslau, well known for his investigations concerning the fossil remains of vegetables, possesses great interest, as offering a general view of the relative distribution of these remains.

Formations.	Families.	Species.
Lower Palæozoic System:		
Grauwacké, silurian, or formations older than the carboniferous series, including the Devonian series, and the oldest coal or culm beds, - -	8	52
Permian system, or Upper Palæozoic:		
Carboniferous limestone, - - - - - -	3	3
True coal measures of Europe and North America,	18	816
Lower new red sandstone, Permian series, containing, among others, a few species common to the carboniferous era, - - - - - - -	4	39
Magnesian limestone and kupfer schiefer, chiefly marine fucoids, Permian system, - - -	3	19
Gres bigarré, Bunter sandstein, - - - - -	8	32
Triassic period; or Lower Secondary:		
Muschelkalk, - - - - - - - -	2	2
Keuper marls, marnes irisées, - - - - -	8	52
Middle and Upper Secondary:		
Lias, - - - - - - - - - - -	12	75
Oolitic series, . - - - - - - -	9	159
Wealden formation, - - - - - - -	8	16
Lower cretaceous beds, - - - - - -	15	59
Chalk, - - - - - - - - - -	1	3
Lower Tertiary.—Monte Bolca beds, - - - -	4	7
Other lower, tertiary, - - - - - - -	10	120
Middle and upper tertiary.—Miocene and pliocene,	52	327
Unknown geological position, - - - - -	4	11
	169	1792

Recapitulation.	Families.	Species.
Older Palæozoic rocks below the coal measures, - - - - - - - -	11	55
Coal measures, - - - - - -	18	816
Newer Palæozoic or Permian system, above,	15	90
Triassic and secondary formations, - -	55	366
Tertiary, - - - - - - -	66	454
Unknown, - - - - - - -	4	11
Fossil plants,	169	1792

Summary of M. Gœppert's numerical distribution of Fossil plants. —The following table presented by Sir R. T. Murchisson at the meeting of the British Association, in 1845, embodies the same facts as are already announced in detail above.

Palæozoic rocks, - - - - - - -	52
Carboniferous group, - - - - - -	819
Permian, - - - - - - - -	58
Triassic, - - - - - - - -	86
Oolitic, - - - - - - - - -	234
Wealden, - - - - - - - -	16
Cretaceous, - - - - - - - -	62
Tertiary, - - - - - - - -	454
Unknown, - - - - - - - -	11
Total, - - - - - - -	1792

It was further stated, that the number of fossil plants known to M. Adolphe Brongniart, in 1836, was 527. In the new list they amount to 1792! and it is seen that the carboniferous group contained more than half the known species of fossil plants; a remarkable circumstance, when it was considered that the great herbivorous *land* quadrupeds had no ascertained *existence* before the tertiary period.*

For a notice of the flora and fauna of the amber forests of the countries bordering on the Baltic, our readers are referred to the head of Prussian Pomerania, in this volume.

MICROSCOPIC OBSERVATIONS ON THE STRUCTURE OF COAL, LIGNITE AND PEAT.

Among other collateral subjects of interest, tending to throw light on the age, the history and the composition of coal, the mode of investigation through the agency of the microscope, is not altogether inappropriate.

Mr. Hutton, of Newcastle, has instituted a series of examinations of the substance of coals, through the aid of the microscope.

Professor Phillips addressed some observations to the British Association, in 1842, on this new test.

In consequence of the facilities afforded for polishing coal, and of examining it by means of transmitted light, some progress has been made in this mode of investigation.

By the process of combustion another method had suggested itself, for making apparent to the eye the vegetable tissues of which certain coal plants were composed. In the ashes of Staffordshire coal,—a variety not strictly bituminous or caking,—Mr. Phillips was impressed with the analogy they presented to the combustion of certain sorts of *peat*, of a laminated texture; and their microscopic examination showed abundant traces of a vegetable character.

* Report of the British Association, 1845.

In some anthracite ashes furnished by Sir Henry De la Beche, vegetable tissues were also found; and the same fact is also visible in the ashes of the Pennsylvania anthracites.

A paper was read to the Geological Society of London, January 9th, 1833, entitled "Observations on Coal," by W. Hutton. The author was led to this subject by pursuing the method of microscopic examination, so successfully employed by Mr. Witham; and from these observations much interesting information has been acquired, respecting the fine, distinct reticulation of the original vegetable texture, still discernible in the various species of coal, and showing the presence, in the Newcastle coals, of cells which are filled with bituminous matter, extremely volatile.

Another system of cells was discovered, different from the others, which he conceived was adapted for containing gas. These supposed gas cells are found empty, and of a circular form, and in groups which communicate with each other; each cavity having, in its centre, a small pellet of carbonaceous matter. The author establishes a clear distinction between these two classes of cells; for the anthracite of South Wales contains the gas cells, but is quite free from those which, in the other coals, are filled with bituminous matter. The anthracite of South Wales affords a free disengagement of inflammable gas when first exposed to the air.*

Additional light is thrown on this subject by a paper of M. Link, of Berlin, "on the origin of coal and lignites, according to microscopic observations."†

The professor remarks that there still prevail two different opinions relative to the origin of coal. The one sustains the view that it is a turf, peat or marsh of the primitive world; the other that it consists of the trunks of forest trees which have been brought together and here buried.

Ordinary peat consists of earthy matter penetrated by the roots or radical fibres of vegetables, with here and there some portions of leaves. This earthy part is composed of the cellular tissue of plants, whose structure has been so flattened by pressure, that it is often impossible to recognise them.

A second and better description of peat is sold at Berlin, under the name of *tourbe de linum*, which consists of cellular tissue, compressed in exceedingly thin laminæ.

A third variety, dug in Lower Pomerania, has acquired the appearance of fossil wood; being compact, and its fracture conchoidal and bright; yet still containing parts which resemble the debris of leaves. There remains no trace of ligneous structure. Some portions of this peat become partially transparent when plunged in olive oil; and still more so when they are coated with rectified oil from coal tar.

By observing a similar process with regard to coals, we are enabled to render a great portion of their parts transparent. It has, in this way, been found that the lignite or brown coal of New Granada, and

* Proceedings of Geol. Soc. of Lon. vol. i. 415.
† Annales des Mines, vol. xvii. p. 593. 1840.

the coals of Newcastle, of Bridgewater, Saint Etienne, and Lower Silesia, present a structure analogous to peat, and particularly to that of the compact tourbe de linum.

In these coals M. Link did not observe a ligneous structure, resembling that of solid wood.

The coals of Upper Silesia have enabled us to make, by means of calcination, a comparison with wood charcoal, particularly with that of birch, pine, and palm—the *bactris spinosa*. Calcination has restored to the cells or vessels all their distinctness, but did not effect any change in the pores or openings.

It would appear, then, that the fibrous coal which covers more or less the compact coal of Beuthen, in Upper Silesia, resembles burnt charcoal, seeing that its compact portion is peaty. All these coals belong to the most ancient formations.

The Muschelkalk coal in Upper Silesia, is turfy, but that of Diester, in the lias, appears to approach to wood.

The coal of the Quadersandstein of Quedlinbourg, exhibits evidently the wood of conifera.

The lignites of Greenland, in which retinasphalt occurs, are peaty in structure, as are those of Meissner, in Hesse.

In those of New Granada, the wood of the palm is discernible by means of the microscope.

In those of numerous positions in Germany can be traced the wood of conifera; while among those lignites which belong to the dicotyledones, but not to the conifera, may be ranked the Surterbrand, the Bersteinholz, the lignite of Meissner, and that of Brohlhale on the Rhine.

M. Göpper, professor at Breslau, has also pursued similar researches, with interesting results;* and has determined with great precision the character of many lignites in Prussia. Among the additional localities of lignites, which contain wood of the family conifera, and genus *pinus*, are those of Siegen in Westphalia; of Friesdorf near Bonn; of Salzhausen in Wetteravia; near Konigs-Bergen-Prusse, and in Hungary.

We cannot conclude this part of our subject without adverting to the investigations of Dr. Mantell therein. We regret that our limited space forbids us to extract more than the following passage from one of his latest publications.

"Although the vegetable origin of all coal will not admit of question, yet evidence of the original structure is not always attainable. The most perfect bituminous coal has undergone a complete liquefaction, and if any portions of organization remain, they appear as if imbedded in a pure bituminous mass. The slaty coal generally preserves traces of cellular or vascular tissue; and the spiral vessels, and the dotted cells, indicating coniferous structure, may readily be detected by the aid of the microscope, in chips or slices. In many examples the cells are filled with an amber-coloured resinous sub-

* Annales des Mines, vol. xviii. p. 448.

stance: in others the organization is so well preserved, that on the surface exposed by cracking from heat, vascular tissue, spiral vessels, and cells studded with glands may be detected. Even in the white ashes left after the combustion of coal, traces of the spiral vessels are discernible by a high magnifying power. Some beds of coal appear to be wholly composed of minute leaves, or disintegrated foliage; for if a mass be recently extracted from the mine, and split asunder, the exposed surfaces are found covered with delicate pellicles of carbonized leaves and fibres, matted together; and flake after flake may be peeled off through a thickness of many inches, and the same structure be apparent. Rarely are any large trunks or branches observable in the coal; but the appearance is that of an immense deposit of delicate foliage, shed and accumulated in a forest, (as may be observable in existing pine districts,) and consolidated by great pressure, while undergoing that peculiar fermentation by which vegetable matter is changed into a carbonaceous mass."*

Professor J. W. Bailey has communicated an article in the American Journal of Science and the Arts, on some microscopic examinations which he has instituted, of the ashes of anthracite coal. He observed that on the surfaces of partly burned laminæ of coal, vegetable structure could be readily detected, and that often the tissues were presented in a state of unhoped-for preservation.

These specimens, the description of whose beauty and perfection can scarcely be exaggerated, present all the original markings of the vessels with a distinctness which leaves scarcely anything to be wished for.

They may be examined either as opaque objects, in which the silex appears in relief against the black coal, and shows the form and markings of the tubes very finely; or still more satisfactory results may be obtained by melting some inspissated Canada balsam upon a plate of glass, and, while in a melted state, applying it to a surface of the coal upon which the ducts had been previously found to exist. When the balsam has hardened the coal may be taken off, and it will be found that it leaves, fixed upon the balsam, a thin layer of silica, containing perfectly preserved dotted vessels, which, when viewed as transparent objects, are nearly as distinct in their markings as if freshly obtained from a recent plant.

Among other inferences, derived from his early examinations, Professor Bailey draws the following:

1. That almost every layer of coal is composed of vegetable matter, which still retains very distinct traces of the original organic structure, and which, consequently, proves that it could never have been reduced to a homogeneous pulp.

2. That the plants which chiefly contributed to form the mass of the coal were not the ordinary dicotyledonous or monocotyledonous plants, but they more probably belonged to the acotyledons, among which the ferns and lycopodiaceæ present similar vascular bundles.

* Mantell's Medals of Creation, vol. i. p. 92.

Mr. A. Brongniart, however, has decided, with Lindley and other investigators, to remove the calamities, the sigillaria, stigmara and lepidodendrons from the monocotyledons, and group them with the dicotyledons.

The presence of bitumen, and the consequent swelling and partial fusion of the ordinary bituminous coal, in making these experiments, render it difficult to obtain, from that species of coal, the tissues in the perfection in which they may be found in anthracite.*

Carbonization of Wood.—Dr. Mantell has treated at length on this interesting subject in his "Wonders of Geology." In a more recent work he remarks,—"that the structure and composition of a plant affected its carbonization there can be no doubt; for in the same layer of stone, [in the calciferous grit of Tilgate forest,] the stems resembling palms, *Endogenites*, invariably possess a thick outer crust of coal; while the stems and roots of the *Clathrariæ*,—plants allied to the yucca, or dracæna,—have not a particle of carbonaceous matter, but are surrounded by a reddish brown earthy crust.

The nature of the stratum in which the plants were imbedded, must, of course, have also influenced the bituminous fermentation. Vegetable remains, when interposed between beds of tenacious clay, by which the escape of the gaseous elements, set free by decomposition, was prevented, appear to have been most favourably situated for their conversion into lignite or coal. Experience has shown that although the true coal-measures are only found beneath the saliferous formation [of England], the production of good combustible coal is not necessarily restricted to any period or series of strata; but may occur wherever the local conditions were favourable to the complete bituminization of beds of vegetable matter. In fact, the productive coal-fields of Buckeburg, in Hanover, are situated in deposits of the Wealden epoch."†

Coniferous Fossil Wood in the newer coal formation of Nova Scotia.—For a knowledge of these fossil trees we are indebted to Mr. Dawson. According to his relation, at a particular level, in the lower part of the newer coal strata, calcareous petrifactions of coniferous wood are very abundant, in some instances appearing to have belonged to extensive rafts of drift-wood. A bed of sandstone, containing one of these petrified rafts, is well exposed on the shore between Cape Malagash and Wallace Harbour, and is there associated with a bed of gypsum, and a thin layer of limestone containing a few marine shells of species found in the lower carboniferous rocks.

In the bed of coniferous wood at Malagash, the structure of many of the trunks has been very perfectly preserved; and slices exhibit, very distinctly, polygonal discs on the walls of the cells, like those of the genus *Araucaria.* On comparing them with others from different parts of Nova Scotia, and New Brunswick, Mr. Dawson found that the species of coniferous trees most abundantly found in the coal

* Silliman's Journal, May, 1846.

† Mantell—Wonders of Geology, pp. 373, 688. Medals of Creation, 1844. Vol. I. pp. 89–90.

formation of Pictou and Cumberland counties have the structure of Araucarian pines. On the weathered ends of trunks of Araucaria, in the sandstones at Pictou and near Wallace, rings of growth are often very apparent. In some instances, the layers of yearly growth having separated in the progress of decay, as is often seen in recent wood, they have left vacant spaces, occupied, in the fossils, by calcareous spar. In a transverse slice the rings of growth can easily be seen by the naked eye. They do not exceed in width those of vigorous individuals of many recent coniferous species, but their limits are much less distinctly marked than in any coniferæ now growing in this climate.

It is perhaps worthy of notice, that the alteration effected from the original structure of these calcareous fossils, consists merely in the filling up of the cavities of the cells with carbonate of lime, and in the carbonization of their walls. When fragments are exposed to the action of diluted hydrochloric acid, the calcareous matter is removed, and a flexible carbonaceous substance, retaining the form of the fragment, remains. This residual woody matter burns like touchwood, and leaves a very little white ash.

Coniferous wood is not unfrequent in the nodules of iron-stone, included in the great coal-bed at the Albion mines. More rarely they afford fragments with the structure of stigmaria.

Stigmaria. At the extremity of Malagash Point, Mr. Dawson discovered in a bed of shale, a fossil stump of a tree, having connected with it, roots with regular scars like *stigmaria.* A portion of one of the main roots, ten inches in length, was seen to be attached to the stump, and other portions appeared in the surrounding clay. The trunk exhibited an external coaly envelope or bark irregularly corrugated: its stony cast showed, indistinctly, alternate smooth and rough vertical stripes, and internally it possessed an eccentric core, probably corresponding with that of the roots, and having large transverse prominences, which appear to have been connected with fibres or bundles of vessels, whose remains extend onward and downward through the outer part of the cast.

Artisia or *Sternbergia.* Fragments of plants of this genus are frequently found in the sandstones of the Pictou coal-field; usually in beds which also contain *calamites.* They are in the state of stony casts, always invested with a thin bark or coating of lignite, whose outer surface is smooth and without transverse wrinkles. Mr. Dawson saw none with any trace of roots, leaves, or fruit, or even of a conical termination: all were cylindrical fragments, and so similar in their markings, that they may have belonged to one species.

Transversely ridged stems, of a character very different from the above, are occasionally found in the carboniferous beds of this province. They are stony casts, having irregular and often large transverse markings, and enclosed in a thick coat of lignite or fossil wood. Transverse sections showed cellular tissue apparently with medullary rays, and much resembling the wood of coniferæ. These last are referred to casts of the pith of trees. Those previously mentioned apparently belonged to a plant having a very large pith and a comparatively thin woody envelope—in short a gigantic rush-like plant, perhaps leafless and nearly cylindrical, like some modern species of *iuncus*.* In this view Mr. Bunbury fully concurs, and recommends an adherence to the name *Artisia* given to these bodies, rather than that of *Sternbergia*, which name belongs to a genus of recent plants very different from these fossils.†

Coal vegetation of Frostburg in Maryland.—There are some details of the fossil forms at the Frostburg mines, deserving note, in an article in the Quarterly Journal of the Geological Society of London, in May, 1846, by Mr. Bunbury. These beautiful plants are figured and named as

1. *Pecopteris emarginata.* [*Diplazites emarginatus* of Goppert?]
2. *Pecopteris elliptica.*
3. *Danæites asplenioides.* (Goppert.)

With these fossil ferns Mr. Bunbury describes the following less rare plants, which are collected at Frostburgh by Mr. Lyell.

4. *Neuropteris cordata*—very abundant, and certainly identical with the English plant.
5. *N*——— gigantea?
6. *Cyclopteris ?*
7. *Pecopteris arborescens.*
8. *P*——— *abbreviata.*
9. *P*——— (?)
10. *Lepidodendron tetragonum.*
11. *L*——— *aculeatum.*
12. *L*——— (?) resembling in its markings the *Sigillaria menardi* of Brongniart.
13. *Sigillaria reniformis?*
14. *Stigmaria ficoides.*
15. *Asterophyllites foliosa.*
16. *A*——— *tuberculata?*
17. *A*——— *equisetiformis ?*
18. *A*——— undescribed, but said to be found in the "middle coal," near Manchester.
19. *Artisia* ——— ?
20. *Calamites nodosus.*
21. *C*——— *dubius ?*

Mr. Bunbury remarks that the very striking similarity between the coal plants of North America and those of Europe makes it probable that a similar kind of climate also existed in both countries at that era; and whatever conclusions we may arrive at, in relation to the carboniferous period in the one continent seems equally applicable to the other. Nothing, he continues, that has yet been ascertained relative to the coal formations of either continent seems at all

* Dawson on Nova Scotia Coal plants. Quarterly Journ. Geol. Soc. London, May, 1846, p. 132.

† Bunbury, Ibid., p. 138. Also Mr. Dawes, on Sternbergiæ, Ibid., p. 139.

inconsistent with the suggestion of Mr. Lyell,* touching the climate of the period in question.

This view is, that the climate was then characterized by excessive moisture; by a mild and steady temperature, and the entire absence of frost; but perhaps not by intense heat. It is admitted, indeed, that our materials for the foundation of this theory are perhaps somewhat scanty; being, chiefly, the general character of luxuriance of the carboniferous vegetation; the great abundance of ferns: and the presence of large leaved monocotyledonus plants of a tropical or sub-tropical aspect: for with regard to the sigillariæ, stigmariæ, asterophyllites, calamites, &c., their real affinities are, he thinks, too doubtful to allow us to found any arguments on them.

That extreme heat is not necessary to the existence of a very luxuriant and quasi-tropical vegetation, is sufficiently clear from Mr. Darwin's interesting observations on Chiloe and other islands of the southern temperate zone.† Chiloe, situated in the 42d degree of south latitude, enjoying little summer heat, and subject to perpetual rains and mists, is covered, as he states, with forests of extraordinary density, and the luxuriance of the vegetation is such, that it reminded him of Brazil. Large and elegant ferns; parasitical monocotyledonous plants, and arborescent grasses, reaching to the height of thirty or forty feet, are abundant. Indeed, in the southern hemisphere generally, owing to the equable climate produced by the great proportional extent of sea, tropical forms, both of vegetable and animal life, range much farther from the equator than in our hemisphere. It appears very probable that the climate of the northern temperate zone, during the epoch in which the coal measures were formed may have been similar to that now existing in Chiloe and the adjoining parts of South America.

Still, considering that the principal coal-fields of England are situated from 13° to 15° farther north than that of Frostburg, the close resemblance of their vegetation is very striking. The absolute identity of some species is not perhaps so remarkable as the very great general similarity of the whole; for those among the Frostburg plants, which cannot be satisfactorily identified with the British species, are, in every instance, very closely allied to them. We should not find so great a degree of resemblance on comparing the recent floras of two regions separated by so many degrees of latitude, whether in Europe or North America. If we may reason at all as to climate, from the fossil vegetation of a country, we must suppose that the climate varied less rapidly with the latitude that it does at present.

In concluding this valuable paper, the writer suggests, that the plants, of which we now find the remains embedded in the carboniferous strata, may probably be but a very small proportion of those which, at that time, flourished on the earth. If, as seems to be now most generally believed, the coal beds are derived from the vegeta-

* Travels in North America, vol. i. p. 148.
† Darwin's Journal, 2d edit., p. 242.

tion of ancient swamps or lakes, existing in the very localities now occupied by such beds of coal, we could not expect to find in them the remains of other plants than such as grew in those bogs, or lakes, or swampy forests, or immediately around them; together perhaps with some which might be washed into them by occasional inundations. May there not have existed at the same time, in other parts of the world, [nay, perhaps at no very great distance from the carboniferous regions,] great tracts of country, indeed whole continents, in which the local circumstances were unfavourable to the preservation of vegetable remains, and of which, consequently, the flora is wholly lost to us?

I think, therefore, that we ought to proceed with great caution in theorizing with respect to the vegetation and climate of the carboniferous era. I believe that the preponderance of ferns in the flora of the coal measures, together with the other characteristic of the fossil vegetation of that period, affords, to a certain degree, good evidence respecting the climate of those particular regions in which the coal measures occur; but we should not be justified in extending our inferences farther. Those parts of Europe and North America, in which the coal-fields were accumulated, may have existed, at that time, in the state of islands, like those of the present Pacific ocean: but it would be rash to infer, as M. A. Brongniart seems disposed to do, that no extensive continents at that time existed in any part of the globe. If in all departments of geology, it is necessary to advance with caution, and to avoid dogmatism and rash generalizations, it is more especially necessary in the department of fossil botany, where so much of the evidence we possess is fragmentary and imperfect.*

MISCELLANEOUS NOTES AS TO COAL AND FOSSIL VEGETATION.

In continuation of this subject, we proceed to advert to the results of some observations which have been made of late years by individual naturalists. Were we to incorporate in this work the facts, theories, and speculations which have been discussed at different times on the coal subject, we should occupy at least an entire volume. There are many excellent treatises embracing these topics, which the reader, if seeking more information, may consult to advantage. The few notes we add here are inserted with little regard to classification.

The discovery, in 1839, during the progress of excavating a part of the route of the Manchester and Bolton Railway, within the limits of the Lancashire coal-field, of numerous fossil trees of the family sigillaria, standing in a vertical position, with their roots embedded in a thin coal seam, gave rise at the time to much discussion. Mr. Hawkshaw described these trees in two communications to the Geological Society.† These trunks were wholly enveloped by a coating

* Bunbury on Fossil Ferns. Quarterly Jour. Geol. Soc. No. 6, p. 82.
† Proceedings Geol. Soc. Lon., Vol. III. p. 139; and 269, 1840.

of friable coal, varying from one-quarter to three-quarters of an inch in thickness. Their internal casts consisted of shale, traversed beneath the place of the bark by irregular longitudinal flutings, less than one-quarter of an inch broad and about two inches apart.

Mr. Bowman communicated a paper on the same subject. He is opposed to the drift theory in accounting for coal beds, because they would have been intermixed with more earthy matter than is now proved to be the case in coal; and because they could not have maintained that singular uniformity of thickness and character throughout so many square miles, and such extensive areas that we find prevails in the coal measures; as an instance of which the author cites the thin seam below the Gannister or Rabbit coal, which extends in a linear direction thirty-five miles. It is much more rational to suppose, that the coal has been formed from plants which grew on the areas now occupied by the seams; that each successive race of vegetation was gradually submerged beneath the level of the water, and was covered up with sediment, which accumulated till it formed another dry surface for the growth of another series of trees and plants, and that these submergences and accumulations took place as many times as there are seams of coal within the confines of each basin.

Mr. Bowman proceeds to the examination of the phenomena presented by the fossil trees discovered in the railroad excavations above referred to by Mr. Hawkshaw. He describes, generally, the markings on the internal casts of the trees. The only indications of scars which he could find, his practised eye recognized to be those of a sigillaria.

From a careful consideration of the phenomena presented by the fossils, the author is convinced that they stand where they originally flourished; that they were not succulent, but dicotyledonous, hard-wooded, forest trees; and that their gigantic roots were manifestly adapted for taking firm hold of the soil; and, in conjunction with the swollen base of the trunks, to support a solid tree of large dimensions, with a spreading top.

With reference to fossil trees in general, and especially to those near Manchester, Mr. Bowman proceeds to show; 1st, that they were solid, hard-wooded, timber trees, in opposition to the common opinion that they were soft or hollow; 2d, that they originally grew and died where they have been found, and consequently were not drifted from distant lands; and 3d, that they became hollow by the decay of their wood from natural causes, similar to those still in operation in tropical countries, and were afterwards filled with inorganic matter, precipitated from water.

The author states his reasons for believing that these were solid timber trees. In soft monocotyledonous trees, their stems never expand laterally, but are as thick when only a few years old, and a foot high, as when they attain the height of sixty or one hundred feet. Their roots, also, instead of being massive and forking, generally present a dense assemblage of straight, succulent fibres, like those of an onion or a hyacinth.

Mr. Bowman then combats the view generally entertained, that fossil stems, with perpendicular furrows, as in the sigillariæ, were succulent or hollow plants. He showed by specimens of recent dicotyledonous wood from New Zealand, that, both upon the bark and on the naked wood, longitudinal ribs and furrows, as regular as those on sigillariæ, were displayed; proving, therefore, that these characters are not incompatible with a dicotyledonous structure. By sliced and polished specimens of the bark of one of these fossil trees, he showed evidence of coniferous structure, proving, also, further, their dicotyledonous character. We note this decision with the more particularity, since M. Brongniart at the same time had asserted that "no wood of dicotyledonous plants, properly speaking, have been found in the coal-fields,"* but has since materially changed his views on that point.

The roots of these trees are fixed in what is now a seam of coal nine inches thick. Mr. Bowman infers that one hundred years must be the minimum of time which would be required for the production of the vegetable matter out of which the nine inches of coal were produced; and he estimates that the thickness of the solid coal is equal to about one-third that of the vegetable matter out of which it was produced.†

An instance very similar to this was detailed by Mr. Witham, in a communication to the Philosophical Magazine, entitled, "On the vegetation of the first period of the world, during the deposit of the transition and coal series." The author illustrates by a diagram the fossil stems of sigillaria, which occur beneath the main seam in the great Newcastle coal-field, at one hundred and fifty yards beneath the surface.

The fossil plants stand erect in the sandstone, their roots being imbedded in the ten inch seam of coal below. "These stems, [as shown in the figure,] are truncated after passing through the sandstone, and are lost in the main coal seam; leaving room to believe that they may have formed part of this combustible mass or bed." The saginariæ, the stigmariæ, and the calamites, he observes, do not appear to have been sufficiently strong to have resisted the force of a current of water, but are placed horizontally.‡

Position occupied by Sigillariæ.—The trunks of these trees are found both in the floor and the roof of coal seams; their position commonly being the upper part of the coal and the lower part of its roof. The sigillariæ are arranged by M. A. Brongniart among the conifera; by Dr. Lindley under the name of caulopteris, and by Count Sternberg as syringodendrons. Some discussion and much new light have arisen, and it seems nearly settled that the numerous tribe of sigillariæ are to be removed altogether from the arborescent ferns to the

* Histoire des Vegetaux Fossiles.
† Proceedings Geol. Soc. London, Vol. III. p. 270.
‡ Phil. Mag. January, 1830.

dicotyledon family. M. Brongniart has been able to take the measurement of one of these stems, which was horizontally extended to the distance of more than forty feet; but has rarely had opportunity to examine their height, their general form, and their mode of termination, on a large scale, in the mines.

In Pennsylvania we have had some favourable opportunities for observing and illustrating the position of enormous trunks in the anthracite mines. The Transactions of the American Philosophical Society contain a memoir on the fossil stems of large trees belonging to the family of sigillaria, which occur both in the roof and floor of a coal seam in Dauphin county.* They consist of several species of these trees, which are displayed in a very interesting manner upon the nearly vertical walls of the vein for several hundred feet in length.

The Floor.—As usual in Pennsylvania, the "bottom slate" consists of indurated clay and shale, more or less laminated. This lamination, it may be observed, is principally due to the flattened sheets of enormous sigillaria. Very few of these compressed trunks are of a less diameter than two feet; many of them are three feet; several are four and four feet and a-half wide, and one specimen is at least five feet broad in its flattened diameter. More than a hundred of these are exhibited in the drawing which illustrates the paper referred to. The coal seam had not at the time commenced to be worked; and as its position was approaching to vertical, the gallery of exploration was conducted longitudinally along it, having the floor on the right hand and the roof on the left. Consequently, although several hundred feet in length of walls were exposed on either side, the height denuded was comparatively limited, and afforded little chance for determining the length of the trunks. In no instance was the area of excavation sufficiently extensive to exhibit either extremity of these gigantic stems, notwithstanding that many of them are inclined in such a position as to be exposed for thirty, forty, and fifty feet of their length, without much apparent diminution or tapering upwards, and are perfectly straight.

The Roof.—This is the north or hanging wall of the vein, and consists of coarse siliceous conglomerate of white quartz pebbles. Between it and the coal, and embossed, as it were, upon the surface of the pudding stone, is a very thin coating of clay slate, and an extraordinary assemblage of prostrated trunks of sigillariæ. In diameter they are much smaller than those of the species which form the floor. Instead of being straight like them, these are bent or curved, and some of them appear to be dichotomous, and to possess the characters of *S. elegans*. Such is the scale, as regards height, of these trees, that the extent of cleared space was, as in the floor, inadequate to elucidate their entire development at any point or in any instance.

* Memoir by Mr. Richard C. Taylor, in Trans. Amer. Phil. Soc., Vol. IX. part II., 1845.

One specimen, although laid bare for a length of more than fifty feet, showed no signs of either termination, and looked as if it might have extended thirty or more feet further. Another exhibited sixty-five feet in length, of a flexuous stem, which, apparently, extended at least thirty feet beyond. A third, the most interesting of the group, showed at its base what obscurely seemed to be the root. Near this base the stem was about two feet and a half in diameter. Forty feet up the trunk it measured two feet broad, and continued in about this rate of diminution as far as it was traced. Seventy feet in length of this specimen occur above the level of the floor of the gallery. It was followed, by direction of the author, several feet further, below the floor, and in all was perhaps from eighty to one hundred feet high when growing; but of this, and of the character of that superior termination, we have no present knowledge. It was covered with a bark of anthracite, about half an inch to three quarters or more thick. The interior cast consisted of shale or fire clay.

On applying to this interesting illustration of the ancient flora, Mr. Logan's views as to the universal presence of the stigmariæ in the argillaceous floors of coal seams, and of their absence in the roofs, it was found that in this instance, where a surface of seven or eight thousand feet had been recently denuded, stigmariæ were rare. Only two well defined specimens, but of small size, were observed. One of these was seen in the roof above the coal; the other in the floor, below it: but detached leaves were abundant in the lower shale. Six other species of fossil plants were observed in the roof, and seventeen species in the floor.* As usual in the coal seams of this country, a remarkable contrast appears in the condition of the roof and floor. While the appearance of the floor attested the state of tranquillity under which the mud of the ancient surface had accumulated, and the pressure that had flattened those enormous stems of sigillariæ upon which the coal appears to be based, the roof, on the contrary, exhibits the usual indication of violent action of the waters, in the rolled fragments of subjacent rocks, and in the prostration and drifting of gigantic trees, such as we have described above.†

A few of these prostrate trees are very imperfectly represented, as regards scale and details, in the following figure, which has been reduced from a very elaborate drawing.

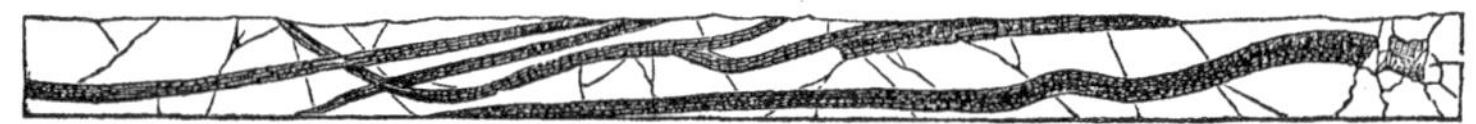

USUAL POSITION OF STIGMARIÆ, IN THE FLOORS OF COAL BEDS.

The existence of beds of Stigmaria, in the slate and fire clay which so generally form the strata, upon which coal seams repose, has been

* Proceedings American Philosophical Society, Vol. III. p. 149.
† Trans. Amer. Phil. Soc., Vol. IX. part II., 1845.

pointed out by various persons; in particular by Mr. Logan, who found it to hold good in the coal-fields of both the European and American continents. It is due to earlier observers, however, to state that this fact had long since been noticed by Mr. Martin, Dr. McCulloch, and others, including numberless working miners.

In a communication to the British Association by Mr. Binney, in 1842, it appears that the workmen in the principal coal-fields in England, more especially that of Lancashire, regard the presence of stigmaria as a favourable evidence of the vicinity of coal.

All the *floors*, with the exception of one rock floor, in the Lancashire region, from the thin coal seams in the Ardwick limestone, to the two seams in the Millstone Grit, a thickness of near sixteen hundred yards, contain Stigmaria ficoides. All the fifteen floors of the Manchester coal-field contain them; and at least sixty-nine beds in the middle and lower divisions of the Lancashire field.

He adds, [a fact we greatly doubt,] that, in all instances of *true floors*, the stigmaria occurs without any intermixture of other plants.

Sir Henry T. De la Beche corroborated the former portion of this statement as regards Glamorganshire, Somersetshire, Yorkshire, Scotland, and Ireland; and said that he had never seen a *workable* coal bed which did not bear out Mr. Binney's conclusions.

Mr. Logan showed that below every regular seam of coal, in South Wales, [and nearly 100 are known to exist there,] is constantly found a bed of clay, so well known to the collier, that he considers it an essential accompaniment of the coal; and only where it ceases, does he give up his expectation of finding coal.

These beds are most strongly marked by containing innumerable specimens of *Stigmaria ficoides.* The stems of this plant, which are usually of considerable length, are said, by Mr. Logan, to lie always parallel to the plane of the bed, and nearer to the top than to the bottom. Portions of the stem of the Stigmaria are found in other parts of the coal measures, but it is only in the underclay that the fibrous processes are attached to the stem, or are associated with it.* The same rule appears to hold good in the coal formation of Nova Scotia, New Brunswick, and the United States.

With regard to the specific plant whose remains have chiefly contributed to form our coal seams, different views have been advocated at times, by naturalists. Without assenting to the doctrine, "that each bed of coal is an ancient Stigmariæ bog,† we think that many other plants united to make up the mass, and that the predominant character of these may, in great measure, be inferred from an inspection of the shales, slates, clays, and sandstones, which occur in close contact with the coal itself. In Pennsylvania we have had abundant opportunity of observing coal seams, whose roofs and floors were crowded with sigillaria, and showed but rarely the traces of Stigmaria, or of those forms which are now ascertained to be the roots of the

* Proceedings Geol. Soc. London, Vol. III. p. 275.

† Proceedings of the American Philosophical Society, May, 1843, p. 182.

Sigillaria itself. In other cases, the prevailing plants of the shales, on which the coal rested, appeared to be Stigmariæ; while the roof contained chiefly Sigillariæ, and Lycopodiaceæ. On the whole we were at one time quite inclined to adopt the view of M. Brongniart, that the mass of coal vegetation was more likely to have been derived from Sigillariæ than from Stigmariæ. The great number of leaves he observes which the Sigillariæ bore along their whole length, and which evidently were disarticulated, and had fallen to the then surface of the earth, announce a life of long duration, and a growth which required a considerable lapse of time.*

On thing appears to be pretty certain,—that the coal-fields exhibit alternate intervals of repose and of energetic action by currents of water—in other words, of a series of epochs of dry land and of inundation. These evidences testify that, after long periods which favoured the quiet growth and accumulation of masses of vegetable matter, they were abruptly terminated; and that this state of things was succeeded by overwhelming currents, which prostrated the forests of Sigillariæ or arborescent ferns; rooted in the ancient surface, and covered them with a debacle derived from older formations, and which we now recognize under the term conglomerates.

During an investigation of the coal beds of Dauphin county, in Pennsylvania, we had ample means of observing, at leisure, these facts: and it was seen, that while the floor of every coal seam consisted of shale, its roof, in the majority of cases, consisted of pudding-stone, whose lower side was impressed and embossed with enormous casts of prostrated Sigillariæ.

Before quitting the subject of coal vegetation, or rather that of the Sigillariæ and Stigmariæ, whose exuviæ are considered mainly to form our coal seams, the progress of discovery in regard to the real nature of those plants, is too interesting to be omitted here. M. Brongniart, after dissecting their stems had arrived at the same conclusion as Mr. Bowman, that these fossils had been wrongly classed with monocotyledonous plants, and from a comparison between the fossil and the stems of those recent vegetables which present the closest analogy, M. Brongniart concludes "that the Sigillariæ constituted a peculiar family of coniferous plants, now extinct, which probably belonged to the great division of gymnospermous dicotyledons. In their external forms they somewhat resembled the *Cacteæ* or *Euphorbiæ*; but, by their internal organization, they were more nearly allied to *Zamiæ* or *Cycadeæ*. The leaves and fruit of these trees are unknown, for no satisfactory connection has yet been established between their stems, and the foliage, and seed-vessels, with which they are collocated."†

But the most important discovery yet announced, relates to the character of the fossil genus *Stigmaria*, which, after all the speculations to which its appearance has given rise, seems to result in

* Brongniart, Histoire des Végétaux Fossiles. For details of fossil vegetation in Great Britain, see England.

† Medals of Creation, Vol. I. pp. 138—140.

determining it to belong to, and, in reality, to form parts of, the Sigillaria itself. Instead of the Stigmariæ being aquatic plants, as it has been customary to consider them, M. Brongniart, author of the elaborate Memoir on the Sigillariæ, "from a careful examination of the internal structure of the Stigmariæ, contended that *they were not aquatic plants, but the roots of the Sigillariæ;* the central axis, or cylinder, bearing a close analogy in organization to the stems of those trees."

"This opinion of the eminent French savant, has been confirmed by the discovery, near Liverpool, in coal strata, of an upright trunk of a Sigillaria, nine feet high, with its roots eight or nine feet in length, still attached and extending in their natural position. *These roots are undoubted Stigmariæ of the usual species, S. ficoides; and the radicles, formerly considered leaves, are spread out in all directions, to the extent of several feet.**

The existence both of Coniferæ and Cycadeæ, which heretofore had been doubted, in the coal measures of the former world, is now established satisfactorily. M. Alex. Von Humboldt assures us that the Coniferæ have not only relationships with the Cupuliferæ and the Betulineæ, by the side of which we encounter them in the brown coal formation, but they are further connected with the Lycopoditæ. The family of the sago-like Cycadeæ approaches the palms in external appearance whilst agreeing essentially with the Conifereæ in the structure of the flowers and fruit. In the coal measures of Nova Scotia, fossil Coniferæ are very abundant, as Mr. Dawson has shown.†

Since the statement of Mr. Binney, respecting the two fossil trees with marked roots resembling Stigmaria, which were discovered at St. Helens in Lancashire, Mr. Dawson has described numerous corresponding instances in the coal-shales of Nova Scotia, and Mr. Bunbury states that the symmetrical quincuncial arrangement of the scars in the Nova Scotia specimens, the presence of the eccentric axis, and the general appearance of the fossils, leave no doubt that they are referable to the supposed genus *Stigmaria.* Dr. Lindley, who seems to have been the first to hint,‡ that Stigmaria might possibly be the root of Sigillaria, compares the dome-like centre and radiating arms of Stigmaria with the roots and base of the stem of *Sig. pachydermata.*§

Mr. Bunbury, in discussing the progress of his investigation into the character of these remarkable extinct forms of vegetable life, observes that the similarity of the vascular tissue of the Sigillariæ to that of ferns is not a sufficient proof of any real affinity to that tribe of plants, since Mr. Brown has ascertained that vessels of a similar structure, constitute the whole of the woody tissue of *Myzodendron,*

* Communication to the British Association, 1843, by Mr. Binney.
† Quarterly Jour. Geol. Soc. of London, May, 1846.
‡ Penny Cyclopedia, art. Coal plants, 1837.
§ Fossil Flora, t. 54.

a genus of parisitical flowering plants, allied to the misletoe, and totally dissimilar to ferns.*

A very satisfactory and characteristic specimen, showing unquestionably that the Stigmaria is the root of the Sigillaria, has lately been brought to light from the Victoria pit at Dunkinfield, in the Manchester coal-field, where, at the depth of 1000 feet, the fire clay, in which the tree was imbedded, underlies the cannel coal. This was first noticed by Mr. M. Dunn, and has since been described by Mr. Binney;† and is now in the collection of the Manchester Geological Society.

The stem of this fossil is unquestionably that of a Sigillaria; exhibiting all the ribs, furrows, and scars of that genus. It is four feet, ten inches in circumference at its base. On the outside is a coating of bright coal, one third of an inch in thickness; very much resembling that found on the *S. pachyderma*. In every respect, Mr. Binney observes, this stem resembles the two trees found in the St. Helen's mine, before alluded to, and also to the Dixon Fold trees, described by the late Mr. Bowman.‡

The roots gradually assume all the true characters of Stigmaria, with depressed areolæ, &c., and have been traced for fifteen feet; at which distance they average about six inches across, without any signs of terminating.

Mr. Binney concludes his description of these fossil trees with the remark, that it seems evident that Sigillaria was a plant of an aquatic nature, from the position of the St. Helen's trees, which were found on the identical spots where they grew, imbedded in a fine silty clay, sixteen yards above or sixteen yards below, or midway between two seams of coal.

Sigillariæ and Stigmariæ of the American Coal-fields.—In confirmation of the last named English observers, and in addition to those discoveries previously cited, by several geologists in British America and the United States, we have more recently had the evidence of many similar fossil trees in the Sydney coal-field of Cape Breton, described by Mr. Brown.§

The sea cliffs on the north-west shore of Sydney Harbour, present an interesting section of the coal measures, and unusual facilities for observation. Fossil trees are seen therein at various levels, but abundantly in a stratum of arenaceous shale, lying almost immediately under the main coal, where, within a space of eighty feet, eight erect trunks are seen, with their roots and rootlets attached to them. They all occur at right angles to the planes of stratification, and belong to the same species; being evidently young individuals, which range from two to sixteen inches in diameter only. Mr. Brown's paper is illustrated by drawings of these fossils as they appeared in

* Mr. Bunbury on Stigmariæ, Quarterly Journal Geol. Soc. of London, Vol. II. p. 136.
† Quarterly Journal of the Geological Society of London, Vol. II. p. 390.
‡ Transactions of the Manchester Geological Society, Vol. I. p. 112.
§ Quarterly Journal Geol. Soc. London, Vol. II. p. 393.

their native sites. Their bark, converted into bright coal, is very thin; it is marked with longitudinal furrows and ridges.

The roots, which are true Stigmariæ, with *rootlets* or [as they generally have been called] *leaves*, spreading out in every direction, are about three inches in diameter at their junction with the stem. Two of these roots, in the plant figured, have been followed to their terminations, where they gradually thinned out to a mere line in one direction, being about three quarters of an inch in width. They are generally thickly studded with tubercles, presenting an imperfect spiral arrangement, and are covered with a thin bark, or coating, of carbonaceous matter.

The leaves or rootlets, varying in length from three to twelve inches, are flattened; being much broader near their junction with the roots than at any other point.

All these circumstances seem sufficient to prove that Stigmariæ and their leaves are, in reality, the roots and rootlets of a class of trees, allied probably to the Sigillariæ.

On the Fossil Vegetation of America—by J. E. Teschemacher—with illustrations in wood.*

An important source of information is presented by the vegetable remains existing in the coal itself; leaving out of consideration those in the shaly roofs and clayey floors of the mines. The Pennsylvania anthracites offer many specimens of these. What is termed charcoal is commonly found in seams and crevices in the coal; and in most of this, the vegetable tissues, although carbonized, are in perfect preservation.

Mr. T. proceeds to describe a series of coal plants from Carbondale, in Pennsylvania.

He considers *Sigillariæ* as the stems of *Filices*, observing, "It seems to me almost impossible not to be convinced, by the arguments of Brongniart, that these are the stems of the arborescent ferns, whose leaves are scattered in such profusion around them,—although I am aware that both Göppert and Lindley have withheld their assent to this opinion."

The present is probably the most fitting place to allude to a late valuable contribution to our knowledge of the internal structure of fossil plants, in the work of A. J. Corda, entitled "Contributions to a Flora of the Ancient World."† The following notices, derived chiefly from the Journal of the Geological Society of London, are selected with reference to their bearing on the families of coal plants which we have been considering.

Sigillariæ. The author differs from M. A. Brongniart, respecting the affinities of these plants. He believes that the Sigillariæ were succulent Dicotyledons, closely allied to the recent Euphorbiæ.

The leaves of the Sigillariæ have been hitherto unknown, except in the single instance of *S. lepidodendrifolia*, as figured and described by A. Brogniart.

* American Journal of Science and Arts, January, 1847.

† Beiträge zur Flora der Vorwelt. Prague, 1845.

M. Corda has discovered the leaves of *S. rhytidolepis*, which bear a strong resemblance to those of *S. lepidodendrifolia*, and other species; and it is very probable that some of the so-called *Lepidophylla*, which occur very frequently in a detached state, in the coal formations, may be the leaves of Sigillariæ.

The author is of opinion, in agreement with a M. A. Brongniart, that the temperature which prevailed during the period of the coal formation, was very high.*

It may be useful to geologists to mention also that the work of M. Corda contains tables, which show the number of fossil plants in each formation of the earth, and the number of fossil ferns in proportion to that of other plants; an enumeration of living ferns, distributed by tribes, and according to the zones of temperature in which they occur; also a list of fossil ferns by tribes, showing the proportion of the fossil to the recent species, known of each tribe: and a table enumerating the arborescent ferns, known in a recent and in a fossil state, and also the *Marattiaceæ*, known in each of these states, comparing them with the total number of ferns, recent and fossil.

As to the class of Plants which form the Coal Vegetation.—According to M. Burat, there are about three hundred species of coal plants recognized; of which five-sixths belong to the cryptogamous vasculaires [vascular cryptogamia?] that is to say, to the ferns, the calamites and neighbouring families. These vegetables form one-sixth of the actual existing flora.

The coal vegetation, which is remarkable for the predominance of the cotyledonous species, is most analogous to that of the existing epoch where developed in certain low and humid islands in the warmest latitudes. The researches of M. A. Brongniart in this vegetation, showing that it resembles that of our equatorial regions in the abundance of equisetaceæ, palms and arborescent ferns, leaves no doubt respecting the origin of the coal, and we can even find direct proofs in the mechanical analysis of certain varieties.

Coal Shales.—We believe it is now generally admitted that nearly every coal seam in the world is imbedded upon an argillaceous stratum, more or less indurated, in every gradation, between soft fire-clay and compact slate. These argillaceous beds are characterized by the abundant traces of the fossil vegetable, Stigmaria, which rule is sufficiently exact, in most instances, to enable the miner, when engaged in exploring for coal beds, to distinguish, with the aid of some other obvious characters, between the shales which form the roof and the floor of those veins, or what is commonly termed their "top slates" and "bottom slates."

In some of the coal-fields of Europe, the "top-slates" or roofs are as much characterized by courses of nodular iron ore, as the fine clay floors are distinguished by their fossil vegetable traces. In the United States, the clay iron ore, although occasionally present, does not seem so abundant as in Europe.

* Quarterly Journal of the Geol. Soc. of London, Vol. II. p. 219.

The carboniferous shales contain but little bituminous matter, in America; and generally speaking, we believe, not in England or Wales. In Scotland we are assured the shale is often so bituminous as to be used for fuel; as at Pitfirrane in Fifeshire, and other places; and it gradually passes into pure coal.*

The upper shales or "top slates" are seldom so regular as the bottom slate. They are commonly thin interposing seams between the coal and the overlying sandstone. In Pennsylvania it is of very common occurrence that beds of coarse gritstone and conglomerate are in immediate contact with the underlying coals; showing that the period of quiet which marked the deposit of mud and clay in which the stigmariæ are imbedded, and that of the tranquil accumulation of vegetable matter which forms the purest coal veins, was abruptly succeeded by a period in which the waters were in a state of tumultuous agitation;—when the trees on the then existing surface were prostrated and buried beneath thick beds of pebbles and gravelly debris.

Origin of Coal.—In relation to this subject, M. Burat has noticed that, as a great number of varieties of coal contain much more ashes than they had in the vegetables from which they were formed; as, in other cases, the ashes are not of the same nature as the vegetable ashes; as, moreover, in a great number of instances, we find very small lines of schist intercalated in the coal; as even selected portions always furnish a considerable proportion, these extraneous matters may have preserved some historical facts in regard to the formation of coal.

On proceeding to the mechanical analysis of these coals, we perceive that they are formed of thin and superposed laminæ, which are composed sometimes of pure, specular coal, yielding scarcely two per cent. of ashes, and sometimes a dull schistose coal, which contains twenty or twenty-five per cent.

The results of these alternations of layers is a diversified structure, in the direction of the stratification, and a series of layers which present frequent traces of vegetable tissue. This analysis demonstrates that the coal is heterogeneous, and is composed of superposed alternations, the one consisting of pure particles which are the result of vegetable decomposition, the other of earthy parts, produced by the action of water, more or less charged with argillaceous matter. The vegetables of which the coal is composed, both M. Brongniart and M. Burat conceive belong chiefly to the small species of the genus calamite; in that respect differing from the larger plants, whose debris is found in the slates and sandstones.

Therefore, from these data, we may conclude that these little alternating bands represent a production and a periodical destruction, like those which might result from the seasons of the year. The brilliant or specular laminæ are the decomposed vegetables of this period; the dull or earthy layers represent a portion of this decomposition

* Nicols's Guide to the Geology of Scotland, p. 64.

mixed with impurities such as may be attributed to the invasion by waters holding argil in suspension, and whose periodical return has been one of the causes of decomposition. To the influence of these immersions must be attributed the effects of erosive currents, which have brought together, at certain points, sometimes thick zones of specular coal, and sometimes beds wherein carbonaceous schist or clay predominates.

Thus, then, the coal-fields may have been produced, in great measure, by the growth, on the spot, of small vegetables, in the manner of peat or turbaries; whilst the larger vegetables may have been drifted from distant and elevated points, when the oscillations of the surface produced the interruption or the renewal of this generative action, in placing the coal surface above or below the level of the sedimentary waters.

This hypothesis explains, not only the formation of the thin and multifold beds, in the basins of the north of Europe, for instance, but they also agree with the generation of the thick and limited basins of the south, and of the great accumulations, like those of Montchanin.

The distinctness of the planes of separation demonstrate that the two principal generators were not susceptible of being confounded; in other words, the deposits of sand and clay were effectuated in the water, whilst the coal, on the contrary, has been produced above these waters. Coal, then, is not, properly speaking, a sedimentary deposit, produced by the transportation of vegetables, or by floating rafts, as has sometimes been supposed; and yet its production has taken place very near to the surface of the water, since it has had frequent penetrations of the two generating influences, which thus accumulated, side by side, products so different.

These views of the origin of coal ought evidently to be extended to all the series of fossil combustibles, which represent the vegetable accumulations of various geognostic periods. The anthracites of the transition formations owe their dry and meagre nature only to the difference in the mode of decomposition, determined by the special conditions of the surface of the globe at that early epoch. It is to be remarked, that if our theoretical ideas of the formation of the globe induce us to attribute this difference to the phenomena of temperature and pressure, which appear to have affected the rocks of the anthraxiferous epoch, this opinion is completely confirmed by the anthracitous state of the combustibles, subsequent to the coal period, which we encounter in the metamorphic formations. We cannot, in fact, doubt that, in the second case, the phenomena of heat and pressure are the modifying causes of beds which originally consisted of coal or lignite.

The tertiary LIGNITES have generally preserved their ligneous tissue so fully, that we can recognize, in many of the fragments, the nature of the constituent wood. The fir, the alder, the beech, and the oak, form the most frequent debris of the lignites of the Alps, and they thus denote a complete change in the vegetation of the earth since the coal period. They are the true fossil forests, which likewise differ from the coal beds by a more circumscribed accumulation, and by a less complete stratification.

In certain exceptional cases, the lignites have a compact structure, almost comparable to that of coal. They then constitute what is called common *jet*. This is the ordinary character of the lignites in the environs of Marseilles; which give rise to an annual production of more than a hundred thousand tons. These lignites form seven perfectly regular and stratified beds, within the tertiary basin, at Fuveau, Crest, Auriol, &c.; the thickness of each of which beds [from one to three feet,] preserves such constancy that it may be recognized by this character alone, in the divers parts of the basin. These beds are comprised between calcareous strata; they are subject to the numerous movements, inclinations, folds, faults, and upheavings which have disturbed the various portions of the tertiary basin.

Among these accidents, there is one which is peculiar to the lignites, and which is known under the denomination of *mouillères.* These consist of portions of beds where the lignite is so fissured and decomposed that it has become very permeable to water. The workings encounter the double difficulty of abundant infiltration, and a production of no value. In their normal state these lignites have much the appearance of coal, but they have not its quality. Nevertheless, in certain positions, in Tuscany, for instance, we find some small beds of a lignite sufficiently perfect to furnish a coke, on distillation.

The general character of lignite is such, that it cannot be considered as possessing a regularity comparable to the coal beds.

Portland Oolite Beds—Contain Zamia, fragments of which are found in the lower calcareous bed of the group—or perhaps in the inferior portion of the cretaceous series. These plants are accompanied by paladinæ or helices, which consequently indicate the passage of fresh water in the seas of this epoch, where are seen the remains of large coniferæ, rooted in the soil, analogous to the *Araucariæ*, now strangers to the present climate of England. But in the midst of these coniferæ we find plants which have a resemblance to the *Cycas* and the *Zamias* of the tropical climates, and also the animal relics which approach to those of the same zone. The dirt bed of Portland, which incloses trees still in place, attests the existence of a vegetable soil, of earth almost dry, which rests upon the marine deposits. This bed has since been recovered by very powerful beds of fresh-water limestone, and then passes under the green sand which follows the chalk.

The Wealden group—incloses various vegetable debris—some of which resemble that of the Portland beds—and we meet with, in place, and in a siliceous state, the trunks of Cycadeas; *Mantellia nidiformis.* With these occur various species of coniferæ, besides the fragments of equisetaceæ and forms of a peculiar species.

Trias, or the *Grès Bigarré*, or copper group. This great formation, which in France has received the name of Trias, because it incloses three principal parts, is composed of deposits of sandstone and marls, of varied colours, which have given to the sandstone the name of Grès Bigarré, [red and white] and to the marls, that of *Marnes Irisées*. The two latter in England are known under the name of the *upper new red sandstone and red marl.*

In this group vegetation has undergone great modifications. The ferns and the gigantic equistaceæ have considerably diminished; while the coniferæ, on the contrary, have become more numerous: plants, analogous to the Zamia, and perhaps to the Cycas, formed at that time an important portion of the flora of Europe; a prelude to the immense development which they made in the succeeding epoch —" *l'époque jurassique*," or lias.

Vegetable debris and combustibles of the *Molasse*.* This tertiary formation occurs above the *Calcaire Grossier*, in the environs of Paris.

The Molasse is very rich in combustible; it encloses the lignites of Languedoc; of Switzerland; the most part of those in Germany, as well as those of Cologne. All the lignites appear to have been principally formed by the *coniferæ*, of which we are able to recognize the tissue, either in the mass of combustible or in the wood which is disseminated in the midst of the various deposits.

It is known, however, that in this formation there are also many dicotyledonous plants, the wood of which is found disseminated here and there; sometimes in a silicified state, clearly exhibiting the tissue peculiar to this class of vegetables, and characterised, above all, by the presence of large longitudinal vessels.

Leaves also exist; often abundantly, even in the clays which accompany the lignites, and in these can be recognized distinctly the characters which the dicotyledones present. Among them are those of the walnut, the maple, elm, birch, &c.

There exist even fruits, which often cannot be distinguished from those which we find at the present day in our climate.

Finally, there are found in this formation the remains of monocotyledonous plants. This wood presents all the structure of the palms; that is to say, an assemblage of ligneous bundles, disposed longitudinally, without regularity, in the middle of a cellular tissue, as in *Palmacites lamennois.*

CUPRIFEROUS LIGNITES.

Vegetable remains under this form present themselves in various geological positions and circumstances, which will be noticed in the progress of this work.

In the provinces of *New Brunswick* and *Nova Scotia* they occur in the regular coal measures. Mr. Henwood has mentioned this interesting fact—that lignites, consisting of ferns and other coal plants occur impregnated with rich vitreous copper ore and coated with

* Cours Elémentaire d'Histoire Naturelle, par M. F. S. Beudant.

green carbonate of copper, on the Nipisiguit, near Bathurst in New Brunswick.* These vegetable remains are, according to Mr. Logan, partly converted into coal, and partly replaced by gray sulphuret of copper. The same occurs in the neighbourhood of Pictou in Nova Scotia, in considerable quantities, and also within the limits of the same coal-field at the Joggins, on the Bay of Fundy. On the Nipisiguit it has even been attempted to work the deposit as a copper mine; but, on account of the irregular distribution of the organic remains, the operations became uncertain and led to the abandonment of the work. This bed is from two to four feet in thickness.†

In *Pennsylvania*, United States, beds of vegetable stems, impregnated with vitreous copper and green carbonate, occur in the shale or argillaceous beds at the base of the Devonian or old Red Sandstone series. In two or three instances, within our own observation, these were commenced to be worked as copper mines, but the quantity of ore was found insufficient for productive operations. The mineral occurs in the form of rich gray sulphuret of copper. So far as our remarks have extended, it is only the terrestrial and not the marine vegetation of this formation that is cupriferous.

In the *State of New York*, cupriferous lignites occur in about the same geological position, in the Catskill mountain series. They consist, like the preceding, of vegetable casts, replaced by gray sulphuret and carbonates of copper.‡

Professor Del Rio mentions certain beds of this character with which he had become acquainted.

In *Russia*, in the carboniferous beds which are considered by Mr. Murchison to be of the same age as the Zechstein of Germany and the magnesian limestone of England. The flora is peculiar to it; and the fossil stems and leaves of plants are very general indications of copper ore, which, in the form of gray oxide and green carbonate, is disseminated through or arranged around them.

The *Kupfer Schiefer of Germany* represents this metalliferous deposit on a smaller scale.§

In the *Tyrol*, in the upper tertiary coal beds of the valley of the Inn—cupreous vegetable fossils occur.‖

Thuringia is remarkable for a cupriferous schist, with lignites and fossil fishes.

In the *Spanish Pyrenees*, Mr. Logan examined, within the coal measures, a bed which presented a combination of coal and gray sulphuret of copper, in the form of vegetable casts. These occupied an eighteen inch seam, cropping out regularly and extensively. It was then worked as a copper mine, and promised a profitable return.¶

In *Ireland*, in a bog on the east side of Glendore Harbour, the

* Mr. Henwood in Trans. Royal Geol. Soc. of Cornwall, 1840.

† Report of the Geological Survey of Canada, May 1, 1845, p. 63.

‡ Mather's Fourth Report of New York Geology, p. 229.

§ Proceedings Geol. Soc. of London, Vol. III. p. 751.

‖ Ibid. Vol. I.

¶ Logan's Report on the Geological Survey of Canada, May 1, 1845, p. 64.

peat was found to be highly impregnated with copper, which was extracted from the burnt ashes.*

In *Scotland* and *England*, some of the beds of the old red sandstone have a green tinge, and the more argillaceous beds are mottled with red and green. The former hue arises from the oxide of iron, the different tints depending upon the amount of iron in the beds and on its state of oxidation, while the green colour is ascribed to the presence of copper. Whether vegetable casts occurs in these beds as in the United States, we have not learned.

TURBARIES, PEAT-BOGS—TOURBIERES, PEAT-MOSSES.

In the succeeding pages will be found copious details, (taken from the foreign portion of the first edition,) in relation to this useful combustible—the most recent deposit, if we may so employ the term, but nevertheless by no means the least valuable, of the class of fuels which we have to bring under consideration. Respecting the origin of these modern deposits, which bear some resemblance to coal-fields, it is not uninteresting to trace the process of their accumulation or development.

Turbaries, formed in depressions of the soil, where the shallow waters constantly remain, are found dispersed, here and there, on the surface of plateaux more or less elevated, or upon low plains, and often follow the direction of the valleys, whose hollows they fill. These deposits sometimes present several beds of the combustible, separated from each other by argillaceous, sandy, or calcareous matters; now and then filled with the remains of aquatic or terrestrial mollusques which still live in the country.†

They only originate under peculiar circumstances. They are formed neither in running waters, nor in deep lakes, nor in the transient pools of water which occasionally dry up. It is only produced in places where the waters stagnate, or are slowly renewed, and have an inconsiderable depth.

The production of peat is principally due to the accumulation of cellular vegetables, which are constantly submerged and which multiply with rapidity; such as the *sphagna*, *confervæ*, &c. To these are added a great number of terrestrial vegetables, which are brought thither by streams, either in their ordinary condition, or during inundations. Frequently, also, we find large trees, which are buried more or less deeply in the moss, and particularly in the lower parts, where they are accumulated upon the sands and clays which form the base. Sometimes these trees appear to be standing, but most frequently they seem to have been broken off on the spot, and thrown down near their roots, which are seen fixed at the bottom of the turbary. In certain cases they are extremely numerous, and seem to indicate entire forests which have been buried in the same spot where

* Jameson's Mineralogy of the Scottish Isles.

† Beudant, Geologie, p. 98.

they grew, before the formation of the peat bog. All these plants conform to the existing vegetation. They consist of resinous trees, of oaks, birch, sometimes the ash, elms, &c. The first are generally the best preserved; they have, especially, maintained all their solidity, and are only blackened: the others, on the contrary, are to a certain extent, reduced to a rotten earth, which falls into powder on drying. We also frequently find the remains of mammiferæ in these peat-bogs, and these commonly belong to animals of the existing epoch. These are the bones of oxen, the horns of stags and roe-bucks, the tusks of wild boars, &c.

Turbaries or peat-bogs are abundant on the surface of the globe, in the cold and temperate regions, and are distributed in basins, like the coal-fields, more or less expanded, at all elevations, and occupy the various depressions of its surface. They are even on the summits of mountains, as in the Alps; on elevated plateaux, as in the centre of France; or in the lowest plains, where they cover somtimes immense spaces, as in Silesia, Prussia, Hanover, Westphalia, and Holland. Details will be found under each of these local heads in this volume.

If the majority of turbaries are formed on the main land, and entirely by fresh water vegetables, there are others which appear to have been deposited in the marshes which communicate with the sea; as the greater part of those in Holland. Some of these deposits consist of wrack or drifted sea-weed and marine plants, such as we still see upon the flat and sandy shores of the ocean, and particularly upon those of Friezeland and Jutland.

It is remarked by M. Beudant, that the hypothesis which assimilates the coal beds to the turbaries is fortified by the different characters which they present. These are, on one side the numerous debris of cellular cryptogamia, which microscopic examination discovers in such combustibles as turf, the trees standing rooted in the middle of the deposits, and the remarkable preservation of the leaves in the schists; on the other the disposition in basins, more or less extended, and isolated from each other, surrounded by the earlier rocks;—all circumstances which seem to indicate pools of water, and marshy places formed in the depressions of an open country. We frequently also observe that a certain number of small independent deposits form portions of a more extended basin; of a species of lake, filled with arenaceous contemporaneous matters, at the surface of which will be formed so many separate heaps of combustible; they are, as it were, inclosed in a species of ancient valleys, along whose length they are disbursed.

Certain deposits of lignite are evidently formed in the same manner as coal, of which they present the same characters, "*allures;*" but there are others which exhibit masses of wood, thrown pell-mell, more or less bituminised, preserving their tissue, buried by chance, in the middle of the sedimentary deposits; reminding us of those which are drifted by great rivers, which deposit them in the lakes, or which are transported to the middle of the seas.*

In France, where every description of fuel is valuable, the working of the turf pits is carefully attended to, and, in great measure, but not entirely, is under the surveillance of government officers.

The usual process of cutting this turf is as follows. When the peat is above the level of the adjacent waters, as it is a substance always soft and easy to be cut, it is worked by digging small trenches with a succession of steps or grades of elevation, whose height is that of the spade which cuts them, say about one foot. These steps are separated by a breadth of at least three feet, upon which the workmen walk in file, one after the other, taking off from each side a series of prisms of about five inches in thickness. These prisms are immediately collected by the porters, *chargeurs*, who follow the cutters with wheelbarrows.

To raise thus a line of prisms from the whole length of a step or bench, is what is called raising a *point* of turf. The labourers can follow on the same step, *gradin*, in working out the successive *points*.

The extracted turf is carried to the drying floors, in the driest and best ventilated places in the vicinity. At first they deposit these prisms of turf flat on the ground, like bricks, and superposed to a trifling height; then, when they have acquired sufficient consistence, they are piled in walls open to the day, about three feet in height, which form a series of broken lines, in such a manner as to present solidity, and, at the same time, to permit the air to circulate without the wind being able to upset them. It is only after complete desiccation, that they are able to pile the peat in the form of stacks, which are then thatched with stubble, to prevent deterioration; for if it has not been well dried, it will heat, and if, on the contrary, it attain a point of desiccation too advanced, it will be crushed so as to occasion much waste.

If the peat-bog be again covered by water, there will be a renewal of its original condition, but very often the workmen are compelled to work beneath the surface of the water, after having lowered its level by every possible means.

The consistence of the turf being very slight when first withdrawn from the water, they employ, in extracting it, implements called *louchets*, whose forms are designed to increase the adhesion of the cutting surfaces to the matter cut. The common louchet is a spade with a lateral wing or flange, making an angle with the surface. With a single cut, this tool can detach a prism of turf whose angular surface facilitates the raising. Other louchets carry a fork with a spring, which is designed to press the prism of peat against the surface of the blade.

In Bavaria, towards the sources of the Mein, the peat-beds are from six to twelve feet thick. The turf is mossy, and contains

* Beudant, Cours Elémentaire d'Histoire Naturelle, p. 115.

numerous buried and decomposed trees; among whose remains we are still able to recognize many existing species.*

It may be seen from the numerous facts which have been accumulated from the foreign portion of the first edition of this work, how far they sustain a theory which supposes a zone or belt of coal vegetation around the earth.†

A difficulty here presents itself at the outset, by reason of the comprising under one common denomination of coal, deposits of very different ages. It is true that carboniferous formations appear, at intervals, in almost every quarter of the habitable globe, but the more recently produced coals and lignites have no apparent conformity with the arrangement of the true coal beds.

The greater part of the basins of true coal is decidedly limited to the space between the Tropic of Cancer and the Arctic circle. But the coals of later epochs,—those from the oolites up to the tertiary periods, obey no such law of arrangement. They are found in both hemispheres, extending almost from pole to pole, and crossing the range of the old coal formations almost at right angles.

Thus we have detached coal deposits of later origin than those of true coal and we have occasional accumulations of tertiary lignite or brown coal southward as low as S. lat. 50°, and as high northward as N. lat. 70°, embracing the extreme accessible points upon our globe.

There is an immense range, although with many interruptions, extending in a north-west direction, over nearly half the circumference of the globe, from New Zealand, Australia, Borneo, Siam, Ava, and Burmah, and across Hindostan, and by the Caspian and Black Seas, across Europe, even to the Baltic.

We are by no means certain, in many cases, of the relative ages of what passes under the ordinary denomination of coal, and besides many extensive deposits have received no scientific examination. But we know, for instance, that brown coal exists as far to the southward as Kerguelen's Land, and at each extremity of North and South America and Asia, and of Africa, at the Cape of Good Hope, and Algeria; throughout Europe, and on both shores of Greenland. Lignite, apparently of the same age, stretches, at intervals, through 125 degrees of latitude, and along both the American continents, from the Straits of Magellan to the Arctic Ocean.

We need not repeat here that these newer coals are at once distinguished by their inferior calorific power; while the naturalist recognizes them by their geological associations and by the peculiar animal and vegetable races which characterise the epoch of their formation.

The subject of peat and its various uses being of much importance, the Editor has inserted the following abstracts of the numerous re-

* Burat, Géologie appliquée, p. 380.

† An Essay on Organic Remains, by Thomas Gilpin, Philadelphia, 1843.

ports contained in the former edition under the heads of Ireland, France, Bohemia, Prussia, &c.

IRELAND.

Turf, (English;) *Peat*, (Scotch;) *Torf*, (German;) *Tourbe*, (French;) *Turbary*, (a peat bog,) from the Latin.

So excellent, plentiful and cheap is this description of fuel in Ireland, that our sketch would indeed be incomplete did we omit to mention it. Its abundance and accessibility to the vast mass of the poorer classes, renders it of no small importance among the natural resources of the island. Not only is it the common fuel of the poor in the interior, and indeed of all classes in some districts, but it is also brought in barges by the grand canal, and consumed to a great amount along with, or instead of coal, in the capital itself.*

So extensive is the supply of peat in Ireland, that it has been estimated to occupy one-seventh of its entire surface. One of the mosses of the Shannon is described by Dr. Boase to contain one hundred and fifty square miles.

The supposed deficiency of good coal in Ireland is less felt as regards domestic than manufacturing purposes. Mr. John Classon, 1845, has stated that Ireland has two canals, running from Dublin, through 2,000,000 acres of turf bog. He mentions, among other instances of the value of this combustible, that a distillery company, by the judicious management of a bog, had their steam power for half the cost it would have been for coals; and were, at the same time, making an estate of reclaimed land for themselves.†

The red peat bogs, which form so remarkable a feature of this country, are chiefly comprised in the great central plain of Ireland. Unlike the English mosses, they are rarely level, but undulating; and in Donegal, there is a bog which is completely diversified with hill and dale.

These bogs consist of moist vegetable matter, containing a great deal of stagnant water; and, after heavy rains, fogs, &c., sometimes burst, and inundate or overwhelm the adjoining country.‡

At the meeting of the British Association in 1842, Mr. Griffith illustrated the mode in which he considered the coal measures had been formed, by describing the general condition of the peat bogs of Ireland. They appeared to occupy basins which had originally been lakes, but the peat moss had grown up to the level of the water, and afterwards, by capillarity, rose twenty or thirty feet higher. The bases of these bogs consisted of clay, covered by a layer of peat, which is composed of rushes and flags. Above this is another bed of peat, closely resembling cannel coal, possessing a *conchoidal fracture, and hard enough to be worked into snuff boxes.* It yielded twenty-

* History of Fossil fuel.

† Ibid., January 3d, 1846.

‡ McCulloch's Geographical Gazetteer. Also his Statistics of the British Empire, Vol. I. p. 357.

five per cent. of ashes, and contained a large proportion of oxide of iron.

The bed was covered with black peat, containing branches and twigs of fir or pine, oak, yew, and hazel—only the bark being left; and where whole trees occurred, the roots were entirely gone. The surface was formed of ordinary bog moss, (*sphagnum*,) nearly white. The whole amount of peat in the bog to which Mr. Griffith referred would, he thought, form a coal seam at least three or four feet thick.*

We have seen a recent statement to the effect that the area of peat land in Ireland is now partly diminished; some of the bogs being reclaimed and converted into arable land, and others are exhausted, drained, or dug out.†

County Clare.—The bog of Douragh, eastward from the Fergus, affords the principal supply of turbary to Ennis and Clare. The bogs in this district abound in timber. A fir tree, measuring thirty-one to thirty-eight inches in diameter, by sixty-eight feet in length, was some time since raised from a bog near Kilrush. The mode of finding bog-timber is rather remarkable. It is ascertained that the dew does not lie on the part of the bog immediately above a tree, as it does elsewhere. Its position can thus be easily ascertained before the dews rise in the morning, when the finder, after probing with a bog-auger, to ascertain whether the wood be sound, marks the spot with a spade, and proceeds to raise the timber at his leisure.‡

The series of extensive bogs in the central part of the island, though separated from each other, have received the common name of "the Bog of Allen."

They vary infinitely in wetness, also in depth, composition, &c. They rest, in general, upon a stratum of blue clay, based on limestone, and are invariably above the level of the sea. Their greatest elevation, however, does not exceed four hundred and eighty-eight feet; the mean elevation being two hundred and fifty feet.

The drainage and cultivation of these extensive portions of the surface of Ireland, have long been regarded as objects of great national importance, and frequent attempts have been made to show that they might be effected at no very great expense. But there are few examples in any part of the island, and those under very peculiar circumstances, of successful bog cultivation. The attempts hitherto made to drain the bogs in Ireland, have not been very advantageous; and even had they succeeded, it is doubtful whether they would have produced any considerable return. It is, indeed, by no means clear, supposing them to be quite dried, that they would not, in most instances, be rendered still more worthless than they are at present.§

These bogs are, however, not without their value. They supply the inhabitants extensively with their fuel. In those parts, indeed,

* Meeting of the British Association in 1842.
† Mining Journal, October, 1845.
‡ Penny Cyclopædia, art. Clare.
§ Wakefield's Account of Ireland, p. 105.

where bogs are scarce, they are the most valuable properties in the country. In not a few localities, they have been wholly cut out; and where this is the case, and other bogs are not easily accessible, the inhabitants have sustained great privations from the want of fuel.*

		Acres.
Cultivated land in Ireland, - - - -		14,603,473
Unimproved mountains and bogs, - - -		5,340,736
		19,944,209
Lakes, - - - - - - -		455,399
Out of this aggregate, coal, more or less, is estimated to extend beneath,	coal,	1,881,600†

The parliamentary commission to inquire into the nature and extent of the several bogs in Ireland, and the practicability of draining and cultivating them, reported in 1814, that "the extent of peat soil in Ireland exceeds 2,831,000 English acres, of which at least there are of flat red bog,

	1,576,000	acres, the most valuable.‡
Of mountain bogs, on the surface of the uplands,	1,255,000	do.
Total, - - - -	2,831,000	do.
Total area of Ireland, -	20,399,608	do.

Mr. Bicheno remarks, that "the rainy climate of Ireland, and the wet occupations of the people, with the nature of their food, make a fire more important to them than to most others; and, in fact, is frequently the substitute for clothing, bedding, and, in part, shelter. Had it not been for the bog, the measures taken in former times to extirpate the nation, might probably have succeeded: but the bog gave them a degree of comfort upon easy terms, and enabled them to live under severe privations of another kind."§

Mr. Griffith,|| from his own observation during twenty years, states an example of peat bog having grown at the rate of two inches every year;—an instance, probably, of excessive growth, under peculiarly favourable circumstances, yet valuable in its direct testimony to the fact that bog, fitly circumstanced, still continues to grow with undiminished vigour.¶

* McCulloch, Gazetteer.
† McCulloch's Statistics of the British Empire, Vol. I. p. 329.
‡ Fourth Report, 28th April, 1814.
§ Ireland and its Economy, p. 28.
|| Mr. Griffith, in the "Bog Reports."
¶ Ordinance Survey of the County of Londonderry, 1837, p. 7.

We have seen an estimate in circulation, by which it is shown how important to Ireland are her peat bogs, in furnishing a valuable fuel, independent of her deposits of anthracite and bituminous coal. According to this calculation, the space occupied by bog, in Ireland, is 2,830,000 acres.

If, however, the quantity capable of being made into turf be taken as low as 2,000,000 acres, and at the average depth of three yards, the mass of fuel which they contain, estimated at 550 pounds per cubic yard, when dry, amounts to the enormous sum of 6,338,666,666 tons.

Taking, therefore, the value of turf as compared with that of coal, namely, as 9 is to 54, the total amount of turf fuel in Ireland, is equivalent, in power, to above 470,000,000 tons of coal; which, at twelve shillings per ton, is worth about £280,000,000 sterling, or $1,335,000,000, U. S.

Species of Lignites found in the Irish Peat bogs.—With respect to the trees which are so frequently found in the Irish bogs, Mr. Aher remarks: "Such trees have generally six or seven feet of compact peat under their roots, which are found standing as they grew; evidently proving the formation of peat to have been previous to the growth of the trees."* In the bogs in the vicinity of Londonderry, according to the report of the Ordinance Survey, in 1837, the fact above stated may be verified in relation to *fir trees*, the lowest layer of which is underlaid by from three to five feet thick of turf. Not so, however, with *oaks*, as their stumps are commonly found resting on the gravel at the base or on the sides of the small hillocks of gravel and sand, which so often stud the surfaces of bogs, and have been aptly called "islands" by Mr. Aher, and "hummocks" by the Americans of the south. It is a remarkable fact, although very common, that successive layers of trees or stumps, in the erect position, and furnished with all their roots, are found at distinctly different levels, and at a small vertical distance from each other.†

We have seen that the bogs contain two important families of trees,—the resinous or coniferous trees, which grew in successive layers or tiers upon the ancient surfaces of peat; and the hard-wooded, non-resinous trees, which grew upon the gravel at the original base. Of the former, the prevailing tree was the common Scotch pine or fir, *pinus sylvestris*;—of the latter, the oak, *quercus robur*, prevailed.

We may be permitted to note here, that in a "Notice of a Submarine Forest in Cardigan bay, North Wales," the author remarks on the occurrence therein of the pinus sylvestris, although the Scotch fir is now excluded from the native flora.‡

MISCELLANEOUS NOTES ON PEAT FORMATIONS.

Mr. Jameson has a remark that we must not overlook; that peat

* Mr. Aher, in the "Bog Reports."

† Ordinance Survey and Report of the County of Londonderry.

‡ Rev. James Yates:—Proceedings Geol. Soc. London, Vol. I. p. 407.

is peculiar to cold or cool climates; and thus nature has provided the constant means of supplying, through this source, the necessities of the people who occupy those climates, and who continually require fuel.

In Scotland, it is observed that the peat at the bottom of a mountain is more decomposed than that which occurs at its top; and that the lignites found in turf bogs or mosses, are more sound upon the summit of a mountain than at its base. It is also observed that the peat of the south of England is more decomposed than that of the north of Scotland: and the peat of France has more of the coaly appearance than that of England.

As we advance towards the warmer climates, vegetable matter is more rapidly decomposed, until, at the tropical regions, the putrefaction of animal and vegetable matters is so rapid, that it prevents the formation of any body of the substance and structure of peat.*

Cupreous Peat and Lignites.—We add one curious fact in connection with peat. In 1812, there existed a bog on the east side of Glendore harbour, in Ireland, which was so much impregnated with *copper*, that forty or fifty tons of dried peat, when burned, yielded one ton of ashes, containing from ten to fifteen per cent. of copper.

A parallel case recently occurred in North Wales, where a solution of copper, which was let loose by accident upon an adjacent peat bog, affected and impregnated the vegetable fibre in preference to the accompanying soil. Mr. Murchison conceives, with regard to the dissemination of copper through the vegetable matter, or its arrangement around the thicker branches of the fossil plants in the thin coal beds of the Zechstein of Permin, Russia, that they attracted around them the cupriferous matter contained in the transporting currents.

Silicified or calydonized Peat, is noticed as occurring in Iceland, with hyalite or opalized mosses, by M. Eugené Robert.†

Uses and adaptation of Peat to various economical purposes.

Charred, coked, or carbonized Peat.—This substance can be charred, and rendered fit to be used like charcoal, in cookery and other domestic purposes, in the same way as wood or coal is charred, and in much less time.

For ordinary purposes, it is charred by some families on the kitchen fire, thus:—take a dozen or fifteen peats, and put them upon the top of the fire, upon edge. They will soon draw up the coal fire, and become red in a short time; after being turned about, once or twice, and they have ceased to smoke, they are sufficiently charred, and may be removed to the stoves. By following this plan you keep up the kitchen fire, and have, at the same time, with very little trouble, a supply of the best charred peat, perfectly free from smoke; and the vapour is by no means so noxious as charcoal made from

* Jameson's Mineralogy of the Scottish Isles, p. 152.

† Bulletin de la Societé Géologique de France, tome XI. p. 350.

wood. Peats, charred in this way may be used in a chafer, in any room, without danger arising from the vapour.*

For the production of *gunpowder*, many varieties of peat are superior to the charcoal of dogwood and alder.

For *gas*, its properties have been tested in Dublin, Paris, and Plymouth; yielding about the same as the Newcastle coal, but its light is superior in brilliancy and power.

For *pavements*, when combined with an artificial asphaltum, composed of carbonate of lime and coal tar, it forms a solid and elastic road, superior in many respects to native asphaltum. The tendency of this artificial asphalte to crack and break, is counteracted by the strong fibre of the turf; which, if added to the chalk and tar, while warm, acts as a binder when the mass is cooled, and obviates its brittleness.†

Analysis of Irish peat, of an inferior quality, from the Bog of Allen; made with a view to ascertain its calorific power, by Mr. C. Cowper, of London. The experiment was pursued by the litharge test, recommended by Berthier. This consists in mixing a given weight of the fuel with a sufficient quantity of litharge, and heating it in a crucible; the heating power is in proportion to the quantity of lead reduced.

By Mr. Cowper's experiments, the following comparative results were obtained, being averages of six or eight experiments each.

10 grs. of good Newcastle coal gave - -	284 grs.
10 grs. of oven coke, - - - -	302 grs.
10 grs. of common peat, - - -	144 grs.
10 grs. of same, coked in a crucible, - -	259 grs.

"The foregoing analysis is founded upon a well known fact, that the quantity of heat, generated during the combustion of any fuel, is in exact relation to the quantity of oxygen consumed in the process. Hence, in order to ascertain the relative calorific power of different kinds of fuel, it is only necessary to ascertain the quantity of oxygen which each consumes in burning."‡

These experiments show that seven tons of peat coke are equal to six tons of good coal coke.

Professor Everitt's experiments, similarly conducted, show that

10 grs. of peat coke, picked surface, gave -	277 grs.
10 grs. of peat coke, lower strata, gave -	250 grs.
10 grs. of pressed peat, gave - - -	137 grs.

Application of Peat to Iron Making.

It has been asked, can peat be advantageously used in the manufacture of iron? Generally speaking, the answer has been in the

* Loudon's Encyclopedia of Agriculture, 1831, p. 747.
† Farmer's Magazine, Vol. XVII.
‡ Byrne on Compressed Peat, Boston, 1841, p. 11.

negative. Yet experience proves that in such a matter we ought not to pronounce an absolute opinion.

The history of the making of iron with bituminous coal or coke, is the most striking instance of this truth. The English forge-masters maintained with all the energy of conviction, that pit-coal could never be used in the fabrication of iron; and they treated with ridicule all who made such attempts. We have witnessed the triumphal results, and its universal and successful application. So also with the employment of anthracite or stone coal in the process of iron making. It had baffled, for a long series of years, every attempt to employ it; and but a short time ago it was pronounced so surrounded with difficulty as to be impracticable. We now see that it is managed with equal or even more facility than with the bituminous coal.

We shall show, in the progress of these pages, that it is not only practicable to employ peat as the fuel in fabricating iron, but that at the present moment, it is absolutely in full operation on an extensive scale; not only in high furnaces, but in puddling and refining; in cubilot and in reverberatory furnaces; in forges; in fact, in nearly all the processes of iron manufacture.

The countries where this combustible is so employed, on a large scale, are France, Bohemia, Bavaria, Westphalia, Wurtemberg, and in several adjacent provinces; thus settling this question in the only way it ought to be answered; practically and successfully.

Table of Analysis of Peat, both in the Raw and the Carbonized State.

By whom analyzed.	Carbon, Per cent.	Vaporizable matter, Per cent.	Cinder, Per cent.	Locality and Description.
David Mushet, Esq.,	25.20	72.60	2.20	Scotch peat, raw state.
Dr. Kane,	61.04	37.13	1.83	Bog of Allen, Ireland.
M. Marcher,	65.00	22.00	13.00	Carbonized peat,
	37.00	48.00	15.00	Raw state.
M. Debette,	67.00	30.00	3.00	Bohemia, carbonized.
M. Berthier,	38.00	28.00	17.40	carbonized?
	24.40	70.60	5.00	Wurtemberg, raw state.
Dr. C. T. Jackson,	21.00	72.00	7.00	Maine, U. S., raw state.
M. Sawge,	22.00	69.70	8.30	Ardennes, France, raw.
M. Diday,	9.00	58.00	33.00	Basse-Alpes, France, raw.

In a pamphlet on this subject, republished by Mr. Alex. S. Byrne, he remarks that charcoal iron is the best known at present in the markets; and that such is its value and superiority, that large quantities are annually imported into England from India and China, and sold at the enormous price of £36 [$173] per ton. Mr. Byrne contends that *Peat Coke is of still greater value than the best charcoal*, and that in the manufacture of iron it stands unrivalled as a fuel.

When properly compressed, two tons of peat coke occupy no more space than one of charcoal: consequently, where intensity of heat is an object, a much higher temperature can be obtained from peat coke than from the hardest and closest charcoal.*

Professor Everitt's investigations of the common Lancashire peat show that in regard to comparative specific gravity—

Compressed peat possesses - - -	1.160
Less pressed, - - - - -	.910
Peat coke, hard pressed, - - - -	1.040
Less pressed, - - - - -	.913
Charcoal from hard woods - -	.400 to .625

Hence it appears that the coke prepared from compressed peat is nearly double the density of ordinary charcoal. In common practice it is estimated that 100 lbs. of charcoal occupy the same space as 200 lbs. of coke. Peat coke would occupy, weight for weight, the same space as common coke.

Professor Everitt adds, that "where bulk of stowage and high intensity of heat are important considerations, the peat coke is superior to wood charcoal." Moreover, the density of peat coke, by means of a stamper press and the use of heat, can be carried up to 1.359, which increases the comparison in its favour.

The admixture of peat, even in its natural state, with common coke, in smelting iron, materially improves its quality; in some instances changing the pig metal from the state of "white iron" to that of "gray iron," technically called "foundery."

Good peat is shown to be preferable to any other fuel, not only for the process just mentioned, but in welding, and for softening steel plates, &c.

For the finer iron works, turf and turf-charcoal are known to be better than wood charcoal. Dr. Kane shows that the precious Baltic iron, for which from £15 to £35 per ton is given, could be equalled by Irish iron, smelted by Irish turf, for £6, 6*s*. per ton.†

From another source we learn that iron, manufactured with peat fuel, is more malleable than Swedish; and that tools made from it are of a superior quality. It has been doubted whether peat can be used in the puddling furnace, but with a diminished produce; yet the working of iron by this fuel is known to improve its quality, and the welds, especially, are superior to those made with coal.‡

It has been proved that, after peat has been well carbonized, it may be employed in puddling and reverberatory furnaces, and forges. As to its use in blast furnaces, peat, which is the lightest of all coals, would consequently seem to be the least fitted for the reduction of ores.§ But even this difficulty has, in great measure, been surmounted in the high furnaces of Germany.

M. V. Lamy has made a series of experiments to determine the

* Observations on the uses and advantages of compressed peat, by Alex. S. Byrne, 1841, abridged from the American Repertory.

† The Industrial Resources of Ireland, by Robert Kane, M. D., 1844.

‡ Mining Journal, Dec. 6th, 1845.

§ On the applicability of peat to manufacturing iron—Mining Review, 1840, p. 46.

quantity of heat given out by peat, in burning, compared with other combustibles: the results are as follows:

One kilogramme, or 2¼ lbs., of the varieties of fuel mentioned below, evolved of caloric the following parts:

Wood charcoal,	75 parts.	Bituminous coal,	60 parts.
Coal-coke,	66 "	Charred wood,	39 "
Charred peat,	63 "	Dry wood,	36 "
Raw peat,	25 to 30 "	Wood with ¼th moisture,	27 "

Thus, as regards charred peat, or turf charcoal, it appears preferable to coal in the manufacture of iron, and is almost equal to wood charcoal.

Compressed peat, dried in a furnace, could be used with decided advantage in a puddling furnace.* In fact, it is already in extensive use, and the results have been very carefully investigated by men of science, as well as of profound practical attainments, more especially in the iron districts of the Austrian Empire.

Prepared Turf or Peat for steam purposes, and in various processes of working iron and steel.—A patent has been obtained by Mr. Williams, managing director of the Dublin Steam Navigation Company, for a method of converting the lightest and purest beds of peat moss, or bog, into the four following products: each of which possesses very valuable properties.

1. A brown combustible,—solid—denser than oak.
2. A charcoal, twice as compact as that of hard wood.
3. A factitious coal.
4. A factitious coke.

Mr. D'Ernst, artificer of fire-works to Vauxhall, has proved, by the severe test of coloured fires, that the turf charcoal of Mr. Williams is twenty per cent. more combustible than that of oak. Mr. Oldham, engineer of the Bank of England, has applied it in softening his steel plates and dies, with remarkable success.

But one of the most important results is, that with ten cwts. of the factitious coal, the same steam power is now obtained, in navigating the company's ships, as with seventeen and a half cwts. of pit coal, alone; thereby saving thirty per cent. in the stowage of fuel. What a prospect is thus opened of turning to admirable account, the now unprofitable bogs of Ireland; and of producing, from their inexhaustible and reproductive stores, a superior fuel, for every purpose of arts and engineering!†

From the experiments of Mr. Le Sage, charred ordinary turf seems to be capable of producing a far more intense heat than common charcoal. It has been found preferable to all other fuel for case-hardening iron; tempering steel; forging horse-shoes; and welding gun barrels.‡ Since turf is partially carbonized in its native state, when it is condensed by the hydraulic press, and fully charred, it

* Mining Review, quoting l'Ancre, p. 53.
† Ure's Dictionary of Arts and Manufactures.
‡ Repertory of Arts, Vol. V.

must evidently afford a charcoal very superior in calorific power, to the porous substance generated from wood by fire.

It was announced, a few years ago, as an important fact, that the steamers plying between Limerick, Clare, and Kilrush, in Ireland, were using peat for their fuel. The Garry-Owen, steamer, has made the passage between Kilrush and Limerick, fired with turf, [although in the midst of a coal region,] in three hours and twenty minutes. We have been recently told that the Shannon steamers mostly use it, and that it is growing into use in mills and factories.*

Turf forms an important article of transportation, to Dublin, &c., on the Grand Canal. In 1831, 48,000 tons were conveyed on this canal.

The city of Londonderry receives its supplies of turf from the countyof Donegal.†

We continue our notices, derived from various sources, of the importance of this heretofore neglected species of combustible.

The Editor of the Mining Journal of London, in an able article, of December 20th, 1845, warmly advocates this subject, with reference to Ireland. He remarks that among the numerous resources which nature has placed in such profusion throughout Ireland, there is, perhaps, none to such a prolific extent, or in comparison, so little valued or turned to the uses for which it is so eminently calculated, as peat or turf.

From Dr. Kane we ascertain that the light turf, which is so much burned, weighs 500 lbs. per cubic yard; while the most dense varies from 900 to 1100 lbs., being about half the average weight of coal. Thus, furnaces, to burn a similar weight of coal and turf, would require double the volume of the latter.

We have alluded, in a previous page, to Mr. Williams's prepared turf. His method of preparing it, is as follows. When freshly cut, the fibre of the peat is broken up, and the mass is placed between cloths and pressed by a powerful hydraulic press; which condenses it to one-third of its original volume, and to three-fifths of its weight, through the loss of moisture.

This condensed peat, when carbonized, gives a fine coherent coke, containing little ashes, and amounts to thirty per cent. of the weight of the turf. Its density is greater than wood-charcoal, and it can be manufactured for 20*s*. per ton.

That this combustible can be successfully employed in iron works, in the puddling and second fusion, in the re-heating and rolling of the metal,—in fact, in all the operations which are effected with coal in England, is practically demonstrated in the furnaces of Ransko, in Bohemia, and of Köenigsbronn, in Wurtemberg, the details of which appear in the succeeding extracts.

A work has been published, 1845-6, by Mr. Mallet, of Dublin, "on the artificial preparation of turf," showing the immense advantages

* The Industrial Resources of Ireland, by Robert Kane, M. D., 1844.

† Ordinance Survey of Londonderry, p. 200, Vol. I.

to Ireland are, or rather might be, its peat-bogs. We have no room here for details of his experiments. He considers the best method of taking turf, instead of cutting it into sods, is, to work it up like mortar, and thus to break the fibre, and mould it into bricks; which are afterwards kiln-dried.

M. Goldenberg has reported on the successful employment of gas, obtained from peat, for the refining of iron and the puddling of steel, in various parts of Germany.

He considers that the successful result of this method will be of the greatest importance to the whole of northern Germany, where extensive beds of turf and lignite prevail.

In their solid state, these combustibles have, heretofore, been of little use in the manufacture of metal; but being reduced into gas, they become a great resource to those districts.

The same process would not be less beneficial to France, which possesses some very rich peat bogs, and scarcely turned to use.

It is also adopted in Sweden, where coal is scarce.

Iron manufactured through the agency of Peat.

Ireland has heretofore paid one million sterling annually to England for iron. It is now contemplated that she will not only keep at home this amount of capital, but will even receive a much larger sum from England for a description of iron which the latter cannot do without, and which, for want of peat or charcoal, she cannot manufacture.

It is expected that Ireland will henceforward have it in her power to supply this iron on better terms than it can be imported from the Baltic, and, consequently, that the large sums now paid for foreign iron, will be spent in the former country, to the great benefit of its population.*

We can find but little further space to cite with reference to the employment of the peat of Ireland the facts and opinions with which intelligent persons have recently furnished us. Among others is an able pamphlet by J. W. Rogers, pointing out a mode for the permanent employment of the overplus labouring population of that country. He suggests the employing them in preparing different kinds of fuel from the immense bog districts. The writer "has been in the habit of having peat charcoal prepared for smith's use, infinitely in preference to any coal," and states that, "if within the reach of the manufactories of iron, at the price at which it can be produced, no other fuel would be used." We add the following extract from this work:—

"Charcoal of peat has been found, by analysis, to possess almost identical qualities with wood charcoal. Prepared, as it hitherto has been, however, it is more friable, and therefore more fitted for many purposes—such as the working of iron, manufacture of gunpowder,

* Editorial remarks, Mining Journal, July 11th, 1846.

&c., &c., and also as a fertilizer—the great value of which is not known in this country; but peat charcoal is quite capable of being prepared so as to obtain a density, little if at all inferior to wood charcoal."

The calorific value of peat coke may be commercially averaged as equal to coal coke. The evaporative powers of the two are nearly equal; but peat coke has the advantage of freedom from sulphur. Its superiority is decided when used for the following purposes:—

For the working of malleable iron.

For melting unmalleable or cast-iron.

For all descriptions of brass and copper work; and

For the smelting and general manufacture of iron from the ore.*

Mr. Rogers has followed up this interesting subject by a pamphlet entitled, "Appeal for the Irish Peasantry," in relation to the employment of the Irish peat. He remarks, that "when compressed coke is carbonized, it gives a fine coherent coke, which contains very little ash, and amounts to about thirty per cent. of its weight. The density of this coke is greater than that of charcoal—being found to range from 0.913 to 1.040." The objection, as regards ships of war in action, that the splintering is so great, that this material cannot be safely employed, is met by the assertion that the evil entirely arises from one cause—that of iron being now made solely by sulphureous fuel. Iron made with the peat charcoal will not splinter.

It is remarked in the Mining Journal, that the smiths in the country surrounding Dartmoor travel twenty or thirty miles to get the peat for charring, and that horse-shoes made with it are known to produce almost double price.†

FRANCE.—Official accounts of the amount of turf raised in France and the number of workmen to whom the process gives employment are annually published; but owing to the difficulty in procuring accurate statements, and to the irregular time and manner of working the turbaries in the communes, some uncertainty always prevails as to the details. In some districts the turf is solely applied to the domestic purposes of the inhabitants. In others on the contrary, the exploration of the pits give rise to considerable works; and this fuel furnishes a supply to various important industrial establishments, such as sugar-houses, distilleries, dye-houses, steam-engines and boilers, and kilns for lime and plaster; and it is even employed, although to a limited extent, in certain iron works. The quantity of peat annually raised, is greatly on the increase in France, Austria, Bohemia, Bavaria, Styria, Wurtemberg, and other parts of Germany, in consequence of recent improvements in its application to the smelting and fabrication of iron.

Peat of Les Landes; employed for manufacturing Iron. Iron works of Ichoux in Les Landes.—According to M. Lefebvre,‡

* Mining Journal, October 31st, 1846.

† Ibid., January 30th.

‡ Annales des Mines, 1839, tome xvi.

the proportions which result from the operations at the refining and puddling furnaces, and forge operations at these works, chiefly through the use of peat, are as follows: 114 kilog. pig iron produce 100 kilog. of bar iron, with 93 kilog. peat, and 52 kilog. wood.

116 kilog. pig iron produce 100 kilog. of bar iron, with 93 kilog. peat, 37 kilog. wood, and 9 kilog. coal.

The peat of Ichoux contains two and a half times more ashes than the peat of Köenigsbronn, in Wurtemberg, there also employed in iron making. In 1842 the establishment in Ichoux was the only one in France in which peat was employed for converting cast iron. There are, however, a great number of forges in the vicinity of the turbaries, which are able to procure this combustible at a small price; and with the example of Köenigsbronn before us, which has been regularly and satisfactorily conducted, during many years, by M. Veberling, we can no longer permit ourselves to doubt the advantages, which are presented, by the employment of peat, in the fabrication of iron.

Iron making by means of Gas obtained from Peat.—This is now practised in France, Germany, and Sweden.

We find it stated in the London Mining Journal, April 1st, 1854, that a French chemist, M. Lallamand, has lately invented a method of making Paper from peat. For particulars of the process we refer our readers to the Journal.

PRUSSIA.—Peat is in very extensive use in Prussia, in Bavaria, and Wurtemberg. At Berlin, and its environs, it is employed in almost all the workshops; and on account of its application to the production of Gas, its consumption is regularly augmenting. The price and qualities of turf, differ greatly in one locality from another. In the north of Germany, the value of the stere, or cubic metre of peat, varies between 1 fr. 30 cents, and 3 frs.

Gas employed in refining Iron.—The gas of the high furnaces in Germany, has been satisfactorily introduced, employing for this purpose, combustibles of an inferior quality; such as peat, lignite, and even wood. At Magdesburg, in the Hartz, not only iron is refined, but steel is fabricated, and possesses all the characters of a good quality. It is expected here, to effect an economy of 50 per cent., in the combustible, inasmuch, as fifty francs worth of wood, converted into gas, will give a result, which they have never yet been able to obtain, with less than a mean quantity of charcoal of double the cost.

The good result of this method, when perfected, will be of the highest importance to northern Germany, which possesses immense deposits of turf, and lignite. In their solid state, these combustibles have been of little service in the fabrication of metals; but reduced to gas, they will become a great resource in those countries.

The same process, will be no less useful to France, which possesses very rich turbaries, of which little use has yet been made. Wood for the fabrication of gas, not requring to be carbonized, will equally become a source of more economical and advantageous employment. The beneficial results, obtained by the use of gas in refining iron, as much from the economy of the fuel, as from the smallness of the loss, and the amelioration of the quality—render it, speedily desirable, that the forge-masters should apply themselves with ardour to the study of this process, and introduce it in the iron works.*

BOHEMIA.—*Iron works of Ransko, in Bohemia, with Peat for fuel.*†

High Furnaces.—These works are situated at the south-west extremity of Bohemia, and belong to the Prince of Diétrichstein. They consist of two high furnaces and two cubilots, which are worked with a mixture of turf and charcoal. There are also several refining fires; the establishment comprising four hundred workmen.

The turf is brought from the turbaries situated some leagues from Ransko. It is there dug in bricks or oblong pieces; of which the three dimensions are 35—16, and 13 centimetres [$=13\frac{3}{4}\times 6\frac{1}{4}\times 5$ inches Eng.] These bricks are exposed in piles to the air, during the fine season, where, in drying, they contract nearly to one third; so that when they are carried to the iron works their three dimensions are there found to be about $7\times 3.5\times 2.4$ inches. A cube metre [$= 35\frac{1}{4}$ Eng. cube feet,] contains 590 of these bricks.

In general, these peat bricks are not employed until one year after having been dug; and it is considered good to wait even a longer time. They are stored under the sheds attached to the high furnace, and are, of course, sheltered from the rain. The fuel receives no further attention or preparation. It was at first proposed to use it in the carbonized state; but as regards this particular quality of peat, the carbon obtained was not found much more advantageous, practically, than the peat itself, and it became too expensive.

They next essayed to dry it in kilns, by means of the waste heat or flame of the high furnace. In time this was also abandoned; because it required immense apparatus to dry all the turf required for consumption at the works; and because this operation is always dangerous, the peat catching fire with great facility; and, finally, because the advantage acquired on one side, would scarcely compensate the expense of manipulation on the other.

In France and in Wurtemberg, they have essayed, several times to compress the peat, to discharge the water, and to condense the combustible matter into the same volume; but experience has shown that this operation is costly and difficult to execute, on account of the elasticity of the peat. Besides much of the combustible substance escapes with the compressed water. On this account they employed, at Ransko, non-compressed turf, simply dried in the air.

Two varieties of peat are used here, the distinctions of which are pointed out by M. Delesse. One of these weighs 400, the other 587 lbs. English, the cube metre of $35\frac{1}{4}$ cube feet English. They cost at

* Association Allemande, faits commerciaux, No. 241, 245.

† Report of Mr. Delesse, in Annales des Mines, abstract.

the iron works 1*fr.* 34 [= 13*d.* Engl. = $0.26 Amer.] per stere = 35¼ cube feet Eng. The weight and cost per stere of the different species of charcoal, employed in the high furnaces with the peat, are as follows:

	Fr. Kil.	Eng. lbs.	Cost at the works.			
Charcoal, resinous wood,	125 =	275 =	4*fr.* 14*c.* =	3*s.*	4*d.* =	$0.80
Charcoal, hard wood, heavy,	213 =	468 =	5 49 =	4	5 =	1.06
Charcoal, employed,	143 =	314 =	4 40 =	3	6 =	0.84

The price of a volume of charcoal is thus more than triple that of an equal volume of peat; it will, therefore, be advantageous to exchange, as soon as possible, the charcoal for the peat.

The ore smelted here is clay-iron stone, of moderate quality, and the fuel is, generally, turf and charcoal mixed. In the making of a ton of iron are employed,—Turf, 34 cwt. 3 qrs., costing 8*s.* 9*d.*; Charcoal, 30 cwt. costing 24*s.* 7*d.*; together, £1 13*s.* 4*d.* Producing iron of the very highest character.*

At Schlakenwerth in Bohemia, near Carlsbad, is a high furnace, which works with a mixture of charcoal of wood and peat charcoal.

The peat is raised upon the plateaux of the Erzebirge, at more than 1000 metres elevation, and its exploitation is only practicable during two months in the year. They carbonize it in the same manner as wood, in circular piles, and obtain a very dense charcoal, which, on an average does not contain more than five per cent. of ashes.

The stere of peat charcoal weighs 300 kilogrammes = 660 lbs. English; and the stere of wood charcoal weighs only 141 kilogrammes = 310; used in equal quantities in the high furnaces. The analysis of the carbonized turf is as follows, on the authority of M. Debette: fixed carbon, 67; volatile matters, 30; ashes, 3; total, 100.

Cubilot Furnaces.—A mixture of equal parts of peat charcoal and wood charcoal is employed in the cubilot furnaces of Bohemia, with heated air. Consequently, in the cubilot, one volume of peat produces absolutely the same effect as one volume of charcoal. See also an article "on the applicability of Peat to manufacturing Iron," in *L'Ancre* and on the same subject in Mining Review, 1849.

We have taken much pains, in the foregoing valuable practicable statements, to reduce the German and French weights and measures to those of England; and also to exhibit the prices both in French, English, and American currency. The results are thus made intelligible to our readers.

KINGDOM OF WURTEMBERG.—*Peat employed in Reverberatory and other Furnaces in Wurtemberg.*—At Köenigsbronn, they execute with peat alone, the refining, and the second fusion of the pig metal;

* Mining Journal of London, December 29th, 1845.

its puddling, the reheating of the lumps, and rolling the bars and plates; in fine all the operations which are made with coal in the English forges. The works are under the care of Mr. Veberling.

The peat is of three kinds, as follows:

1st. *Peat of Dattenhausen.*—Fibrous or consisting of interlaced filaments, its colour varying from dark yellow to brown.

Peat in Iron making.—Comparative weight and volume of a brick of each kind.

After drying in the air.	Weight in grammes.	After desiccation in the kiln.		
Value in cubic centimetres.		Volume in cubic centimetres.	Weight in grammes.	
Yellow kind . . . 1304	258	994	231	Ashes 3½ to 4 per cent.
Brown kind . . . 799	218	611	196	

2d. *Peat of Günzburg.*—Compact; having an earthy aspect; colour deep brown, often passing to black; ashes, 6 or 7 per cent.

3d. *Peat of Wilhelmsfield.*—Dark brown; resembling straw to a certain extent. Weight of ashes 5½ to 6 per cent.

	Volume in cubic Centimetres.	Weight in Grammes.
Before desiccation in the kiln, - -	813	265
After, . - - - - - - -	703	231

This species is first dried in the air, at the place where it is dug. The bricks are placed upon a floor, and are turned from time to time. At the end of eight or ten days they are collected in little piles between which the air circulates freely; and three weeks after, if the weather has not been too rainy, they can be transported to the iron works to be further dried in kilns; the description and details of which we cannot follow here, and which bricks are either heated by means of the waste heat of the furnaces or by ovens constructed for the express purpose; or by the union of both. These turfs after being thus artificially dried, absorb anew the moisture of the atmosphere. It is therefore necessary to store them in places which are as dry as possible. However, the quantity which they will thus absorb is so small, that they remain several months and even a year in the storehouses without losing their applicability to metallurgic uses.

Of these three species of peat that we have enumerated above, the proportionate diminution of their weight and volume when dried is as follows:

	1st.	2d.	3d.
Diminution of volume, - - -	0.24	0.10	0.135
Diminution of weight, - - -	0.10	0.19	0.12

Cost of 1 kilogramme or metrical quintal [= 220 lbs.] delivered at the iron works of Itzelberg, fr. 1, 29 c. = 1*s*. 6*d*. = $0 36; being about $3 50, or from 13*s*. to 15*s*. per ton; the distance from Königsbronn being 2 kilometres (= 1⅓ miles.)

M. Berthier's analysis of the peat of Königsbronn is as follows:

Carbon,	24.40
Volatile matters,	70.60
Ashes,	5.00

It is employed without admixture of other fuel, in the refining, puddling and reverberatory furnaces.*

BAVARIA.—*Employment of Peat in the Iron Works of Weiherhammer.*—This peat is procured from the numerous tourbieres of the Fichtelgebirge, which are worked during the fine season, and the turf is left to dry for six months; then it is stored, but is not employed in the iron works until a year after it has been dug. The peat is of good quality, compact, heavy, yet containing no more than from 3½ to 5 per cent. of ashes.

At the Weiherhammer works are two puddling furnaces, one of which is generally in activity. The puddled iron is converted into bars in the ordinary charcoal forges, or in a chafing (réchauffer) fire, which is fed with peat alone. As the peat which is dried in the air produces with difficulty a temperature high enough to remelt the iron, the combustion is hastened by means of a forced current of air. This air, furnished by the blowing machine of the refining furnace, the remelting of the pig metal is effected with the greatest facility. The result of these operations is as follows:—To produce 100 kilogrammes of bar iron = 220 lbs. English; fuel required, all peat, 2,416 stère = 85.32 cubic feet, English; pig metal employed, 128 kilogrammes = 281 English lbs. These proportions are equivalent to 1 ton and 621 lbs. of pig metal, and 868 cubic feet of peat, to make 1 ton (2,240 lbs.) of bar iron.†

HOLLAND.—Holland possesses no mines of mineral coals. As some reparation for this privation Nature has furnished her with inexhaustible supplies of peat. In a compressed state, peat approaches more closely in economical value to coal than is usually supposed. It has been successfully employed as a substitute for the latter, both in Europe and America, in iron works. For the ordinary domestic purposes of the poor, as we have witnessed in Holland, Scotland, Wales, Ireland and England, the pungent quality of the smoke forms the chief objection to its use. This complaint obviously arises from the imperfect application of the fuel, as formerly prepared.

It has even been found that gas, for lighting, can be produced from

* Sur l'emploi de la tourbe dans la mettallurgie du fer, par M. A. Delesse. Annales des Mines, tome ii. 1842, p. 758.

† Sur l'emploi de la tourbe dans la metallurgie du fer, par M. A. Delesse, Annales des Mines, 1842. 1st Edition, 539.

it. As long ago as 1683, J. J. Beecher published an account of his having not only produced gas in England from common coal, but in Holland, from peat or turf.*

ORGANIC REMAINS IN THE CARBONIFEROUS PERIOD.

Insects.—Professor Agassiz remarks that, "with regard to insects, their existence has been already ascertained in the coal formation, which, in my opinion, is much more intimately connected with the palæozoic than with the secondary formations, by the whole of its organic characters."

Entomostraca, of small size, abound in certain coal formations, and they are found after that period in a multitude of deposits.

Trilobites, which are unquestionably the most ancient type of the class crustacea, appear under the strangest and most varied forms, from their first occurrence in the most ancient palæozoic formations. This type, however, does not go beyond the period of the coal formation, when it is replaced by gigantic Entomostraca, which are in some degree the precursors of the *Macrura.*

Fishes.—"When I commenced the publication of my researches on fossil fishes, I was acquainted with no species more ancient than that of the coal formation, and even with a very small number of these. Now, not only is the list of species and even of genera proper to these formations considerably increased, but the more ancient deposits are daily increasing more and more the number of types to add to our catalogues. The strata of the Devonian system, and those of the Silurian system, have in their turn furnished a contingent, which continually goes on increasing."

We cannot here resist the desire to pursue our quotations from the same Professor's Fossil Fauna of the precursor of the great carboniferous formation, the old red sandstone, which also contains the most ancient deposits of coal that are yet known. "The ichthyological fauna of the old red sandstone appears in such extraordinary and fantastical forms, that the most trifling remains of the beings which lived at that epoch, cannot fail to interest the attention of the naturalist. In no other formation do we find an assemblage of fishes, deviating so strikingly from all that we are acquainted with in our own days. The study of no other fauna requires so many years before we become sufficiently familiarized with its types to venture to classify them, and fix their relations to those of other creations.

Comparisons with the remains of anterior formations would have been impossible; because it is in the old red sandstone that we meet, for the first time, with a complete ichthyological fauna. The Silurian formation, it is true, contains some remains of fishes; but hitherto they have been so rare, and the number of species so limited, that it may be safely affirmed that it is only with the Devonian formation that fishes have really acquired some importance among other fos-

* History of Fossil Fuel, p. 405.

sils; or, at least, that the part they performed in nature becomes appreciable."

"What first strikes one, on studying the ancient deposits is, that fishes are the only representatives of the branch vertebrata which exist in the old red sandstone, or even in the coal formation; in so much that we have a good right to call the epoch when these formations were deposited, *the reign of fishes.*

"The consideration that the fishes of the old red sandstone really represent the embryonic age of the reign of fishes, has even been with me a powerful motive to undertake the examination of these ancient animal remains, as my first *monograph*, forming a continuation of my *researches;* since it was here there existed evident facts to prove the truths of this great law of the development of all living beings."

In concluding the introductory article, from whence these few brief but comprehensive passages have been selected, M. Agassiz remarks, that viewing this assemblage of fossil fishes of the old red sandstone, as a simple group of divers, but contemporary species, and apart from all systematic considerations, we are struck with the great diversity which the species really present. "Who would have expected that we should ever find, in spaces so limited as those which have hitherto been explored, above a hundred species of fossil fishes, in the Devonian system alone; that is to say, in a stage of our formations which was believed a few years ago to be confined to the British Islands, and to which, in consequence, only a local value was assigned; and yet, all other things remaining equal, the ichthyological fauna which this formation contains, is as considerable as that which inhabits the coast of Europe; and even although the species of the old red sandstone do not belong to so great a number of families as the living species, they are not less varied in their forms and general aspect, nor less curious in their external characters and organization, nor less different from each other in size, and the degree of locomotive power with which they were doubtless endowed."*

Foot-marks discovered in the coal-measures of Pennsylvania.—In Vol. II. of the proceedings of the Academy of Natural Sciences of Philadelphia, 30th of December, 1845, is an account of fossil foot prints in the sandstone of the coal measures of Westmoreland county, Pennsylvania, by Dr. A. T. King. Those particularly described are reptilian foot-marks, and occur about three miles from Greensburg, and others at Derry, twenty-seven miles from the same town, which seem chiefly to have been made by *ruminant mammals.*

These sites have subsequently been visited by Mr. Lyell, and form the subject of a preliminary article, in the Quarterly Journal of the Geological Society of London.†

* From Professor Agassiz, "Monographie des poissons fossiles du vieux grès rouge." Article in Edinburgh New Phil. Journal, July, 1846, p. 17.

The number of species of fossil fishes, in the entire series of formations, are now known to M. Agassiz, to be not less than two thousand.

† Journal, Vol. II., p. 418, 1846.

The stone on which the Greensburg impressions occur, is a sandstone which rises up from beneath the well-known and widely extended main or Pittsburg ten feet coal seam, whose outcrop is worked in this neighbourhood. The slabs of sandstone are separated by layers of a fine unctuous clay, such as would be admirably fitted to receive the most delicate and faithful impressions of the feet of animals treading upon it.

Twenty-two of these Cheirotherian impressions were discovered by Dr. King, on the under sides of the sandstone slabs, standing out in relief. They occur in pairs; each pair consisting of a hind and fore foot. There are two rows of these tracks which are parallel, or have been formed the one by the right fore and hind feet, the other by the left; the toes turning one set to the right, and the others to the left; and the distances between the successive footsteps being about the same throughout.

Mr. Lyell concurs with Dr. King as to the authenticity of these foot-marks, and conceives that an important truth has been brought to light, through the exertions of the latter gentleman;—that the land on which forests of Sigillaria and Lepidodendron grew, gave support also to large air-breathing quadrupeds. Few geologists, he observes, will now be prepared to believe that this single species or genus of reptiles, or that one class only of vertebrated animals, had possession of the islands and continents, on which so widely-extended and magnificent a vegetation flourished.

With regard to the other supposed impressions of various animals, they appear to be artificially formed; probably by the Indians who occupied the country, and occur under entirely different circumstances to the reptilian tracks near Greensburg. Dr. King agrees with Mr. Lyell in abandoning as spurious all the imprints except those of the large reptile. These reptilian tracks occur in one locality only; no others have yet been found in the same place, nor under similar circumstances elsewhere.

Respecting the traces of organic forms, other than those of vegetables, in the coal formation, we are precluded from entering into details which do not strictly comport with the plan of this work. The shales and argillaceous ore-beds of the coal measures, in most coal-fields, exhibit numerous remains of *conchifera* and *mollusca*. In several instances traces of *fishes* also occur, as we have previously noticed.

In the newer coal formation of Nova Scotia, Mr. Dawson discovered scales of fishes, and traces of shells. But the most interesting discovery in that quarter, is the foot-marks of unknown animals, impressed upon the sandstones. They appear to be those of birds, such for instance as are left by the common sand piper when running over a firm sandy shore. The foot-marks of another animal were subsequently observed, and in frequent instances these were partially obliterated by *rain-marks*. Many beds are represented as *rippled*, rain-marked, or covered with worm-tracks, all indicative of a littoral origin. The footsteps of another animal, considered to be a reptile by Mr. Owen, were observed by Mr. Logan. This detection of animal tracks on the coal measures, is announced as the first instance we have obtained of the probable existence of air-breathing land animals, at any period anterior to the new red sandstone.

Dr. A. T. King, in 1845, discovered, as we have already remarked, undoubted reptilian impressions of footsteps in the coal measures of Pennsylvania, proving, as subsequently observed by Mr. Lyell, the existence of large air-breathing quadrupeds, on the same soil which produced the forests of Sigillaria and Lepidodendron.

In relation to these interesting indications of the early inhabitants of the earth, we may be allowed to cite an eloquent authority. "It is strange that, in a thin bed of fire clay, occurring between two masses of sandstone, we should thus have convincing, but unexpected, evidence preserved concerning some of the earth's inhabitants, at this early period. The ripple-mark, the worm-track, the scratching of a small crab on the sand, and even the impression of the rain drops, so distinct as to indicate the direction of the wind at the time of the shower,—these, and the foot-prints of the bird and the reptile, are all stereotyped, and offer an evidence which no argument can gainsay,—no prejudice resist,—concerning the natural history of a very ancient period of the earth's history.

But the waves that made that ripple-mark have long ceased to wash those shores; for ages has the surface, then exposed, been concealed under great thicknessess of strata; the worm and the crab have left no solid fragment to speak to their form or structure; the bird has left no bone that has yet been discovered; the fragments of the reptile are small, imperfect, and extremely rare. Still, enough is known to determine the fact, and that fact is the more interesting and valuable from the very circumstances under which it is presented."*

SECTION IV.

MINING CASUALTIES AND PROVIDENT INSTITUTIONS.

On the mining casualties or accidents, and on the provident institutions, relief funds, benefit societies, *caisses de prévoyance*, *caisses de secours*, and similar institutions which have been established for the relief of working miners, in the principal coal-producing countries.

During the preparation of the present work, we had collected numerous statistical facts on a branch of our subject which appeared fraught with unusual interest, namely, that of the casualties to which

* Ansted's Picturesque Sketches of Canada.

the coal miner's occupation is especially subjected, and the means which in late years, have been adopted to afford him aid under the many attendant circumstances of privation, sickness and distress.

We had originally distributed these notes under their local heads, but soon perceived that that arrangement was not likely to prove the most useful or convenient ; and that the whole matter would be more appropriately disposed in a distinct section. The topic had acquired additional interest in proportion to the accession of information, until it appeared to us that, in a philanthropical sense, few were more entitled to our calm consideration. By no government, probably, has its investigation been carried to a more praiseworthy extent than by the Belgian, and with this conviction, no apology seems necessary for adverting to the opinions and experience of some of her most enlightened official writers.

It will be borne in mind that these investigations are especially directed to the case of the operatives engaged in the extraction of mineral fuel, and not in the mining generally of the metals. There appears to be a wide difference in the character of the two classes of employment. Each has its contingent difficulties, each its attendant dangers, but superadded to these are the peculiar, the instantaneous, the uncontrollable risks, in the daily operations of the coal miner. Of all descriptions of subterranean undertakings, it is conceded, that of coal mining is accompanied with the most frequent dangers to the workman; and the most appalling of these dangers arise from causes over which he possesses the smallest control, and which do not attend the extraction of the metalliferous ores. It is this sad experience and the urgent necessity for alleviating its calamitous results, which have called into exercise the aid of the economist; has awakened the sympathy of the philanthropist, has appealed to the aid of the rich and the protection of the powerful, and has united, in common cause, the proprietor, the explorer, and the working labourer.

Influenced by considerations suggested by these and some other obvious circumstances, we have concentrated under one section, and proceed to exhibit in the following preliminary chapter, the data we have collected on the subject of mining casualties and miner's provident institutions, commencing, as we feel bound, with those of Belgium.

BELGIUM.

On the 19th December, 1841, M. Desmaissières, minister of public works, made a report to the king, on the provident or relief funds, "*caisses de secours*," of working miners, established in Belgium.* We proceed to trace the substance of that excellent report, with the addition of some subsequent notes from the papers of M. Auguste Visschers† and others.

* Rapport sur les caisses de prévoyance en faveur des ouvriers mineurs.

† Notice sur l'établissement, en Belgique, de caisses de prévoyance, Bruxelles, February, 1843.

The creation of private "*caisses de secours*," in the vicinity of the Belgian collieries, dates only from the commencement of the present century.

With the enlargement of coal mining undertakings arose more frequent casualties among the workmen and increased demands, on very inadequate resources, to alleviate the consequent distresses. The aid afforded to the sick and the wounded at this period is stated to amount to almost nothing.

It was in consequence of a series of appalling accidents and deplorable loss of life in the mines of Belgium, principally in the department of Ourthe, in 1812, that the attention of the imperial government was attracted to these events, and to the means of ameliorating them. By a decree of the Emperor Napoleon, 26th May, 1812, the first "relief fund" was founded. By another imperial decree, 3d January, 1813, regulations were established concerning a subterranean police. At the entrance of the allied armies the relief fund or chest ceased to exist, and the Netherlands government did not consent to its re-establishment.

The casualties to which we have alluded were chiefly these.—On the 10th January, 1812, sixty-eight miners perished in the coal pit of Horloz; victims of the fire-damp. The 28th of February, following, twenty-two workmen were buried in the waters of the mine of Beaujonc. Hubert Goffin, a common workman, saved, by his courage and presence of mind, seventy labourers, who were buried under ground five days and nights. For this act he received the order of the legion of honour.

When the distressing catastrophes in the mines of Cockerill and of L'Esperance, March, 1828, and August, 1829, occurred, the government of the Netherlands granted six thousand two hundred florins; at the same time public charity and the treasury of the mining companies united to alleviate much of the suffering. Seventy-two workmen had perished by these two accidents; eleven others were wounded; but the warning was not yet sufficiently solemn.

On the 3d August, 1831, thirty-six workmen perished, victims of fire-damp, at the colliery of the *Grande veine du bois d'Epinois.*

Upon the 26th June, 1833, twelve workmen fell by the same cause, at the mine of *Petit Forêt.* The following 8th of August, thirty-eight miners perished by an inundation in the coal pit of *Monceau Fontaine.* The 31st of the same month, an eruption of water caused the death of thirteen more at the coal pits of *Sartes.* On the 16th April, 1834, fire-damp caused the death of nine workmen at the mine of *Poirier.* The 18th April, 1835, fifteen workmen lost their lives in consequence of fire-damp, at the coal pit of *Trien-Kaisin.* Sixth of December following, fifteen miners perished by the same cause at the coal pit of *Kessales;* five others were dreadfully wounded. On the 16th May, 1836, an inundation destroyed twenty-nine workmen, in the coal pit of *Sainte-Victoire.* Fourteenth June, following, twenty-two workmen fell victims to the detonation of carbonated hydrogen gas, at the colliery of *Grand-Buisson.* Sixty workmen perished, choked, or burnt, on the 22d of June, 1838, at the coal pit of *L'Esperance* at *Seraing*. The 8th of April, 1839, the "*grisou*" fire caused the death of fifty-five miners, at *Horloz.*

Of the solitary cases of violent death we have no separate record before us, during this period, and many victims fell, isolated without the public remembering to compassionate and aid their families.

In some of the disasters we have recorded, the royal munificence, the treasury of the state, or private subscriptions, came to the succour of the parents and relatives of the victims. But the great majority remained without any assitance.

We extract the following table from official documents, concerning the accidents that have happened in the mines of the kingdom, from 1821 to 1840, inclusive.

General Cases.

Mining Divisions.	Number of accidents.	Number of Workmen.		
		Killed.	Wounded.	Total.
1st division, Province of Hainault . . .	693	878	440	1318
2d " Provinces of Namur and Luxemburg,	80	62	30	92
2d " Province of Liége, ...	579	770	442	1182
Total casualties in the Kingdom, . .	1,352	1,710	882	2,592

These cases may be subdivided under eight heads, whereby we are enabled to show the nature of the casualties, their frequency, and the mortality attending them.*

Nature of the accidents between 1821 and 1840.	Number of cases.	Number of Workmen.		
		Killed.	Wounded.	Total.
Falling in of the roof, of stones, coal, &c. . .	389	334	114	448
Divers accidents in the pits, ...	262	232	53	285
Ascending or descending by ropes or chains, .	226	261	50	311
Divers causes, ...	146	106	68	174
Fire-damp, ...	130	505	472	977
Ascending and descending the ladders, . .	95	73	30	103
Explosions by powder, ...	75	31	79	110
Inundations, ...	29	168	16	184
	1,352	1,710	882	2,592

Average number of workmen employed, in the Belgian coal mines.	From 1821 to 1830,	25,980
	From 1831 to 1840,	31,500
	Mean of 20 years,	28,740
	Year 1842,	39,277

* Rapport au Roi, Statistique de la Belgique, 1842, p. ci.

The cases of fire-damp, [detonation of carbonated hydrogen gas,] form the most murderous, if not the most frequent of these accidents. Below is a summary of those explosions that occurred during the period from 1821 to 1840.

Fire-Damp.

Mining Divisions.	Number of accidents.	Number of Workmen.		
		Killed.	Wounded.	Total.
1st.—Province of Hainault, ...	70	211	244	455
2d.—Provinces of Namur aud Luxemburg, .	2	1	3	4
3d.—Province of Liége, ...	58	293	225	518
Total in the Kingdom, ...	130	505	472	977
Or 37 per cent. of the whole, . . .		25.16	23.60	48.85

Thus, within twenty years, thirteen hundred and fifty-two serious accidents have taken place, and two thousand five hundred and ninety-two victims have perished, or have been grievously wounded or maimed. This forms on an average one hundred and twenty-nine persons a year, in a population that may be placed at about twenty-eight thousand persons. Nine hundred and seventy-seven individuals have fallen victims to fire damp alone. But the seventeen hundred and ten miners who perished during this time, had wives and children, left in want and misery. In valuing at four, the number of unhappy creatures, dependent for their subsistence on these victims, and who were abandoned without resources, we shall have an amount of six thousand eight hundred and forty suffering beings, whose misfortunes result from the working of the coal mines.

In November, 1841, a dreadful explosion took place in the coal mines of P. Felix, Hainault, at a depth of 1450 feet causing the death of thirty miners.

In May, 1845, another disastrous explosion of fire-damp occurred in the bottom of a coal pit, at *Boussu*, near *Quirrain*, where no less than one hundred and forty out of two hundred miners, who were at work at the time, lost their lives.

An explosion of fire-damp took place in a colliery near Mons, March 22, 1847, at a time when fifty men were below. Of these, twenty-six were killed, and the remainder were all, more or less, wounded seriously.

The recently published report of the Belgian mines, from 1840 to 1844, enables us to complete, so far, our table of the number of workmen who were killed or wounded by *explosion of fire-damp*.

	Number of cases.	Number of Workmen injured.		
		Killed.	Wounded.	Total of victims.
From 1821 to 1840, ...	130	505	572	977
1840 to 1844, ...	64	122	180	302
In twenty-four years, ...	194	627	652	1279

Mining Accidents—from the "Compte Rendu de 1839—1844."

First division of Mines—Province of Hainault.—During the period from 1840 to 1844, inclusive, the working the mines of this province has occasioned 572 grave accidents, and caused to perish 291 workmen, and wounded 494 others,—785 victims.

The mean number of workmen employed in the mines of the first division during this period, was 27,512. It appears, therefore, that for each thousand miners, there were twenty accidents, and twenty-eight victims, of which eighteen have received wounds, and ten have been deprived of life.

Second division—Provinces of Namur and Luxemburg.—From 1840 to the end of 1844, fifty-two accidents occurred. The number of victims was sixty-six, of which thirty seven perished, and twenty-nine received serious wounds.

The average number of miners employed during this interval, was 2450. Thus, for every thousand workmen, there were twenty-one accidents,—eleven persons wounded, and fifteen others killed; that is to say, twenty-six victims.

Third division—Province of Liege.—One hundred and fifty-one accidents, killing 218, and wounded 57 : total 275 victims.

The average number of miners working in this district was, 10,932. Consequently, for every thousand workmen, were fifteen accidents, five persons wounded, and twenty killed; that is to say, twenty-five victims.

General Review of the Accidents which happened in the Coal Mines of Belgium from 1840 *to* 1844, *inclusive.*—Number of accidents, 775; deaths in consequence, 546; severely wounded, 580,—total number of victims, 1126.

Average number of miners employed at this period 40,894. This is nineteen accidents, fourteen persons wounded, and thirteen killed; total victims, twenty-seven for every thousand.

Table of the Nature of these Accidents, arranged in the order of their frequency or number.

	Number.
Falling of stones, crushing of roof, &c. - -	271
Divers causes, - - - - -	159
Accidents in the pits, - - - -	120

	Number.
From ropes and chains, - - - -	68
Fire-damp, coup de feu, "*grisou,*" - - -	64
Falls from ladders, - - - - -	47
Explosion of powder - - - -	33
Inundations, - - - - -	13
Total, - - -	775

In the table below the same accidents are classed after the *order of their importance.*

Nature of the Accidents.	Number of Workmen.		Total of victims.
	Wounded.	Killed.	
Fire-damp,	180	122	302
Fall of stones, crushing in, &c.	158	134	292
Various causes,	107	64	171
Various accidents in the pits,	45	93	138
From ropes and chains,	27	72	99
Falling from ladders,	30	19	49
Explosion of powder,	27	12	39
Inundations,	6	30	36
Totals,	580	546	1126

In pursuance of this momentous subject of coal mining accidents, the recently published report of the mining operations between the years 1840 and 1844, inclusive, developes some interesting statistics. It is seen that while the production of coal in the kingdom has increased twenty per cent., the number of working miners has only augmented seventeen per cent.

In the same lapse of time the total number of these disasters augmented thirty-nine per cent., and that of the victims have increased only nineteen per cent. This augmentation of the number of victims bears principally upon the wounded, which has increased fifty-two per cent., while the number of killed has diminished to fifteen per cent.

Nature of the Accidents.	In every 1000 Workmen killed or wounded, the result is thus:		
	Period from 1821 to 1840.	Period from 1835 to 1839.	Period from 1840 to 1844.
Fire-damp,	377	378	268
Falling of rocks, stones, coal, &c.	173	193	259
Divers accidents,	177	185	274
Falls from ropes, chains, ladders, &c.	160	159	132
Explosion of gunpowder,	42	38	35
Inundation,	71	47	32
Totals,	1000	1000	1000

The foregoing table, derived from the "Compte Rendu de 1839–1844," is arranged in the order of importance of the several classes of accidents, at separate periods of time, and calculated by the actual per centage in every thousand workmen.

The increase or decrease in each class of accidents is rendered sufficiently apparent without further comment. We would only notice that, contrary to prevailing opinion, founded on the increased depth of the mines, the loss of life by fire damp has remarkably declined. The accidents attributable to the crushing in of the roof, the falling of stones, coal, &c., has more than proportionately increased within the same period of time.

The magnitude of the catastrophe that occurred at the coal works of *L'Esperance*, at length attracted serious attention. During the years 1839, 1840 and 1841, the subject of establishing relief institutions in the mining provinces was advocated by the ministry, and eventually decided by several royal decrees. The archives of the ministry contain several propositions which were made upon this subject. M. *Auguste Visschers*,* the present director of the administration of the Belgian mines, published an article which attracted much notice, and which was reprinted in 1843. It is entitled, "Notice of the establishment in Belgium, of Provident Institutions, caisses de prévoyance, for the benefit of the Working Miners."†

In the subjoined note, A. will be found the titles of many of the publications which appeared in Brussels, in relation to this subject, by distinguished writers. To these works we will refer those of our

* To whom the author of the present volume is personally indebted for valuable documents on this interesting branch of statistics, and from which he has not hesitated to make copious extracts.

† Literally, "foresight chests;" the object being not merely to afford relief, but to encourage in the mining population habits of foresight. Mining Review, Vol. XX. p. 167, 171.

Note A. *Caisses de prévoyance.*—The following publications treating more or less directly upon this subject, have been printed in Brussels of late years.

I. *Rapport sur les Caisses de prévoyance en faveur des ouvriers Mineurs*, presented to the king by M. Nothomb, Minister of Public Works, 24 June, 1839.

II. *Rapport sur les Caisses de prévoyance en faveur des ouvriers Mineurs*, presented to the king by M. Desmaisières, Minister of Public Works, 19 December, 1841.

III. *Rapports Annuels des Commissions Administratives des Caisses de prévoyance en faveur des ouvriers Mineurs, instituées dans les Provinces de Hainaut, de Liége et de Namur;* several years.

IV. *Rapports sur les Institutions de Bienfaisance du Royaume;* official reports in 1825, 1826, 1827, 1828, &c.

V. *Essai sur les Moyens d'améliorer le sort des ouvriers*, by Count Arrivabene, 1832.

VI. *Des Caisses d'épargne et de leur Influence sur les Classes Laborieuses*, by M. Ducpetiaux, 1831.

VII. *De la Condition Physique et Morale des jeunes ouvriers et des Moyens de l'améliorer*, by the same, 1843.

VIII. *De l'establissement de Caisses de prévoyance en Belgique, en faveur des ouvriers Mineurs*, by Auguste Visschers, 1839 et 1843.

IX. Eleven publications and reports upon the same subject, in relation to local establishments in the provinces of Liege, Hainault, Namur and Luxemburg, and the arondissements of Mons and Charleroi,—1839, 1840 and 1841.

The generous bounty of the king, appreciating the acts of heroism and courage to which the accidents in the collieries frequently give rise, has expressly instituted a medal of recompense in favour of working miners. The royal decree of the 19th October, 1840, determined the form and model of this medal, divided into two classes, [gold and silver.] Both of them bear on one side the effigy of the king, and on the reverse the insignia of the miner's profession, with these words inscribed upon the exergue:

ACTE DE DEVOUEMENT, RECOMPENCE NATIONALE.

readers who desire more detailed information on a matter of no ordinary interest.

We have only to add in this place a short resumé of the general plan and condition of these useful institutions, chiefly on authority of the reports of M. Desmaisieres and M. Visschers.

The organization of the *caisses de prévoyance* in the five subdivisions of the Belgian coal basins, is the same throughout. The statutes are approved by the king: the governors of the provinces preside over the administrative commissions, which are composed of "exploitants" and master workmen, and render annual accounts to governors. The resources for these institutions are derived from deductions from the wages of the men, equal to one-half per cent., and from contributions of the mine owners to the like amount. Each year, since 1840, the legislature has voted about 42,000 francs; three important societies contribute at least 5000 francs; the provincial council at Hainault annually votes 6000 francs: to these may be added the funds derived from endowments, and from the donations and bequests of individuals. Independently of the temporary relief afforded to the widows, orphans, and dependents on the deceased, the benefits are still further extended by furnishing the means of instruction to the children. Thus, the institution, in providing for the moral wants of living generations, continue to ameliorate the future conditions of the working miner. The benefits are not limited solely to the alleviation of the physical necessities.

The beneficial effects of this system are best evidenced by the practical working since its introduction. On the 1st of January, 1842, the proportion of mining establishments, [exploitations,] associated on the foregoing principles, and the number of workmen who had enrolled themselves as members, were as follows:

	Exploitations.	Working miners.
Affiliated exploitations, forming societies,	210	31,971
Establishments not yet associated, -	160	7,306
Total in the kingdom, - - -	370	39,277

Hence we perceive that the affiliated coal establishments of Belgium amounted to fifty-seven per cent. of the whole number, and the workmen attached to the provident societies were not less than eighty-one per cent. of the aggregate mining population. This is the best indication of the general approbation, by the miners themselves, as well as the owners and lessees of the collieries, of these institutions, throughout Belgium.

"Thus, happily," concludes M. Dermaisieres, "have these humane projects been most nobly brought about, by the influence of the proprietary, by the wisdom of the government, and by the parental solicitude of the sovereign."

The working miner, left to himself, has not the foresight, and does not possess the influence necessary to bring to a good issue such projects as these. It is then for the manufacturers and for the civil administrations to set on foot the establishment of beneficent and relief funds. Modern philanthropy has nobly pleaded the cause of the workman. What is important above all, is to protect them against the reverses which continually threaten industry, in all the gigantic extension to which it has reached. It is not enough to provide for his health—for his comfort—he ought to be habituated to reflect as to the future. Once accustomed to do this, and theworkman will become more moral; because he will be persuaded that his condition will be ameliorated.

GERMANY.*

The mining art was early diffused through the states of Germany. Various edicts granted privileges, or what were then called franchises, to the cities of the mineral districts.

In the greater part of these ordinances we perceive "dispositions protectrices" to the workmen; particularly the assurance of certain aid to himself and to his family, in cases of accident.

The ordinances of 1524 and 1538, made for the mines of Hartz, [Hanover,] assured to the wounded labourer, besides medical aid, the enjoyment of his pay, for eight weeks, if the working company made profits; but only during four weeks if it lost. Hence we observe that it was the mining company on whom the expense devolved.

A similar ordinance, of the 22d July, 1564, made in the electorate of Trèves, reserved a certain weekly sum from all the workmen's wages, towards these objects. This is the earliest edict which makes mention of a reserved fund, introduced by fixed regulation.

An edict of the margrave of Brandebourg, 20th Oct. 1599, bestowed franchise and privileges on the city of Tarnowitz, in Silesia. Art. I. of this act founded a common fund, to be supported by moneys retained from the pay of the working miners. Its object was to contribute to the foundation and construction of churches and schools, and at the same time to afford Christian assistance to the wounded workmen; or, in case of death, to their widows and orphans.

We will not here enumerate all the ordinances prescribed in favour of the mining workman, and the establishment of common funds. Similar institutions exist even in SWEDEN, for the working forgemen, sick or wounded; each owner of forges, every master forgeman, contributes. The simple workman bears a reserved amount equal to the half of that which is contributed by the master forgeman.

Germany presents us, from an early period, an example of two institutions by which we might profit. 1. Mutual insurance funds for poor *mines:* 2. Relief societies—"caisses de secours"—for the *miners.*

The first of these institutions is especially useful in the infancy of the art of working the mines; but when,—extraction having attained considerable amount,—the production tends to exceed the require-

* Abridged from the "Notice sur l'établissement, en Belgique, de Caisses de Prévoyance, en faveur des Ouvriers Mineurs." Bruxelles. 1843.

ments of consumption, is all insurance between the mines, all association for works of general utility superfluous? The system of insurances against risks of every kind has only been developed within a few years, in Belgium and in France. Mines are penetrated with the necessity of remedies against the evils of unlimited competition; against the disorders which it has tended to produce. Even in Belgium, already, one of our financial societies has suggested the formation of a species of institution, [syndicat,] for the industry of the high-furnaces and forges. These ideas should not be lost. Public riches as well as private fortunes suffer from the disorders of excessive competition; of an imprudent excitement, given to the productive forces. The history of late years should serve us as a warning.

"Caisses de secours" for poor miners, wounded or sick, at length exist in Belgium, but the government alone could not have established them. In Germany, the development early given to the "exploitation" of mines; the important number of workmen devoted to this branch of industry; the revenues which the princes derived from it; the influence which they enjoyed in these mine operations, by virtue of the principle which attributes them to the sovereign, [droits régalien,] have induced the depositories of power to regulate all that which concerns the extraction from the mine, the duties, and the relations of masters and servants. Sacrifices were at first alone imposed on the working companies; subsequently the workmen were called on to contribute. The princes granted subsidies or privileges to the "chests;" in many of the mines free action was reserved to the benefit fund.

These institutions were regulated by some suitable persons, chosen by the officers of the prince. The funds were inclosed in boxes having several keys. Sometimes also the workmen bore a part in the directing commission.

The working miners were not, in Germany, and are not now, abandoned by their masters: the authorities are careful to provide for their necessities. These chests were sometimes very rich. According to Jars, the revenues of the *Caisse* of the poor miners of the department of Freyberg, amount annually to 24,000 livres: about the year 1757, the capital, invested at five per cent. interest, was 32,646 livres.

THE PRUSSIAN STATES.

As regards legislation over the aid afforded to the workmen, in mining casualties, the articles 214 to 220, of the general code of the Prussian states, were formed, in order to generalize and reproduce the various local statutes. It is unnecessary to cite them here.

Finally, public authority has recently sanctioned regulations for the established provident institutions, in favour of working miners, in *Rhenish Prussia.*

GREAT BRITAIN.

What the wise direction of public authority has established in Germany, the spirit of association, the sentiment of individual independence, the habit of calculation and of observation, have consecrated in Great Britain. The associations of provident institutions, of saving, of insurance, of charitable, friendly and benefit societies and clubs, in this country, have been clothed with the popular character, always visible in all its institutions. However, the patronage of the higher classes is not refused. It is probable even that these establishments have been originated by the masters or by the mining companies; but these parties have placed themselves in the back-ground of the picture. The charitable or friendly societies have become now part of the customs of the English people. The soil of Great Britain is covered with them.

We have consulted the documents relating to the benefit societies, or those of mutual assistance, in several parts of the United Kingdom. In general, although the donations of distinguished patrons, or those of the proprietors, are welcomed, the major part of the funds is supplied by the workmen; not by means of a voluntary assessment, but by virtue of statutes to which they submit on entering into the establishment.

These institutions participate in the character of insurance societies, but they present this peculiarity, that they are not, to any one, the object of lucre or of speculation. They possess the defects of the societies of mutual help; inasmuch as, in general, they apply only to a small number of individuals. But the wisdom with which the funds are guarded, the prudence which they exercise not to encroach upon the reserve, show that the inconveniences are at least but slight. The workman knows that the "chest" is only maintained by his contributions; he knows that the funds cannot be diverted, and he makes no complaint, in any case, of the insignificance of the aid he receives in proportion to the sacrifices that he has made.

Nevertheless, the funds are, ordinarily, sufficient, and in affinity with the wants of the members. The proprietor of the English mines interests himself in the lot of his workmen. He takes pride in seeing them well-ordered and economical. For his own advantage, he constructs, for the use of his work-people, habitations convenient to the seat of their operations. He gives them, sometimes, dwellings gratuitously. He founds schools for the children; he furnishes a place for a common library. He contributes to the stock for mutual assistance, placed under his patronage; he holds the funds, and pays the interest on them.

In England especially,—and the attempts to reform the poor-laws demonstrate it,—they seek to avoid the inconveniences of those institutions which are solely charitable or purely helpful. The superior classes, so enlightened in this kingdom, interfere in these institutions only to facilitate their operation. The government, whose action ordinarily remains latent, limits itself to the publication of the precise formula for the regulation of the various societies of insurance or benefit.

These societies when they have acquired some extension, are very careful to solicit and secure legal sanction. An advocate of the crown is appointed to review the rules of the associations which aspire to be incorporated. The acts of Parliament, 10 Geo. IV. and 4 and 5 William IV., fix the course to be pursued, and the final sanction is accorded by the magistrates of the county.

The numerous philanthropic societies in Great Britain second the tendency of the English people to profit by the benefits of co-operation. Association, in the times to come, will produce such wonders as we owe, in the order of physics, to the accumulation of steam, or to electricity. It is a lever or powerful spring, which till now has been employed but imperfectly; but which, well directed, will be the principle of prodigies which the future will disclose.

The English workman is, in general, better instructed, and is in easier circumstances, than those of Belgium. He not only has a love for his profession, but entertains a great respect for his superiors and for the laws. The habit of economy, the advantage he finds in it, the pride which the sentiment of his power and good conduct gives him, contribute to strengthen these moral ties. We speak not now of the workers in the great manufactories: reduced to the state of paupers; ill fed; exposed to every privation. But the working miner is, in this kingdom, in a more favourable position than the Belgian miner.

There are two traits of character in the English workmen that we must not lose sight of:—the care that they take to provide a suitable and religious burial for their deceased comrades, and the importance which they attach to the education of their children.*

The picture thus presented by M. Visschers, of the condition of the English miners, is drawn by a friendly hand, and perhaps may be considered somewhat flattering.

To the foregoing liberal views of this philanthropist, we proceed to note some prominent statistics on the miners' Benefit Societies, and on the casualties of coal mining in England; a country which has perhaps a greater interest in these subjects than any other, being the largest coal producer, and employing a more numerous population in its extraction, than the rest of world united.

The continuance of voluntary subscriptions to the innumerable provident societies of the mining districts, proves the prevailing reliance on their efficacy in times of emergency; while the almost universal enrollment, as contributing members, of the class of operatives especially interested in the result, attest the estimate which has been formed, by the working miners themselves, of the salutary influence of those associations. Local instances, we are constrained to admit, may be cited where abuses have existed; where the system has been rendered less operative for good by defective arrangements; by erroneous calculations at the outset; or by occasional improvidence in

* Notice sur l'établissement de Caisses de Prévoyance. M. A. Visschers, 1843, p. 18.

the management; such, for instance, as has been shown by a Parliamentary Report of the South Staffordshire coal-fields; a district which has acquired a lamentable notoriety, for the habits and the moral and social condition of its mining population. But the general working of the relief funds and provident societies, throughout the length and breadth of the land, is satisfactory: creating habits of foresight and economy—compulsory probably at first—and, above all, estimable in bringing opportune succour to the maimed and the sick, and relief to the infirm; in providing support to the survivors of those frightful accidents which so often occur; in securing decent burial to the dead, and, in affording consolation to the families of such as have unfortunately perished.

The drawback on the utility of the ordinary country clubs, seems to consist in their local operation and restricted character; in the limited and fluctuating nature of their resources. Often based on erroneous data; frequently originating with, and conducted by, the uninformed; isolated in all respects;—they want the power and uniformity, almost amounting to nationality, which the coherence of the Belgian confederated *exploitations*, guaranteed by the solemn sanction of the government and laws, seems to assure to the individual societies of which they are made up.

In England there are no public institutions to supply the deficiencies of the country club system. It has been even considered better to leave the supposed evils to be corrected by the interested parties themselves. Moreover it is contended that the extension of especial public protection and relief in favour of one class of operatives, is incompatible with strict justice towards numerous other classes who also pursue hazardous occupations; such, for instance, as the sea service, in which 2,000 British sailors are annually estimated to perish by shipwreck; which appears very little to exceed the number of the killed and disabled miners.

Violent deaths, which occurred in 55 mining districts of England and Wales, in the year 1838:—

By falling down shafts, - - - - - -	63
Breaking of ropes, - - - - - -	1
Ascending and descending, - - - - -	10
Drowned, - - - - - - - - -	22
Falling of stones and coals, - - - - -	97
Explosions of gas, - - - - - -	88
Explosions of gunpowder, - - - - -	4
By trams and wagons, - - - - - -	21
By various injuries, - - - - - -	43
Total, - - - - - -	349

Mining Casualties in the South Staffordshire Coal District.—This district has been recently investigated by the "Midland Mining Commission," and forms an important part of their able Report, drawn up by Mr. Tancred. He remarks, "I come now to, perhaps, the most distressing part of my subject, on which I have to present details which I am persuaded must shock the feelings of all who read them; I allude to the frightful amount of accidents and loss of life which is day by day leaving the fatherless and widows to lament the sudden loss, in the midst of health and vigour, of those on whom they depended for support.

"I should hardly have been disposed to investigate this subject so minutely, had I believed that such a destruction of human life was a necessary and inevitable accompaniment of the working of the thick-coal seam; for in this by far the most danger is incurred. On the contrary, however, I shall have the consolation of proving that such is not the case, and shall produce instances in which a gratifying contrast to the general course of things is exhibited."*

The writer proceeds to quote the records of the "General Registry Office," which furnished the following appalling results.

Table of the Deaths of miners in the Dudley Coal-field, in $5\frac{1}{2}$ years, viz. from July 1837 to December 1842, and the proportion of such deaths as result from accidents, in eleven parishes, whose population in 1841 was 221,018. We may observe that in this population there appears to be no registry or estimate of the total number of miners, so that we are deficient in the means of comparing the results with other districts.

Death of miners, of 15 years and upwards, in $5\frac{1}{2}$ years, - - - - - - - - -	1122
Of the above number killed by accidents, - -	610
Proportion per cent. [being 54.3 killed out of every 100 deaths,] - - - - - - - -	54.3
Average age of miners at their deaths, years, - -	$36\frac{2}{3}$

Thus every miner has more than an equal chance of being killed, in pursuing his occupation.

According to a return by Mr. Best manager of the large works of the British Iron Company, the proportions of casualties, in the Netherton colliery, were, in 1842, or rather for 45 weeks in that year, as follows:

	Men employed.	Accidents.	Of which were fatal.	Accidents. per cent.	Deaths. per cent.
In the thick coal pits,	82	59	4	72	5
In the thin coal and ironstone pits,	92	67	0	73	0
	174	126	4		

As during this year, 1842, the works were inactive for seven weeks, on account of the strike of the men, if we take the proportion for the entire year or 52 weeks, the result shows a total of near 146 accidents sufficiently serious to prevent men from working, out of 174 mines.

* Midland Mining Commission, First Report, 1843, p. liv.

Mr. Best adds, that in the same year was paid, to the sick colliers, miner's widows and orphans in that establishment, upwards of £560=$2,721.

Mr. Smith, manager of the property of the Earl of Dudley, employing 1054 miners, furnishes the following statement also for the year 1842, or for 45 weeks only.

Thick coal colliers, - - - - -	429
Thin coal and ironstone miners, - - -	290
Limestone miners, - - - - -	325
	1054 men.

Medical relief and pensions paid to wounded and superannuated miners and widows, £960 9*s.*=$4,658.

From the details of the mining casualties in this region, it appears that none have been ocasioned by inundation, and very few by explosions of fire-damp. This gas appears to be not engendered by the Staffordshire coal so abundantly as in most other fields; the men usually working with open candles.

The reporter goes on to state the remarkable circumstance that, with so great a number of frightful accidents, constantly occurring, there is nothing in the shape of a hospital in the whole mining district, with the exception of a few in-door patients at the Wolverhampton Dispensary. All other cases, requiring pecular skill, must be sent to Birmingham.

A serious case of explosion occurred on the 18th August, 1845, at Tividale, near Dudley, when twenty miners lost their lives. Among other cases may be added that at Round's Green colliery, near Oldburg, by which twenty lives were lost, on the 17th November, 1846, leaving fourteen widows and one hundred and two orphans destitute.

Benefit Clubs in the Dudley or South Staffordshire Coal-field.— The report from which we have last quoted, examines into the nature of the associations among the miners of this district, for the relief of the members in sickness, and for their burials, and allowances to their widows. "These institutions, so beneficial in themselves, and so well calculated, if properly regulated, to counteract the habitual improvidence of the workmen, and to compensate, in some small degree, for the absence of a wealthier class living amongst them, are, by the perverse ingenuity of interested parties, converted into one of the numerous means by which the hard-earned wages of the miner are transferred from his pocket to the till of the public house."

We cannot enter here into the details which appear in the pages of the report. It is evident enough that the practical working of the system is greatly in need of amendment; and, indeed, occasionally is productive of injurious consequences. The evidence shows that these clubs are always held at public houses, and are promoted by the publicans for their own benefit. By reason of the abuses of the system, and of the appropriation of the funds to drinking and unnecessary expenses, the results are far less beneficial than they might be.

The "*friendly societies* or *sick clubs*," are very numerous, and are established on various principles. They engage with a medical man to attend the members during sickness; and he is paid from 2*s*. to 4*s*. [fifty cents to one dollar] per annum, for each member enrolled. It is a a general rule, that if a member continues a charge on the sick fund for twelve months at one time, he is reduced to half pay for life, and allowed to follow his employment if he is able.

There are also "*Odd Fellows and Lodges*," established on the same principle as the clubs, with the addition of the ceremonies and feasts, but as these are attended with considerable expense, the steady mechanics prefer the common "*sick clubs*."

The "*Field Clubs*" are confined entirely to the miners, and afford medical attendance and sick-pay only during illness from accidents occasioned by the work. The payments to these clubs are compulsory, and the employer always stops the contributions, out of the wages of the men. The miners are very generally in a sick or "*life club*" in addition to their "field club."

Mr. Simkiss, from whose evidence we extract the foregoing notes, adds that "all the clubs, or societies, in this neighbourhood, must be founded on erroneous principles, as they are of short duration. The oldest I can find have not been in existence more than seventy years; and by far the majority do not last one-third of that time." For some years, when new members are joining, and the funds are consequently increasing, they appear to be in a prosperous condition. But when the original members grow old and become a serious charge on the funds, young men look out to join younger clubs. The original club first becomes stationary; then as the deaths increase, the funds decline, the members diminish; and, after struggling for awhile, they ultimately divide what little stock is left among the few survivors. Thus, those who have, during life, contributed regularly, with an assurance of provision in their old age, are now, when they want the most, left with nothing to depend upon but parochial relief.

The miners and mechanics, generally, make no provision (with the exception of clubs,) for the wants and infirmities of old age. There are very few who ever make any deposits in the *savings' bank*, and the instances of miners becoming depositors, are so few, that we may almost say there are none of that class."

Besides the voluntary clubs above mentioned, established by miners themselves, there is a regular system of relief for men wounded or killed in the service, and also to their widows.

In the thick-coal mines, [ten yard coal] the custom is for the owner of the mine to allow 6*s*. a week to the wounded; and 1*s*. 6*d*. a week to the widow of any man killed; together with 1*s*. a week to each child which she may have under the age of ten. The men on their part, make a special collection at every weekly pay, for any wounded man, or widow of a fellow workman, of as much as will make up another 6*s*. weekly.*

* First Report of the Midland Mining Commission, p. li.

The most mischievous sort of clubs, both to the finances and the morals of the miners, are what are called "*money clubs;*" being subscriptions to certain funds, which are very abundant in the South Staffordshire coal district, all which money is spent in drink at public houses. By the evidence of the resident clergy, the regulations and conduct of some of the clubs here are, "terribly destructive of the morals and savings of the workmen."

It is due to the clergy of this district to state that they strenuously exerted themselves to induce the adoption of clubs conducted on sounder principles, as well as other institutions calculated to remove the evils complained of. Here are "*Provident Societies,*" for savings and for relief in sickness, in connection with Sunday and day schools. There are also "*Wesleyan Clubs,*" and "*Clothing Clubs;*" the latter have a beneficial effect in inducing habits of saving. To these we may add the "*Dorcas Societies,*" consisting of charitably disposed ladies, who hold stated meetings and make articles of clothing which are sold to the poor at half the price of the materials.

A series of articles on Benefit Societies, by Dr. Beard, was published in the "People's Journal" in 1847. From these we learn in detail how extensive and almost universal are the failures of the English Benefit associations, arising from defective management, and from the erroneous structure of their respective constitutions.

In the "Odd Fellow and Friendly Societies" the scale on which they have failed—and, unless great changes are introduced, will fail—is, according to Mr. Nelson, fearfully large. The Rev. Mr. Sherman lately stated at a public meeting in Liverpool—'Mr. Ansell had told him of two thousand societies having been submitted to him in three years, whose affairs were proved to be altogether insolvent.' "But," observes Dr. Beard, "there is another kind of failure: the staff breaks under the hand of the poor sick, aged man, the first time he leans on it. Benefit societies, in numberless instances, *do not* afford the needful aid."

We trust there needs no apology for calling attention through the medium of the foregoing passages, to the defective construction of associations which are designed for the most useful and beneficial purposes.

The extent of their failure can scarcely be fully ascertained; for the sufferers are in humble life; are scattered up and down in society and have no sufficient means of making their injuries known. A committee of the House of Commons is the only resource by which benevolent men could acquire some knowledge of the number of these failures, and of the sufferings they have entailed."*

Cornwall.—"The diseases of miners," forms the subject of a paper by Sir Charles Lemon, addressed to the Royal Institution of Cornwall; and also another by Dr. Barham. They contain some very important statistical statements respecting the deaths and diseases among the mining population of Cornwall, and a series of compari-

* The People's Journal, July, 1847.

sons between the mining districts of Cornwall and the coal regions of Staffordshire, Northumberland, &c.*

Miners' Club.—It does not appear that provident associations, of much utility or permanent character, prevail in the mining districts of Cornwall, and the want of a better system there has been frequently deplored. Various plans for the establishment of hospitals, mining schools, and beneficial societies, founded on an adequate scale and based on correct principles, have, from time to time, been ineffectually advocated in Cornwall. That of a general miners' club is the last.

In 1846, efforts were made to engage the public sympathy in favour of this association, and an earnest appeal was made to the lords and adventurers to give their countenance to the plan, and to support the wishes of the great body of miners. The project of a "Miners' Society" had previously been abandoned for want of the requisite co-operation of the influential classes.

It was urged that necessity and policy required the establishment of a provident association in every district, or of branches emanating from one general society or club, which should be based upon such principles as should amply provide for the necessities of the labouring miner, when, from the various risks, inseparable from his hazardous avocation he should be incapacitated from labour.

If, as we infer from the address of some of the advocates for this general club, the thirty thousand miners are to contribute their full share of the annual funds, and the lords and adventurers, and benevolent persons to supply the other moiety, as in the case of the Belgian "*caisses de secours*," the project appears to be unobjectionable; for it has been well proved, that no plan works well as a merely charitable institution giving gratuitous aid; no project is successful in teaching the importance of foresight and timely economy, which does not comprise the contributions of the working miners themselves, and constitute them joint guardians of the funds destined to relieve their future wants.

Mining Accidents in Great Britain.—The Mining Journal, January, 1844, published a list of four hundred and eighty-three deaths and accidents noted in its columns during the previous eight months.

Out of one thousand one hundred and twelve deaths of colliers only among that class of population, reported by the Midland Mining Commission, no less than six hundred and ten arose from accidents; by which it appears there is a frightful advance of mining mortality. The editor conceives that "the number of lives sacrificed annually, cannot be less than two thousand five hundred, exclusive of the numerous cases recorded in which severe injuries have been received, resulting ultimately in the loss of lives."†

It is due to Mr. English to state that he has for a considerable time past done his utmost towards the humane object of establishing institutions for the relief of the sufferers by accidents in mines, and

* Mining Review, August, 1839, and 1841.

† Mining Journal, January 20th, 1844.

has let no opportunity pass of awakening sympathy in favour of that unfortunate class. We fear that there is too much truth in the following severe remark:—

"England is justly proud of her numerous charities, her hospitals for the sick and maimed, her asylums for the aged and decayed members of society, and her institutions for the support and protection of the widow and orphan. But, with shame be it spoken, a country indebted, in a great measure, for her position to her mineral riches, cannot reckon, amongst her numerous charitable establishments, one which is devoted to the maimed or aged collier or miner, nor a [public] fund wherewith to support the widow and the fatherless who may be bereaved of their natural protector by accidents in mines."*

The same gentleman, in a petition to the House of Commons, dated 28th January, 1846, stated that the loss of life in mines and collieries within the preceding year, was upwards of a thousand individuals.

A printed statement has subsequently appeared, wherein it is shown that in twelve cases alone, in thirty-two years, there was sustained a loss of seven hundred and twenty-three lives in the Durham and Northumberland coal-field, chiefly by EXPLOSION in the pits. We add the details below.

Date.	Localities.	Lives lost.	Causes.
1812, May 25th,	Felling,	92	Explosion.
1815, May 3d,	Heaton,	75	Inundation.
do. June 2d,	Newbottle,	57	Explosion.
1821, October 23d,	Wallsend,	52	do.
1823, November 3d,	Rainton,	59	do.
1835, June 18th,	Wallsend,	102	do.
1839, June,	St. Hilda,	32	do.
do. June 23d,	South Shields,	51	do.
1841,	Wellington and Thornley,	41	
1843,	King pit,	28	
1844, Sept. 28th,	Haswell,	95	do.
1845, August 21st,	Jarrow colliery,	39	do.
	Cases of death,	723	

A petition was presented to the House of Commons in 1843, by the pitmen of the Tyne, the Wear, and the Tees, in which they state that within the preceding twenty years upwards of *seven hundred* pitmen, the friends and companions of the petitioners, had been miserably destroyed in the Durham and Northumberland mines, by explosions of inflammable gas, and that others met the most fearful deaths from various other causes; that these explosions have always been traced to the want of sufficient ventilation, permitting the accumulation of the gas in such masses that, when set fire to, it explodes with sufficient force, sometimes, to blow men up a shaft six hundred

* Mining Journal, Vol. XIII. p. 391; also, 31st January, 1846.

feet deep as if from the mouth of a cannon, and to shake the solid ground similarly to an earthquake. They state that, knowing the Davy lamp is liable to fire an explosive mixture under certain circumstances, they cannot rest satisfied with their lives being secured by an imperfect instrument, easily deranged, and which at the moment of the greatest danger brings on the mischief it is intended to prevent, and on the supposed safety of which has been based the modern practice of carrying foul underground workings to a most dangerous extent.

The petitioners, who assembled to the number of fifteen thousand, at their meeting, suggested that the only way of working the mines with security, would be by sinking two shafts at the "winning," and as the work extends making additional shafts. The mine would then be thoroughly ventilated, the coal more easily worked, and the petitioners secured from these terrible accidents.

Mining Casualties in the North of England.—The lists we have given of the loss of life, chiefly by explosions in twelve cases alone, in the counties of Durham and Northumberland, by no means exhibits the entire number of deaths there from that cause.

The subjoined statistical table shows that they comprised in eighty years, between 1756 and 1836, the destruction of one thousand four hundred and twenty-seven miners. The cases of explosion more than one hundred in number, were attended by the loss of one thousand three hundred and one lives, out of this complement of one thousand four hundred and twenty-seven. A large extension to this catalogue might be made by the addition of the cases since 1836, in fact, amounting to many hundreds.

Period.	Causes.	No.	Total deaths.	Annual deaths.	
1756 to 1800.	By explosions,	305	311	7⅔	
	By inundations,	6			
1800 to 1815.	By explosions,	332	424	26½	
	Inundations,	74			
	Bursting of steam-boiler, locomotive,	18			
1816 to 1836.	By explosion,	664	692	34½	About 21,000 persons employed above and below ground.
	Inundation,	3			
	Suffocation,	7			
	Falling of stones,	6			
	Bursting of steam-boilers,	12			
			1427		

These tables can only be usefully compared with each other, and with the results of other mining districts, when we know the number of workmen actually employed at those times and places respectively. According to Mr. Buddle, the foregoing list of deaths does not comprise those which result from the ordinary casualties of life.*

* M. Piot in Annales des Mines, Vol. I., 1842.

The following statement has been published of six cases of fire-damp in the Jarrow colliery, on the Durham side of the Tyne, and the number of deaths they occasioned.

In 1817, 1st explosion,	- - - -	6 killed.
1820, 2d "	- - - -	2 "
1826, 3d "	- - - -	42 "
1828, 4th "	- - - -	8 "
1830, 5th "	- - - -	42 "
1845, 6th "	- - - -	39 "
In twenty-eight years,	- - - -	139

The attention of the government has been attracted to the formidable nature of these explosions, and in recent important cases, it has nominated commissions, consisting of gentlemen of science and experience, to such as in the cases of the Haswell and the Jarrow explosions, and have directed a searching investigation to be instituted into the causes which led to these catastrophes.

The quality of the deleterious gases of the Jarrow, the Hebburn and the Gateshead collieries, was examined in 1846, by Mr. Thomas Graham, and the Mining Journal of June 16th contains an article by that gentleman, "on the composition of the fire-damp of the Newcastle coal-field," and the result of his investigation. From this paper it appears that the gas of Killingworth colliery, near Jarrow, where the great explosion of 1845 took place, issues from a fissure in a stratum of sandstone, and has been kept uninterruptedly burning, as the means of lighting the horse road in the mine, for upwards of ten years, without any sensible diminution in its quantity. At the Gateshead colliery, also, the gas is collected as it issues, and is used for lighting the mine, while at the Hebburn colliery the gas ascends from a bore made down into the Bensham coal-seam, which is highly charged with gas, and has been the cause of many accidents.

We add to the table in the foregoing extract, a recent incomplete return of the numbers of miners who have perished in the Durham and Northumberland coal mines in the last forty-two years:—

			Total deaths.	
From 1803 to 1821.	From explosion,	105	229	
	Inundated by water,	75		
	Choke damp,	9		
	Boiler bursting and other causes,	40		
1821 to 1843	From explosions,	732	821	About 990 cases of death.
	Falling stone, choke damp and other causes,	89		
1844 to 1845	Explosions, in two cases only, exclusive of other accidents,		134	
Total killed in 42 years in one district, besides numerous cases which have been omitted,			1184	

This appalling account of loss of life in this class of working men has, it is said, led among other causes to the association of colliers in the north of England, called "the Union," which lately comprised 60,000 persons.

This association, it appears, has other objects besides those contemplated by the Belgian provident institutions, or the "caisses de secours" of the French mines. One object aimed at is the facility it affords for enabling large bodies of operatives to strike for rise of wages, &c., or to consolidate the interests of an important and numerous class in the community.

It is asserted that there is not a colliery in the kingdom in which the men are not daily and hourly exposed to similar fatal accidents as are recorded above, which cannot be wondered at, considering the bad ventilation, and the extent of the underground operations, where in some mines sixty or seventy miles of passages have been cut.

Dr. Barham has communicated an article on "the accidents and diseases of miners," more especially directed towards those of Cornwall. He institutes an interesting comparison between the number of deaths in the Cornish mines of copper and tin, and those of the coal districts. The chances of violent deaths in the latter greatly preponderate.

Thus, there were in the Tyne and Wear district, in the forty years from 1800 to 1840, 1480 deaths from accidents, out of a mining population of 21,000 persons, men and boys, of whom five-eighths worked underground.

Out of the 1480 deaths in the collieries, 1325 deaths, or nine-tenths of the whole, were caused by explosions or inundations—accidents to which Cornish miners are rarely subjected.

From official returns given by the Register-general, we are furnished with materials for comparing the mortality among miners with that in other classes of the community. By the census of 1841, the number of males, of twenty years and upwards, employed in the coal mines, as well as those of salt and the metals, was 124,667. Among these, the violent deaths registered in the year 1840 were 498. The only employment which was equally fatal was that of the navy and merchant service. The relative proportions of deaths in an equal number, are as follows:—

In the navy and merchant service, - -	4006
In the mines, - - - - - -	3939
In the agricultural population of England, -	1221*

Frightful as is the foregoing statement of the mortality in the northern coal-field mines, we derive some consolation from perceiving that it falls very short of that in the Belgian coal-field. The results are interesting.

General Cases.—In Belgium, out of 20,000 miners, the total number of deaths was 1710 in twenty years, prior to 1841, averaging

* Mining Journal, January 1, 1841.

85.5 per annum, or thirty to every one thousand miners employed, annually.

In the Durham and Northumberland coal-field the total of deaths in the mines was 990 in the twenty-four years prior to 1846, averaging 41.28 annually, in 21,000 miners, or 19.65 to every 1000 annually.

To render the comparison more exact, we find that the number of deaths in Belgium, proportionate to 21,000 miners, (out of 28,000) is 64.12 per annum.

The Durham and Northumberland coal-field, average of 24 years prior to 1846, 41.28 per annum. Do., average of 20 years prior to 1837, 34.60 per annum.

Thus the number of general cases of death, in a corresponding number of miners, is 55 per cent. greater in Belgium than in the English northern coal-field. When, however, we analyze the causes of these deaths, we observe that the proportions are reversed, and the fatal cases of *fire-damp* are far greater in the Newcastle than in the Belgian coal-field.

Explosions.—Belgium, 503 killed out of 28,000 miners, or 377 out of 21,000, average 18.85 per annum in 20 years.

Northern coal-field, out of 28,000 miners, or 866 out of 21,000, average 36.09 per annum in 24 years.

The mortality by fire-damp being greater in the English coal-field by 31 per cent. than in that of Belgium; or thus, annual deaths by explosions in the Belgian coal-mines, 0.89 out of every 1000 persons employed. In the Newcastle coal-field, 1.72 out of every 1000 persons employed.

Our data is somewhat too scanty to pursue these comparisons far. A statement of the number of miners killed, from various causes, in fifteen years in the basin of the Loire, in France, shows the deaths to be in the proportion of 1 in 100 persons employed. This ratio, if correct, is enormous. That of Belgium averages 1 in 327. The Newcastle coal-field, 1 in 508.

Forest of Dean.—Royal Commission of Inquiry into mines.—In 1842, ample reports were made by the chief commissioner of the Gloucestershire mining district. He says that this woodland and mining region, although comprising an area of only 22,000 acres, is so much isolated in its character and local customs, that it presents a field of more than common interest.

The employment of females in the mines and collieries is happily almost unknown in the forest. Boys, however, and those often of a very early age, are employed in considerable numbers, as the thinness of the seams of coal requires the labour of mere children, from their very limited height.

From the evidence adduced in the commissioners' report, it is proved in some of the forest mines, that the subterranean roadways or passages are so small, that even the youngest children cannot move along them without crawling on their hands and feet; in which unnatural and constrained position they drag the loaded carriages or *hods* after them. And yet, as it is impossible, by any outlay compatible with a profitable return, to render such coal mines fit for human beings to work in, they never will be placed in such a condition. Consequently, they never can be worked without inflicting great and irreparable injury on the health of children.

From the peril arising from the destructive influences of malaria and inflammable gases, these mines seem, in a great degree, happily free; and the accidents from explosions are of rare occurrence. The excellent attention given to the system of ventilation, adopted in the Forest collieries, in fact, affords a very general protection from the fatal effects also of carbonic acid gas, or choke-damp.

Staffordshire.—Five lives were lost by an explosion in the Yew-tree colliery, Sedgely, 23d March, 1847.

June 2nd, 1847, eight men and three horses were killed by an explosion of carburetted hydrogen, at Gerard's Bridge colliery, near St. Helen's. In the same month, by an explosion, in Croft Pit, near Whitehaven, four lives were lost. In the same month were nine persons killed by explosion at Kirkless Hall colliery, about two miles from Wigan; besides which there were eight or ten others who were not expected to recover, and about twelve others less seriously injured. Also in the same month, near Wigan, two persons killed, and at Felling colliery, near Gateshead, six miners killed by explosion or fire-damp.

Yorkshire coal-field, XX.—Fifteen persons lost their lives by fire-damp, Nov., 1841, at Barnsley. At Huddersfield, three explosions in 1841. On the 5th of March, 1847, an explosion of carburetted hydrogen took place in the Great Ardsley main colliery, near Barnsley; 95 men were working in the pit at the time, 66 of whom were instantly killed, several died subsequently, and only 10 escaped unhurt.

At Beeston, near Leeds, 17th May, 1847, an explosion led to the death of nine miners.

Lancashire coal-field, XVIII.—Haydock colliery, near Newton.—On the 5th Nov., 1845, an explosion took place, whereby nine persons were killed, and ten others so dreadfully mutilated, as to be unable to survive, with the exception of one.

In the Moyston colliery, ten lives were lost and seven wounded by fire-damp, in 1840, and six persons burned and five hurt in May, 1846.

An explosion from fire-damp in a colliery near Preston took place on the 24th of November, 1846, and on the same day another occurred at Coppell colliery, Standish. Twelve lives were sacrificed in these two cases.

In the same month, by an explosion at Chorley, eight persons were instantaneously killed.

March 18th, 1854, a dreadful explosion occurred in Avley mine, of the Ince Hall Coal Company, near Wigan, by which 120 men lost their lives.

In the same mine, sixty lives were lost by an explosion in March, 1853.

SCOTLAND.

The ordinary casualties of mining occupations prevail here; but that arising from fire-damp does not appear to be so common.

Two explosions took place in 1845, in the Victoria colliery, near Nitshill, Glasgow, but without loss of life.

Ameliorations in the habits and condition of the mining population have taken place, within a few years. Amongst these, none, perhaps, is more important than the prevention of the employment of females in the coal mines, both of Scotland and in some English districts. At the time of the passing of Lord Ashley's Act, in 1842, there were no less than 2400 females in the coal pits of Scotland; seven hundred women in those around Wigan; many in Staffordshire, &c.

SOUTH WALES.

It is ascertained that the loss of life by fire-damp is not less frequent, although on a smaller scale in this coal-field, than in the highly bituminous coal basins of the north of England. Those which occur are in great measure limited to the bituminous portion of the Welsh basin. One of the most important of these accidents was an explosion in the Duffryn colliery, whereby 29 miners were killed, on the 2d August, 1845. Minor cases, of the death of from two to ten persons, are less rare, and scarcely a week passes without a case of explosion.

In January, 1844, twelve persons were killed by this cause, at Dinas colliery, and several accidents from the fire-damp took place in other collieries. Three miners were destroyed at Nantyglo, in July of that year. In 1845, a good many accidents occurred, by explosions: at Patricroft, at Swansea, at Mynydd Newydd colliery, four deaths. In May, 1846, a severe case of explosion at the Risca colliery, and in the following month, eight persons were burned at Homfray's colliery, Tredegar.

The employment of females in the mines is or was prevalent in South Wales, but it is hoped the degrading practice is diminished.*

Benefit societies, for the relief of sick and wounded miners, or for their families, are numerous throughout the mining regions, and are productive of considerable good.

CONDITION OF THE MINING POPULATION OF GREAT BRITAIN.

This has been the subject of investigation for some years, and annual reports have been made to government. Difficulties, abuses, and grievances, under which the working miners and their families suffered, have been diligently investigated and pointed out, and reme-

* Royal mining commission of inquiry into mines.

dies have been suggested and acted upon. We cannot here enter into these details. It is evident that remarkable differences in the habits, morals, and comfort of the same classes existed in different mining regions of Great Britain. The causes of these discrepancies or contrasts have been traced to their sources, and placed before the public. The general social condition of this class of population, we have every reason to know, has been greatly ameliorated by means of these investigations.

Some colliery districts, it is well known, have always maintained a more moral, a more respectable and intelligent population, than others. We have no means of classifying these, even were it desirable to do so. Some have been more prominent than others, as we have shown in relation to the Dudley coal-field, and formerly some of those in Scotland; but in all, we have the satisfaction of stating great improvements have taken place, of late years.

Thus, we read, in a recent article, that "not a little of the success of the Coalbrookdale coal and iron works, must be attributed to the great attention paid to the religious and moral training of the workmen, and the care bestowed on their physical condition. Excellent schools are provided for the children, and lecturers are occasionally engaged to instruct the adults. The training and education of the children, the aids for mental improvement, offered with no niggard hand to the operatives—from 3000 to 4000 in number—have rendered the work people of Coalbrookdale a very superior class to those usually employed in the mines and forges."*

Education of the youth at the collieries, through the untiring agency of the benevolent proprietors and the exertions of the clergy, is making rapid progress in many districts. At the Low-moor Iron Company's colliery, near Bradford, where, in 1841, only two in ten could read, out of 1100 employed, there were, in September, 1845, out of 494 boys, between ten and eighteen, at work, 411, or 83 per cent. who could read.

Condition of the Mining Population of Great Britain in 1847.—By an act of Parliament, passed in 1842, commissioners were appointed "to inquire into the operations of that act, and into the state of the population in the mining districts." Four reports, between 1844 and 1847, have been presented by these commissioners, under the provisions of what is generally known as Lord Ashley's act, and refer to portions of England, Scotland, and Wales. We have extracted freely from a portion of these.

An excellent article on these reports, but more especially on that which relates to Scotland and the North of England, appeared in the North British Review for November, 1847. No apology, we trust, is necessary from us, for introducing a sketch of the article to which we refer, particularly as it supplies some information in which we were otherwise defective.

On the authority alluded to, some facts as to the condition of the

* London Art Union, 1847.

colliers in the north are detailed, which might almost appear incredible.

"Whether it may have arisen from the nature of the employment underground, or whatever may have been the original cause, we shall not wait to determine; certain it is that, till about the commencement of the present century, colliers were kept in a state of perpetual bondage, and from the first moment of their existence were considered as belonging to the property which gave them birth. Without the permission of the proprietor, they could not receive employment in any other place. In fact they were held to be part and parcel of the establishment for carrying on the working of the coal; and if it happened to be let, they were specially described in the lease, and transferred to the lessee, in the same manner as if they had been a number of horses. When the legislature passed measures for the benefit of the community generally, the colliers were expressly exempted from the privileges which such measures conferred. Even in the well-known *Habeas Corpus* act it was declared, '*that this present act is no ways to be extended to colliers and salters.*'

"In 1775, an act of the British Parliament was passed, which declared that colliers and salters were to be no longer 'transferable with the collieries and salt-works;' but upon certain conditions, which were then deemed 'reasonable,' they were to be gradually emancipated and set free, and others prevented from coming into such a state of servitude. But the act of 1775 does not seem to have operated satisfactorily; and in 1799 another act was passed, which completely freed colliers from the bondage in which they had been previously held, and placed them on a footing of equality with the other labourers of the kingdom."

Another evil, of great magnitude, which had long existed in Scotland, and which presented the greatest obstacle to the improvement of the condition of the mining population, was unquestionably the employment of females under ground.

The arrangements were such, that the labour of the man who worked or hewed the coal was wholly unproductive without the assistance of his wife or daughter, whose occupation it was to carry it away; and unless trained to it from their infancy, it was totally impossible for females to engage in such employment.

In a very interesting book, entitled "A General View of the Coal Trade of Scotland," published by Mr. Bald, in 1808, a graphic description of the work performed by a female "coal-bearer" is given. We are compelled to abbreviate the details. "The collier leaves his house for the pit, about eleven o'clock at night, [attended by his sons, if he has any sufficiently old,] when the rest of mankind are retiring to rest. Their first work is to prepare coals, by hewing them down from the wall. In about three hours after, his wife, [attended by her daughters, if she has any sufficiently grown,] sets out for the pit, having previously wrapped her infant child in a blanket, and left it to the care of an old woman, who, for a small gratuity, keeps three or four children at a time, and who, in their mother's absence, feeds them with ale or whiskey, mixed with water. The children who are a little more advanced are left to the care of a neighbour.

The mother, having thus disposed of her younger children, descends the pit with her older daughters, where each having a basket of suitable form, lays it down, and into it the large coals are rolled; and such is the weight carried, that it frequently takes two men to lift the burden upon their backs; the girls are loaded according to their strength. The mother sets out first, carrying a lighted candle in her teeth; the girls follow, and in this manner they proceed to the pit-bottom, and with weary steps and slow, ascend the stair till they arrive at the hill or pit-top, where the coals are laid down for sale; and in this manner they go on for eight or ten hours, almost without resting.

We have seen a woman, during the space of time above mentioned, take on a load of at least 170 lbs. avoirdupois; travel with this 150 yards up the slope of the coal below ground; ascend a pit by stairs 117 feet, and travel up the hill 20 yards more to where the coals are laid down. All this she will perform no less than twenty-four times, as a day's work.

The whole distance, thus loaded, during each day, was	5016 yds.
And the unloaded distance, - - -	5016
Total of the dailywork, $5\frac{3}{4}$ miles, or - -	10,032 yds.

In those pits which are so deep as to prevent the women from carrying the coals to the surface, the distance which they bring their loads to the pit-bottom may be stated at 280 yards. This journey they will perform thirty times with the weight above-mentioned, in the space of ten hours; so that the journey performed each day is, loaded, 8400 yards; not loaded, 8400 yards. Total length 16,800 yards, or more than $9\frac{1}{2}$ miles. The perpendicular ascent of the slope of the coal being 700 yards."

This is the testimony of Mr. Bald, who has been for half a century at the head of the mining of Scotland, and who has done more than any other man, not merely to improve the method of working, but to elevate the character of the worker.

We are assured that this is no exaggerated statement. It is utterly impossible for language to convey to a stranger anything like an adequate idea of the immense toil which those poor women had to undergo. It was reckoned nothing extraordinary, at a Lothian colliery, for a woman to carry on her back from 35 to 40 cwt. of coal each day, a distance of between 300 and 400 yards; the greater part of the road being not higher than $4\frac{1}{2}$ feet, and, in some cases, a considerable portion of it covered with water.

The reviewer, with perfect justice remarks, that it is certainly something very remarkable that, in the vicinity of the most polished city in the kingdom, and for the purpose of supplying it with an important necessary of life, there should have been in existence, until as it were yesterday, one of the most offensive and disgusting systems of slavery that ever disgraced a civilized country!

On the 7th of May, 1842, Lord Ashley, with whom the commission originated, rose in his place in the house of commons, and moved "for leave to bring in a bill to make regulations respecting the age and sex of children and young persons employed in the mines and collieries of the United Kingdom." In introducing the subject, this philanthropic nobleman said—

"It is not possible for any man, whatever be his station, if he have but a heart within his bosom, to read the details of this awful document, without a combined feeling of shame, terror and indignation. But I will endeavour to dwell upon the evil itself, rather than on the parties that might be accused as, in great measure, the authors of it. An enormous mischief is discovered, and an immediate remedy is proposed; and sure I am that if those who have the power will be as ready to abate oppression as those who have suffered will be to forgive the sense of it, we may hope to see the revival of such a good understanding between master and man; between wealth and poverty; between ruler and ruled, as will, under God's good providence, conduce to the restoration of social comfort, and to the permanent security of the empire."

After describing the measure in detail, Lord Ashley concluded in this striking and beautiful language:—

"Is it not enough to announce these things to an assembly of Christian men, and British gentlemen! For twenty millions of money you purchased the liberation of the negro, and it was a blessed deed. You may this night, by a cheap and harmless vote, invigorate the hearts of thousands of *your* country people; enable them to walk erect in newness of life; to enter on the enjoyment of their inherited freedom, and avail themselves, (if they will accept them,) of the opportunities of virtue, of morality, and of religion. These, sir, are the ends which I venture to propose: this is the barbarism that I seek to remove. The house will, I am sure, forgive me for having detained them so long; and still more will they forgive me for venturing to conclude, by imploring them, in the words of holy writ, 'to break off our sins by righteousness, and our iniquities by showing mercy to the poor, if it may be a lengthening of our tranquillity.'"

We have been informed, that during the delivery of Lord Ashley's speech, the house of commons was a perfect calm—not a whisper was heard. The simple announcement of the injuries inflicted, the sufferings endured, the degradation and ignorance prevailing, made such an impression in the house that many a stout heart melted, and tears were shed, where seldom tears had been shed before.

The act passed in the autumn of 1842. It has now been five years in operation, and we will venture to affirm that no measure was ever passed which so fully realized all the expectations of its supporters, or so completely refuted all the objections of its opponents.

Notwithstanding the imperative character of the act of parliament which forbids the employment of females underground, the law has been evaded in some districts, especially in the Welch collieries, and it seems singular that these violations of the law invariably originate with the females themselves, for whose especial benefit Lord Ashley's Act was introduced.

According to the report of the commissioner appointed to inquire into the state of the population in the mining districts, in 1850, it appears that notwithstanding all the precautions taken by several of the owners and managers of the large collieries and iron works in Monmouthshire, Brecon and Glamorganshire, females were again at work below ground. Great difficulty was experienced in detecting them, as they always concealed themselves and got into hiding-places when the pits were searched by the agents.

Judging from their own admissions, it is probable that not less than 200 females have been employed in some of the principal works, during the last year. A considerable proportion of these were of the tenderest years—11 to 13—the rest quite young women. The habit, therefore, was re-commencing with an entire new generation.

Among the many things required to aid in correcting the sensual habits, and enlightening the minds of the mass of the population of this district, the better training of the female portion of it is considered by all thoughtful persons as one of the first and most pressing.

What, however, can be expected from those who are allowed to begin life in an employment of this kind, and to continue it until they become wives and mothers?

The employment, underground, of boys under 10 years of age, has also been resumed in several works, in the Welch district. So large a proportion is a plain proof that the practice has been resumed without check, and that another generation of colliers and miners is growing up, whose minds are becoming indurated by the same process of early removal from school and employment in the pits, at the age of seven or eight.

It is but justice to the managers and mineral agents of some of the principal companies to state that they have endeavoured to stop the practice, and have caused boys under the legal age to be strictly excluded from their pits, in the hopes of their parents keeping them at school until the age of ten.*

We are happy to learn by the London Mining Journal, April 1st, 1854, that from the suggestions of Mr. Tremenheere, about four years ago, with a view to hold out an inducement to parents to keep their children at school two or three years longer than they previously had done, a distribution of prizes offered by the iron and coal masters took place at the St. Thomas National School, Dudley. The prizes were £4 each to fifteen boys, and £3 each to thirty boys and girls, besides smaller gifts to other children.

* Mr. Tremenheere's Report in 1850.

FRANCE.

Mining Casualties in the Coal Basin of the Loire or St. Etienne.

Years.	Workmen employed.	Number of Workmen. Killed.	Wounded.	Total.	English Tons (of 10,146 met. quin.) produced.
1817	1,825	18	27	45	382,625
1818	1,915	14	37	51	387,352
1819	1,927	20	16	36	328,200
1820	1,945	20	38	58	374,390
1821	2,038	19	33	52	397,920
1822	1,958	25	26	51	415,131
1823	2,259	32	16	48	445,113
1824	2,514	12	19	31	548,567
1825	2,814	21	21	42	503,341
1826	2,708	26	30	56	552,216
1827	2,738	17	11	28	616,653
1828	2,190	28	11	39	656,490
1829	2,970	46	16	62	614,684
1830	3,029	30	29	59	673,400
1831	3,053	30	10	40	625,486
	36,879	358	340	698	7,521,568
Average of 15 y's.	2,458	23.86	22.66	46.52	Av'ge to each miner 204 an'ly.

Of the foregoing list of casualties occurring in the fifteen years, from 1817 to 1831, the immediate causes were as follows:—

Crushes or falling in of the ground,	299=20.	annually.
Inflammable gas or fire-damp explosions,	179=11.92	"
Falling of rocks, timber, &c.	220=14.60	"
	698=46.52	"

Proportion of deaths, 1 in 100; proportion of accidents, 1 in 52.

It is remarked that Mondays are the days of the week on which the greater part of the accidents happen.

The casualties recorded above which occurred in a working population averaging during the fifteen years only 2,458 miners, seem to be very disproportionate to the number employed; being at the rate of 1 in 155, annually. In the foregoing statement the deaths by fire-damp at St. Etienne, during this period, form only one-fourth of the total number; while the falling in of the ground in the subterranean works has been the principal cause of nearly three-fourths of these accidents.* The safety lamp of Sir Humphrey Davy was introduced here in 1825.

In 1845, a scientific commission was formed in France, to inquire into the causes of fire-damp in the French mines, and to suggest the best means of preventing them.

Most of the mines in this department possess local *caisses* or funds

* This preponderance of the number of accidents in the Basin of the Loire, from the falling in of the ground or roof in the subterranean works, seems to be of local occurrence. In other districts, those arising from explosions of gas are the most disastrous.

for the relief of the families of workmen; this resource has often proved inadequate; but the owners of the mines and the inhabitants of the district have generously vied with each other in relieving the families rendered needy by these accidents. The government has also relieved many sufferers.

By a circular of the under-secretary of state for the public works, addressed to the prefects and engineers of mines, they are directed to furnish, sometime in January of each year, an account of the accidents which may have occurred in the mines and quarries during the preceding year; distinguishing the number of workmen employed in each department and the number of killed and wounded, the cause, &c.*

We have elsewhere observed that the daily employment of the coal miner was accompanied with far greater risk of life than that of the workman engaged in the extraction of metallic ores, and consequently that it was the more expedient to provide the means of meeting or alleviating the accumulated casualties of the collier's life. From a statement before us, it appears that, in France, the proportion of coal miners killed or seriously injured, annually, has amounted to one in one hundred and forty-four, while in the mines of metalliferous ores, the proportion is only, on an average, one in four hundred and twenty five.

We have shown that in the basin of St. Etienne alone six hundred and ninety-eight coal miners have been killed or wounded in the space of fifteen years; that is to say, forty-six per annum; but this represents one-third of the coal production of France. It has been ascertained that the general average of deaths in the coal districts of France, occasioned by mine accidents, is between ninety and one hundred, for an extraction of coal amounting to three millions of tons, about equal to that of Belgium, where the average deaths are as much as one hundred and twelve annually.†

The great sources of accidents which occasion sudden death in coal mines, are obviously the irruptions of water and the explosions of the fire-damp; the last being the most frequent and active. The fall of rocks, the *eboulements* or crushing, which seems at first sight, to constitute the principal elements of danger, enter only into a fraction of the sum total of disasters. The proportion of killed and wounded by fire-damp in Belgium, between 1821 and 1842, is about 38 per cent. of the total number of accidents.

These distressing occurrences have by no means diminished of late years. On the contrary, they appear to increase in all the coal mining countries of the world, in the ratio of the increased depth to which the working of the collieries are carried.

On the 23d March, 1847, an explosion took place in one of the coal pits of La Grame, in Alsace. Out of thirty-six workmen who were in the pit at the moment, twenty-four instantly perished, and the other twelve were seriously burned.

* Annales des Mines, tome vii., 1845.

† Géologie appliquée, par M. A. Burat, Paris.

An imperial decree, 26th May, 1813, founded, in the department of Ourthe, a *caisse de prévoyance*, in favour of the poor colliers. But this institution, even in the same department, had not taken deep root; at Liege it did not survive the fall of the empire.

A royal ordinance, 25th June, 1817, established at Rive-de-Gier [Loire,] a common "foresight chest" for the working miners of this basin; the association also had no long duration.

The two institutions, of the empire and the restoration, failed from the same cause. They were unable to resist the times; because they were not rooted in the economical habits of the workmen. They were almost exclusively benefit societies; the providence of the workman, which ought to have formed their principal characteristic, was there only an accessory.

The generosity of the proprietary had little utility, because they helped the unfortunate *unproductively*. The Belgian societies are not so: the gifts of the mine explorer and the subsidies of the government have resulted in stimulating the clubbing of the workmen, and of doubly interesting them.

The principal mine establishments of France usually possess local *caisses de secours*, in favour of their workmen; but no common association unites together several of these mines. These "chests" are organized and administered somewhat after the manner of those at the Belgian collieries.

Insurance associations multiply throughout France; they protect the assured against the principal disasters which can await him. The philanthropic societies of Paris, of Nantes, and of Mulhouse, contribute much to spread similar institutions and the germs of order and economy, in the working class.

The miners of France are still without public provident societies. It is more difficult to organise such associations in this kingdom than in Belgium; either because the miners are more scattered, or because the dangers are less. But it is not the less useful for the working class to see these imperfect distributions replaced by permanent institutions.

In continuation of the subject, we notice here the progress made to form somewhat similar institutions to those of Belgium. Among others in France, to which we shall refer, is the "relief chest" or fund, "*caisse de secours*," for the benefit of the working miners in the department of Ariége.

In 1842, a report was made by the minister, M. Teste, Secretary of State for the Public Works, to the king of the French, in which a project was submitted to his majesty for the establishment of a "*caisse de prévoyance*" at the mines of Rancié, in Ariége. The minister remarks, that these institutions, so useful to the whole working class, are especially needed by the miner, who is, by the nature of his employment, exposed to daily danger, and who can with great difficulty, by his own individual economy, assure himself of resources against the casualties to which he is subject.

The administration has long been aware of the influence which this kind of establishment would have over the well-being of this class of workmen.

A decree of the Emperor Napoleon, in 1813, prescribed, for the department of the Ourthe, the formation of the relief chest for the miners of that basin.

A similar institution was created in the department of the Loire, by royal ordinance, in 1817.

The advantages which were anticipated from these beneficent projects were not immediately realized. We cannot here act by force. They required the free assent and co-operation both of the proprietary of the mines, and of the operatives or workmen. This indispensable junction of the will of the whole often occasioned serious obstacles. Individual resistances fettered that which was the interest of all to extend.

In Belgium they have been more happy. The workmen in the five subdivisions of the coal territory of that country, have concerted together to organise relief funds and annuities; they have drawn up their statutes in a form which has met the approval of the government, and secured the benefits of its protective influence; and an association is thus firmly established which comprises the various concessionaries and workmen.

Belgium, in this respect, is in a most favourable position. In France, the works are, in general, more dispersed. With some exceptions, the working miners do not present in France, as with her neighbour, a population concentrated upon certain points; devoted exclusively, from father to son, to working in the mines; and having those habits of fraternity from which results a powerful and moral bond.

At the same time, it is but just to say, that several of the mines of France offer examples of provident or relief funds, which have been successfully established.

One of the largest establishments in the kingdom, the iron mines of Rancié, is placed in circumstances where these ameliorations can be immediately realized. These circumstances are detailed to the government by the minister. Although they are of an interesting character, we are compelled to omit them. Finally, he solicits the sanction of the king to the meditated institution, and thereby to confer on it that consistence and firmness which is essential to its well-being.

On the 25th May, 1843, a royal ordinance was issued to create a relief fund, "*caisse de secours*," in favour of the working miners of Rancié; and, by a series of articles, the plan of the institution and of the government, and the duties of the members, are detailed. The relief fund comprehends aid to the sick, the aged, and infirm miners, and affords assistance to the widows and children of deceased workmen. The plan has met with the general approbation of the community for whose benefit it was designed, and the cordial support of the mayors of the eight communes of the valley of Vicdessos; which communes were declared by royal ordinance, in 1833, concessionaries of the said mines of Rancié.*

Existing in the midst of the elements of destruction, the working miners have not been able to escape from superstitious impressions, and, on this head, we find that in mining countries, widely separated from each other, there exists a similar belief, which attributes most of the accidents to local or evil spirits who, in order to defend the subterranean treasures against the encroachments of man, oppose to his progress the waters, the gases, the fallings down, &c. There results a real evil from these superstitions: they afford a pretext for carelessness, already too prevalent among miners, and for neglecting to take the necessary precautions. Accidents are, moreover, multiplied with the extension of the subterranean works; and, while seeing men exposed all their lives to risks so terrible, there is no company or government that remains unmoved or has not sought to prevent them; at first by wise prescriptions, and then to mitigate their effects. The establishment of benefit societies—*caisses de secours*—is one method generally adopted. A fund created by a reserve from the wages of the workmen and other persons employed, and from the eventual profits of the society, is distributed among the wounded, the widows, and the children of those who fall. These funds are administered by a council, presided over by the administrator or manager, and of which the engineer, the cashier, and several master miners delegated by the workmen, form a part. This council regulates the number of pensions and the time which they ought to continue, according to the nature of the accidents and the position of the injured individuals, as to means of subsistence.†

A fact of some interest, when viewed in connection with the subject of these pages, has been elicited by the returns made to the French government relative to the savings banks of France. It appears that, in 1845, 123,000 workmen of different trades, were depositors in these banks; and that, out of this number, no less than eighty-one thousand were miners.

Considering that the mining population of France is not as 1 to 20 of the working classes, it would appear from these statistics, that miners are here more economical and prudent than any other section of the labouring community.

UNITED STATES OF AMERICA.

Bituminous coal works.—With the exception of the Richmond coal-field, there are few works where the subterranean operations are sufficiently extended to render them dangerous from the presence of fire-damp.

In March, 1839, an explosion took place in one of Heth's pits, Chesterfield, Virginia, by which a number of lives were lost.

These mines are from 400 to 700 feet deep, and are almost entirely worked by slave labour.

* Annales des Mines, 1843, Vol. III. p. 923.

† Burat, p. 610.

In 1844, another explosion took place in one of the Black Heath pits, while four Englishmen and eight negroes were in it.

1854, May 14th, another terrible explosion took place at the "English Coal Pits," in Chesterfield county, by which 20 men were killed.

Anthracite Collieries.—Some few of the deepest workings in the Pottsville district have been so far troubled with fire-damp, as to require some caution on the part of the miners. Explosions on a small scale occasionally occur, but we have no very serious cases to record, up to the present time; although it is evident that, as the mines are deepened, the risk and danger increases, and it will require greater circumspection hereafter.

In February, 1847, seven lives were lost by an explosion in Spencer's mine, near Pottsville.

Benefit clubs and associations for relief in times of sickness prevail in this mining district, after the English method.

Miners' Asylum.—A project was suggested through the columns of the Miners' Journal of Pottsville, in 1840, for the relief and support of such miners as became disabled by accident from pursuing their dangerous avocation. It contemplated the raising of a fund sufficient to erect a building for the reception of sick and wounded miners, and an annual contribution towards its endowment. The plan was originated in charitable and philanthropic motives, but was somewhat crude in its conception; and, as it proposed to be maintained from resources derived from a tax on the purchasers of coal, consumed at a distance and in great measure by other states, while the mining community was to be relieved from all the burthen, it was not likely to secure public favour. It has been thoroughly proved that no project of this sort is successful in its results, in which the expenses are not, to a certain extent, defrayed by the miners themselves; or, as in Belgium and in many instances in England, supported at the joint expense of the owners and the workmen, and carried on under their joint control and supervision.

In Pottsville, the great centre of the mining population of the anthracite region of Pennsylvania, we may infer that the moral condition of the working class, particularly of the rising generation, is carefully looked after, from the establishment of so many schools for their education. It was ascertained, in 1842, that there were then in Pottsville the following schools and number of pupils then receiving instruction in that place, which but a very few years ago was a barren and profitless wilderness.

	Pupils.	Teachers.
Sunday schools, - - - -	1137	150
Eight public schools, - - -	472	8
Private schools, - - - -	479	15

Out of the population then amounting to but 4500 souls, brought together from various parts of the globe, we thus find eleven hundred children receiving the benefits of a Sabbath school education, and nearly one thousand children who attended the public and private schools during the week days.*

Mauch Chunk, Anthracite District.—We have been favoured by the kindness of the Lehigh Coal Company with the following synopsis of the mining and general population, the schools of instruction and places of worship within the limits of the Company's coal operations, in the year 1846, all of which had scarcely commenced to exist twenty years ago.

The mining accidents are not very frequent: they have probably averaged three lives lost a year.

Details.	Summit Mine and vicinity.	Farm and Old Tunnel.	Nesquehoning mines.	Mauch Chunk.	Total.
Total mining population, - - - - - -	1,350	90	756	300	2,496
Working men employed at the mines, - -	650	30	300	100	1,080
Sabbath schools, English, - - - - - -	2		1	3	6
" " Welsh, - - - - - -	1				1
Common or public schools, - - - - -	4		2	4	10
Methodist Church, - - - - - - -				1	1
Roman Catholic Churches, - - - - -	1		1		2
Presbyterian Churches, - - - - - -	1			1	2
Episcopal Church, - - - - - - -			1	1	1
Division of Sons of Temperance, - - -				1	2
Division of Odd Fellows, - - - - -				1	1
Beneficial Society, - - - - - - -				1	1

And a population, at Mauch Chunk, of 1500 persons dependent on the coal operations, exclusive of those engaged in transportation and the coal trade, &c., here and elsewhere.

We have brought together in the foregoing pages a small portion of the details which, were we writing on no other subject than the moral and physical condition of the working population in the coal districts of various countries, would be very inadequate to our purpose. But our scope is too limited to admit of further extension in this branch of inquiry, however interesting it may be.

In the sketch before us, we have had a two-fold object: first, of glancing at the innumerable casualties attending the miner's life; at his moral and physical disabilities: second, at the means which have been taken by the philanthropic and the benevolent, in conjunction with his own exertions and the triumphant assistance of education, to ameliorate his condition, to compensate as far as possible for unavoidable privations, and to elevate him above a position which, as we have seen, is too frequently one of extreme degradation. It will

* Miners' Journal.

be regarded, with pleasure, that the efforts made in his behalf by the good and the influential of all these countries have not been profitless; and that the impulse thus given at the commencement of the work, has been seconded by the majority of the workers themselves. This is the true, the rightful working of the system; and the beneficial results already appear on every hand, where so many interests co-operate to relieve the wretched, to inform the ignorant to elevate the debased, to reform the improvident, to encourage the industrious, the successful issue appears inevitable; it is, indeed, already manifested in the improved condition of those, to whom these benevolent exertions have been directed.

To those who have perused the sickening details recorded in the Report of the Midland Mining Commission, and the almost incredible evidence of the condition of the South Staffordshire mining population, so late as 1842, the prospect of any amendment of such a lamentable state of society must be welcome. To those, again, who apprised, for the first time, of the degrading and demoralising employment of thousands of females in the coal mines of England and Wales, and moral Scotland, the interposition of the legislature, sanctioned by the approbation of the wise and humane throughout the empire, must afford cause for rejoicing in the progress of a reform so needful. The beneficial effects of education and of the Sunday school system, of which we have ample evidence both in the New and the Old World, have had the most happy influence on the industrial classes and on none more than the youth of the mining communities. The noble example set by the Belgian government in supporting the provident associations of the miners in that country, will be followed by other European states, and is already in progress in France.

NORTH AMERICA,

COMPRISING

1. THE UNITED STATES—including,
2. STATE OF OREGON,
3. STATE OF UPPER CALIFORNIA.
4. STATE OF TEXAS.
5. NEW MEXICO.
6. MEXICO.
7. BRITISH AMERICA.
8. RUSSIAN AMERICA.

UNITED STATES OF AMERICA.

AREA, exclusive of Texas and Oregon, estimated at 2,300,000 square miles.

	Population.	Slaves.
Population in 1840, - -	17,063,353	
Of which there were Slaves, -		2,487,355
Population in 1845,* - -	19,914,362	

The total area of this country, including Oregon, Texas, and California, is set down in 1851 at 2,981,123 square miles.†

The census report of Mr. Kennedy makes it amount to 3,230,572 square miles.

Population in 1850,‡	
Free, - - - - -	19,987,573
Slaves, - - - - -	3,204,347

Population in 1853, is estimated at 25,000,000§

Weights and Measures.

Ordinary estimate of Bituminous Coal—28 bushels = 1 ton of 2240 pounds.

Occasionally it has been customary to allow 30 do = 1 ton of 2240 pounds.

But, we also find it stated, in the west, at 26½ do = 1 ton of 2240 pounds.

At Richmond coal pits the common measure is 5 pecks to a bushel.

The same coal put on board at Richmond, is 4 pecks to the bushel.

The Richmond coal bushel at the pit's mouth, is said to weigh 90 pounds = 24 bushels and 80 pounds to 1 ton.

* American Almanac, 1847.

† According to Colonel Abert, Chief of the Topographical Bureau, Washington, D. C.

‡ Hunt's Magazine.

§ Boston Post, August, 1854.

The four peck bushel weighs 72 pounds, and the ton contains 31 bushels and 8 pounds.

In the south, bituminous coal is sold by the barrel, weighing 172⅓ pounds. There are therefore, 13 barrels to 1 ton of coal.

In the anthracite trade the prevailing standard is by the ton of 2240 pounds.

Occasionally, in retailing, the ton is only 2000 pounds; it is so quoted at New York, Cleveland, &c.

On the State Canal, and the Tidewater Canal, the toll is levied per 1000 pounds of coal.

Foreign bituminous coals are, or were, commonly sold by the chaldron of 36 bushels. A chaldron of these coals weighs 25½ cwt.

A bushel, measured when dry, weighs 84 or 85 pounds; but in Pennsylvania, in Ohio, at Cleveland, and several other places, the bushel is equivalent to 80 pounds.

What used to be sold under the denomination of a Newcastle chaldron, weighed 2 tons and 13 cwt.

The Nova Scotia chaldron is 1½ tons, or 3,360 pounds of 42 bushels; but the measurement yields 48 bushels.

The Boston retail chaldron is commonly 2,500 pounds, but sometimes 2,700 pounds.

The tariff duty was levied on the chaldron of 2,880 pounds, or 36 bushels of 80 pounds each.

A corde of wood 8 × 4 × 4 feet = 128 cubic feet.
The Austrian corde of wood is 88¼ do.
40 feet of round timber } = 1 ton.
50 feet of hewn timber }

1 pound	= 0 kilogrammes	.4535
1 barrel of flour	= 98 do.	.391
1 bushel	= 35 litres	.236
1 yard	= 0 mètres	.9144

1 Mexican or Texas *Vara*, represents 33⅓ English inches, = mèt. 0.847 French.

United States Currency.

1 dollar = 100 cents, = 4*s.* 1¼*d.* = 5*fr.* 35*c.* [4*s.* 16*dec.*]
1 cent, = 0*fr.* 5*c.* 35.
Par value of 1 United States dollar in London, 47½ pence.
£1 sterling = $4.84 nearly, = 25.89*fr.* to 25.76, legal value.
1 shilling, English, $0.24*c.* 20.
1 crown " 1.21.
The value of the 5 franc piece is fixed by Congress at 93 cents.

Spanish American Currency.

1 hard dollar, = 100 cents, = 4*s.* 3*d.* = 8 reales.

Previous to the 31st July, 1834, the American eagle contained 270 grains of standard gold, viz. 247 grains pure, and 23 grains alloy. By the Act of Congress of that date, the weight of the eagle was reduced to 258 grains, of which 232 are pure gold, and 26 alloy. In consequence of this alteration, the sovereign, or pound sterling, that was formerly worth $4.57 cents, is now worth $4.83.8 cents.* Under the present American system, it is believed that gold is over-valued from 1¼ to 1½ per cent.

Money.

"The mode still adhered to by many of quoting exchange between the United States and London, is both *obscure* and *absurd;* as the premium or discount is founded upon the *false*, or nominal *par* of $4.44, instead of the true par of $4.86." In the calculation of duties at the United States custom-houses, since 14th July, 1832, the value of the pound sterling is fixed by law at $4.84.

The banks receive and pay out sovereigns at $4.85.†

Among the "Documens sur le Commerce extérieur," of France, published in June, 1843, occurs the following passage relative to the pound sterling. "In 1842, an Act of Congress, of the 27th July, fixed the legal value of the pound sterling at 4.84, [=25*fr.* 89*cts.*] for the conversion, in American currency, of prices to foreign agents, and for payments into the American treasury. Previously, this value was $4.44 [=23*fr.* 75*c.*] a rate that the American treasury had substituted, in 1840, for that of $4.80, [=25*fr.* 68*c.*] established by law of the 2d March, 1799."‡

The weight of the Spanish and American dollar is 416 grains troy.

100 dollars, therefore, are equal to 211,742 Sicca rupees.

or, deducting the duty, to 207,508 Sicca rupees.§

The dollar is a legal tender, at the price of 4*s.* 4*d.* English currency in the East Indies.

Value of Foreign Moneys,

AS TAKEN AT THE CUSTOM-HOUSE IN NEW YORK, IN 1846.‖

Barcelona and Catalonia livres, - - - $0.35¼

* Bicknell's Gold Chart. Also, Moore's "Philadelphia Price Current."

† New York Journal of Commerce, October, 1839.

‡ In continuation of this subject, we insert the following illustrative note:

A COMMERCIAL ABSURDITY.—The current quotations, as 7, 8, or 9 per cent. premium for exchange on England, which we see in the newspapers, do not mean a premium on the par value of the pound sterling, but on a fictitious valuation of the pound which prevailed in this country a century ago, when the States were colonies. For example, the pound sterling, or gold sovereign, is to day worth $4.85 in Wall street, which is about the par value as established by law of Congress. A thousand of them would be worth $4850. The current rate of exchange on England in Wall street is now about 9 per cent. premium, as the phrase is, for bills payable in London or Liverpool. But the premium is not on $4.85, the par value of the pound, nor yet on the pound sterling, but it is on $4.44, the old colonial value of the pound. For example, A. B. buys a bill of exchange for £1000 on England, from C. D., at 9 per cent. premium; he pays $4844.44 for it. Suppose he gave a thousand sovereigns for it, at current value, there would be a balance in his favour, so that, in reality, the rate of exchange on England, instead of being 9 per cent. against us, is in our favour, because bills can be obtained cheaper than gold. Of course, then, there is no object in sending gold to England. Hence the absurdity of this ideal mode of dealing in exchange on England, which is still kept up by our merchants and newspapers.

§ Hand Book of India, 1844, p. 65.

‖ Williams's Statistical Companion, 1846.

16

Brabeint florin, - - - - -	0.34
Bremen dollar, - - - - -	0.78
Bengal, Calcutta and Bombay sicca rupee, - -	0.50
China tale, - - - - -	1.48
Crowns of Tuscany, - - - -	1.05
Denmark Rix dollar, - - - -	1.00
Ducat of Naples, - - - - -	0.80
Dutch florins or guilders, - - - -	0.48
English pound sterling, fixed by law, - -	4.84
French francs, - - - - -	0.18¾
" livre, - - - - -	0.18½
" five franc piece, - - - -	0.93
Genoa livre, - - - - -	0.18¾
Halifax pound, - - - - -	4.00
Hamburg Rix dollar, - - - -	1.00
Leghorn dollar, - - - - -	0.90¾
Louis d'or, or Rix dollar of Bremen, - -	0.78½
Portugal milreas, - - - - -	1.24
Russian rouble, - - - - -	0.10½
Spanish rial of plate, - - - -	0.10
" of vellon, - - - -	0.05
Saxon dollar, - - - - -	0.69

System of Weights and Measures

IN GENERAL USE, IN RELATION TO COAL, IN THE UNITED STATES OF AMERICA.

In our progress through this section of our work, it has been our endeavour, when reporting on given amounts of coal, whether anthracite or bituminous, to render them in one uniform standard. We have effected this, at the cost of much extra labour, and have brought out the results in a common denomination—that of WEIGHT. We have done so under the conviction that, sooner or later, the principle must be universally complied with; whatever may be the present prevailing provincial customs, or the ordinary usages of trade.

Whether 2240 be the most scientific or appropriate number of pounds to constitute one ton, or otherwise, it certainly possesses the convenience of general adoption, in the principal coal-producing countries of the world, in our times; and of being employed by the great maritime nations, among which Great Britain and the United States stand pre-eminent.

In this country, there can be no sufficient reason assigned why bituminous coal should almost invariably be sold in bulk—that is to say, by *measure*—and anthracite by *weight;* or that the former should be calculated in the eastern ports by the chaldron; in the southern by the barrel; at the mines by the bushel; and, although much more rarely, by the European chaldron and the ton:—which ton is generally 2240 lbs. weight at the place of production, and 2000 lbs. at the place of consumption; especially in the eastern ports.

Without altogether discarding the old denominations, in our returns, we have, at least, accompanied them by other tables, representing an uniform standard of weight. We have not limited the process to American returns, but have applied it to every country. In relation to the United States, we have done so the more readily, because we hoped thereby to be instrumental in terminating the highly objectionable system of buying and selling a mineral substance like coal by measure; whether that measure be a peck, a bushel, a barrel, or a chaldron; for we have all these varieties, and all are equally indefinite, equally liable to abuse, and equally disadvantageous to both buyer and seller.

In Europe,—in all the great coal-producing and coal-buying countries,—it has long ago been demonstrated, after very full investigation in all its bearings that there exists no fair and equitable system, suited alike to the buyer and seller, the miner, the producer, the transporter and the consumer, except that of *weight*.

Local usages and peculiarities are always sources of embarrassment in commercial transactions. We feel their influence in this country daily.

In Pennsylvania, for instance. In the tariff of tolls on bituminous coal and anthracite,—fixed officially and annually, and to be received on the State and Tide-water canals and railroads,—the article, mineral coal, is charged per 1000 lbs. weight. Now, as in Pennsylvania there is no such weight recognized by the producer or by the trade, for the reason already assigned, that the bituminous coal is sold by measure and not by weight, this new denomination applied to the article on its transit merely, is obviously a source of inconvenience to more parties than one.

So also in relation to anthracite; for as all which is transported on these canals and railroads is mined by the customary ton, of 2240 lbs.; conveyed to the landings by the ton; freighted by the ton, and are bought and sold by the same weight: the departure from a universal practice of the trade by the interposition of the 1000 lbs. standard, instead of the genuine ton, not only occasions unnecessary trouble, both to the payers and the collectors of toll, but interposes an uncalled for difficulty in one branch of the trade.

Neither is the rule so general as to demand this interposition, on the score of conformity; for instance, the coal which has descended the Tide-water canal, and has there paid toll per the 1000 lbs. weight, on passing into the Delaware and Chesapeak Canal, pays its toll on the ton of 2240 lbs.

Again, in relation to the movement of coal in neighbouring states, we may remark on the want of uniformity in the system of weights, which circumstance interferes with the means of acquiring correct statistical information. Thus, the bituminous and anthracite coals which pass down the Schuylkill navigation, in Pennsylvania, are returned by the large ton, while all that pass over the New York canal are returned by the small ton, of 240 lbs. less; and yet the same coal was imported into the state by a different scale of measure, both that which was shipped at Cleveland or Erie, for Buffalo, or that which came from Pennsylvania by the Tioga railroad.

In Pennsylvania, the Union Canal,* the Schuylkill Navigation, and the Lehigh Navigation, as well as the Delaware and Chesapeake Canal, levy their tolls, and arrange their freights, by the standard ton adopted by the trade: thus employing a different system to that used on the state works.

We will proceed to point out some of the extraordinary discrepancies which prevail in relation to the coal trade; and which, considering we have heard for years past that great attention has been paid at Washington to the establishment of a national system of weights and measures, our readers would not suspect was still remaining in full operation.

Towards the commencement of the coal trade in Pennsylvania, even anthracite was calculated by the bushel. In the vicinity of Pottsville and Wilkesbarre, in those times, leases of mines were granted, the lessee in the former place paying two cents a bushel.†

In some parts of the bituminous coal-field, thirty bushels have been supposed to be equivalent to a ton in weight; in others twenty-eight bushels for gross weight and twenty-five bushels for minimum weight; and we have also heard of twenty-six bushels as representing the ton. For a long time, the usages of the trade, as regards anthracite, assigned twenty-eight bushels as the equivalent of one ton of Lehigh coal, thirty bushels of Schuylkill, and thirty-three bushels of Lackawanna coal. It is needless to point out the utter worthlessness of a system, if system it can be called, so vague, so utterly incorrect and unphilosophical; a practice which operated, so long as it was pursued, equally to the prejudice of the producer and consumer. Yet, it will scarcely be credited, the early returns, during several years, of Lehigh anthracite, were made in bushels.‡

There are numberless and insuperable obstacles to making weight and quantity synonymous terms. In point of fact, there is so much guess work—so much uncertainty, in assigning a standard of weight against a given bulk—which bulk is, of itself, entirely unsettled in real practice, that the consequence, not unfrequently, is that bituminous coal, in the large way, is not really measured at all. The present custom observed is this: Contracts for the coal which descends the Pennsylvania State Canals, are generally made by the ton of 2000 lbs. weight: considered equivalent to twenty-five bushels, each bushel being estimated to weigh 80 lbs. But what is termed the gross ton of twenty-eight bushels, which is supposed to represent the 2240 lbs., is in as frequent use; one being as often used by the shippers from the Allegheny mines as the other.

On the western rivers twenty-eight bushels represent a ton. On the Union Canal, of Pennsylvania, the liberal allowance of thirty

* The returns of bituminous coal passing on the Union Canal, used sometimes to be made in bushels and sometimes in pounds—latterly in tons.

† Pamphlet on the Coal and Iron business, Poughkeepsie, 1828.

‡ Dr. James's History of Pennsylvania Anthracite, Memoirs of the Historical Society of Philadelphia, Vol. I. 1826.

bushels to each legal ton is made and accounted for. Yet, in this case, the coal conveyed on the Union Canal is specifically heavier than that of the western rivers, in the proportion of 1.350 to 1.230 spec. grav. The official returns to Congress comprehend all bituminous coals throughout the Union, be their specific gravities what they may, at 80 lbs. to the bushel, and twenty-eight bushels to the ton. In Michigan, the coal business is conducted by the bushel measure.

At the coal pits at Chesterfield, near Richmond, Virginia, the coal trade adopted as the standard, five pecks to the bushel; weighing ninety pounds. Consequently each ton of 2240 lbs. actually contains only twenty-four bushels and 80 lbs. over. This measurement, it must be stated, solely applies to coal at the pit's mouth. At the terminus of the railroad, twelve miles from the mines, another system commences. Here, at Richmond, where the coal is shipped, the orthodox bushel is four pecks. This bushel, therefore, weighs 72 lbs., and the ton is now represented by thirty-one bushels and eight pounds over, instead of twenty-four bushels ; notwithstanding which the sales in Boston and New York are made by the chaldron of thirty-six bushels, or by the ton of twenty-eight bushels. At Baltimore, twenty-eight bushels.*

In the southern ports, in Pensacola, Mobile, and New Orleans, another and peculiar standard prevails for the sale of bituminous coals; and we find that an indeterminate measure of capacity, called a barrel, prevails. Thirteen of these barrels constitute one ton; each barrel, whatever be the specific gravity of the coal, being calculated to hold a quantity which corresponds with two and a half bushels. This coal was purchased at the mines by the bushel.

With regard to the customs of the trade in the eastern ports: in Boston, foreign bituminous coal is imported and sold by the chaldron; American bituminous coal is generally sold by the bushel; and anthracite is purchased at the rate of 2240 lbs., and retailed at 2000 lbs. the ton. Sometimes the returns are given in tons, sometimes in bushels, sometimes in chaldrons, and one denomination being occasionally mistaken for another, we need not wonder at the singular discrepancies in the published statements of the coal trade there. In Philadelphia, anthracite, both wholesale and retail, is always sold by the legal ton of 2240 lbs.† In New York and Boston, the ton is only 2000 lbs., thus gaining six tons on every cargo. On the Reading Railroad, a ton of coal is 2240 lbs., but a ton of merchandize is only 2000 lbs.‡

Nova Scotia coal is imported, in some quantity, into Boston; always by the chaldron. But what constitutes a chaldron seems a matter of somewhat arbitrary character. It sometimes is fixed at 3000 lbs. weight; sometimes at 2928 lbs.; but most frequently at 3360 lbs., or one ton and a half. The tariff duty is customarily levied on the chaldron of 2880 lbs., or thirty-six bushels; while the

* Baltimore Report, Nov. 16, 1843.

† In 1853, the coal proprietors agreed to fix the retail measure of a ton at 2000 lbs.

‡ Reports of the Reading Railroad Company, January 13, 1845, p. 15, and subsequently.

retailer sells a chaldron, which is sometimes 2500 lbs. and sometimes 2700 lbs. weight. The Nova Scotia chaldron of $1\frac{1}{2}$ ton, should contain forty-two bushels, of 80 lbs. each; but the custom of the trade, we are informed, raises the admeasurement to forty-eight bushels. In like manner, the ton is rated at thirty-six bushels, instead of twenty-eight.

Amidst all the intricacies of these returns and dealings, it is very difficult to get at the real quantity and prices of foreign imported coals; as the number of chaldrons purchased at the place of production, materially differs from that on which duty is paid, and from that which is sold to or by the retailer.

The registration of imports of Virginia coal in Boston and New York, is by the number of bushels only. At Sydney and Pictou, the mine or colliery measure is thirty-six bushels to the ton, even measure, or twenty-four bushels heaped measure.

While on this subject it may not be altogether out of place to note that complaints have been made in Pennsylvania respecting the irregularity prevailing in relation to the weights of other substances besides coal. We have recently observed an article in a Philadelphia paper to the following effect: "It is a singular fact that our measures for grain are larger than those of New York, Boston, or Baltimore. This deviation from uniformity is greatly complained of by our country dealers and farmers, who ship to Philadelphia, from Maryland, Delaware, and Virginia. A very large quantity of grain is shipped from those States to New York and other ports, which would come here, were it not for the inconveniences arising from the falling short of the measure. Some years since, Congress passed a law providing for uniformity of weights and measures throughout the Union. But we regret to say that our old standard of dry measure still holds its place."*

The same writer alleges that equal ground of complaint exists in relation to the measurement of bark, and the same difficulty with respect to the retail measure of charcoal, in Philadelphia, has been lately settled by municipal legislation.

In the lead region of Missouri the present standard of weight appears to be on the 1000 lb. weight, as in Wisconsin and Iowa. Formerly, the custom prevailed of 108 lbs. to 1 cwt., or 2160 lbs. to the ton.

In Pennsylvania, the weight of a ton of iron is local and arbitrary. Thus, we are informed by an experienced iron master, of Centre county, the number of pounds usually assigned as a ton vary according to the following scale:

Iron Ore,	Sometimes taken at	2240 lbs.	
	and sometimes at	2480	
Manufactured Iron,	Pig iron universally	2240	
	Blooms and Puddled Iron	2464	The ankoney or double gross ton.
	Nails	2000	

* The North American, January 10th, 1845.

In the tariffs of the United States customs, in every case the ton is required to be of the weight of 2240 lbs.

Among many other irregular and uncertain customs of local weights and measures, we extract the following from our notes.

In Kentucky, corn is measured by the barrel, which is five bushels of shelled corn. At New Orleans, a barrel of corn is a flour barrel full of ears. At Chicago, lime is sold by the barrel, and measured in the smallest sized cask of that name that will pass muster. A barrel of flour is seven quarters of a gross hundred, (112 lbs.) which is the reason of its being the odd measure of 196 lbs. A bbl. of tar is 20 gal., while a barrel of gunpowder is only a small keg holding 25 lbs., and of cotton, a *bale* is 400 lbs., no matter in what sized bundles it may be sent to market.

Ere we terminate this article, we will advert to two or three facts that have come to our knowledge respecting the uncertainty of any standard of measurement, after long experience, that can be adopted as a substitute for weight, in the sale of coals. For instance, 1 bushel of English coal, *measured when dry*, weighs from 84 to 85 pounds. The American bituminous coals are commonly averaged at 80 pounds per bushel. The same English coal, *if measured when wetted*, paradoxical as it may appear, the weight will be found not so great. The fact is proved, conclusively, that in the dry coal the small particles run to fill up the cavities, making the whole almost a solid mass: whereas a bushel of wet coals only closes up the hollow cavities; the fragments clog together, and the whole do not weigh so much as the dry coal of the like admeasurement.

With regard to the increased measure acquired by breaking up coal, it was commonly proved by the trade, that that which in the large or coarse state measured *five bolls* (say tons or chaldrons,) when broken up, fine, in the hold of the ship after delivery on board, measured *nine* bolls.

As to the continuance or toleration of the system of *heaped measure* for coals, we trust that an end will ere long be put to what has, with perfect propriety, been termed "a barbarous custom." A commercial author Mr. McCulloch, observes that "all articles that may be sold by heaped measure, *ought* to be sold by weight. In Scotland indeed, the use of heaped measure was abolished *above two hundred years since.*" The French, Belgian, Prussian, Austrian, Spanish, Portuguese and nearly all other European nations, adopt well ascertained weights for the purchase and sale of coal, and not measures of capacity. Throughout Hindostan coal is always sold by weight. In the Indian countries north of the Nerbudda river, there is no dry measure of capacity, and everything is, therefore, sold by weight. This appears also to be the case in most of the Nizam's districts, adjoining those of Ahmednugger. The introduction of a system of measure into the Decan, seems to be of a late date.*

In the English act of 5 and 6 of Will. IV. 1835, are the following important provisions:

* Martin's Colonial Statistics. Appendix iv. p. 143.

It abolishes all local or customary measures.

It prohibits "the mischievous practice" of *heaped measure.* All bargains, sales, and contracts, made after the passing of this act, by heaped measure, shall be null and void: and every person who shall sell any articles by heaped measure, shall be liable to a penalty not exceeding 40*s.* for every such sale.

From and after the 1st of January, 1836, all *coals*, slack, culm, and cannel, of every description, *shall be sold by weight, and not by measure*, under a penalty of 40*s.* for every such sale.

All articles [except gold, diamonds, &c.] shall be sold by standard *avoirdupois weight*, of 14 lbs. to the stone, and 8 stone to the hundred weight, and of 20 such cwts. to the ton = 2240 lbs.

"The fact, that so monstrous a system should have been persevered in for more than a century, sets the power of habit, in reconciling us to the most pernicious absurdities, in a very striking point of view. Happily, however, the nuisance have been at last abated.*

The United States is the only coal country, of importance, in the world, where the practice remains uncorrected.

The duty on foreign coke and culm, prior to the modification of the tariff in 1846, amounted to 60 per cent., on its wholesale market value, English, at the principal coal shipping ports, in 1846.

That on foreign bituminous coal was from 70 to 90 per cent., on shipping prices aboard, which is, among the highest duties, payable on any imported article, under the operation of the tariff of 1842.†

A drawback is allowed on foreign coal exported from the United States, in such cases, for instance, where it has been landed and placed in depot, for the use of the British steamers. This law was confirmed by the Senate of the United States, in January 1840.‡

The amount of re-exported coal, on which the duty was remitted, was 11,364 tons, in 1849, valued at $45,957.

We have no official return of the amount of American coal annually exported into Canada from the lakes. (See p. 243.)

The act of Congress, passed July, 1846, rescinded the tariff of 1842, and substituted a modified one, which took place, December 1, 1846.

The annexed table is founded on the Report of the Secretary of the Treasury to Congress, Dec. 3, 1845.

* McCulloch's Commercial Dictionary, p. 294.

† Letter of the Secretary of the Treasury to Congress, December 16th, 1844.

‡ Hazard's U. S. Commercial and Statistical Register, 1840. Also Commercial Reciprocity—Hunt's Mag. Vol. X. pp. 358 and 526.

American Tariff of Duty on Foreign Coals.

Import duties on foreign bituminous coals brought to the United States. In all cases the duties are collected upon the ton of 2240 lbs.

	Year.	Duty per bushel heaped.	Per ton of 28 bushels.	Per Chaldron of 36 bushels.	Average selling price at New York, per Chaldron or Ton.	Rate per cent. on value.
		Cents.				
TARIFF.—4th July, . .	1789	2	$0 56			22
10th August, .	1790	3	0 84			33
2d May, . . .	1792	4½	1 25	$1 65	$19 00 Chaldron,	50
7th June, . .	1794	5	1 40	—		56
1st July, . .	1812	10	2 80	3 60		112
By act 27th April, continued to 1824,	1815	10	2 80	3 60	20 to 23 do.	112
	1816	5	1 40	1 80	12 to 15 do.	56
By act 2d May, continued to 1833,	1824	6	1 68	2 16	10 to 11 do.	67
Compromise act of . .	1833	5	1 40	1 80	8 to 14 per ton. proposed to be gradually reduced.	56 to 40
August 30th, ...	1842	6¼	1 75	2 70	7 16 per ton. 5 56 do.	70 to 90
Proposed Tariff in . . .	1844	6 to 5½	1 00	1 50	defeated and abandoned,	50
1st December, 1846, . .	1846	an advalorem duty of 30 per cent. equiv. to 0 45			7 00 per ton.	30

*British Export Duties.**

On coals shipped from the ports of Great Britain to foreign ports, from the year 1835 to 1842.

To foreign countries in British ships, 10 per cent. *ad valorem.*

To " " in foreign ships, { 4*s.* per ton, for large coal. 2*s.* per ton, small coal and culm.

To British possessions in British ships, free.

These duties were abolished in 1845; no duties received since March, 1845.

There is no *import duty* on American bituminous coals or anthracite, brought into the Canadas, or any part of British America.

As regards Louisiana, by a treasury letter, July 3, 1821, the trade of Louisiana is placed on the same footing as that of the United States of America, by the government of Great Britain.

* Official Documents of Great Britain.

Gross Importations of Foreign Coals,

From Great Britain, British America, and all other places, into the United States, both in American and in foreign vessels, from 1789 to 1853, inclusive, showing their declared value, the tariff, and the amount of duties received thereon ; the commercial year ending on the 30th June, annually.*

Years.	Bushels.	Tons.	Official Value.	Duties Received.	Tariff.
			Dollars.	*Dollars.*	
1789†	107,810	3,850			2 cents per bushel.
1795	125,357	4,477		8,338	5 cents do.
1800	330,041	11,787		25,150	
1805	498,543	17,805		25,810	
1810	392,857	14,030		19,907	
1814	19,367	691	War.	War.	10 cents do.
1815	98,398	3,514	Peace.	Peace	
1820‡	673,711	24,061		53,685	5 cents do.
1825	722,255	25,795	108,527		6 cents do.
1830	1,640,295	58,582	204,773	98,417	
1835	1,679,119	59,972	143,461		
1837		153,450			
1839		181,551			
1840	4,560,287	162,867	387,238	273,610	
1842	3,962,610	141,526			
1843		41,163	116,312		
1844		87,073	203,681	133,845	30 per cent.
1845	8,543,327	85,776	187,962	130,221	
1846		156,853	336,691	254,149	
1847		148,021			
1848		196,168	426,997	128,099	
1849		198,213	382,254	114,676	
1850		180,439	361,855	108,557	
1851			478,095	143,429	
1852			405,652	121,695	
1853		231,508	490,010	147,003	

From this table of gross *importation* have to be deducted, in order to show the actual *consumption*, the annual quantities taken out of storage at the depots for the Atlantic steamers, &c. and re-exported. This re-exported quantity is about one-twenty-fifth part of the entire importation.

There are other statements of the importations of English coal, differing from the foregoing, but we take the following table from the official returns of the British exports. The reduction since 1842, as compared with the seven preceding years, is attributable to the operation of the tariff of that year ; the imports from England have not materially changed in the aggregate, and those from Nova Scotia were about the same in 1847 as in 1842. At the same time the importations of English and colonial coal into Boston have increased since the tariff of 1842.

* Register's Office, Treasury Department, 1850.

† Statistical Annals of the U. S. of America, by Adam Seybert, M. D., 1818.

‡ Report of the Secretary of the Treasury, December 3d, 1845, and subsequently, to 1853.

§ Reading R. R. Report, 1854.

Coal.—Balance of Imports and Exports.

Table of British and colonial bituminous coals, culm and coke, received in the United States, chiefly New York and Boston, from 1822 to 1853, deducting the coal re-exported, and showing the *consumption:* in tons of 28 bushels.

	Years.	British and Colonial tons of 28 bus.	Value.		Years, ending June 30.	British and Colonial Coals retained in the U. S.	Coals of Great Britain only, according to the Parliamentary returns.		
							Tons.	Declared value.	
			Dollars.			*Tons.*		*Sterling.*	*Dollars.*
Paying duty 6 cents per bushel.	1822	34,672	139,790	Duty 6 cents per bushel.	1836	108,432	30,220	£17,080	71,052
	1824	27,314	111,541		1837	153,450	46,574	29,252	121,683
					1838	129,083	57,175	27,949	116,267
	1826	34,647	145,262		1839	181,551	52,930	40,013	166,454
	1828	32,364	104,292		1840	162,867	77,559		
					1841	155,394	52,273		
				$175 per ton.	1842	141,450	68,407		
	1832	72,978	211,017		1843	41,163			
					1844	87,073	29,232		
	1834	71,626	200,277		1845	85,776	58,381		
					1846	156,853	57,903		173,290
				30 p. c.	1847	148,021			
					1848	196,168			
					1849	198,213			
					1850	180,439			
					1851				
					1852				
					1853	231,508			

Average Prices of Foreign Imported Coals,

At the ports of shipment, according to the official valuations, returned to the United States.

Years.	Per ton of 2240 pounds Foreign Coal.	Years.	Average price per ton of 2240 pounds Foreign Coal.	Wholesale prices at Philadelphia, per ton of Anthracite.
	Dollars.		*Dollars.*	*Dollars.*
1822	4 00	1836	2 27	
1824	4 08	1838	2 40	
1826	4 19	1840	2 37	5 50
1828	3 21	1841		5 00
1830	3 50	1842	2 68	4 25
1832	2 90	1843	2 83	3 50
1834	2 79	1844	2 72	3 37½
		1845	2 62	3 50
		1846	2 41	4 75

The fourth column shows the average price, at the place of shipment, of all descriptions and qualities of foreign coal, both English and colonial. The colonial coal is shipped at a lower price than the European.

The amount of Nova Scotia and Cape Breton coal imported into Boston during the ten years from 1835 to 1845, was 314,565. In 1845, 33,628 chaldrons of this coal were imported into Boston, or 42,035 tons. In 1846, 26,851 tons only. In 1849, 34,531 chaldrons; 1850, 32,486; 1851, 30,183.

Condensed View of the Importation of Foreign Coal into the United States.

The following statement is compiled from both the American and the British parliamentary returns, and, although incomplete, will probably furnish the most approximate view of the actual importation of coal into the United States; in tons of 2240 lbs.

In framing this statement, we have endeavoured to rectify various discrepancies in the returns, but not always satisfactorily.

Years.	British Coal, from the Parliamentary Returns.*	Colonial Coal, from the U. S. Returns.	Gross Importation. Bushels and Chaldrons reduced to tons.	British Coal re-exported for the use of the Steamers.†	Gross Value of Imported Coal.	Average value per ton, at the Shipping Ports.
	Tons.	*Tons.*	*Tons.*	*Tons.*	*Dollars.*	*Dollars.*
1801			21,027			
1810			14,031			
1820	average of 10 years annually.		34,205	150	108,527	4 08
to						3 80
1830					204,773	3 50
1831	15,103	21,406	36,509			
1832	42,210	41,934	83,144		211,017	2 90
1833	28,512	63,920	92,432			
1834	39,855	51,777	91,632		200,277	2 79
1835	19,585	40,387	59,972		143,461	2 39
1836	30,220	78,212	108,432		244,995	2 27
1837	46,574	106,876	153,450			
1838	57,175	71,908	129,083		308,591	2 40
1839	52,930	128,621	181,551			
1840	77,559	85,951	163,510		387,238	2 37
1841	52,273	103,121	155,394			
1842	68,407	73,114	141,521		380,635	2 68
1843	10,917	64,186	75,103	8,557	116,312	2 83
1844	29,832	57,241	87,073		203,681	2 72
1845	58,381		85,776	11,364	187,962	2 60
1846	57,903	95,330	156,853		336,691	2 41
1847			148,021		370,985	2 50
1848			196,168		426,997	
1849			198,213		387,370	
1850			180,439		361,855	
1851					478,095	
1852					405,652	
1853			231,508		490,010	

* Parliamentary Tables of Revenue, Commerce, &c.

† Treasury Reports U. S., December 3d, 1845, to 1853.

Tariff Duties on Coals Imported.

The question of how far the coal trade of the United States required protection from an external competition, has, at various times, been the subject of public discussion among the parties interested.

As may be seen in the preceding table, various modifications have been introduced, from time to time, in the scale of duties on imported coals.

The Committee of the Senate of Pennsylvania appointed to investigate the subject of the coal trade, reported March 4, 1834; and, among other important points, stated that they were led to the consideration of the question, "whether the bituminous coal of Pennsylvania can be brought into general use, east of the mountains, for manufacturing purposes, and be transported to the eastern markets upon such terms as to supersede the use of foreign coals?"

The report proceeds to notice the effect that duties on foreign coal had heretofore produced on the sale of American coals, in the markets on the sea board.

In 1815, when the duty on foreign coals was three dollars and sixty cents, the price in New York was twenty-three dollars the chaldron, of thirty-six bushels.

From 1816 to 1823, inclusive, during which time the duty was one dollar and eighty cents, the average price was about eleven dollars.

From 1824 to 1834, the duty was two dollars and sixteen cents, and the average price about fourteen dollars.

For the twenty years prior to 1834, the average price has been about twelve dollars and fifty cents; and, therefore, it has not varied in proportion to the tariff; nor does it appear to have been influenced by the rates of duty—for, in 1821, when the duty was one dollar and eighty cents per ton, the price of coals was fourteen dollars; and, in 1830, when the duty was two dollars and sixteen cents, the price was only eight dollars. The difference in price, it would therefore seem, has been produced by other causes.

From 1824 to 1834, the duty was one dollar and eighty cents a ton, and the average price during the same period, was about ten dollars; yet in the latter year it declined to five dollars and fifty cents, and five dollars per ton.

In 1842, when the duty was one dollar and seventy-five cents, the average price in New York was seven dollars and sixteen cents per ton.

In 1844, with a duty of one dollar, the price was five dollars and fifty-six cents. In 1846, with an ad valorem duty of thirty per cent., or about forty-five cents per ton, the price was seven dollars.

The authors of the Report observe, that there are other causes which co-operate in influencing prices, more than the tariff. The price, heretofore, seems to have been governed, almost entirely, by the scarcity or the demand for fuel.

For ourselves, we think that inferences drawn from the state of the markets at any period reaching further back than the last fifteen or twenty years, are of very little avail, and indeed ought to be discarded, as unsound. Previously to this period, the United States was not a coal producing country. Its fuel was the timber of the forest; it supplied no coal for domestic use in the eastern cities, and consumed but an insignificant amount of the foreign coal; which amount has been decreasing for the last twelve years, even with a diminished tariff. These duties could have exercised no influence on the prices, or upon the supply of anthracite from Pottsville, or of bituminous coals from beyond the Alleghanies; because, in point of fact, neither of them had reached the seaports, previously to the time of which we speak.

There can be no foreign competition now feared in relation to anthracite; and, probably, very little in relation to any substitute for that fuel, in the shape of bituminous coals—foreign or domestic. We do not think that the bituminous coal of Pennsylvania will ever find an extensive sale on the seaboard. The use of anthracite for all domestic purposes is so firmly established, that no other quality will henceforth find admittance into our houses. It is for the home consumption of the interior, and eventually the countries lying north of the great lakes, that the northern coal-fields of the United States may look for increasing markets. For the same cause, there never can be any large demand or sale for foreign bituminous coal at the sea-ports, except, perhaps, near the northern frontier; because its uses are, from the causes we have specified, much more limited on the eastern borders of America than in most parts of the world.

That these are not newly or hastily adopted opinions, may be seen from an article on the same subject, published by the writer in 1840. After reviewing the position of the Pennsylvania anthracite trade from 1820 up to 1839, and that of the foreign importation of bituminous coal along the Atlantic sea-board, during the same period, the author remarks, "It appears, that in a similar ratio as the consumption of this admirable fuel—the Pennsylvania anthracite—increases, so does the importation of foreign or English coals diminish; and the remark even extends itself to the diminution of Richmond bituminous coal, during a given term of years."

"The anthracite trade of Pennsylvania is decidedly on the increase; while the bituminous coals, borne coastwise or imported, have either remained stationary or have furnished a diminished supply."*

The seven subsequent years of the Boston coal trade has proved that the writer's views were not far from the verified results. The comparative business done in the years 1840, 1847, 1850 and 1853, are as follows:

* Report to the Dauphin and Susquehanna Coal Company, by R. C. T., May 1, 1840.

	Tons in 1840.	Tons in 1847.	Tons in 1850.	Tons in 1853.
Pennsylvania Anthracite received,	73,847	258,093	294,675	362,441
American Bituminous Coal,	3,299	4,554	2,265	
Foreign " "	49,997	65,203	42,532	62,110

Thus while the anthracite importation has increased three and a half times, that of the foreign and the American bituminous coal, has been almost stationary.

American Coal Exported from the United States in 1849-50, *to the British Dependencies.*

	Tons.	Value.
1846—To the British West Indies,		$765
1849-50.		
To British American Colonies, -	1102	4,549
To British West Indies, - -	- -	10,090
To Canada, - - - - -	9076	36,813

Preliminary Sketch of the Coal Fields and Coal Trade of the North American Continent, as at present known to us.

The substitution of mineral coal, or of any other combustible than the timber of its indigenous forests, whether as a domestic fuel, or for the manufacturing purposes of an increasing population, had its origin in America scarcely more remote than the memories of the living generation. In Pennsylvania, the anthracites, which now number by millions of tons, their annual production, were unknown to the community twenty years ago, and had then but commenced to find their way into the dwellings of the wealthier inhabitants of our maritime cities.

It will be remembered, that nearly the whole area of the great basin of the Mississippi, the valley of the Ohio, and the western slope of the Alleghany mountain or Appalachian range, embracing the great central coal-field hereafter to be described, was—although geographically subdivided into several states and territories—until after the middle of the eighteenth century, in the partial occupation of Indian tribes. Until about a quarter of a century ago, this immense coal area, taking the country at large, was held to be of small value, even by the civilized successors of the aborigines. The purchases made, at sundry times, by William Penn and his family, and subsequently by the proprietaries, did not embrace any portion of the anthracite districts until 1749, or of the Allegheny bituminous coal region of Pennsylvania, until the year 1768. The acquisition of these coal-fields in no respect influenced the arrangement between the parties; and, to this day, the supply of that description of fuel to the seaboard, is insignificant, when compared with the magnitude of the source from whence it is drawn.

By the terms of the treaty of 1768, which was the last purchase made by the proprietary, they became possessed, with a small exception, of the whole superficial area of the bituminous coal land of Pennsylvania:—that is to say, the entire country between Lycoming creek, the north branch of the Susquehanna, and the head waters of the Alleghany river, down to the Ohio, for *the sum of ten thousand dollars.* The presence of coal, in certain places, became known about this time; for we have seen maps, of the dates of 1770 and 1777, which, among other places, marked the site of "coal mines" on the Ohio side of the river. In 1785, the first tract which was secured on account of the value of the coal upon it, within the new purchase, was patented near Clearfield; and the first ark load of coal descended the Susquehanna from thence. In 1828, the first cargoes of coal from the Alleghany coal-field at Karthaus,* reached Philadelphia and Baltimore; but the distance from market was found too great, and the means of transportation too imperfect, to hold out any hope of a profitable coal business.

The eastern margin of the Alleghany coal-field has been approached in two places from the seaboard by Pennsylvania canals; and, in a third, by the Chesapeake and Ohio Canal in Maryland. It will also be traversed by the Central Railroad. The supply from these sources to the sea-coast is not large; being of course regulated by the limited demand for this species of fuel, of which, anthracite, everywhere, has precedence. It is chiefly in request for gas-works, forges, blacksmiths' use, and for certain industrial purposes. Wherever the two species of coal can be obtained on equal terms, or are equidistant from their centres of production, anthracite maintains its indisputable supremacy. It does not appear probable that Pennsylvania will ever acquire a large market to the eastward for her bituminous coal, so long as her anthracite fields remain unexhausted. For precisely the same reason, it seems to us equally improbable that the anthracite coal trade with the eastern cities will be perilled by the existence of bituminous coal-fields in New Brunswick, Nova Scotia, and Cape Breton. Where there is anything like an equal choice, the demand for Pennsylvania and Virginia bituminous coals will continue small at the eastern ports, owing partly to the causes named, and partly to the heavy cost attending its transportation. Only one hundred and seventy-five tons of Pennsylvania bituminous coal are reported to have reached Boston, during the year 1846, while the supply of the same description of coal from Virginia and foreign countries also simultaneously decreased. But there is another, a better, and a vastly more extensive market, to which Pennsylvania, Ohio, and probably Michigan, may ultimately look for the disposal of their bituminous coal. We refer now to the whole of the countries bordering upon the North American lakes; embracing a

* This name is pronounced *Cart-house* in Pennsylvania.

large portion of the Canada frontier, now rapidly filling up with settlements, and all the opposite portions of the United States. These, in good time, it seems to be very certain, will more than compensate for the loss of a monopoly of two or three ports nearest to the coal-fields of Nova Scotia. Besides the demand for domestic and manufacturing uses, it is probable that much coal will be needed for smelting the copper ores of Lakes Huron and Superior.

From the Cumberland and Frostburg angle of the Alleghany coal-field, there seems a better prospect for the bituminous coal trade, than from Pennsylvania, although, probably, the larger part of the amount produced will be consumed in iron works on the spot. The conveniences for transporting the Cumberland coal to Washington and Baltimore, and for a certain export trade, added to the very high intrinsic value of the coal itself, and the comparative remoteness of the anthracite districts, will secure to this region a fair share of the bituminous coal business.

Surveys of vast bodies of land were made in western Virginia and Kentucky, in 1795, and even previously to the revolution; but we perceive no evidence that those lands possessed any other than surface value, or that the presence of seams of coal, if known, conferred any additional value upon them. Even at the present moment, we know that enormous areas yet remain untouched, and that the time is not yet arrived when they can be estimated beyond their mere agricultural prices. It is only along the flanks of the principal streams, such as the Ohio and the Kanawha, that the vegetable fuel has, in some measure, given place to the mineral combustible.

At the period of running the boundary line between North Carolina and Virginia, in the year 1728, the narrator describes the country as yet in its pristine state of savageness. Settlements extended no further west from the Atlantic than a hundred miles, and the remainder was still the home of the Indian, and the feeding ground of the Buffalo.* It was towards the close of that service, although the western extremity of the line was still left unfinished, that the expedition found itself amongst unknown mountain ranges, and were still remote from the eastern flank of the great Alleghany coal field, and from that wild, bordering, elevated country of Tennessee, of which we yet know but little beyond the late reports of Dr. Troost.

South-west from hence, through the state of Tennessee, and far into the then French province of Louisiana, the coal-field has received but partial investigation. That part which terminates in Alabama, was, but a few years ago, a part of the Cherokee country, from which the aborigines were then removed.

Along its course through Kentucky, Tennessee, and Alabama, many iron works have been established within a few years; but the insular position of this part of the region is unfavourable to the transportation of coal to distant markets.

The last mentioned state constituted the easternmost part of the

* History of the dividing line betwixt Virginia and North Carolina, by Colonel Wm. Byrd, Esq., of Westover, 1728.

original territory of Louisiana, and the mountainous portion of it yet retains its wilderness character. Even so late as 1800,—what in 1817 constituted the separate territory of Alabama,—its entire population did not exceed 2000 persons.

We have traced, in few words, the great Alleghany coal-field in its progress through eight of the Atlantic States. With the exception of a small area towards the north, it lies so remote from the seaboard, that it is not probable that much coal will find its way in that direction. On the south, its best market is the cities on the Gulf of Mexico; and, already, great progress has been made in railroads and inland navigation with that view. These cities now pay enormous prices for coal that has descended the Ohio, the Mississippi, the Tennessee, and the Cumberland rivers, by very long and circuitous channels.

There are no reliable returns of the quantity of coal which descends to these markets, nor of the ordinary production of the states. Western Virginia, in 1840, returned about 300,000 tons as her annual production. In the Ohio division of the Alleghany coal-field, coal was known previously to 1777, since it is marked on Captain Hutchin's map of that date, although not mined until many years after. In 1840, it returned 125,000 tons as her share of the annual production; a small yield, certainly, for a highly favoured district of 11,900 square miles.

But Ohio has since found a northern market for her coal, through the port of Cleveland, as Pennsylvania has done through her port of Erie; and hence, through the lakes to the Canadas, and the countries which border the great lakes. The present export of coal from thence, is about equal to one-half of the entire production of Ohio seven years ago; and nearly equals the whole consumption of the state in 1838. Erie also received 70,000 tons of bituminous coal, in 1847.

The Maryland division is one of the smallest of the Alleghany districts; yet as its coal is probably the best in America, there is no doubt but it will contribute a large quota of coal, and much iron, to the Atlantic ports, by means of the railroads and canal, now in full operation.

We close our circuit of the Alleghany coal-field by returning to western Pennsylvania. Although in 1840, this section returned 415,000 tons, the quantity was evidently much under-rated; and we cannot assume the amount now annually raised at less than one million of tons. One fourth of this quantity descends the Ohio river; one-half is consumed in Pittsburg and in the establishments around that great manufacturing city; the remaining fourth is consumed in the interior.

Let it be borne in mind, that all this business has sprung up within the memory of persons now living. In the year 1753, there was, probably, no white man living within the limits of the present city of Pittsburg; and, in 1775, only a few cabins were standing there.

Yet, in our day, three-fourths of a million of tons of coal are annually received there, and the extent of the iron manufacture is so great as to confer upon the place the title of "the Birmingham of America."

Not more than 40,000 tons of bituminous coal annually pass through the state canals, eastward. This is the maximum of the present demand for this description of coal. The largest portion of this is deposited at the iron works along the route. A very small quantity passes through the Union Canal. The remainder is either shipped at Havre de Grace, or is conveyed in boats to the Delaware.*

Another outlet besides that of Erie, for the northern section of the Pennsylvania coal-field, has been in existence several years. By the Tioga Railroad the coal from the little isolated coal-basin of Blossburg passes into the State of New York, and thence as far as Lake Ontario on the north and Albany on the east. The detached coal-basin near Towanda will, when the meditated improvements are completed, furnish another portion of the State of New York with mineral coal. From the port of Erie, above mentioned, more than 20,000 tons of Pennsylvania coal are annually shipped for exportation to Canada, &c.

Such is the great Alleghany coal-field, whose outline and resources in mineral fuel, we have thus traced. It is impossible to contemplate its gigantic proportions, and its enormous, yet almost untouched resources, without being struck with the magnificent field which it presents for future enterprise.

There is a small detached basin of semi-bituminous coal, lying in Pennsylvania, to the east of the Alleghany Mountain. This locality is called the Round-top Mountain. The coal at present is only employed for the consumption of the neighbourhood, on account of the want of means for transportation.

We pass now to the great depository of anthracite in Pennsylvania; the only one, in fact, of material value on this continent. Here we have the most interesting assemblage of isolated coal-basins that the world has yet produced, or the geologist investigated. We can only now advert to them with extreme brevity; but here we can afford to be brief.

The physical features of this anthracite country are wild; its aspect forbidding; its surface broken, sterile, and apparently irreclaimable. Its area exhibits an extraordinary series of long parallel ridges and deep intervening troughs. This group of elongated hills and valleys consists of a corresponding number of axes, all or nearly all of which range in exact conformity to the base of the Alleghany Mountain. When viewed from the latter, they bear a striking resemblance to those, long rolling lines of surf, wave behind wave, in long

* In 1852, the amount of bituminous coal in Pennsylvania brought to market is supposed to be about 1,300,000 tons.

succession, which break upon a flat shore. A century ago a large portion of this region had received, upon the maps, the not unapt title of "The Wilderness of Saint Anthony."

Three-fourths of a century after, when the greater part of this area was still in stony solitude—when this petrified ocean, whose waves were sixty-five miles long, and more than a thousand feet high, remained almost unexplored,—a few tons of an unknown combustible were brought from thence to Philadelphia, where its qualities were to be tested, and its value ascertained.

But the miner has entered into this Wilderness of Saint Anthony,—and canals have penetrated it,—and railroads have traversed it;—basin after basin of this combustible has been discovered in it;—tract after tract has supplied productive collieries in it;—until, in a single year, [1847] it had furnished the surprising amount of three millions of tons; [or an aggregate of near nineteen millions of tons of anthracite within the last quarter of a century;] and 11,439 vessels cleared from the single port of Philadelphia, in that season, loaded with a million and a quarter of tons, for the service of the neighbour-bouring states.

Such then is the anthracite region, and such its rapid progress in production. To Pennsylvania, in relation to the future, its value, in connection with the corresponding advance of her manufacturing industry, surpasses the power of computation.

Some detached spots in the states of Rhode Island and Massachusetts, have, from time to time, furnished a small and irregular supply of anthracite. The presence of this coal, indeed, was known many years before the Pensylvania anthracite was first mined. Numerous efforts have been made, from time to time, to explore the coal beds in these places, but they have generally ended in failure; owing, as we conceive, to the disturbed and contorted structure and metamorphic character of the enclosing rocks.

In attempting to introduce a fuel so difficult of ignition, at a period when the only coal known and used was of the fat bituminous variety, imported from Europe, and when no adaptation for burning it had been matured, it is not remarkable that the semi-crystalline anthracite of Rhode Island should acquire a bad reputation. Notwithstanding this, and its proverbial unfitness for all purposes of combustion, we believe in its intrinsic excellence—in which belief, both practice and the result of its analysis fully bear us out—and that there exists no better anthracite in the world. But as we have before suggested, the unusually modified nature of the inclosing rocks, the tortuous and schistose characters which they now assumed, and in which the most experienced geologist might fail to detect the representatives of the coal measures, seem to present insuperable difficulties in working the combustible to advantage.

Before quitting the Atlantic slope, we have to notice a bituminous coal-field, small in area, yet rich in the abundance of its coal, and most favourably situated in regard to facility of transportation. Almost in the centre of Eastern Virginia, between N. latitude

37° and 38°, lies the little basin of Richmond, or Chesterfield. This field contains the oldest worked collieries in America. Mr. McClure described it more than thirty years ago; and it was, apparently, for years, the only point in the United States where bituminous coal could be procured, and shipped coastwise. The amount of this export trade does not appear to have ever been large. The official returns show an increase from 48,000 tons in 1822, to 142,000 in 1833, and then annually diminishing to 65,000 tons in 1842. In the southern part of this field, however, new collieries have been opened, which sent down about 50,000 tons of coal in 1847.

Crossing from hence to the northwest, we have next to mention a coal-field of no mean size, yet at present known but to few, even within its immediate limits. It occupies an area of from three to five thousand square miles, in the centre of the peninsula of Michigan; communicating readily with all the great lakes. The amount of coal mined here, is very trifling at present, the beds being quite thin; and this country is so recently occupied, that as yet there has been no demand for this description of fuel. But it seems destined, ultimately, to be of importance, as the use of coal becomes general, and as the timber of the forest decreases. It possesses remarkable geographical advantages; being the only coal-field in a vast extent of country. As population flows into the countries bordering on the lakes, this Michigan coal-field, although not so productive as some others, cannot but become, ultimately, of considerable value.

Passing now to the southward, we enter the great Illinois coal-field, which occupies an extent nearly equal to that of England; yet the State has but recently commenced to make use of the coal with which nature has so bountifully provided her. Except in the vicinity of the larger towns and rivers, the business of mining coal here has made but small progress.

The existence of this combustible was proved by the French explorers at an early period. It was certainly known to Father Hennepin in 1679, almost a hundred years before the Pennsylvania coal was discovered, and is marked on the map which illustrates his journal. He points out a "cole mine" above Fort Crevecoeur, on the Illinois river, near to the site of the present Ottawa. He further states, that in this country, then occupied by the Pimitoui or Pimitewi Indians, now Peoria, "there are mines of coal, slate, and iron; and several pieces of fine red copper, which I have found, now and then, upon the surface of the earth, make me believe that there are mines of it."* This is the earliest notice, on record, of the existence of coal in America; and the same may be said of the bouldered masses of native copper, which we know to have been drifted from their original sites, only discovered but recently, on the borders of Lake Superior. At this period, viz. from 1680 to 1698, and subsequently,

* Map and description of a large country newly discovered in the Northern America, situated between New Mexico and the Frozen Sea; together with the course of the great river Meschasipi. By Father Lewis Hennepin, Missionary Recollect and Apostolic Notary. 1698.

the Illinois river formed part of the main route from the French missions on the Niagara, by Lake Karegnonde, now Huron; by Lake Illinouach, or the Lake of the Illinois, now Lake Michigan, to the Mississippi, and thence down to its very mouth.*

We confess to entertaining strong feelings of interest in the description of these newly explored countries, by the good missionary fathers of those days; among whom stand conspicuous the names of Hennepin, Gabriel, Zenobe, Marquette, and, a few years later, of Father Charlevoix, and others. They were men of observation, yet of simple lives—messengers of peace and good will—mediators between the native savage and the white invader. Here, in these remote missions, they erected the first Christian altars, and planted the first germs of civilization; they shared the perils of exploring unknown regions, and they were the intelligent chroniclers of the times;—the faithful and simple narrators of those hazardous voyages.

The records that these religious men have bequeathed to us, form the most interesting, the most valuable statistical memorials that we possess of the aboriginal state of the interior of this Continent. It were a service not unworthy of some man of leisure, to collect together, ere they are totally lost, these and many other illustrations which distinguish the progress of French discovery, and which so especially belong to the history of this Continent.

On occasion of the peace of 1763, Colonel Croghan was sent by the British government to explore the country adjacent to the Ohio river, and to conciliate the Indian nations. This officer was captured by a party of Indian warriors, and carried up the Wabash river, through the Illinois country. At a point on this river, apparently nearly where Williamsport or Covington now stands, the author states that "on the south side of the Ouabache runs a high bank, in which are several fine coal mines." This is the earliest notice of coal in this part of the coal-field.†

Respecting the coal region of Missouri,‡ we have very incomplete information. The geological distribution of the formations there are ill defined. It will, no doubt, be found, on attentive examination, that the coal measures exist in numerous patches or detached areas, whose boundaries are influenced by the physical configuration of the country, and that the coal series, although scattered at intervals over a great surface, do not cover, at any point, any very large areas; for as the carboniferous strata approach so nearly to the horizontal position, and, moreover, not being of much thickness, a large proportion of productive coal land has been removed by the erosion of the rivers, and similar causes of denudation and excavation. This is also apparently the case in relation to the Illinois coal-field, last men-

* The same line, from Lake Michigan to the mouth of the Illinois river, also formed the northwest boundary line of the country which was ceded by the French to Great Britain, in 1763.

† Printed from the original journal, in the Monthly American Journal of Geology, 1831, Vol. I. p. 257.

‡ See extract from Dr. Owen's Geological Report, published in 1852, of Iowa, Wisconsin, and Minnesota, and the Coal Measures of Missouri.

tioned. The patches of coal formation are scattered all across the State of Missouri, and appear at intervals over a wide tract of country; stretching through a part of Arkansas into the territory now occupied by Indian tribes, and thence, for an unknown distance, towards the southwest, apparently into Texas. The earliest notice of the existence of thick seams of excellent coal in the Osage country, bordering the river of that name, we believe, was furnished by Captain Pike, when on his exploring expedition, in 1806.

Of Texas, there exists at present no geological map or description. Both coal and anthracite are described as existing at the head waters of the Trinity, the Sabine, and some other rivers; but this country is, as yet, in too unsettled a state, to encourage mineral explorations, to any extent. It appears that coal, in detached areas, as in Missouri, crosses Texas and enters Mexico. An important coal-field which crosses the Rio Grande, in N. latitude 27° 30′, near Loredo, has very lately been examined by an officer attached to the United States army of invasion. This coal formation bears away into the interior of Mexico, in a southwest direction, by Guerrera, on the Salado river, where it is of good quality. Coal is also found in the provinces of Oajuca and Vera Cruz, and in abundance at Tehuantepec; doubtless geologically newer than the true coal.

We can scarcely speculate here on the existence of bodies of true coal, along the Pacific slope of this Continent. In some cases brown coal has been mistaken for the older coal. In the midst of the Rocky Mountain range, Captain Fremont discovered coal beds of dubious geological age; intermediate, apparently, between the brown coal and the true coal. A coal bed of considerable thickness has lately been noticed in the Raton mountains, east of Taos, and on the head waters of the Canadian river.

The scope of the present sketch leads us now to notice one of the most interesting geological phenomena in the new world. We refer to that enormous range of brown coal, apparently of the tertiary period, which follows the eastern flank of the Rocky Mountains, from near Mexico even to the Polar sea. Nature has, indeed, worked on a truly gigantic scale. We see here a deposit of brown coal, so extensive that the magnitude of its proportions is far from being defined; yet enough is known to show that it exceeds, in longitudinal range and breadth, all others on the present surface of our planet. So far seems to be established, that, allowing liberally for interruptions in continuity, supposing that any such exist, it occupies thirty-five degrees of latitude, or near 2500 miles, following its oblique range; and has a maximum breadth, at N. latitude 48°, of four hundred miles; the whole area, as near as we can venture to compute, being 250,000 square miles, or one hundred and sixty millions of acres—more than twice the size of Great Britain. Compared with this, the largest coal-fields in the world are absolutely small.

Should it prove that the coal, which has been traced at no very distant intervals, westerly from Mackenzie's river to the Icy Cape, by Point Barrow, and into Behring's Strait,—along the north coast of Russian America,—is also of the same geological age as that which ranges parallel with the Rocky Mountains, we might add twenty degrees more to the thirty previously mentioned; while, at the same time, the oblique direction of the latter adds five degrees more to the total range.

Turning to the southward, after an uncertain interval of twenty-five degrees, we find ourselves again on coal strata, apparently of similar age to the northern zone just described, and occupying about two thousand five hundred miles more. At certain points along the Pacific side of the southern Mexican provinces, from the Isthmus of Tehuantepec to that of Panama; and then, with a few interruptions, continuing all down the western side of South America to the Equinoctial line; and thence to Lima; and again appearing on the coast and adjacent islands uninterruptedly from Valparaiso to below Chiloe Island, and even through Patagonia, at least as far south as 50° lat., a belt of brown coal formation and tertiary strata, borders the Pacific, or skirts the Andes. At all of the points which have been successively named, investigations into the quality of this coal and its fitness for the purpose of steam navigation, have, for some years past, been made, and the details will be furnished in the appropriate place.

Commencing our admeasurement near the Isthmus of Panama, in N. lat. 10°, and descending to S. lat. 50°, the traces of a tertiary formation, containing lignites, and fossil wood, are reported to extend, almost continuously, for four thousand miles. We are within reasonable bounds when we assume two thousand five hundred miles, as the extent, with occasional interruptions, in which brown coal or carbonized wood is traceable.

Thus there exists, ranging nearly with each other, but separated by a great breadth of unexplored ground, two apparently contemporaneous belts, 2500 miles long each, extending through both continents to points at least one hundred and twenty degrees asunder,—namely, the Frozen Sea or perhaps the Icy Cape to the north, and southern Patagonia to the south. We think we cannot be far in error if we assign five thousand miles, out of eight thousand four hundred miles, to this remarkable coal formation. We do not know if this statement be entirely new. Were it not supported extensively by good geological authority, and by a series of facts and observations which will scarcely be called in question, we should hesitate extremely ere we gave it circulation.

Returning once more towards the north, coal is mentioned as occurring near Monterey, in California. Petroleum and asphaltum, and perhaps anthracite, occur in this parallel. Captain Fremont discovered coal, probably brown coal, in the centre of the Rocky Mountain chain, at an elevation of 6820 feet above the sea. Another coal-field was found by Captain Fremont, in his recent expedition, in N. lat. 41½°, extending from 110° to 111° W. long. Both of these deposits appear to be about the Oolitic age. There is a coal range at an elevation of more than 7000 feet, in the great mountain range, east of Santa Fé, in about N. lat. 37°; but of what geologic age we know not.

In Oregon we have had notices both of coal and lignite, from various explorers. Good coal is reported in Vancouver's Island, in Queen Charlotte's Island, Millbank Sound, and other points on the western border of British America. Passing round by the undetermined coal-beds of Russian America, there appear to be various scattered points within the Arctic circle, where coal has been discovered by our enterprising navigators. At Prince Regent's Inlet, at Byam Martin's Island, and Mellville Island, both true coal and brown coal were obtained.

Within the arctic regions other considerable bodies of the newer coal are known to exist, particularly at Disco Island, Hasen Island, and on both sides of Greenland.

All these northern coal localities seem scarcely more than mere objects of geological interest; for, in regard to their practical value, all that is known to us at present is their mere existence. Still, it can never, surely, be held as a matter of no importance, the fact of this local distribution of mineral combustible, throughout regions which have no timber, or even shrubs, to serve the purpose of fuel.

Canada and the territory west of it, for a vast space, contains no known deposit of coal. This country is destined to be tributary to the states of Ohio, Michigan, and Pennsylvania, for her future supply of mineral fuel, which can be transported at a very cheap rate through the chain of inland lakes.

New Brunswick, Nova Scotia, Cape Breton, and Newfoundland, make up, by the prodigious expansion of the coal formation in that quarter, for the deficiencies of the upper province.

At what period mineral coal first attracted attention, and was applied to the service of the original colonists, does not appear. The business of the General Mining Association, sole lessees of the enormous coal-fields of Nova Scotia and Cape Breton, did not commence until 1827, but the coal of Cape Breton has been worked for sixty years at least. In the south-western part of the province of New Brunswick, the mining of coal seems to have been commenced a little earlier, but the returns exhibit a meagre amount of business.

In Newfoundland coal is mentioned at an early period; and, in a climate not particularly adapted to the growth of timber for fuel, the substitute was gladly welcomed.

The coal trade of Nova Scotia and Cape Breton, is of comparatively modern date. It furnishes supplies to the cities on the St. Lawrence, and a few thousand tons annually find their way, with or without commercial interposition, to the ports of Boston and New York, where, however, the use of anthracite is so completely established, especially for domestic use, as to exclude, in great measure, the smoky bituminous coals, come from where they may.

The coal-fields of British America, although possessing iron ore in great abundance, have no iron works established within their limits. It is no wonder, indeed, when we see regions so highly favoured by natural resources and advantages, neglected, or sacrificed to the paralizing influences of an imbecile monopoly.

In the foregoing outline we have traced, with such brevity as the subject permitted, the prominent features of those vast depositories of mineral coal which nature has so bountifully distributed over the American continent. We have shown, that the mining of this coal, the establishment of an important coal trade, the employment of the fuel in industrial arts, in steam engines and steam vessels, on railroads and canals, in blast furnaces, and iron works, and factories; in fact, its application in a thousand forms, is but of yesterday's growth. We have shown, too, that a portion—immeasurably the larger portion—of these prodigious areas of coal formation, has still no appreciable value, but continues at present wholly unappropriated. It will, doubtless, long remain in reserve, for the service of other generations.

Still, if we measure the future by the past—and we feel assured that we may safely do so in a vastly accelerated ratio, taking Pennsylvania as an example, and admitting the surprising increase of population as an essential element in the calculation—the production and conversion of iron and coal, with all their attendant and ever increasing uses, together with the influence they cannot fail to exert upon agriculture, involve results of which we have now but a remote perception. We cannot but think that the close of the present century will witness an advance in the industrial resources of the country, and a consequent extension of domestic prosperity such as it may be presumptuous, at the present moment, on our part, to anticipate.

In corroboration of the above opinion, we conclude this chapter with an extract from the late message of the governor of Pennsylvania, January 16th, 1852, giving a startling estimate in anticipation of the immense wealth of the anthracite coal fields.

"The whole amount of anthracite coal mined and taken to market in 1840, was 867,000 tons. In 1852, the products will reach near 5,000,000 of tons, being an increase in twelve years of 600 per cent. This rate of augmentation up to 1870, would give the startling production of over 45,000,000 of tons, and yielding at the present Philadelphia prices, the sum of $180,000,000, being more than treble the present revenues of the whole United States."

IRON MANUFACTURE AND TRADE OF THE UNITED STATES.

	Pig Iron, tons.
1810,* 153 furnaces,	54,000
1828,	130,000
1829,	142,000
1830,	165,000
1831, 939 furnaces, forges, &c.	191,536
1840, 804 do. 795 bloomeries and forges,	347,700
1845, 540 do. 950 do.	777,600
1846,	765,000
1847,	800,000
1850,	400,000
1853, estimated in a local Journal, at	1,000,000

A writer in the "Merchant's Magazine," March, 1845, gives the following estimate of the production of iron in the United States at that period:

	Tons.
540 blast furnaces, averaging 900 tons each, per annum,	486,000
950 bloomeries, forges, rolling and slitting mills, yielding of bar, hoops, &c.,	291,600
950 and of blooms, castings, machinery, stove-plates, &c.,	151,500
The market value of these, in 1845, was $33,940,500.	
Quantity of bar, hammered, pig, scrap, and sheet-iron, and steel imported,	92,077
Value of the same, $7,794,110.	

From a sketch of the American iron trade and production by Mr. Feuchtwanger, in November, 1847, it appears that the quantity of pig-iron produced in the United States in 1810, was 53,908 tons. In 1847, above 500,000 tons.

The value of manufactured iron and steel *imported* into the United States, during the year ending June 30th, 1846, *paying duties ad valorem*,	$4,023,590
Paying specific duties,	4,463,739
Total,	$8,487,329

	1844.	1845.
Value of American iron and manufactured iron exported,	$716,332	$845,017

* Chiefly taken from "Domestic Production of Iron, by H. C. Carey," in a volume entitled "Documents relating to the manufacture of Iron, on behalf of the Convention of Iron Masters, 1849."

MANUFACTURE OF IRON IN THE UNITED STATES, FROM THE CENSUS OF 1850.

Pig-iron.

Capital invested in manufacture, - -	$17,348,000
Raw material consumed, - - -	7,000,000
Cost of labour, - - - -	5,966,000
Value of products, - - - -	12,740,000
Hands employed, - - - -	20,458

Cast-iron.

Capital invested in manufacture, - -	$17,456,000
Raw material, - - - -	10,346,000
Labour, - - - - -	7,000,000
Value of product, - - - -	25,000,000
Number of hands, - - - -	20,507

By another account, according to Mr. Kennedy's census tables, there were in 1850,

Forges, furnaces, &c., in the United States,	2190
Employing capital, - - - -	$49,258,006
Male hands, - - - -	57,021
Consuming raw materials, worth - -	27,049,753
And producing, tons, - - -	1,165,544
Of wrought, cast and pig-iron, valued at -	*$54,604,006

Import, Production and Consumption of Iron in the United States in 1851.†

	Tons.
Importation, - - -	341,750
Production, - - -	413,000
Consumption, - - -	754,750 or $69\frac{8}{4}$ per head.

Value of Exports of Iron from the United States.

Different kinds of Iron.	1849.	1850.	1851.	1852.
Pig, bar, and nails, ...	149,358	154,210	215,652	118,624
Castings, ...	60,175	79,318	164,425	191,388
All manufactures of iron, . .	886,639	1,677,792	875,621	1,993,807
Total, ...	1,096,174	1,911,320	1,255,698	2,303,819

* United States Gazetteer, 1854, published by Lippincott, Grambo & Co.
† Secretary Corwin's Report.

Condensed Table of the Imports of the various description of Iron into the United States each year, ending June 30th, constructed from the report of the Secretary of the Treasury.

Different kinds of Irons.	Years.	Tons.	Assessment per ton.	Value.
			Dollars.	*Dollars.*
Pig iron, ...	1843	3,783	12 46	248,251
	1844	14,944	13 42	500,522
	1845	27,510	18 40	506,291
	1846	24,187	20 24	489,573
	1847	27,955	19 25	554,486
	1848	51,632	15 79	815,415
	1849	105,632	13 60	1,405,613
	1850	74,874	12 69	950,660
	1851			787,524
	1852			935,957
	1853			1,528,031
Hammered iron, ...	1843	6,254	52 37	327,550
	1844	18,876	47 99	872,157
	1847	14,911	54 90	854,708
	1848			975,214
	1849	10,598	49 61	525,770
	1850	14,706	50 64	744,735
	1851			900,026
	1852			1,302,809
	1853			627,675
Bar iron rolled, ...	1843	15,757	32 45	
	1845	51,188	33 05	
	1847	40,183	53 23	
	1848			3,679,598
	1849	173,457	34 93	6,060,068
	1850	247,951	29 83	7,397,166
	1851	254,310	29 00	7,324,283
	1852			8,568,317
	1853			15,402,776
Scrap and old iron, ...	1843	157	17 43	
	1845	5,847	20 48	
	1847	1,693	22 27	
	1848			140,037
	1849	9,450	15 28	144,424
	1850	10,104	16 03	161,981
	1851			112,029
	1852			102,292
(The column of Duties is omitted.)	1853			145,059

*Value of Iron Imported into the United States.**

Years.	Value.	Duty.
1844	2,395,760	1,607,113
1845	4,075,142	2,415,003
1846	3,660,581	1,629,581
1848	7,060,470	2,118,141
1849	9,262,567	2,778,770
1850	10,864,680	3,259,404
1851	10,781,312	3,234,094
1852	18,843,569	5,632,484
1853	27,015,364	8,104,609

* Report of the Secretary of the Treasury, 1853.

In 1852 New York received of manufactured iron from Britain, 135,299 tons, Boston 24,414 tons, and Philadelphia 12,024 tons.

Table of the tons of Iron of all kinds imported from Great Britain to United States.—The following figures exhibit the number of thousands, leaving out the fractional hundreds.

Years.	Tons.	Years.	Tons.	Years.	Tons.
1820	8,000	1830	21,000	1840	72,000
1821	9,000	1831	41,000	1841	112,000
1822	15,000	1832	45,000	1842	107,000
1823	13,000	1833	62,000	1843	38,000
1824	11,000	1834	47,000	1844	102,000
1825	13,000	1835	63,000	1845	68,000
1826	12,000	1836	91,000	1846	
1827	21,000	1837	54,000	1847	
1828	22,000	1838	78,000	1848	
1829	17,000	1839	85,000	1849	315,000

Wire Cables for Mines and Inclined Planes, for Tiller Ropes, &c.—See much practical data collected under this head on iron in Pennsylvania, &c.

Casualties of Miners.—Provident Institutions, "*Caisses de Secours,*" Relief Funds, &c., see details under this head, in a preliminary chapter.

As regards the relief and support of aged or disabled miners in the United States, particularly in Pennsylvania, it is but justice to the editor of the Miner's Journal, of Pottsville, to state, that he has sought on several occasions to attract attention to this very desirable object. The casualties to which this class of useful operatives is continually exposed, calls for some provision for the aged, the infirm, and the injured; and for occasional relief in distressing cases, to their bereaved families. All mining countries have perceived the necessity of adopting measures which shall effect these benevolent objects, in behalf of a population whose employments peculiarly and perpetually expose them to the most distressing calamities.

The countenance of the state government would not, of course, be withheld from "The Miners' Provident Institution," but it is obvious, and has been decided in every well regulated mining region, that the burden and the management of such institutions as are here suggested, must be jointly borne by, and emanate from, the two most interested parties—that is to say, the proprietors of the minerals and the workmen themselves.

The community, as experience has shown, will not consent to be taxed for the relief of one class of operatives, however strong their claims. All other classes of persons pursuing hazardous occupations, would view such a measure as an act of injustice to themselves. Above all things to be avoided is the conversion of benevolent institutions, however well conceived or modelled, to any thing like local or political influence.

In England, the operative miners have held back from such institutions, under the impression that the amount of their subscriptions would be so much deducted from the poor-rates, and, consequently, that their contributions would in reality prove a bonus to their employers, rather than a benefit to themselves. In the coal regions of the American states, no such objection can be urged, as the rates for the support of the poor are extremely trivial: and it seems most just and fitting that the operatives should, as in France and Belgium, have a share through their representatives, in the management and appropriation of the funds to which they have contributed their portion.

At the commencement of this work, we devoted some space to a consideration of this subject, and we conclude by referring the reader to that article, which abounds in facts of extreme interest.

Statistics of United States.

We make a few extracts from the Boston Post for August, 1854, which contains some valuable information relating to the vastness and riches of the United States.

"The extent of its sea-coast, exclusive of islands and rivers, to head of tide-water, is 12,669 miles. The length of ten of its principal rivers, is 20,000 miles. The surface of its five great lakes is 90,000 square miles. It contains within its limits the longest railway upon the surface of the globe,—the Illinois Central,—which is 731 miles in length. Annual value of agricultural productions, $2,000,000,000. Capital invested in manufactures, $600,000,000. Value of the gold of California, $100,000,000 per annum."

Railroads in the United States.

	Miles.	*Cost.*
In 1845–6, there were in activity 113 public and private railroads, whose aggregate length was - - - - -	4870	$127,417,758
Average cost per mile, $26,932=£5,564.		
In 1846–7, completed lines, 133, -	5703.12	
		Capital.
In 1847, by "Doggett's Railroad Guide," there were completed, - -	5740.00	$122,525,937
		Cost.
1853, total number in operation, 264, - - - - - - - - -	17811	$508,588,038
In 1854, the number of miles of railroad in operation within the limits of United States,* - - - - - -	20000	$600,000,000

* Boston Post, August, 1854.

Canals of the United States.

	Miles.	*Cost.*
Cost of fifty-seven canals, up to 1845, length, - - - - - - -	4102	$113,934,163
1854, length of its canals, - -	5000	
Amount expended on the canals of the United States is about - - - -		$150,000,000

Lines of Magnetic Telegraph.

	Lines.	*Miles.*	
At the end of 1847, finished,	18	2311	in operation.
	7	2586	under construction, nearly finished.
	10	3815	to be completed in 1848.
Total,	35	8712	
		1768	projected.
Total telegraphic conductors,		10,480	

At the commencement of 1852, the aggregate number of main and branch lines in the United States were about 100.

	Miles.
Completed and in operation, - - - - - - -	27,000
In construction, - - - - - - - - -	10,000
Extent of lines throughout the world and in operation, -	40,000
Great Britain has - - - - - - - - -	4,000
America, - - - - - - - - - - -	27,000

Summary.

Aggregate of the 57 canals of the U. States, in 1845,	4102 miles.
" of Railroads finished in 1847, - - -	5740 "
" of Lines of Magnetic Telegraph, - -	8712 "
Total, - - - - - -	18,554 "
Aggregate of the canals in the U. States, in 1854,	5,000 miles.
" of Railroads, - - - - - -	20,000 "
" of Magnetic Telegraph, - - - -	27,000 "
Total, - - - - - -	52,000 "

Registered, Enrolled, and Licensed Tonnage owned in the principal States.

States.	1845.	1846.	1847.	1848.	1849.	1850.	1851.	1852.	1853.
New York, .	625,875	655,695	747,124	733,077	796,491	835,867	931,193		1,149,133
Massachusetts,	524,994	541,520	577,520			313,192	342,936		
Maine, · .	320,059	358,123	374,354						
Pennsylvania,	147,812	148,058	183,007		188,087	206,497	222,428	229,443	
Louisiana, .	170,525	181,258	213,539						

Comparative view of the Registered, Enrolled, and Licensed Commercial Tonnage of the United States, exclusive of those engaged in the Fisheries.

Years.	Tons.	Years.	Tons.
1815, - -	1,368,127	1848, - -	3,154,042
1820, - -	1,280,166	1849, - -	*3,334,015
1830, - -	1,191,776	1850, - -	†3,535,454
1840, - -	2,180,764	1851, - -	‡3,772,439
1845, - -	2,417,002	1852, - -	4,138,441
1846, - -	2,562,084	1853, - -	§4,407,010
1847, - -	2,839,046		

Tolls received on State Works—Canals and Railroads.

Tolls collected by the canals and railroads on the transportation of merchandise for the internal trade of the country, exceeded $25,000,000 in 1853.||

	1845.	1846.	1852.
New York, - -	$2,620,532	$2,764,121	$3,179,145
Pennsylvania, - -	1,196,979	1,295,494	1,938,574
Ohio, - - -	495,313	630,770	¶693,675
Indiana, - - -	46,279	53,930	—

Vessels built in the United States and annually employed, from 1815 *to* 1852.

Years.	Ships.	Brigs.	Schooners.	Sloops and Canal Boats.	Steam-boats.	Total No.	Total Tonnage.
1815	136	224	680	274		1,314	154,624
1820	21	60	301	152		534	47,734
1825	56	197	538	168	35	994	114,997
1830	25	56	403	116	37	637	58,094
1835	25	50	302	100	30	507	46,238
1840	97	109	378	224	64	872	118,309
1845	124	87	322	342	163	1,038	146,018
1850	247	117	547	290	159	1,360	272,218
1851	211	65	522	326	233	1,357	298,203
1852	255	79	584	267	259	1,444	351,493

This table is merely designed to show the progress of ship building, at intervals of five years.

Steam Vessels.

The number of steamboats built and annually employed in the

* Including those engaged in the fisheries.

† 269,773 tons in the Whale, Cod, and Mackarel Fisheries.

‡ Including 138,014 tons in the fisheries; and 521,216 enrolled tonnage employed in steam navigation.

§ Including 159,831 engaged in the fisheries; and 514,097 enrolled tonnage employed in steam navigation. Report of Secretary of the Treasury.

|| Report of Secretary of Treasury.

¶ American Almanac, 1854.

United States, between the years 1823 and 1850 inclusive, is 2690.

Registered and enrolled steam tonnage, Dec., 1851, of the United States, amounted to 583,607 tons; in 1853, 604,617 tons.

Number of steamboats plying from Philadelphia to different points on the Delaware and its tributaries, and to New York. 1844—35 boats, consuming 45,000 tons of anthracite.

Steamboats on the western waters. 1846—1500 boats, whose tonnage was 145,311 tons.

Steamboats on the Lakes. 1846—80 boats, whose tonnage was 54,486 tons. 1851—180 boats, whose tonnage was 212,000 tons.

War steamers. 1846—11 boats.

1853.—From A. Guthrie, Esq., steamboat inspector, the following statement is derived:—*

No. of Boats.

Philadelphia, 1853, - - 60.
New York, - - - 92. Considered far below the number.

The navy of the United States contained in 1852, 11 ships of the line, mounting from 74 to 120 guns.†

Twelve first class frigates, mounting 44 guns each.
Two second class, 36 guns each.
Four brigs, of ten guns each.
Three schooners, of 1 and 2 each.
Five steam frigates, 6 to 10 guns each.
Four first class steamers, 1 to 10 each.
Seven second class, and five store ships and brigs.

Foreign Commerce of the United States in 1847.‡

	Cleared for foreign ports.			Arrivals.			Value of Exports and Imports, including Specie.
	Vessels.	Tons.	Crews.	Vessels.	Tons.	Crews.	
American,	8,102	2,202,393		7,730	2,101,359		Imports, $146,545,638
Foreign,	6,268	1,176,605		6,499	1,220,346		Exports, 158,648,622
Total, .	14,370	3,378,998	165,792	14,229	3,321,705	163,889	$305,194,260

Value of imports in 1851, 216,224,932.—Including specie.
Value of exports in 1851, 218,388,011.

* Hunt's Magazine, June, 1853.
† U. S. Gazeteer, 1854, and American Almanac.
‡ Secretary of Treasury's Report.

Foreign Commerce of the United States in 1852.

	Cleared for foreign ports.			Arrivals.			Value of Exports and Imports, including Specie.
	Vessels.	Tons.	Crews.	Vessels.	Tons.	Crews.	
American,	8,887		117,033	8,964		115,846	Imports, $212,613,282
Foreign, .	10,438		97,722	10,607		100,821	Exports, 209,641,625
Total, .	19,325			19,571			$422,254,907

Imports and exports continued for the fiscal year ending June 30th, 1853, including specie.

Imports, - - - - - - $267,978,647
Exports, including specie, - - - 230,452,250

Total, - - - - $498,430,897

During last fiscal year, 1853–4, the aggregate value of foreign imports is estimated at - - - - $315,000,000
Revenue from customs in 1853–4 was - - 64,224,189
" " " 1852–3 was - - 58,500,000

* Washington Correspondent of the Courier and Inquirer, August, 1854.

THE FOLLOWING STATEMENT

Was prepared with a view to show the progress and prominent authors of the State Geological Surveys, the names of *the principal contributors to geological discovery in the United States*, and, also, of those of British America and other portions of the American continent, chronologically arranged. It will be a matter of regret if we have omitted any names that are entitled to appear upon our list of principal workers in the geological field; but we fear that it is unavoidable.

UNITED STATES.	William McClure, Geology of the U. S. (1809,) 1817, 1822; Prof. Cleaveland, 1816; C. Lyell, 1845; Ed. de Verneuil, 1847.
MAINE.	Chas. T. Jackson, M. D., Survey of the State, appointed in 1836, 3 Reports, 1837, 8, 9; Prof. Cleaveland.
NEW HAMPSHIRE.	Dr. Jackson, nominated 1840, reported 1841.
VERMONT.	Prof. C. B. Adams, State G. Survey, appointed 1845, with Prof. Hitchcock; Prof. Emmons, 1844.
MASSACHUSETTS.	State Survey: Prof. Hitchcock, appointed 1830, published report 1833, 700 pages; re-surveyed by the same in 1837; final report in 1840; Dr. C. T. Jackson, in 1838, 1840; W. C. Redfield, 1841; Prof. Hitchcock, 1845; C. Lyell, 1845.
RHODE ISLAND.	L. Vanuxem, 1825; Dr. Meade, 1820; State G. Survey: Dr. C. T. Jackson, appointed 1839, reported 1840; C. Lyell, 1845; Prof. Emmons, 1844.
CONNECTICUT.	State G. Survey: Prof. C. U. Shepard, report 1837; also, Prof. J. G. Percival, report 1842; Prof. Hitchcock, 1841; W. C. Redfield, 1841; W. W. Mather, 1834; Dr. Barratt, 1845; J. D. Whelpley, 1845.
NEW YORK.	State G. Survey: Profs. Emmons, Mather, L. Vanuxem, L. C. Beck, T. A. Conrad, and J. Hall, appointed in 1836, 5 annual reports, final report in — vols. with geological map; Van Rensselaer, 1825; Amos Eaton, 1820, 1824, 1830; W. C. Redfield, 1841; Prof. Dewey, 1845; R. C. Taylor, 1847.
PENNSYLVANIA.	Cist, 1821; P. A. Browne, 1825, 1831–2; Dr. G. Troost, 1826; G. W. Carpenter, 1828; Silliman's Journal, 1830; G. W. Featherstonhaugh, 1831; R. C. Taylor, 1832, 1843; Packer's Report, 1833–4; Prof. Silliman; W. E. Logan, 1842; State G. Survey: Prof. H. D. Rogers, commenced 1836, 6 annual reports, 1836 to 1842; W. R. Johnson, 1839, 1841; Dr. R. Harlan; Mining Journal, Pottsville; M. Chevalier, 1839.
NEW JERSEY.	L. Vanuxem, 1822–8; Dr. S. G. Morton, 1828–9, 1834; State G. Survey: H. D. Rogers, ordered in 1835, first report 1836, final report, 1840; C. Lyell, 1842–5; W. C. Redfield, 1841–3; W. Lonsdale, 1845.
DELAWARE.	Dr. S. G. Morton, 1828, 1834; State G. Survey, 1837: Prof. J. C. Booth, two annual reports, one final report.
MARYLAND.	H. Hayden, 1820; State G. Survey: Dr. J. T. Ducatel, commenced 1834, seven annual reports; P. T. Tyson, 1837; T. A. Conrad, 1830–6, 1841; Prof. Silliman, 1838; L. Vanuxem, 1841.
VIRGINIA.	State Survey: Prof. W. B. Rogers, appointed 1835, six annual reports; Dr. S. P. Hildreth, 1835; R. C. Taylor, 1834; T. G. Clemson, 1835; Prof. Silliman; Prof. Bailey; C. Briggs; H. C. Lea, 1843.

NORTH CAROLINA.	State G. Survey: Prof. Olmstead, 1824-5; W. C. Redfield, 1841; Prof. E. Mitchell; J. T. Hodge, 1841-3; C. Lyell, 1842; T. A. Conrad, 1843; R. C. Taylor, 1845: Prof. E. Emmons, State G. Survey, 1851.
SOUTH CAROLINA.	L. Vanuxem, 1826; C. Lyell; M. Tuomey, State G. Survey, 1848; Dr. R. W. Gibbes, 1845-8.
GEORGIA.	J. R. Cotting, State Survey, commissioned 1836, report 1841; J. H. Couper; C. Lyell, 1842.
WEST FLORIDA.	J. L. Williams, 1827.
EAST FLORIDA.	T. A. Conrad.
ALABAMA.	Prof. M. Tuomey, State Survey, commenced 1847, reported 1850; I. Lea, 1833; Prof. Brumby, 1838; T. A. Conrad, 1832; Dr. R. Harlan, 1841; C. Lyell, 1845-6, Dr. J. H. Kain, 1845; W. S. Porter, 1827; R. W. Withers, 1828, 1845; S. G. Morton, 1833; A. Jones, 1834; C. U. Shepard, 1834; C. S. Hale, 1848.
MISSISSIPPI.	B. L. C. Wailes, 1845; Dr. M. W. Dickerson, 1845; T. A. Conrad.
LOUISIANA.	Brackenridge, 1814; Darby, 1818.
ARKANSAS.	T. Nuttall, 1819, 1821; H. R. Schoolcraft, 1818; Major Long, 1820; G. W. Featherstonhaugh, 1834; A Geological Survey recommended by the Governor, Nov. 1846.
TENNESSEE.	State G. Survey: Dr. G. Troost, 1831 to 1846, 8 reports; T. A. Conrad, 1835; Dr. D. D. Owen.
KENTUCKY.	D. Trimble, 1836; State G. reconnoissance: Prof. W. W. Mather, 1838; W. Cooper, 1832; C. Lyell, 1842.
OHIO.	Dr. S. P. Hildreth, 1835; State G. Survey: Prof. W. W. Mather, assisted by Dr. S. P. Hildreth, Prof. J. Locke, Prof. J. C. Briggs, J. W. Foster, 2 reports, 1837, 1838; Prof. Briggs, 1838; Prof. J. Hall, 1843; Dr. J. S. Taylor, 1845; Charles Whittlesy, 1848.
MICHIGAN.	Schoolcraft, 1821; State G. Survey: Dr. D. Houghton, 1838, 4 reports to 1845; Dr. C. T. Jackson, 1845; numerous Reporters of Surveys in the Copper Region of the north.
INDIANA.	State G. Survey: Dr. D. D. Owen, 1837, 2 reports, 1837-8; Prof. J. Locke; Prof. J. Hall, 1843.
ILLINOIS.	Father Hennepin, 1680; Colonel Croghan, 1763; Dr. D. D. Owen, 1839, 1844; Prof. C. U. Shepard, 1838; Prof. J. Locke; Prof. J. Hall, 1843.
MISSOURI STATE.	Captain Pike, 1805-6-7; Bradbury, 1809-10-11; Brackenridge, 1814 Major Long, 1819-20; H. R. Schoolcraft, 1818; Dr. Daubeny, 1838; T. G. Clemson, 1838; J. N. Nicollet, 1841; J. T. Hodge, 1842; G. W. Featherstonhaugh, 1835-6; Prof. C. U. Shepard; T. Nuttall, 1819; Major Cass, 1820.
MISSOURI TERRITORY.	Lewis & Clarke, 1804-5; Capt. Pike, 1805; Major Long, 1819-20; Dr. James, 1819-20; H. D. Rogers, 1834; J. N. Nicollet, 1838 to 1841, reported in 1843, with his large map; G. W. Featherstonhaugh, 1834-5; Brackenridge, 1814; E. Harris, 1845; Lieut. A. R. Johnston, 1845.
WISCONSIN.	Dr. D. D. Owen, 1839, 1844; Prof. J. Locke, 1840, 1842; Major Long, 1820; Dr. D. D. Owen, 1852.
IOWA.	J. T. Hodge, 1842.
NEBRASKA.	Dr. D. D. Owen, 1852; Prof. Jos. Leidy, 1852.
OREGON AND UPPER CALIFORNIA.	Capt. Fremont, 1843-44; J. D. Dana in Wilkes's Exploring Expedition, 1841; Prof. J. Hall, 1845; Prof. J. W. Bailey, 1845.
TEXAS.	W. Kennedy, 1844; Lieut. B. P. Tildon, 1847; Dr. F. Rœmer, 1848.
NEW MEXICO.	Don Manuel Alvarez, 1847; Lieut. Col. Emory, 1847.
UTAH.	Capt. H. Stansbury; James Hall, 1852.
MEXICO.	A. Von Humboldt, 1809, 1823; M. Chevalier, 1835; Joseph Burkart 1836; Lieut. B. P. Tilden, 1846-7; Lt. G. W. Raines, U. S. N., 1848.

BRITISH AMERICA.

NEW BRUNSWICK.	Dr. A. Gesner, 1843–5; Henwood, 1840; Capt. Bayfield; W. E. Logan, 1845; J. W. Dawson, 1845.
NOVA SCOTIA.	Messrs. Jackson and Alger, 1841; Halliburton; Dr. A. Gesner, 1840, 1842–3–5; J. W. Dawson, 1843–5; R. Brown, 1843; W. E. Logan, 1842–5; C. Lyell, 1843.
CAPE BRETON.	The Abbé Raynal; R. Brown, 1845; J. W. Dawson, 1845; C. Lyell, 1842.
NEWFOUNDLAND.	Mr. Jukes; Sir R. H. Bonnecastle.
CANADAS.	Dr. Bigsby, 1819–20; Provincial G. Survey: W. E. Logan, 1842–4–5; A. Murray, 1844–5; Capt. Bayfield.
NORTHWEST TERRITORY.	Capt. Franklin, 1820–1–5–6–7; Dr. Richardson, 1820–1–5–6–7; Sir A. Mackenzie, 1789, 1792; Mr. Hearne, 1769; Dr. Bigsby, 1821, 1824; Capt. Parry, 1819–20–4; Capt. Ross, 1830–1–2–3; Mr. Isbister, 1845; Capt. Beechy, 1825–6; Capt. Back, 1833; Lieut. Wilkes; Capt. Fremont, 1843–4; Messrs. Simpson and Dease, 1837–8; Edw. Harris, 1845.
GREENLAND.	Capt. Scoresby; Capt. Clavering.

THE ALLEGHANY OR APPALACHIAN COAL-FIELD.

COMPRISING WHAT IS FREQUENTLY DENOMINATED THE GREAT CENTRAL BITUMINOUS COAL RANGE OF THE UNITED STATES.

WE may add, also, that in some of our most ancient topographical maps, this vast range was formerly known as the "ENDLESS MOUNTAINS."

Some of our cotemporaries following up the suggestion of Mr. Darby,* prefer the term Appalachian to that of the Alleghany coal range, and, apparently, with some reason. That able geographer was no doubt governed by good evidence, in favour of its adoption in its generic and most extended sense, while admitting the word Alleghany in what may be termed its specific or local application; thus employing the one to designate an entire system; the other as an integral part of that system. We think that it was so meant by Col. Long in 1831,† and by one or two cotemporary geologists. More than a century before this, we remark the use of the generic term, "the vast Appalachian range of mountains," by Col. Byrd, in his lively narrative of the running of the dividing line between North Carolina and Virginia, in 1728; which diary did not appear in print until 1841.‡

Our cotemporaries of the Virginia and Pennsylvania geological surveys, we believe, have adopted the general scope of this designation, yet with some modification; conferring, if we mistake not, the term Appalachian on the magnificent central and elevated region, within whose borders yet slumber in undisturbed darkness, untold millions of acres of coal and iron.

It seems to us, however, that the entire Appalachian system of the geographers was intended to comprise a widely extended series of mountains, some of which are far removed, by space as well as by geological structure, from that which is generally designated throughout its course, the Alleghany coal-field. The Alleghany is therefore a coal-range, *par excellence*, which the Appalachian certainly is not. Hence appears the peculiar fitness of the term "Alleghany coal-field."§

Col. Byrd's reference to the Appalachian mountains seem to indicate merely the Blue Mountain and those parallel ranges which

* Darby's Geography of the United States.
† Monthly American Journal of Geology, 1832, p. 347.
‡ Westover Papers. Petersburgh, Va. 1841.
§ As to the orthography of Alleghany or Allegheny, we know of no standard; any more, in fact, than there is for all the other Indian names. It appears to be commonly written in the first manner, and here we follow Darby.

stretch co-extensively with the main escarpment of the great coal-field.

We would, therefore, in the following pages, venture the use of this phrase, wherever, in its local application, it is unmistakable: for our conviction is, that whether we choose to adopt it or not, it will ever continue a term in general acceptation—a name imperishable as that of the Alps.

Let us add here, that we find it marked upon our maps, of every age, and in all parts of its course, from the southern borders of New York state, even into Tennessee. We even see the name "Alleghany Mountains," prolonged eight hundred miles further northward, as far as Cape Gaspé, and the gulf of St. Lawrence.*

According to the scheme of Darby, the entire Appalachian system occupies one hundred and twenty thousand square miles; Messrs. Rogers estimate it at 130,000. Of this great area, the bituminous coal-field of the Alleghanies occupies about one half.

In the primitive maps, which have come under our notice, no part of this range is recognized as essentially Appalachian, beyond the northern limit of Alabama; the country, first of the Appalachians and since of the Cherokees. Daniel Coxe, an admissible authority in geographical nomenclature, shows us in his description of the "province of Carolana," A. D. 1722, that this prolonged range of mountains in Alabama, formed a part of the royal grant of 1630, half a century before its supposed discovery by the French, and was named the "Palachean Mountains."† It has generally been admitted, that the warlike Indian tribe of that name, for the most part occupied the rich low ground that bordered on the gulf of Mexico. They were there first found by the Spaniards, in 1528: and there, in the "Appalachee country," subdued only as late as 1702, they have left their traces in corresponding local names. The Appalachicola, and the Appalachie rivers, yet mark their origin, as does the bay of Appalachie, where Narvaez first landed in 1528. The map of M. G. De L'Isle, in 1720, shows the site of the town of the Appalaches, in 1540, a little south of the present city of Mobile.‡

In Father Charlevoix's map, also, in A. D. 1720, republished in English in 1766,§ the Appalachian mountains appear, extending to about N. lat. 35°, which again is the northern limit of Alabama, as previously adverted to, and nearly corresponding with the north boundary of what was then South Carolina, ceded by France to Great Britain in 1763. The same mountains were assigned by Father Hennepin and Mons. La Salle as the boundary of Louisiana, and were so considered by the Spaniards and previously by the French.||

* Martin's Statistics of the Colonies of the British Empire, pp. 147—151.

† Description of the Province of Carolana, by the Spaniards called Florida, and by the French La Louisiane. By Daniel Coxe, of New Jersey, 1722.

‡ Carte de La Louisiane, par Guillaume De L'Isle, de l'Academie Royale des Sciences. Amsterdam, MDCCXX.

§ "A map of the British Dominions in North America, as settled by the Treaty of Peace, 1763."

|| Brackenridge's Louisiana. 1814, p. 26.

Mr. Jefferson, who wrote his "Notes on Virginia," in 1781, ascribes the origin of the term to the Indian tribe of Appalachies, who resided on Appalachicola river. "Hence," he observes, "the mountains giving rise to that river, and seen from its various parts, were called the Appalachian mountains; being, in fact, the end or termination only, of one of the great ridges passing through the continent. European geographers, however, extended the same, northward, as far as the mountains extended; some giving it, after their separation into different ridges, to the Blue Ridge, others to the North Mountain, others to the Alleghany, others to the Laurel Ridge; as may be seen in different maps. But the fact, I believe, is, that none of these ridges were ever known by that name to the inhabitants, either native or emigrant, but as they saw them so called in European maps."*

In the authorities we have cited, we think we have perceived evidence of the actual extent and limit of the Appalachian area: but we do not object to the adoption of the name, on a far wider scale, if it can be advantageously introduced. We desire only to retain for the coal range, a name for which we confess we entertain great affection.

Certain of our American archæologists do, indeed, go a little further; they are so impressed with the fitness and comprehensiveness of the term, that the New York Historical Society has propounded, in all seriousness, the adoption of ALLEGHANIA for the name of the whole Union.†

We now turn to more direct geological considerations. The Appalachian system as contemplated by Darby and his successors, comprises a vast series of parallel ridges, in advance, to the east and south-east, of our Alleghany region, and includes not only the bituminous coal-field of the latter, but nearly every known American rock formation, from the new red sandstone down to the granite. It comprehends the whole carboniferous group; the anthracite basins of Pennsylvania, and the bituminous coal-field of the Alleghanies; the subjacent Devonian system; and, beyond this, the entire Silurian series and palæozoic rocks, and finally the primitive group.

In physical geography, the arrangement is wholly unobjectionable:—as applied to geology, it seems to be too indefinite, and suggests the subdivision of its members.

The distinguishing geological characteristics of the Appalachian system have been traced by the Messrs. Rogers, with a masterly hand.‡ Every one who seeks to know something of the physical to-

* Jefferson's Notes on Virginia, p. 26. 1784.

† A Report of the Historical Society of New York, dated 31st March, 1845, contains the following resolution: "That the name of '*Alleghania*' be recommended as the best, considering that it is derived from the grandest and most useful natural feature, common to the whole country; an eternal type of strength and union; stretching from the Gulf of Mexico to the great lakes;—that it is associated with the most interesting portions of our history; and that, in adopting it, we should restore to the land one of the primordial titles of the aborigines."

‡ On the Physical Structure of the Appalachian chain. Trans. of the Association of American Geologists. Vol. I. p. 477. 1843.

pography of the most valuable and remarkable portion of the American continent, cannot fail to derive advantage from the perusal of this lucid exposition.

THE ALLEGHANY, APPALACHIAN, OR ENDLESS MOUNTAIN COAL-FIELD,

DESCRIBED ACCORDING TO ITS EIGHT PROVINCIAL DIVISIONS.

THIS magnificent coal-field traverses eight of the principal states in the American Union. In the greater part of these states, geological surveys have been in progress for a number of years; and periodical or preliminary reports have, from time to time, been submitted to the legislatures of those states respectively, by the surveyors. In but a very few of them, however, have the final details of the survey; the engraved and geologically coloured maps; and the various essential illustrations, incidental to the whole work, been published. The design, therefore, of these geological investigations, has but very partially been carried out, on account of the alleged expense; a failure which is much to be regretted. We may add, that our own labours in preparing this part of the present volume, would have been greatly curtailed, had those geological surveys been brought to maturity.

DIMENSIONS.

The greatest length of the entire coal-field, measuring along its centre, - - -	750 miles.
The greatest breadth, - - - -	173 "
The average or mean breadth, - - -	85 "

By the computation of Professor Mather, the area, in round numbers, is 50,000 square miles.* In calculating the actual size of this region, we have comprised within its limits those rocks which, by every geologist, are associated closely with the coal; that is to say, the sandstones, conglomerates, shales; the argillaceous slates, and occasionally the intercalated limestones, which combine to make up the series usually called the coal measures or carboniferous strata. From this maximum superficies, we have made no deduction for accidental areas, occupied by inferior formations; such, for instance, as are brought to the surface by one or several anticlinal axes. We

* Second Geological Report of Ohio, p. 7. 1838.

have reason to know that the aggregate of these interposed areas is considerable. Neither have we made allowances for the numerous cases of denudation; for the partial removal of large areas of coal strata;—for the wide and deeply indented valleys; or for the innumerable ravines which cut away the productive strata, leaving large unprofitable areas. It is unusual to enter into such details; yet if they were investigated and computed, their aggregate would bring down the result of the available or productive areas to an unexpected degree. We know of vast bodies of so-called coal lands, within this field, that have scarcely the tenth acre really productive in that combustible, through a combination of the circumstances just alluded to. Our estimate of the Alleghany coal-field is, for this reason, an extreme one, yet we believe it is strictly correct as to the general superficies.

Taking, therefore, for our principle of admeasurements, the areas in question in their most enlarged sense,—the gross rather than the net returns, if we may so speak,—we find the areas of the provincial divisions of the Alleghany bituminous coal-field, to be as follows, in round numbers:

	States.	Area of the whole State. McCulloch.	Area of Bit's Coal Strata therein.
		Square Miles.	*Square Miles.*
I.	Alabama, . . .	51,770	4,300*
II.	Georgia, . . .	58,000	150
III.	Tennessee . .	45,000	4,300
IV.	Kentucky, . .	40,500	9,000
V.	Virginia, . . .	64,000	21,000
VI.	Maryland. . .	13,950	550
VII.	Ohio, ...	44,400	11,900
VIII.	Pennsylvania, .	47,000	15,000
	Total, . . .	364,620	65,300

NOTE.—The older returns of the respective areas of the States give larger results than those published by S. A. Mitchell, in 1836, and subsequently. These returns, therefore, make the aggregate of the seven States in the adjoining column, 350,449 square miles, instead of 364,620 square miles; the former being probably the more accurate.

Since completing this computation, we have observed that Prof. H. D. Rogers† has calculated that the superficial area of the Alleghany coal-field "upon a moderate estimate, amounts to sixty-three thousand square miles."‡

Were we to make the deductions for unproductive areas, for erosions of strata, and for such coal-beds as are never likely to be reached by the miner, it would perhaps be a liberal estimate to rate the workable area of the whole at forty thousand square miles.

* Prof. Tuomey says, that the productive coal measures of Alabama are known to occupy an area of 5,300 square miles, excluding the counties of De Kalb, Morgan, and Lawrence, and the region north of Tennessee river, "because they have not yet been sufficiently explored to ascertain whether they belong to the productive carboniferous rocks."—1850.

† Trans. Assoc. American Geologists and Naturalists, Vol. I. p. 436.

‡ In a recent outline of the Geology of the Globe, by Prof. Hitchcock, he states that the great Apalachian coal-field extending from New York to Alabama is 720 miles in length, and covers 100,000 square miles.

Even on this principle, we have here in this field, no less than 25,600,000 acres of productive coal: an enormous aggregate, of whose ultimate value no present estimate can be formed—no array of figures or of words can adequately portray.

It is beyond the scope of human vision to contemplate, in our day, the results associated with these millions—the industrial facilities, the wealth, and power, and influence at home and abroad, which they must inevitably confer upon the future inhabitants of the country.

LOUISIANA.

It has been very recently announced in the south, that bituminous coal has been discovered *in situ* on the Iberville river, and in sufficient quantity to supply the ordinary demand for coal in this part of the country. Should the report prove correct, of which we entertain some doubts, the existence of the coal formation, or an insulated portion of it, prolonged in the range of the main coal-field of the United States, to a point so near the Gulf of Mexico, is not alone a highly interesting geological fact, but a very important one as regards the coal statistics and the mineral resources of the southern portion of this country.

Another report [1847] is of a species of coal, found, it is said, on the shores of Lake Bistineau, on Red river, and which is proved to be adapted for the forge or grate. This lake is bordered by a sandstone, over which fossils are described to be deposited, and fossil wood: tertiary lignite?

Again, another announcement [1847] is of coal at Lake Borgne, below Lake Pontchartrain; probably carbonised wood.

The bulk of the coal which is consumed along the southern shores of Louisiana and Alabama, is not derived from resources existing within the latter state, but has descended from remote places high up the Mississippi and the Ohio rivers, or by the circuitous navigation of the Tennessee, the Cumberland, and other rivers, which originate to the northward of Alabama.

For want of railroad and canal facilities for transportation to the south, the exports of north Alabama, amounting to about one-fourth the products of the state, are now transported along the rivers we have mentioned, to New Orleans, a distance of from fifteen hundred to seventeen hundred miles.*

The bituminous coal which arrives at New Orleans and Mobile, is

* Report on the Alabama, Florida, and Georgia Railroad.—*Campbell.*

sold by a local measure called a barrel, and not by the ton weight, or by the bushel or the chaldron, as in other parts of the United States. Thirteen of these barrels are estimated to be equivalent to one ton.

In a table showing the receipt of the principal articles from the interior, during the year ending August, 1852, with their estimated average and total value.

	Amount.	Average value.	Dollars.
Western coal,	850,000 Bushels.	50 cts.	$425,000

The following statement from the local commercial returns, exhibits the number of barrels, of western bituminous coal, sold in the port of New Orleans, annually; the fiscal year commencing on the first of September. The prices are very fluctuating: we have seen coal quoted as low as 75 cents, and as high as $1 25 per barrel: the average is not far from 80 cents; average for 1847, is 75 cents per barrel; 1853, $1 50 per barrel. The increased demand is very great of late years, as seen in the table below.

Years.	Barrels.	Years.	Barrels.	Years.	Barrels.	Years.	Barrels.
1830	40,800	1838	99,220	1843	255,568	1847	356,500
1832	50,000	1840	99,915	1844	227,788		
1834	65,100	1841	121,233	1845	281,600		
1836	85,328	1842	140,582	1846	262,800		

We add here, from a late report of Colonel Albert, of the United States Topographical Engineers, a few notes on the internal commerce of New Orleans, in 1846.

Value of commerce of western rivers with New Orleans,	$9,737,354
Exports and imports of New Orleans, in 1842, officially,	50,566,903
" " " " 1846, "	62,206,719
Natural internal water-courses in communication with New Orleans,—miles, - - - - - -	16,674
Population of internal water-courses in communication with New Orleans,—persons, - - - - -	8,877,456

From the Custom House entries for the whole year ending June 30th, 1851, the number of vessels arriving from sea to the port of New Orleans, were 2266—tonnage, 910,855.

Included in the arrivals are 412 foreign vessels from foreign ports, with a total measurement of 185,386 tons.

Commerce of New Orleans, 1851–1852.

The value of products received from the interior since 1st Sept. 1851, is $108,051,708 against $106,924,083 last year, 1852.

The value of the exports of American produce for the year ending

June 30th, 1851, according to the Custom House records, was $76,344,569 against $81,216,925 last year.*

I. ALABAMA DIVISION.

BITUMINOUS COAL AREA, 4300 SQUARE MILES.

THE fossil coal plants of Alabama, were, subsequently to the date of Mr. Lyell's paper, submitted by him to the inspection of Mr. Bunbury, who identified several of these vegetable remains with well known European fossils. Out of sixteen species thus examined, one half agreed with the plants of the old carboniferous formation in Europe, and the rest belong to genera which are common in the English coal measures. Thus, at the distance of nearly five thousand miles, [the broad Atlantic now intervening], we observe a new proof of the wide extension of a uniform flora in the carboniferous period.†

The *manufacture of Iron* is steadily increasing in Alabama. Vast quantities of ore occur in Tuscaloosa county.

COAL-FIELD OF TUSCALOOSA.

Under this title Mr. C. Lyell has described the southern part of the Alleghany coal-field.‡ We derive the following extract from that account as published in Silliman's Journal.

The city of Mobile is supplied with bituminous coal for fuel and gas, chiefly from this coal-field, by means of the Tombecbee river; a navigation of more than three hundred and fifty miles. M. Lyell, had at first suspected from various circumstances, that this deposit might be related to the Richmond coal in Virginia, which has been shown to be of newer date than that of the Alleghany range. This impression was, however, entirely removed on inspection of the district in question.

The coal seams are worked in open quarries, where the outcrops of several seams are dug successfully, the quality being good. They are covered with beds of the ordinary black, carbonaceous shale; full of impressions of more than one species of calamite; with Pecopteris and Neuropteris, Sigillaria and Lepidodendron, and occasionally Stigmaria. The perfect identity of these coal plants with those of Europe, of Ohio and Pennsylvania, was recognized. They also dif-

* De Bow's Review.
† Proceedings of the Geol. Soc. August 1, 1846.
‡ Silliman's Journal, May, 1846.

fered essentially from the vegetable remains that are most abundant and characteristic in the newer coal-field of Richmond.

The strike of these coal beds of Alabama, on the Warrior river, and to the eastward, is north-east and south-west; agreeing with the general direction of the Alleghany mountains, of which they are, geologically speaking, a southern prolongation, and are bent into anticlinal and synclinal ridges, similar to those of the Alleghanies.

The carboniferous strata here appear to come into direct contact with the cretaceous rocks. The productive coal measure seen by Mr. Lyell on his tour, consisted of the usual sandstones, shales, and clays, with seams of coal; the thickest seen by him being about four feet; but a ten foot seam has been discovered further to the north than the localities he visited. This carboniferous formation is many hundred feet thick; succeeded by a great series of gritstones; and thence passes downward into thinly laminated sandstones and dark slates.

Under this carboniferous group lies a limestone formation, with much intermixed chert and hornstone. In the pure limestone, which is fetid, no fossils occur; but in some of the associated siliceous beds, fossils abound, apparently those of the mountain limestone. Throughout the entire range of the inferior limestone, occurs an enormous mass of brown hematite. "From the accessibility and richness of this ore, its proximity to the coal-field, and to the navigation of the Tombecbee river, I can hardly doubt that, like the coal itself, it is destined, at no distant day, to be a source of great mineral wealth to Alabama."

Mr. Lyell adds, that the fossil plants of Alabama, situate in latitude 33° 10′ north, form a subject of peculiar interest; being apparently the extreme southern limit to which the peculiar vegetation of the ancient carboniferous era has yet been traced, whether on the western or the eastern side of the Atlantic.

On this point, however, we believe we shall be able to show an extension of the true coal formation much lower south than the Tuscaloosa coal, at 33° 10'. Coal and anthracite are reported at a number of points in Texas, stretching in a southwest direction across the headwaters of the Trinity, the Brassos, the Colorado, and other rivers which empty into the Gulf of Mexico, and coal even crosses the Rio Grande into Mexico, below the latitude of 28°. More than this, we know that coal is worked at Guerrero, on the river Salado, for the use of the American steamers on the Rio Grande, in N. latitude 27° —being six degrees lower than Mr. Lyell's extreme southern limit.

The first Biennial Report of Professor Tuomey, Tuscaloosa, 1850, enables us to give by permission the following valuable details, as well as the accompanying map, which we have adapted from his geological map of the State.

These coal fields extend from the northeast part of the State in a southwest direction, becoming much wider in their course. The coal rests upon a 'millstone grit' of quartzose conglomerate, and white siliceous sandstones, and with them, lies unconformably upon the silurian rocks. Towards the south, the carboniferous limestone and a superimposed sandstone resembling the millstone grit, are interposed. "The carboniferous rocks are either horizontal, or inclined at a very small angle, while the silurian rocks are always either vertical, or dipping at an angle which is seldom less than 40°."

The coal measures are composed of various beds of coal, shale, and conglomerates, capping the mountains of the northeast (mostly north of the Tennessee river) in scattered patches. The coal does not repose in basins, but in long depressions corresponding with the general northeast and southwest axis of the formations. The Cahawba and Coosa fields dip south of east, whilst towards the Warrior river it is in the opposite direction.

The following section (omitting the details of the valley limestones) exhibits the Warrior coal-field on the left, and that of the Cahawba on the right, each resting upon the grit.

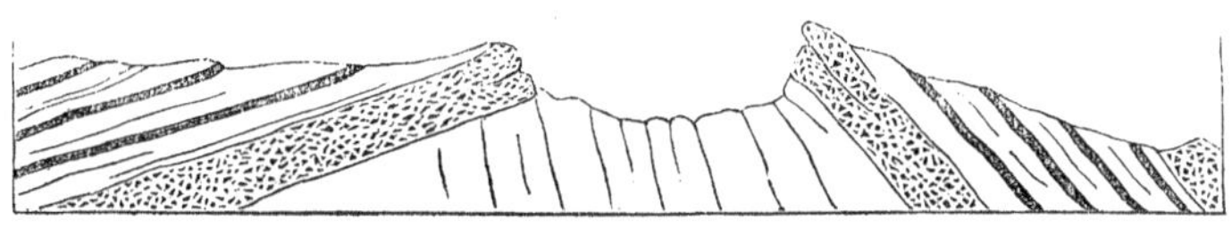

The Coosa coal-field lies along the Coosa river, several tributaries of which traverse it. The coal is good, and free from slate. Mr. Tuomey mentions two beds, "each five feet clear coal." It has been taken down the Coosa and sold at forty cents a bushel at Montgomery; and it is within a convenient distance from iron ore and limestone.

The Cahawba coal-field is considered a prolongation of the preceding. It contains coal of good quality, and strata from six to ten feet thick are mentioned. Two of the latter are worked. In connection with this field Professor Tuomey mentions a bed of argillaceous sandstone with marine fossils, which "prove very clearly that the waters of the ocean once rolled over the beds of coal on the Cahawba."

The Warrior coal-field disappears at its southern extremity under the tertiary beds of Tuscaloosa, where coal is mined for the use of the place. The dip is very slight throughout this field, which brings the coal to the surface in many places by denudation. The coal is taken from the seams exposed along the water-courses, or from the bottom, when a stratum has occurred hard enough to resist the action of the stream.

"I witnessed here the novel process in the art of mining, namely, *diving for coal.* A flat boat is moored parallel with the joints, and near the edge of the coal; long wedge-shaped crow-bars are driven into the seams by means of mauls manœuvred by the men in the boat. When a ledge of about two feet is loosened in this way across the seam, the men take the water, and dive two or three together, according to the size of the masses to be brought up, and lift the coal bodily to the surface, and place it in the boat. As an improvement on this simple process, a crane is rigged on the boat, and a chain slipped round the blocks of coal, raises them into the boat. I have seen, in this manner, masses raised, that weighed 800 or 1000 pounds. The coal thus raised is free from all shale and other impurities, for as the coal parts along the bands of shale, the latter are left behind. Notwithstanding the primitive appearance of this method of raising coal, it is, nevertheless, under favourable circumstances, and where the water is not too deep, one of the cheapest modes in practice, and with the addition of a diving dress, I am inclined to think that in no other way could coal be raised at an expense so moderate."*

Mulberry Fork coal in Walker county. "A very remarkable change takes place in the character of the coal on this branch of the Warrior; it becomes very much cleaner and harder; although it breaks up into small pieces, it rarely produces much slack; and as it comes from the bed, it resembles screened coal. The horizontal partings of the seams often exhibits casts of plants that give a dull lustre to the coal, which makes it appear almost slaty. This, of all the coal in the State, will best bear transportation, on account of its superior hardness."†

The introduction of coal for economic purposes in Mobile was attended with as many difficulties arising from prejudice, as that of anthracite in Philadelphia; and notwithstanding the importance of this branch of industry, it can hardly be considered as established. In 1849, there were about 200 persons engaged in the business during August, September, and October, (when the water was low,) and but three mines were worked below the surface of the earth. The editor, whilst travelling in Alabama, in 1852, found that large tracts of the coal formation remain as public land, and can be taken up at the usual government price. They have been neglected principally because the best beds require capital to open them properly, and because the transportation depends upon the river navigation. The time must come when Mobile will be the centre of a great coal trade on the Gulf of Mexico.

* Report, p. 87.

† Report, p. 89.

II. GEORGIA DIVISION.

We possess no details of this small angle of the coal formation. It probably does not comprise more than one hundred and fifty square miles of bituminous coal area, and was, until a few years back, owned by the Cherokees.

III. TENNESSEE DIVISION.

This part of the Alleghany range, occupies an area of 4,300 square miles, the greater part of which consists of the elevated local group known as the Cumberland mountains.

The geological survey of this State has, during many years, been confided to the charge of Professor Troost, who has communicated to the General Assembly of Tennessee, a series of eight periodical reports, between the years 1831 and 1846. Those which more especially refer to the coal region, are the third, dated October, 1835, and the eighth, dated in November, 1845, but they contain very few practical and economical details respecting mineral coal.

It is understood that West Tennessee has received some investigation from Dr. D. D. Owen, and that the Messrs. Rogers have made some reconnoissances in East Tennessee; but as none of these have, we believe, been published, we are precluded from the advantage of citing the geological results ascertained by those indefatigable investigators.

From the maps which accompany the State annual reports, we perceive that the boundaries of the coal formation are singularly irregular; occasioned by the numerous projecting spurs of the Cumberland mountains, forming alternate bays and promontories, on their western flank. These maps are exceedingly coarsely executed, and do not afford the details we could have desired. We are unable to collect much statistical information from the published reports and maps, respecting the number or thickness of the several coal seams in this section of the great coal-field. In the first report they are somewhat vaguely alluded to as "several outcrops of horizontal strata of great extent." In no instance, in that report, are the respective thicknesses of coal beds recorded; nor can we form any opinion of the amount of the production or consumption of coal, within the district. The quality of this coal is, however, spoken of as excellent. Analyses of two specimens are furnished by the geologist, whence it appears that the coal here approaches in character to the semi-bituminous variety in Pennsylvania and South Wales, and that it possesses only from 14 to 17 per cent. of volatile matter.*

The following section we have compiled from data furnished by Professor Troost:

Fig. 1.

Section of the Tennessee Coal-Field, across the Cumberland Mountains.

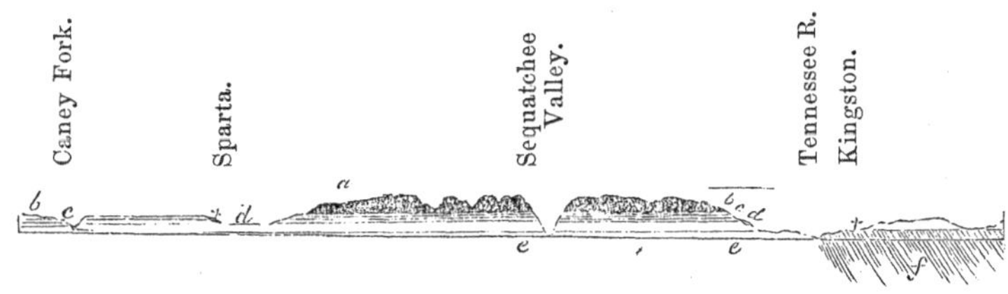

Scale 20 miles to an inch. 75 miles.

a a Coal Measures, almost horizontal.
b b Cherty Sandstone, with Iron Ore.
c c Bituminous Shale.
d d Mountain Limestone—Oolitic above.
e e Old Red Sandstone.
f Upper Silurian rocks, probably broken by parallel axes.

It is evident that much remains to be done in Tennessee, in the way of geological elucidation, and the development of the coal and iron of the state. We believe, however, that very little progress has yet been made in coal operations, or in any branch of mining industry, and, consequently, that opportunities for examination in so wild a region, were rarely afforded to the geologist. He observes, that the deposits of coal in southern and central Tennessee, are evidently of great extent, and are, as yet, only partially brought to light; while those that are best known are only slightly penetrated. Most of them, in fact, have done nothing further than merely to furnish the fuel for the blacksmiths of the surrounding country.

Notices of a great many coal seams appear in the eighth report; but the continuity of these has been so little made out, that there are, at present, no means of determining their number and continuity, so as to be able to recognize them in other parts of the region. The geologist enumerates several beds, of two feet, three feet, and four feet, in thickness; one of six feet, another of eight feet; one of twelve, and a large one of twenty feet. These are all described as good coal, with various qualities and adaptations.

The reporter concludes with the observation, that there exists an inexhaustible treasure of this combustible; which, if once the means of transportation are established, by a railroad, will make Middle Tennessee entirely independent of the uncertain water communica-

* See the Tables of Analysis at the end of this work.

tion which is generally unavailable here during the summer season, and may be the means of transforming this portion of the state into a manufacturing country.*

In the vicinity of the Cumberland mountains a considerable quantity of coal is consumed in the iron works. Some is also transported to distant iron works, and another portion descends the Tennessee and Cumberland rivers, and thence, circuitously, by the Mississippi to New Orleans and the intervening towns, and even to Mobile. At the Dover iron establishment in Cocke county, East Tennessee, the proprietors bring from the coal region a hundred thousand bushels per annum.

According to the third report of the state geologist, this coal is shipped from various points, but particularly from near Kingston; from whence it passes down the Tennessee river more than six hundred miles to the Ohio, and thence, more than a thousand miles further, to New Orleans; making a voyage of no less than *seventeen hundred* miles of inland navigation.† From the western margin of the Tennessee coal-field, a certain quantity of coal is sent down the Cumberland river, nearly an equal distance, to its place of destination.‡

In the eighth annual report, Dr. Troost states that, in 1845, only a few loaded boats descended the Tennessee river, some of which reached New Orleans; but as that city is now much more conveniently supplied from the Ohio river, although quite as long a voyage, he thinks it doubtful as to the future descent of the Mississippi river for Tennessee coals. There is a project under consideration, and urged by the state geologist, for making a rail-road from the southern border of Tennessee, near the Georgia line, across the coal fields of the Cumberland mountains, to Nashville; a measure which would greatly favour the interests of a large area of coal land.

PRODUCTION.

We will not quote the congressional return of bituminous coal mined in this state, in 1840, because it is obviously incorrect. The report of the Secretary of the Treasury, on the domestic products of the United States, of the 6th January, 1845, is still less entitled to credit; being yet more meagre and incomplete than the other.

Like the other coal-fields of this country, that of Tennessee has contributed very little of the ore from which iron is smelted. This is owing, probably, not so much to the absolute deficiency of the argillaceous carbonate of iron which usually accompanies the coal measures, but to the greater abundance, if not to the superior quality, of various other kinds of iron ore, which are distributed throughout the country. Those employed in the furnaces of Tennessee are the

* Eighth Geological Report of Tennessee, p. 15.

† Third Report to the Legislature of Tennessee, by Gerard Troost, M. D., 1840, p. 4.

‡ Report on the Alabama, Florida, and Georgia Railroad, 1838, p. 6.

"Red Oxide," and the various "Hæmatites," and brown ores; and being smelted by the aid of charcoal, produce an iron of excellent quality.

NORTH CAROLINA.

We have no authentic knowledge of the bituminous coal in this state, but anthracite to a very small amount is raised here. The return to Congress in 1840, on which, by the by, little reliance can be placed, shows only fifty tons in that year, and seventy-five bushels of bituminous coal.

Bituminous coal occupies a detached basin on the borders of the Roanoke, and near the state line adjoining to Virginia, in Rockingham county.

Another small basin is situated in Chatham county, in the centre of this state, ten miles south of Pittsborough. This bituminous coal is described as occupying a thin bed, and at present is not worked to advantage.

Other insulated patches of coal are stated to exist in the same range, but, to the present time, little or no advantage has been taken of the presence of this invaluable mineral combustible in various districts of this state.

The coal of Chatham county, on the Cape Fear River, can be conveyed down that river in vessels of from 150 to 200 tons burden; and from below the coal-bed, 30 or 40 miles in barges, said to be free from sulphur—some beds have been partially explored.

April 22nd, 1854. We are indebted for the following valuable information on the Coal Formations of N. Carolina, to the politeness of Professor E. Emmons, of Albany, who has recently made a geological examination of this State.

"The coal rocks of North Carolina lie in two separate troughs. Those which constitute the Deep River coal rocks, begin about six miles south-east of Oxford, in Granville county, and passing south-westwardly to Deep River, thence onward nearly in the same direction to South Carolina, crossing the Pee-Dee three-quarters of a mile south of the mouth of Rocky River. The width of the rocks belonging to this series, varies from four or five miles to eighteen. The greatest width which they attain is upon Deep River; and the least just south of Oxford, in Granville county, where the formation terminates in a point. No coal however has been discovered north-east of Haywood, where the formation crosses the river for the last time in its progress northwards. The other trough containing coal and coal rocks, lies upon Dan river, and extends from the Virginian line near Leaksville, to Germantown, a distance of thirty miles. It extends into Virginia also, and I am informed coal has been found on the Virginia side.

"The Deep river coal formation is divided very naturally into five parts. 1. A conglomerate, at the bottom sixty or seventy feet thick. 2. A red sandstone (free stone) both coarse and fine, alternating with soft marly and mottled sandstone, which may be estimated at about 3000 feet in thickness. 3. Black and green slates and shales, alternating with gray fossiliferous sandstones, at least 1000 feet thick. 4. Drab coloured shale and sandstones, which contain some salt. 5. Red conglomerates, shales, and both green or red, and mottled with green, some 2 or 3000 feet thick. The coal is confined to the 3rd division of the series, or to that which is characterized by the presence of black bituminous shales. The number of seams of coal is not satisfactorily determined. Five seams are known in immediate relation to the thickest mass of black coal slate, the latter of which is about 600 feet thick; this upper seam is just below the middle of this slate. It was struck in boring at Egypt, on Deep river, at a depth of 360 feet. The seam was five feet thick. In the same slate, about two miles to the north-east, at a place known as Farnville, the thickness of the coal seam is about six and a half or seven feet thick, with about ten inches of slate, which divides it at the out-crop into two seams. Following the out-crop of the slates south-west, along Deep river, the coal either appears at the surface at several points, or has been discovered by sinking pits. Coal has been discovered for about twenty miles on a northern out-crop, and the indications seem to warrant the expectation that it will yet be discovered both to the north-east and south-west, or as far as the black bituminous slates extend. Measures are now being taken to test the validity of this expectation. The coal is usually bituminous and of excellent quality, having a clean fracture, and comparatively free from smut. It kindles readily and burns with a bright flame, becomes adherent, and furnishes a strong heat. It is also at most of the pits which have been sunk, free from sulphuret of iron.

"The coal is not, however, all bituminous. Anthracite seams occur at two localities, but seem to be local and to occupy an inferior position, or near the junction of the slates with the inferior sandstones. The questions relative to these anthracite seams are not as yet fully settled. No coal has as yet been sent to market, excepting a few tons, to test its value. It is, however, so highly esteemed by smiths in Fayetteville, that it readily sells for forty cents per bushel, and for economy, it is regarded as cheaper than charcoal at five cents per bushel—2000 tons will soon be sent to New York, prior to the completion of the improvements upon Deep river. The coal seams are accompanied with seams of fire-clay, nodular argillaceous oxide of iron, as in all the coal fields of this country.

"The grits are suitable for grindstones, and the conglomerate at the base of the formation is largely employed for millstones. The formations rest unconformably upon the gold slate series. It is from these that the material of coal formations is mainly derived. At the gulf on Deep river, an iron foundry is now being erected, together with machine shops and buildings for the manufacturing the various articles necessary in the common affairs of life.

"The Dan river coal formation resembles, so far as the rocks, sandstones, marls, slates and fossils are concerned, those of Deep river. The coal however is semi-bituminous, or approaches closely some of the Pennsylvania anthracites. The thickest seam yet discovered is only three feet. It is proper, however, to remark, that few explorarations only have been made. The out-crop of coal will probably be found to extend from Leaksville to Germantown. The out-crop of coal on Deep river is known only on the northwestern side. The dip of the coal seams is variable; at some places it is from 5 to 10°. At the Farmville and Taylor seams it is about 20°, or between 15 and 20°, but becomes flatter in their descent. The southeast side of the formation is concealed by the tertiary sands. The area which is estimated to be underlaid with coal, is about 45 square miles. The main seam is supposed to have an average thickness of six feet. The rocks, I believe, are thicker than upon Deep river, and the formation is overlaid with very thick and heavy beds of brecciated conglomerates. The geological age of the North Carolina coal is supposed to be the same as that of Eastern Virginia, and yet it will be perceived that the coal shales are underlaid by heavy beds of sandstone and conglomerates, which do not exist in Eastern Virginia or the Richmond coal field. So also the thick masses of rather angular conglomerates, reposing upon the top of the formation in the Dan river rocks, is unknown there. These several formations and those of Virginia, are not strictly identical. All contain fish, but those of Dan and Deep rivers are specifically different from those of Richmond; they however belong to the genus Catoplerus, of Redfield. Teeth of Saurians are found in the slate, and in the coal are common, and belong to an order or division of Saurians which have bi-concave vertebræ. In my report to the legislature of North Carolina, I stated that I then regarded the evidence of the age of these coal formations as leaning or rather favouring the Trias period. As yet, however, our lights are too few and feeble to determine the question in a perfectly satisfactory manner."

IV. WESTERN VIRGINIA.

VIRGINIA DIVISION OF THE ALLEGHANY COAL-FIELD.

We have previously assigned for the space held by Virginia, no less than twenty-one thousand square miles. Of this enormous area the actual amount of coal land now in profitable operation is comparatively small. Large bodies of land on the western slope of the Alleghany range, descending to the Ohio, still continue unsettled, although there is, at the present moment, a current of emigration setting in that direction, that a strong impetus is evidently given to industry and improvement in a heretofore much neglected district.

Dr. S. P. Hildreth, in an unusually long and elaborate article in Silliman's Journal of Science, a few years ago, furnished numerous illustrative details of the coal strata in the western borders of this state.

This communication made under some disadvantages in a scientific point of view, was quickly followed by a series of reports from the able geologists appointed to that service by the state.* These have placed a large mass of useful geological information before the public; although, as relates to statistics, they are somewhat less copious.

It will be observed, on reference to the table of analysis in our Appendix, which we derive almost solely from the state reports, that the western coal seams are much more bituminous than the eastern. These western coals contain nearly equal proportions of volatile matter and carbon, and belong strictly to the class denominated *Fat; adhesive* bituminous coals.

At Wheeling, and for fourteen miles down the Ohio, the cliff or bank of the river presents an uninterrupted bed of highly bituminous coal about ten feet thick.† This seam with some other smaller ones, constitute Prof. W. B. Rogers's "Upper Coal series," and extends from Pittsburg, southward to Clarksburg, in the parallel of Marietta; and according to Prof. H. D. Rogers does not extend beyond the Guyandotte river.

Along the valley of the Monongahela are several fine beds of coal. One of them distinguished as the "Pittsburg seam," is the ten feet bed before spoken of, which, to some, is known by the name of the "Main Coal" of northern Virginia, and is readily recognized where it passes the Great and the Little Kanawha river, and thence to the Big Sandy river, on the borders of Kentucky.

* The Act passed in 1835; the "Report of the Geological Reconnoissance," appeared in 1836, and a geological survey of Virginia, was officially directed to follow that Reconnoissance.

† Hildreth—in Silliman's Journal, 1835.

The greatest thickness of workable coal is stated to be nine and a half feet at the mouth of the Scott's Run. The second coal seam in importance, is about five feet thick. A third is from three to four feet. A fourth, geologically the highest known coal bed of any value in Virginia, Pennsylvània, and Ohio, is five feet in thickness.*

On the Great Kanawha are large developments of bituminous coal, as may be inferred from the foregoing paragraphs.

The workable coal seams in the upper group, are thus enumerated by the State geologist.

	Feet.
The first, or main seam,	5 to 9
Second "	3½
Third "	5½
Fourth "	7
	25 feet workable and one vein not workable.

Besides beds of limestone, amounting to fifty feet thick.

The middle division, or group, contains five feet coal in three beds, and twenty-four feet of lime stone, in eleven beds.

The lower coal group contains five small seams, whose aggregate is but nine feet, only one bed of which is workable.

It would seem, therefore, that of these thirteen coal beds, having an aggregate thickness of forty feet, four seams, comprising eight, yards of workable coal may be relied upon, through nearly the whole length of the state, as the productive power of Western Virginia.

For the analysis of the coal of these seams, at various localities, we refer to the tables at the end of this book; they are derived from the State Reports.† We regret our not having the advantage of quoting the final report of the scientific gentlemen at the head of this department.

In 1835–6, it was stated that in the salt region of Western Virginia there were ninety establishments, producing one million bushels of salt annually, and consuming 5,000,000 bushels, or 200,000 tons of coal.‡

In 1840, the amount of bituminous coal mined in the Alleghany, or western counties, was returned at 298,698 tons of twenty-eight bushels. The aggregate of coal, at that time produced in all Virginia, was 379,369 tons—which is probably less than the actual amount—would be about equivalent to nine acres annual consumption from the upper group of workable coal beds, in Western Virginia. The number of workmen, or miners then in employ, was nine hundred and ninety-five, and the capital so engaged was estimated at $1,301,855.

These Congressional Reports so abound in discrepancies as to

* State Geological Report, 1840, p. 86 to 92, by Prof. W. B. Rogers.

† Report Geol. Surv. of Virginia. 1840.

‡ Records of General Science, Vol. V. 1836. Also American Journal of Science, Vol. XXXIX. 1836.

materially impair their usefulness. For instance, Pennsylvania, which makes a larger return than Virginia, viz. 415,023 tons of bituminous coal employing one thousand seven hundred and ninety-eight miners, rates the capital invested at only $300,416. The results of that inquiry (Congressional,) are really so vague, that we can venture to draw no inferences from them.

CANNEL COAL.

Near Charleston, Kanawha county, a bed of this description of coal has been subjected to experiment, with satisfactory results. Its range is apparently of considerable extent.

On the Kanawha and its tributaries, five veins of common bituminous coal and two of cannel coal are found, all capable of being worked, and above the level of the river. 1853.

The northern extremity of Virginia corresponds, in respect to number and quality of its coal beds, with those to which we are enabled more particularly to advert in the State of Maryland.

At the position overlooking the Potomac Valley, called Brandt's Mines, there are described five workable beds, from three feet to fifteen or more feet thick each, or thirty-five feet in all.*

IRON MANUFACTURE OF WESTERN VIRGINIA.

This is of growing importance, but our details are scanty.

WHEELING.

The following is a statement of the annual products of the Iron business, for the years 1845, 1846, and 1848, showing the increase of business in that branch.

	1845.	1846.	1848.
Iron Works at Wheeling,	$300,000	$435,820	$500,000
Foundries, - - -	74,000	99,000	150,000
Engine building, - -	60,000	96,000	100,000
Steel factories, - -			90,000
			$840,000

On approaching the eastern margin of the great coal-field, it has been found that the prevailing quality is much less bituminous there than near its opposite margin.

Immediately west of the escarpment of the great Alleghany ridge, in Hampshire county, are parallel and nearly horizontal coal seams, extending along the borders of the Potomac; five of these are of workable thickness, the aggregate being about thirty-five feet thick.

* Charles Kinsey's Letter to the Union Potomac Company.

EASTERN VIRGINIA.

SMALL DETACHED AREAS OF ANTHRACITE AND SEMI-BITUMINOUS COAL.

For some years it has been known that here and there a few patches of coal occur in the mountain ranges which run parallel with the eastern base of the Alleghany mountains. In Berkley county, a few developments of excellent anthracite in a ridge near Martinsburg, led to the formation of a working company, some years ago.* The undertaking failed, it is understood, on account of the deficiency of the supply of coal. In several other counties, along the same range, which corresponds with that of Schuylkill county, similar anthracite traces prevail, although of inconsiderable value.

Other localities occur in this parallel in which coal of the semi-bituminous species appears. The seams of this coal vary from three to seven feet in thickness, in Botetourt and Montgomery counties, in the Little North mountain.† Their analysis, which will be found in the tables at the end of this work, shows them to resemble the coal of the Round Top mountain, in Pennsylvania.

Only two hundred tons of anthracite were returned as the production of Virginia in 1840.‡

RICHMOND OR CHESTERFIELD BITUMINOUS COAL-FIELD.

This is an area which has been longer known and worked than perhaps any other in the United States. The geological map of McClure—the father of American geology—was prepared to illustrate his memoir in 1817. In position and form, as there represented, this coal-field differs in no material respect from the maps of the present day—and thirty years of operations there have not greatly enlightened us as to its details.

Mr. Nuttall, in the Journal of the Academy of Natural Sciences, of Philadelphia, some years ago, furnished a short notice of the vegetable fossils which characterize the shales of this coal basin, and adverts to the vestiges of fossil fishes which had already attracted the attention of previous naturalists.§

Some reference to these Virginia coal plants may be found in Sternberg;‖ but Mr. Nuttall was the first to point out the prevalence

* Memorial of the Baltimore Convention to the Commonwealth of Virginia, 1834.
† Report of the Geological Reconnoissance of Virginia, p. 90.
‡ Congressional Report on the Census, 1841.
§ Journal Acad. Nat. Sciences, Vol. II. p. 36.
‖ Sternberg, Book III. p. 16, 1826.

of certain fossil vegetables, which seemed to confer a peculiar character on the formation, although the value of such a test was probably unknown to, and unsuspected by the observer of those days.

M. Adolphe Brongniart has figured and described some coal plants from hence, considering them as belonging to the true carboniferous period.*

The author of this work, in 1834, addressed a memoir to the Geological Society of Pennsylvania, respecting a portion of this coal-field This paper was accompanied by several diagrams, among which wa a vertical section of one of the deep shafts, the first illustration of thi description, that the coal-field had received.† This communicatio was followed by another from Mr. Clemson, on the analysis of th coal collected from several of the mines on each side of the Jame river.‡

In the same year a brief account of this basin and its coal trade wa given in a Report to the Senate of Pennsylvania.§

In 1839, a concise description of the same district appeared in a able Report of the Committee on a National Foundry, ascribed t the Hon. W. Cost Johnson. It also refers to a document, submitte at a public meeting at Richmond, February 6th, 1838, wherein it i affirmed, "that every cannon foundry in the United States is furnishe with coal from the Black Heath pits, and that other pits supply large quantities to the northern iron factories."||

The geological age of this coal-field has been a subject of some in vestigation, owing to the anomalous character of the beds of shal and sandstone which overlie the coal. These differ entirely from thos of the regular coal series in other parts of the American Continent Those occupied by shale are distinguished by peculiar fossils. Th numerous suite of interstratified rock beds consist of granitoid sand stones, or *psammites;* derived from the destruction and reproductio of the primitive rocks in which this basin is placed. A rock of pre cisely similar appearance, crosses the Schuylkill from seventeen t twenty miles above Philadelphia; the resemblance being so close a to show no distinguishable difference in hand specimens. The sand stones of some coal-fields on the European Continent are of this cha racter. The coal of the basin of Blanzy, in France, occurs in a gneis valley, and alternates with granitoid psammites. That of Fins e Noyant, also in France, reposes upon granite; and that of Ahun con sist of strata which are recomposed from the debris of granitic rock Near Oporto, also, anthracite occurs, interstratified with granitoi psammites, overlying primitive rocks, and covered by chlorite slat In Northern Bengal and Bhotan are similar granitoid sandstone containing brown coal.

During a transient inspection of these strata, in 1834, it seemed t

* Histoire des Végétaux Fossiles, p. 126.

† Trans. Geol. Soc. of Penna. 1835, Vol. I. p. 275; Pl. 16.

‡ Ibid. p. 295.

§ Senate Journal, Vol. II. 1833–4, p. 567.

|| National Foundry Report, Doc. No. 168, p. 41.

the present writer that the series was at least contemporary with the ordinary coal measures. Perhaps the well known presence of the fossil *Calamites Suckowii*—recognized by A. Brongniart himself from hence*—a species common to the anthracite shales of Wilkesbarre, to the lowest bituminous shales of Continental Europe, and to the old coal measures of England and Wales,† contributed to this impression.

Subsequently, the vegetable fossils of these remarkable strata have been maturely investigated. To the fossil fishes have been applied the test of modern science; and the opinion is now settled that we must look to a later period than that of the carboniferous era for the origin of the Chesterfield deposit.‡

In 1843, appeared a memoir "on the age of the coal rocks of Eastern Virginia," by Professor W. B. Rogers, which has thrown additional light on this interesting subject. We will briefly endeavour to convey the author's views and the testimony which appears strongly to sustain them. Abundant evidence, of a satisfactory character, is produced, of the geological peculiarities of the numerous series of beds which overlie the thick deposit of coal here. This bed is of irregular thickness, in consequence of the uneven surface of the primary rock on which the coal was deposited, by which, at certain points, it is only a very few feet, and at others deepens to upwards of forty feet. The author assigns for this coal the same geological age as that of the shales and granitoid sandstones overlying it. The entire group presents, it is conceived, striking analogies, in its vegetable remains, to the oolite coal formation of Brora, in Scotland; of Whitby, in Yorkshire; and of certain other European localities. Some of these plants appear to be specifically the same as the English fossils of that epoch; while the rest are very closely allied to certain species of the same genera found in connection with the oolite coal of Yorkshire and Sutherlandshire.

It is to be regretted that no figures of these fossils illustrate this able paper, the value of which would have been greately enhanced by such essential aids. Nevertheless, the elaborate descriptions of the plants from the coal shales are decisive, in most instances. The fishes, mentioned many years ago by travellers and geologists have been fully investigated and named by Mr. W. C. Redfield, and go far to settle the point.

These facts seem strong enough to justify the referring this coal-field of Eastern Virginia, to a place in the Oolite system, on the same general parallel with the carbonaceous beds of Whitby and Brora,—that is, in the lower part of that group.§

We have to add, in corroboration of these views, that, at a meeting of the British Association, Sept. 1846, Mr. Lyell stated that he

* Histoire des Végétaux Fossiles, p. 126.

† Count Sternberg, Book IV. p. 16, 1826.

‡ Proceedings of the Academy of Natural Sciences, January, 1842.

§ Trans. Association of Amer. Geologists, Vol. I. p. 308. Also the State Report for 1840, p. 36.

had lately examined this coal-field, and had submitted some of the fossil fishes, obtained from, it to M. Agassiz, which he referred to the Oolite period. The fossil plants, likewise procured from hence, were examined by Mr. Bunbury, who considers that they present an assembage which agrees with those found at Whitby in Yorkshire, and therefore of the Oolite period. This coal-field, consequently, is newer than that of the true carboniferous formation.

We must not forget, however, in relation to priority of observation, that Mr. Nuttall had long ago, recognized among the fossil plants of the coal-shales here, the *Zamia* or *Cycas*, and the leaves of one of the *Scitamineæ*, similar to those of *ginger*, and some enormous flaccid-leaved gramineous plant; all of which are characteristic of the Oolite period, although not so applied, at the time, by that intelligent naturalist.*

We rejoice to perceive this triumphant application of the test of organic remains, in determining otherwise very doubtful points as to the age of rocks; a principle which, some years ago, we, with all the partialities of an original disciple of William Smith, almost feared, was not appreciated as it deserved.

Whether the entire body of the coal itself be referable to this epoch is by no means settled. It has been suggested that the fossils above-mentioned, and seen at some of the pits, represent "a distinct formation of coal from the main or true carboniferous formation, and many suppose it a deposit of after date."†

On the 14th of April, 1847, a paper was read before the Geological Society of London, from C. Lyell, on the Richmond coal-field of Eastern Virginia. It is stated that the shells in these coal measures consist of countless individuals of a species of Posidonomya, much resembling the *P. minuta* of the English Trias. The fossil fish are homocercal, and differ from those previously found in the new red sandstone, [Trias ?] of the United States. Two of them belong to a new genus, and one to *Tetragonolepis*, and are considered by Prof. Agassiz and Sir P. Egerton, to indicate the Liassic period.

In the charcoal Dr. Hooker detected vegetable structure, not of Ferns or Zamites, or any Conifer, but perhaps of Calamites.

Mr. Lyell considers this coal as of the age of the inferior Oolite, or the Lias.

The fossil plants of the Richmond coal-field have been also carefully examined by C. J. F. Bunbury, Esq. Fifteen different forms are described; of which, however, only ten are sufficiently well preserved to be determined with the requisite precision. Six of them are ferns; of which three are new species; one of them being identical with a species characteristic of the Oolites of the Yorkshire coast. One species of Equisetum, undistinguishable from that of Whitby; one or two Calamites; two of Zamites. Mr. Bunbury thinks that the Richmond coal-field is of later date than the great

* Journal of the Academy of Natural Sciences of Philadelphia, Vol. II. p. 36.
† Silliman's Journal, July, 1842, p. 9.

carboniferous system, and that it must be referred either to the Jurassic or the Triassic series,—more probably to the former.

Fig. 2.

Section of the Bituminous Coal Field near Richmond, Va.

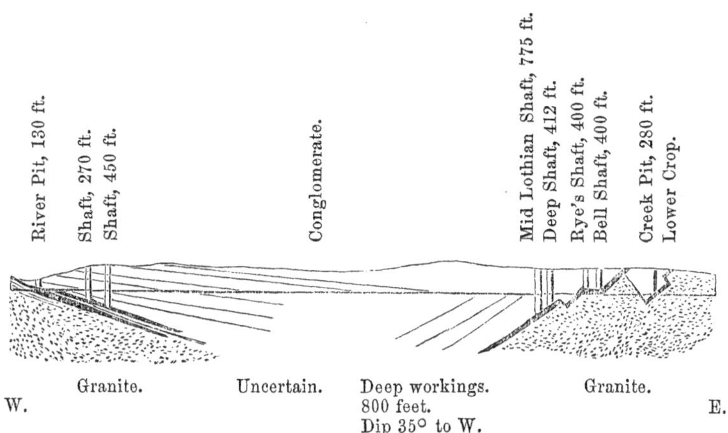

We have seen no satisfactory announcement of the superficial area of this basin, and from local circumstances, it is not very readily defined. It has been considered as thirty or thirty-five miles in length, having a maximum breadth of eight, and an average of five or six miles. The state report of 1840 estimates the length at thirty miles. We believe that we shall not err greatly in assigning one hundred and eighty-five square miles as the extreme productive area of the Chesterfield basin.

There exist, probably, no data by which the depth to which the coal descends in this trough, can at present be ascertained. The outcrop, on the eastern margin, rises at a much higher angle than that on the western. In the former, the shafts are far deeper than any others in America. The vertical section published by the Geological Society of Pennsylvania, in 1835, is by no means the deepest in the district, for there are other pits which are placed at a greater distance from the outcrop of the great mass of coal. Reid's deep pit at Chesterfield, to which we refer, measures four hundred and twelve feet, from the surface to the granite floor. The following summary shows the number of seams of rock, sandstone, shales, and coal, which were penetrated by this shaft. It is to be observed, however, that as the strata were not intersected at right angles, but at the inclination at which they were cropping out to the surface, they are proportionately and individually of greater thickness, as represented in the table, and traversed by the pit, than in strict accuracy, ought to be assigned to them.

The entire series, so far intersected, comprises ninety-four beds. Forty of these, consisting of varieties of carboniferous, micaceous and argillaceous shales, occupy an aggregate of one hundred and thirty-four feet. Fifty-one other strata consist of granitoid psammites, carbonaceous sandstones, white or gray micaceous grits of various degrees of texture, ranging from conglomerates up to schistose sandstone, and comprise a thickness of two hundred and sixty-seven feet. At the base of the group are two or three coal beds with intermediate shales; the whole embracing a thickness of from eleven to forty feet of coal, and even fifty feet, according to the irregularities in the granite floor.

As before stated, there are other shafts which reach the coal at a deeper part of the basin than at this point; and as the operations are carried down the slope of the main seam, the works are necessarily becoming deeper, and the coal is excavated and brought to the surface at a corresponding increase of expense. Up to the present time, we believe, the eastern outcrop alone is that which is almost exclusively wrought, and owing to the steep inclination of the coal measures, not to a greater breadth than from one half to three quarters of a mile. The general course of this eastern boundary is S. 24° W. or thereabouts, and every investigation shows the arrangement to be that of an extremely elongated trough, occupying a hollow in the primary rocks. In the workings examined by the present writer, the coal was only separated from the granite by a bed of about a foot thick, of porphyry.

Some of these deep mines contain a good deal of water, and are consequently attended with expense to keep dry; for as the structure of this trough precludes the possibility of draining and working by the economical system of adit levels or tunnels, all the water, as well as the coal, has to be elevated to the surface by machinery, and generally, by steam power. Other mines are comparatively dry, even in the vicinity of the wet shafts. This fact speaks conclusively as to the dislocated nature of the coal basin, near its eastern margin.

The Maidenhead or Heath's mines are remarkable for their dryness, and no water occurs in any of the workings: such, at least, was the case in 1835, at which time no steam engine was needed for pumping. Some portion of this freedom from water is perhaps attributable to the subsidence of old works.

The Black Heath mines were then on fire, and could not be worked.

The Bell workings had been on fire for twenty-five years, and the fire had advanced to the workings of the Rise shaft. They were all walled up. No flame arose, but a hot, smothering fire continued, year after year, the heat from which was sensibly felt, at some distance.

Only one trifling accident, it was said, had, up to that time, happened from fire-damp.

The Bell and Rise workings, of above four hundred feet depth, belonging to Mr. Mills, gave way on the Christmas eve of 1833. Being a holiday, it was fortunately the cause of saving the lives of a great many miners, who usually worked there, at all hours; and it is remarkable that not a single life was lost on that occasion, although a few workmen were still below, employed in pumping, when the alarm was given. On the succeeding morning, the course of all the underground workings could be traced on the surface, as on a map, by the lines of fissure, running in the same direction, and extending along the ground.

Explosions of fire-damp have been of occasional occurrence in these mines. In March, 1839, an explosion occurred in one of Heath's pits, by which a number of lives were lost, chiefly of coloured miners, to the number of fifty-three out of fifty-six persons, who were then in the mine.

The shaft of the mine was seven hundred feet deep, and the falling in of the earth was so great, that suffocation must have ensued to all who escaped the fire. Explosions, we are informed, occurred several times in the Maidenhead pits, prior to this period, and on these occasions several men were killed and burned.

Great improvement in the system of ventilation have been recently adopted in these deep mines. The first accident, from this cause, took place about 1817: fortunately the explosion occurred when the miners were out at their dinner, about one o'clock, in the day.

In 1841, and preceding years, several accidents by explosive gas occurred in Wills's mines, by which some lives were lost, and several men were severely burnt.

In June, 1844, an explosion took place in one of the Black Heath pits, while four Englishmen and eight negroes were in it. According to the statements of the time, only one person of this party was taken out alive.

May 15, 1854, a terrible explosion occurred at the Chesterfield coal pits, 14 miles from Richmond, Va. Twenty men were in the pit at the time, all of whom were instantly killed but one, who was taken out alive, but dreadfully, and, it is supposed, fatally injured. The pit was over 600 feet deep. Several explosions have occurred heretofore in the same mine, and this last accident was caused by leaks from old damps.

SUBTERRANEAN TEMPERATURE.

Prof. W. B. Rogers has communicated the result of a series of obvations on subterranean temperature, in the Chesterfield mines. The conclusion arrived at is, "That from the *invariable* plane, downwards for many hundreds of feet, the temperature augments at the rate of one degree for every sixty feet of depth."* This result agrees with that recorded by Professor John Phillips, in determining the ratio of descending temperature in the deep mine at Monk Wearmouth, nearly sixteen hundred feet deep. In this case it was also deter-

* Trans. Assoc. American Geologists, Vol. I. p. 538

mined that the temperature increased one degree for every sixty feet of depth.*

In the American Journal of Science and Arts, of July, 1842, is a "Geological and Statistical Notice of the Coal Mines in the vicinity of Richmond," by A. S. Wooldridge, President of the Mid Lothian Mining Company.

The principal mines then in operation, were those of the Maidenhead or Black Heath company, by several shafts which vary in depth from one hundred and fifty to six hundred, and even to upwards of seven hundred feet. The coal from these mines is of good quality, and averages thirty-six feet in thickness.

To the north of these are Wills's pits. A shaft here is about four hundred feet to the bottom; from whence two inclined planes, following the slope of the seam, are conducted, so as to increase the depth of the mine, about three hundred feet more. The coal is here thirty feet thick.

The Gowrie pit is four hundred and sixty feet deep—the coal only six feet. There are various other workings in this region, the shafts of which vary from one hundred and fifty to four hundred feet deep.

The Mid Lothian Company's mines lie to the south of the Maidenhead mines. Coal was struck here in 1839, in a shaft, at the depth of seven hundred and twenty-two and a half feet. The coal was thirty-six feet thick, and the sump below the coal being sixteen and a half feet deep, the entire depth of the shaft is, therefore, seven hundred and seventy-five feet. The coal inclines to the westward at an angle of thirty-five degrees, and in some places is full fifty feet thick, owing to the uneven configuration of the bottom rock, as was observed in other places.

Mr. Wooldridge states that large quantities of inflammable gas are constantly thrown out from the coal in this mine, and great care is taken to prevent the disastrous consequences of an explosion.

Our limits do not allow us to follow the details further. It appears that many new collieries are brought into operation, and a number of others are either exhausted or have failed from pecuniary or other difficulties.

It seems almost certain that the bituminous coal of Richmond, being of an age not older than the Oolitic or Jurassic period, partakes inevitably of the defects of the coal of that period, and can never attain to the rank of the better class of bituminous coals in the United States, any more than the Yorkshire coals of the same age in England. Its chief excellence consists in being a good grate coal for domestic use.

At a meeting of the American Association for the advancement of Science, May, 1854, Mr. H. R. Schoolcraft brought forward a paper on the Dora coal formation of Virginia, which was answered by Prof. W. B. Rogers, a brief summary of whose remarks on the subject we beg leave to subjoin:—

* Philosophical Magazine, Dec. 1844.

"This Dora bed was no new discovery. Twenty years ago it was explored. It lies some 16 miles from Staunton, in Virginia. It was a thin and illusive coal-bed. Rumours of its wealth had often stimulated speculators to their ruin. Companies had often been formed with the intention, first, of working the stock; and second, possibly, of working the coal. The section that he had seen was worthless. He had heard of a seam of it fifteen feet thick. But the rocks are piled up topsy turvy. The older Silurian rock overlaps the carboniferous limestone. The rock that we are accustomed to designate as number two, rides over number ten, and all between have entirely disappeared. In one place the coal seam was four inches, in another four feet thick, and in others you might call it fifteen or twenty feet thick of shale and coal, according to the direction of your measurement. When one falls upon one of these rich pockets, the rumour flies, and the proximity to market makes up a fever of excitement. He did not affirm that it was worthless, but in the lack of definite data, and with no adequate sections of it, he anticipated no large resource of coal from the Dora bed."

PRICE OF COAL AT THE MINES.

The various circumstances attending the quality, locality, &c., affect correspondingly the prices, and render it unsafe to quote any statements, that may be considered as representing an average.

In 1836, it was stated that the coal proprietor could deliver coal at Richmond, twelve miles, at fifteen or sixteen cents a bushel, shipped on board. This appears to be a high estimate when compared with the Alleghany coal which is brought by canals, from two hundred and fifty to more than three hundred miles, and sold in Philadelphia for eighteen cents per bushel.

In 1838, in the National Foundry report, it was stated that "coal could be furnished and pay a reasonable profit to the collier, at ten cents per bushel on the north, and twelve and a half cents on the south, side of James's river, = $2.80 to $3.50 per ton.

In 1846, Richmond coal obtained from twenty to twenty-two cents per bushel, in Philadelphia, which was two or three cents per bushel higher than Alleghany bituminous coal.

The average annual amount of Richmond coal received in the port of Philadelphia, in the six years, from 1824 to 1829, was 124,305 bushels, or 4,143 tons.

The average annual amount imported into Boston during seven years, from 1835 to 1841, inclusive, was 162,552 bushels, or 5805 tons. The entire importation of American bituminous coal into Boston was diminished to little more than 4000 tons, in 1847.

QUALITY.

Some analysis of the Chesterfield coals will be found in the appendix. They were also subjected to the scrutinizing investigation of Prof. Johnson, in 1844. The number and species of American coals* experimented upon were about forty; and we find, from the tables of results furnished in the Report, that the Chesterfield coals, taken from four different pits, ranked as follows:

	Numbers.			
Rank in the order of their relative weights, -	10	27	32	40
" " of rapidity of ignition, -	8	13	16	20
" " of completeness of combustion,	6	12	16	28
" " of evaporative power under equal weights, - -	20	22	26	29
" " of evaporative power under equal bulks, - -	24	27	28	30
" " of evaporative power of combustible matter, - -	22	23	26	33
" " of freedom from waste in burning, - - -	20	25	27	28
" " of freedom from tendency to clinker, - - -	20	26	28	29
" " of maximum power under given bulks, - - -	25	27	28	31
" " of maximum rapidity of evaporation, - - -	1	15	26	39

CLOVER HILL COAL MINES.

In 1845, the Clover Hill Railroad Company constructed a road from a shipping point on the Appomattox river, near Fredericksburg, to what was then a new coal region, but is ascertained to be an extension of that which had been long known and worked, in a north-east direction. These mines are about fourteen miles south of the most southerly, previously wrought, and there is a space of ten to twelve miles between the two regions, in which no mines are yet opened, but in which the coal measures, and in fact, coal, is known to exist in a greater or less extent, although it has never been explored. The general impression, among colliers, is, that the coal is co-extensive from one end of the field to the other, and some even extend their views further south, and place it in North Carolina where the coal is found west of Raleigh. The Clover Hill portion ot the field has been practically opened to commerce only during 1846 and 1847. The coal is not shipped at Richmond, but at twenty miles distant near City Point, at the mouth of the Appomattox river. Consequently it is not included in the Richmond returns, and the amount is additional to the Richmond reports.

We are informed by J. Hopkins, Esq., that since the completion of the road, up to the 1st October, 1847, there has been sent by

* Report to the Navy Department of the United States on American Coals, by W. R. Johnson, 1844.

railroad 2,187,000 bushels of coal, besides some sent by other conveyances; of which 1,592,830 bushels have been shipped for northern or southern ports; the remainder was consumed in Richmond and Petersburg. The business of 1847 is at the rate of 1,500,000 bushels or 53,500 tons, and is increasing. It will probably exceed 2,000,000 bushels or 71,000 tons in 1848.

PRODUCTION.

In 1840, the Congressional return of the annual production of this coal-field was 80,671 tons. By a subsequent return to the Virginia legislature, it appears that the quantity of bituminous coal raised here, between the years 1822 and 1841, inclusive, was forty-nine millions of bushels =1,750,000 tons, being at the average rate of 87,500 tons per annum, for twenty years. The expense of raising this coal was stated to be $1.12 per ton, or four cents a bushel.

Table of Annual Shipments of Virginia Coal.

From Richmond exclusive of the home consumption. The original returns are in bushels, a pernicious custom which should be abolished, but for convenience of reference we have reduced them to tons of 2240 lbs.,* as in all cases where measures are quoted instead of weight.

Years.	Tons.	Years.	Tons.	Years.	Tons.	Years.	Tons.
1822	48,214	1830	91,786	1838	96,428	1841	71,071
1824	59,857	1832	117,857	1839	85,714	1842	65,750
1826	79,214	1834	110,714	1840	78,571		
1828	89,357	1836	110,714				

In 1847, the Clover Hill mines furnished at the rate of 53,000 tons a year, as before mentioned, in addition to that from the Chesterfield district.

Current prices of coal at Richmond, January 1st, 1848.†

Chesterfield coal, 10 cts. to 18 cts. per bushel—$2.80 to $5.04 per ton.
Best Clover Hill coal, 20 cts. " $5.60 "

Importation of Virginia coal into Boston.

1846 - - - - - 183,352 bushels.
1850 - - - - - 63,415 "

* Register of the Treasury. Also Hunt's Merchant's Magazine, Vol. VIII. p. 548.
† Richmond Newspapers.

IRON MANUFACTURE OF EASTERN VIRGINIA.

In the entertaining diary of Colonel Byrd, written between the years 1728 and 1736, but first printed in 1841, occur some curious and interesting details of the iron works of that period. There were in 1732, four furnaces in Virginia; but at that time no forge had been erected in this colony; although a very good one was then in operation for making bar iron, at the head of the bay, in Maryland. "It was feared that the English parliament would soon forbid us that improvement; lest, after that we should go farther, and manufacture our bars into all sorts of iron ware, as they already do in New England and Pennsylvania. Nay it was questioned whether we should be suffered to cast any iron, which they [the English] can do themselves at their furnaces."

Colonel Spotswood, who furnished Colonel Byrd, in 1732, with much practical information, was not only the first in Virginia, but the first in North America to erect a blast furnace [about 1715.] He stated that "they ran, altogether, upon bloomeries in New England and Pennsylvania, till his example had made them attempt greater works. In the latter colony, they have so few ships to carry their iron to Great Britain, they must be content to make it only for their own use. The four furnaces then at work in East Virginia, circulated a great sum of money, for provisions and all other necessaries, in the adjacent counties. They are, besides, a considerable advantage to Great Britain, because it lessens the quantity of foreign bar iron, heretofore imported there, and paid for in silver. On the contrary, all the iron they receive from the plantations they pay for in their own manufactures, and send for it in their own shipping."

Colonel Spotswood also erected an air-furnace at Massaponux, which he brought to perfection, and was able to furnish the whole country with all sorts of cast-iron, as cheap and as good as ever came from England."*

In 1750, a bill was passed in parliament for the repeal of the duties on the pig and bar iron made in the British colonies of America; but the interest of the iron manufacturers in Great Britain prevailed so far, as to add to the bill a clause, prohibiting the erection of any mill or other engine for slitting or rolling of iron, or any plating forge, to work with a tilt hammer, or any furnace for making steel; for it was feared that the colonies might interfere with the manufactures of their mother country.

In 1775, the American war with England broke out, and, at its termination, a new era commenced in the history of the American iron trade.†

* Westover Manuscripts, 1841.

† Scrivenor's History of the Iron Trade.

VIRGINIA.

LIGNITES IN SANDSTONE NEAR FREDERICKSBURG, OF THE OOLITE AGE.

In 1834, the Geological Society of Pennsylvania, published in their first volume a paper, communicated by the author of this work, on the Lignites of the secondary horizontal strata of Fredericksburg, accompanied by six lithographed figures of plants. These lignites are in no place in sufficient abundance to constitute a seam or bed, much less a workable bed; but as interesting specimens of silicified masses of wood, and fragments even of large trees, which reminded us of those of the Portland rock of the South of England; besides an infinite number of impressions and carbonized remains of more delicate varieties of plants, that are not undeserving of a passing notice.

On looking over the imperfectly defined series of these plants, it will be seen that they are all cryptogamous, cellulares, or acotyledones, with the exception of Thuytes; and that they belong to genera some of whose species are distributed abundantly amongst the coal vegetation of all parts of the world. These species, however, appear to be new; that is, they do not belong to the carboniferous period. One approaches to the oolite period, and the consideration given to this group of plants led to the conclusion that they were "perhaps coeval with the oolites."

The large broken masses of silicified wood are, unquestionably, remains of vasculares, or dicotyledonous plants or trees, no member of which, we believe, has yet been observed in our ancient coal vegetation. These resemble, somewhat, the silicified wood of the Portland oolite; and, like them, exhibit no marks of perforation by the teredo.

It must be observed, that all the genera to which we have assigned the fossil plants of Fredericksburg, occur in the oolitic group of Europe. For this fact we have the testimony of M. A. Brongniart; of Saussure, Phillips, Murchison, De la Beche, and many others. Mr. Nuttall has described silicified wood, near the James River, having characters resembling those we have mentioned at Fredericksburg.*

If we mistake not, Professor W. B. Rogers has also satisfied himself that the date of the Fredericksburg sandstone "is referable to that of the oolite."†

The geological and topographical position of this lignite sandstone is immediately beneath the older tertiary formation, and superficially

* Trans. Geol. Soc. of Penna., 1831.

† Proceedings Acad. Nat. Sciences, Philad., January, 1842.

occupies a belt immediately west of it, overlying the primary rocks of Fredericksburg and Petersburg.*

Between Fredericksburg and Richmond, lignite and thin seams of impure bituminous coal, according to the State Surveyor, are of frequent occurrence in these feldspathic sandstone beds, provisionally termed "upper secondary stone," which are, in many places, largely intermingled with dark coloured micaceous slates and bituminous shales.

PETROLEUM.

In the valley of the Little Kanawha, about six miles from the mouth of Hew's River, is a spring from which from fifty to a hundred barrels of petroleum are annually collected. Petroleum also rises in nearly all the wells in the salt region of the Kanawha.

* See also—The Virginia Annual State Report, 1840, pp. 27 and 35. American Journal of Science, Vol. XXIX. p. 86.

V. KENTUCKY DIVISION

OF THE GREAT ALLEGHANY COAL-FIELD.

THE superficial coal area within this State we have computed at 9,000 square miles.

Professor Mather computed it at only seven thousand, which, if we estimate the workable or productive area of coal alone, would be very ample. In 1837, an address, recommendatory of a State geological survey, was made by Mr. Trimble to the Kentucky legislature.* In consequence of this movement, Professor Mather was instructed to make a geological reconnoisance of the state, which was accomplished the following year, 1838. Since that time, no further progress towards a more detailed survey has been made, and our information is less ample than it could be desired.

The congressional return from Kentucky, 1841, shows that she raised in 1840, 58,8167 bushels, or 21,000 tons of coal, an amount far below the actual production.

There appear to be several qualities of coal here. The "main seam," which extends from Pittsburg to Wheeling through Virginia, is said to reach Sandy river at the boundary of this state, but does not pass into it, or extend but a very short distance, southward.

Of Cannel coal, several seams are said to be found on the Kentucky river, and the quality is highly commended.

Nearly all the coal brought into use in Kentucky is reported to be of the description called *Cannel.* It is slightly bituminous, but rarely cakes in burning. Its analysis seems to ally it to the dry or semi-bituminous coal of the Cumberland mountain, described by Dr. Troost.

Mr. Trimble details some experiments made by steam-boats on the Ohio, from which it was ascertained that the daily expense of fuel, when mineral coal was used, was less than one half that of cord-wood.

Four hundred and fifty steamboats, using twenty cords of wood in the twenty-four hours, and running two hundred days per annum, will consume an amount of wood, whose value at $2 50 per cord, would be - - - - - -	$4,500,000
By the use of coal, during the same time, and producing similar effect - - - - - -	1,500,000
Annual saving	$3,000,000

* Hon. D. Trimble; Report on the coal and iron trade of Kentucky, 1837.

The price of this coal at Louisville, 1844, was seven and a half cents per bushel, by the boat load, equivalent to two dollars and ten cents per ton.

Mr. Mather's report (1838) to which we shall now more particularly advert, states that, at that time, at least one million of bushels [35,714 tons] were annually sent to market from the mines on the principal rivers.

He estimates that the coal formations of Kentucky cover twelve thousand square miles, of which seven thousand square miles contain workable coal beds.

The coal is of three varieties—

1. Bituminous—Caking coal.
2. do. but not adhesive.
3. Cannel, or Splint coal—Steam coal.

Besides the million bushels which descend the principal rivers, about two millions more are consumed in the iron and salt works of the state ; thus amounting to 107,143 tons. This shows the fallacy, we have before pointed out, of the census returns, which, two or three years afterwards, when the production was greater, only included 21,000 tons.

The use of coal for steamboats, the reporter urges, is increasing rapidly, and its recommendations, for that purpose, are principally these—

1. It makes a more uniform and more easily regulated fire than wood.
2. The economy in the use of coal, over wood, is three-fifths.
3. The weight of equivalent quantities of coal and wood is as one to three.
4. The bulk do. do. as one to nine.
5. The labour and expense do. putting on board, as one to four.*

The geologist enumerates a great many details and localities within the Kentucky region where coal prevails.

The *Cannel coal*, on the bank of the Kentucky river, occupies a bed of four feet thick, of which about three feet are of this variety, the remainder, or upper part, being common bituminous coal.

The bituminous coal seams of Kentucky appear seldom to exceed three feet thick, and in general are of less dimensions. They are, however, accompanied throughout the entire extent of the coal field, by the valuable mineral, argillaceous carbonate of iron. Mr. Mather's calculation is that it averages one yard thick over the whole 12,000 square miles; equivalent to 38,400 millions of tons: "a quantity sufficient to supply a ton of iron, annually, to every individual in the

* Robert Triplett's Circular.

United States, [the population being then fifteen millions,] for 2,560 years." But we have stated that not more than 7,000 to 9,000 square miles contain workable coal beds. If such an amount of iron ore really exists as three feet in thickness, under the entire area of the coal-field, it far exceeds anything of the kind in any other region we are familiar with in the United States; for rich as the States are in this mineral, in the aggregate, the supply from the carboniferous strata, appears to be but feeble and uncertain; although the deficiency seems amply compensated for in the immense supply of hematite furnished by the subjacent limestone series.

VI. OHIO DIVISION

OF THE GREAT ALLEGHANY COAL-FIELD.

THE superficial coal area within this State we have computed at 11,900 square miles.

At what period the principal deposits of mineral coal became known is uncertain. On the ratification of the treaty of peace in 1763, Colonel Croghan was the first agent deputed by the British government to descend and explore the Ohio, and conciliate the Indian occupiers. His private journal, which was only published in 1831, makes no allusion to coal in this state, but he especially noted the beds of coal on the banks of the Wabash.* It was certainly known shortly afterwards; for in Captain Hutchins' map, published in London in 1777, we observe that coal mines are marked on the western side of the Ohio river.

Occasional notices of portions of the Ohio coal region, and certain local developments therein, have long ago appeared.

In 1835, an elaborate article was published in Silliman's Journal of Science, "on the bituminous coal deposits of the valley of the Ohio," by Dr. S. P. Hildreth. It furnished some useful details of coal operations and statistics, in this and the bordering States. This memoir was illustrated by a great many wood cuts of fossil remains, local sections, and a geological map of the Ohio valley, including parts of Pennsylvania, Virginia, and Ohio. The geological investigations set on foot in those States, by direction of their local governments, have, in great measure, already superseded Dr. Hildreth's memoir. Nevertheless, as the work of an individual explorer, unassisted by the official patronage and the treasuries of those States, it

* Monthly Journal of Geology, Vol. I.

is deserving of honourable mention, as a serviceable contribution to American geology.

It has been, not unfrequently, observed, in relation to the State surveys, that they have more regard to technical and theoretical geology, than to practical and industrial results. In this light, it has been argued, these State surveys have somewhat failed in the utilitarian results expected from them. Perhaps it were scarcely fair to unite all these multifarious duties in the same party. Geologists are commonly occupied with duties sufficiently onerous and laborious in their specific departments, and in investigations over fields heretofore little trodden by men of science, to make much progress in economic and statistical researches. It were better, no doubt, that these distinct subjects of inquiry were divided, or that they should follow each other. On the whole, we think we are not far wrong, in the belief, that the American geological surveys do, in point of fact, contain more details of statistical, commercial and industrial utility, than can be found in the geological reconnoissances and memoirs of any other country.

But to return to that of Ohio. The attention of the legislature having been called to the subject by the governor, a select committee reported on so much of his Excellency's message as related to a geological survey of the State.

Professor W. W. Mather, in association with Dr. J. Locke, and other competent assistants, commenced the survey, and one report of their joint labours appeared in 1837, and another in 1838. As the reports of the former gentleman are always characterized by special attention to economic geology, those of Ohio furnish a large amount of important statistics, from which many of the following notes are derived.

From the magnitude of this coal field, which comprises one-third of the entire area of the State—bordered by the Ohio river for three hundred miles, and intersected, longitudinally and centrally, by the Ohio and Erie Canal—it will readily be perceived that its coal mines must be classed with the most prolific sources of local productive industry.

In the words of the reporter, "it is estimated that about twelve thousand square miles are undoubtedly underlain by coal, and five thousand by *workable beds* of that valuable mineral." This estimate appears to be a very fair one, and precludes all misapprehension as to the available amount.

The physical features of the country are favourable to the working of these horizontal coal strata, by the simple means of adit levels; and it will be long ere the wants of the community call for another system of working, either by steam power, deep shafts, or costly machinery.

"Probably a mean thickness of six feet of coal, capable of being worked, over five thousand square miles, is a moderate estimate of our resources in this combustible."* According to certain data,

* First Annual Report, by W. W. Mather, pp. 5, 6. 1837.

there are now beneath the surface of these five thousand square miles, thirty thousand millions of tons of coal. In the ordinary method of computation, in these cases, we may safely estimate that at least twenty-three thousand millions of tons are available. Could we contemplate a demand for Ohio coal as large as five millions of tons per annum, there will be an annual supply unexhausted until the termination of four thousand six hundred years.

In the second annual report, the author, after revising the geological data which form the basis of this computation, affirms, that from the information subsequently acquired, in 1838, he felt not only justified in sustaining the foregoing statement, but in materially enlarging it, for it had been proved by later investigations, that, in some counties, the coal was from twenty to thirty feet thick, in the aggregate.*

In the official report to Congress, in 1841, it appears that there were raised within the State in 1840, 125,478 tons of bituminous coal; employing four hundred and thirty-eight workmen, and $46,775 of capital.

In 1838, the quantity produced was estimated at 107,100 tons.

The Ohio geologist urged the substitution of mineral coal for the ordinary charcoal, in iron works. It had already been partially adopted by means of a mixture of the two kinds of fuel. The Ohio coal is proved to make excellent coke; and in that state is used in equal proportions with the charcoal. In effective result, it is ascertained, that an increased make of iron occurs; equal, it is affirmed, to thirty-three per cent.

The price of coal here, as elsewhere, fluctuates according to the demand. It is considered to be worth four cents per bushel at the place of production; and after being conveyed one hundred miles, to the ports of Cleveland and Erie, it usually sells for fourteen to sixteen cents per bushel; and, at more distant points, it produces eighteen cents per bushel; equivalent to from three to five dollars per ton, at the places of consumption, according to the distance of transportation; even reaching as high as ten or twelve dollars at New Orleans.

The ordinary charcoal furnaces in Ohio require a command or resource of from two thousand to five thousand acres of woodland, to keep them constantly supplied. Now, with six feet of coal beneath the very ground upon which the furnace is erected, the produce of only half an acre, annually, would be needed, while its surface would still grow timber, or be under cultivation. Such are the different circumstances which attend the two descriptions of fuel.

Like all districts which are covered with primeval forests, Ohio will long continue to make use of wood for domestic use, until it be exhausted, or shall become more expensive to procure than coal.

One hundred and four thousand three hundred and twelve tons of fuel were consumed in the iron works of Ohio, in 1840.

* Second Report, 1838, p. 7.

Raw bituminous coal has at length [1846] been solely employed in a blast furnace in this State, at Poland, on the Mahoning river. This is the first American furnace in which pig-iron has been so made.

At Carr's Run, one hundred and sixty miles below Wheeling, and two hundred and four miles above Cincinnati, an important seam of coal is mined, and this fuel is supplied to the steamers as they pass along the Ohio river. It is dug almost at the water's edge; and, consequently, the cost of transportation is a mere trifle. The price of the coal is generally six cents a bushel. In regard to quality, this is a lighter and drier coal than that of Wheeling; being less bituminous and less adapted to the uses of the blacksmith; but it is better approved for steamboats and for reverberatory furnaces. The same vein is worked on the Kentucky, or opposite side of the river.

*Ohio Coal raised.**

	Tons.
1840	104,312
1843	103,850
1847	181,600
1848	227,104

The principal points of coal mining in Ohio are Tallmadge, Summit county; Pomeroy, Meigs county; Nelsonville, Athens county; and the same points in Stark and Coshocton counties. The amount brought to market from these several points in 1850–51, was as follows: Total in bushels, 6,489,299.

CAPACITY.

Professor Briggs divides the Ohio coal district into two geological series; the higher and the lower group.† The lower series embraces but a portion of the main area towards the east, in Jackson, Scioto, and Lawrence counties. In these he has observed three workable beds of coal. "The aggregate thickness may be safely estimated at from ten to twelve feet." The gross quantity of the coal beneath this area, which is represented to be two hundred and fifty square miles, is computed to be two thousand two hundred and fifty millions of tons; which, upon the ordinary method of computation, and presuming that it is accessible to the miner, may be reckoned about seventeen hundred millions of tons, of available coal. It would be more prudent, however, not to extend the estimate beyond fifteen hundred millions. It is but right to add here, that a more detailed examination, subsequently, led the reporter to enlarge his computation to three thousand millions of tons.‡

In the Ohio second annual report, is an account of the coal of

* Cincinnati Atlas.
† Briggs's First Report, as Assistant Geologist, 1837.
‡ Briggs in Second Report of Ohio, p. 41. 1838.

Muskingum county, through which the river of that name passes. Here are six workable beds; four of which extend nearly thirty miles through the county, and two extend about fifteen miles. The aggregate thickness of available coal being determined to be eighteen feet, the amount of fuel is hence computed at three hundred and fifty-nine millions of tons. Admitting a profit of only twenty-five cents per ton, the result on this comparatively small fraction of the gross productive coal area, is found to be about ninety millions of dollars —$90,000,000.

The Tuscarawas valley and adjacent district, lying north of the preceding, comprehend, according to the authority last quoted, an area of about five hundred and fifty square miles; which on a rough calculation, is considered to be underlaid by an average thickness of six feet of coal. These data, therefore, furnish a gross result of thirty-three hundred millions of tons, of which probably about two thousand millions are attainable.

IRON MANUFACTURES OF OHIO.

The following summary has been compiled from the Railway Record of Cincinnati. In Cincinnati and its suburbs there are not less than 60 Iron factories of the largest sort, which, with their dependent work-shops give employment to full 5000 operatives. The iron ore is found almost entirely east of the Scioto; and occasionally in the form of bog ore in the north. The iron works and Iron produce of Ohio are:—

	Pig Iron.
Furnaces	35
Tons of Iron used	140,610
Pig Iron made, tons	52,658
Bushels of Coal Consumed	605,000
Coke and Charcoal, bushels	5,428,800
Operatives	2,415
Capital invested	$1,600,000
Value of Products	$2,000,000

1853. In the production of pig-iron, Ohio is the second State in the Union, being next to Pennsylvania, the latter producing half the pig-iron in United States and the former one-tenth.

Iron Castings.	Tons.
Factories	183
Pig metal iron and ore used	41,000
Castings made	38,000
Coal consumed, bushels	848,000
Coke and Charcoal "	355,120
Operatives employed	2758
Capital employed	$2,000,000
Value of products	$3,200,000

Wrought Iron.	Tons.
Factories	11
Pig metal used	13,675
Blooms	2,900
Coal consumed, bushels	600,000
Coke and Charcoal	466,900
Operatives	708
Wrought iron produced	14,416
Capital invested	$700,000
Value of products	$1,500,000

The following general view will give the relative standing of the principle States in the manufacture of iron.*

	Iron Works.	Value of Products.
Pennsylvania	631	$20,327,000
New York	401	7,941,000
Ohio	229	6,700,000
Virginia	122	2,450,000
New Jersey	108	1,975,000
Tennesee	81	1,610,000

These States produce more than two thirds the iron ore and iron of the United States.

STATE IMPROVEMENTS IN OHIO.

State Canals and Roads, eight hundred and fifty-two miles, cost $15,283,783. Gross revenue derived from the six canals and other State works in Ohio.—In 1844, $569,676; in 1845, $494,313; in 1846, $630,770.

In 1851, $833,033 received from tolls; 1852, $757,562.

In 1853, Canal tolls	$605,165
National road,	35,354
Maumee road,	10,462
	$650,981

THE LAKE TRADE FROM CLEVELAND.

Since the opening of the canal, from the coal-field to Lake Erie, at the Port of Cleveland, the latter has become an important outlet for the productions of this State.

The following table comprises the collector's annual returns of mineral coal which arrived at Cleveland, *via* the Ohio Canal, and also the amount which was shipped at the port.

* Mining Magazine, 1854.

Years.	Received by Canal.		Shipped.	Years.	Received by Canal.	
	Bushels.	Tons.	Tons.		Bushels.	Tons.
1830		5,100		1845	889,880	31,781
1837		12,269		1846	893,806	31,921
1838		9,298		1847	1,238,622	44,236
1839	140,042	5,000		1848	1,925,451	66,551
1840	167,045	5,065		1849	1,910,474	65,707
1841	479,441	17,122	4,329	1850	2,347,844	83,850
1842	466,844	16,673	2,825	1851	2,992,343	107,134
1843	387,834	15,515	11,168	1852	3,940,749	137,926
1844	550,842	22,035	16,613	1853		

Lake Commerce of Cleveland, Ohio.

Years.	Vessels.	Tonnage.	Men employed.	Exclusive of Steamboats.		Steamboats.		Coal Expor'd.		Value of Exports.
				Arrivals.	Departures.	Arrivals.	Departures.	Tons.	Value.	
1830	15	1,029		213	218				*Dolls.*	*Dollars.*
1835	38	3,962		878	870					
1840	66	9,304		1,344	1,344					
1842	80	8,671		1,418	1,412	1,050	1,050	2,825	7,119	5,851,898
1843	86	9,386	565	1,382	1,432	1,100	1,100	11,168	35,204	
1844	98	11,738	681	1,472	1,522	1,260	1,260	16,613	49,839	
1846										7,040,492

Commerce of Cleveland, Ohio, 1851, was as follows:*

Imports coastwise, - - - -	22,804,159
Exports, - - - - -	12,026,497
Total coastwise, - -	34,830,656
Foreign Imports, - - -	360,634
" Exports, - - - -	284,937
Total Commerce, - -	35,476,327

The population of Cleveland, which numbered in 1850, 17,600, is now put down at 50,000. 1853.

In 1841, out of 1364 arrivals at this port, 437 were from Canadian ports on Lake Erie, and from American and Canadian ports, *via* the Welland Canal.

Out of 1366 departures, 422 were to Canadian ports, and similar places as before named. In 1842, entered from Canada, 356; cleared for Canada, 363.

In 1846 the foreign [Canada] trade of this port was as follows:

* Hunt's Magazine.

ARRIVALS AND CLEARANCES.

	Vessels.	Tons.
In American vessels,	165	12,258
In British vessels,	162	18,759
	327	31,017

Hence it would appear that the trade with Canada is diminishing at this port.

COMMERCE OF THE LAKES ABOVE NIAGARA FALLS.

As bearing collaterally on the progressive advancement in the indigenous production and commerce of the States which border upon the Upper Lakes, we add a few statistical notes.

General Commerce of the Upper Lakes, showing the periodical increase.

Years.	Oservations.	Steamers and Propellers.	Shipping of all kinds.	Tonnage.	Steamboat receipts.
					Dollars.
1825	Tonnage on the Lakes, . . .	1		2,500	
1832	First year after opening the Ohio Canal,			8,552	
1833	Second year,*	11		10,471	229,212
1836	Fifth year, ...			24,047	
1838	Seventh year, ...	15	73	34,277	
1840	Ninth year, ...	43			725,523
1841	Tenth year, ...			41,184	767,123
	On all the lakes, ...			56,252	
1845	Fourteenth year,† . . .	60	380	76,000	
1846	{ Tonnage at the end of the 15th year, .	80	452	91,250	
	{ " on all the lakes. . .			106,836	
1847	In commission, on the western lakes, .	86		113,000	
1851	Tonnage on all the lakes, . .	180		212,000	

There are now steamers on the Western Lakes, of 1140, 1300 and 1705 tons burthen.

PORT OF BUFFALO ON LAKE ERIE.

Number of arrivals from the lakes in 1825, only 200 of all descriptions; in 1846, steamers, 1310; propellers, 200; other vessels, of various denominations, 2,357. Total arrivals, 3857; aggregate tonnage, 912,957 tons.

Value of the property arrived and cleared on the canal at Buffalo, in 1846, $38,214,025.

* Letter on the Lake Commerce, by J. L. Barton. Buffalo, 1846.

† Buffalo Commercial paper, September 9th, 1847.

Coal received at Buffalo from the lake, chiefly from Pennsylvania.

Tons.	Years.		Vessels.	Tons.
995	1845			
4330	1846	Arrivals from Canada,	487	95,879
7716	1847	Cleared to Canada,	492	96,441

*Value of property cleared on the canal at Buffalo, in 1853, - - - - - -	$22,652,408
Value of property left at Buffalo going to Western States and Canada on the Erie canal, - -	64,612,102
	87,264,510

TRADE WITH CANADA.

Total value of the imports into the district from Canada, 1853, was, - - - - -	$392,719
Duties collected, - - - - -	84,943
Value of exports to Canada from the district of Buffalo Creek, - - - - -	992,406

Coal exported from district of Buffalo, 1853:

Tons.		Value.
217	- - - -	$1,636

PORT OF ERIE.

Export of Bituminous Coal received from Pennsylvania:

Years.	*Tons.*		*Value.*
1845, - -	8,507	=	$21,218
1846, - -	21,534	=	53,835

Total receipt of coal at the Port of Erie, by the Erie Extension Canal:

1846,	25,000 tons.	1849,	70,326 tons.
1847,	70,000 "	1852,†	140,000 "

Number and description of vessels built on Lake Erie, during six years, from 1841 to 1846, inclusive: Steamers, 47; propellers, 19; sailing vessels, 185. Total, 251 vessels, having a tonnage of 49,801 tons.

* Hunt's Magazine. De Bow's Review.
† Cleveland Herald.

COAL TRADE OF THE LAKES.

Sources of Supply, Markets, and Prices,—from Notes prepared by Col. James L. Barton, of Buffalo.—" The Lake markets are supplied with the anthracite and bituminous coals. The Anthracite is obtained from coal fields in the Schuylkill, Lehigh, and Lackawanna regions, Pennsylvania. Large quantities are brought to the Hudson river by the Delaware and Hudson canal, and much greater supplies are conveyed from Philadelphia, by sea, to New York. It is brought up the Hudson to Albany. The Erie, Champlain, and Oswego canals are the channels through which this description of coal reaches the Lake country. The state permits coal to pass through the canal at a rate of toll of one half mill per 1000 lbs. per mile. In this market, this kind of coal can be furnished for $7 per ton. It is used by our iron-makers for casting purposes. Bituminous coal, of which large quantities reach the lakes, is obtained from several sources in different states. This coal is made use of here for smithing and fuel; and is sold at $6 a ton. The quantity of Anthracite and Blossburg coal delivered at Buffalo, by the Erie canal, in 1849, was 13,367,595 pounds [5967 tons], and at Oswego, 6,608,422 pounds, or 2950 tons.

Large quantities of bituminous coal are obtained from coal-fields in the north-western part of Pennsylvania, 60 to 80 miles south of Lake Erie. This coal is of a very superior quality, and is extensively used by boats and other machinery driven by steam, as well as for smithing and domestic purposes of fuel. Very large supplies reach Erie from the beds. Bituminous coal is also freely obtained in Ohio, from beds lying along the line of the Ohio canal, and eastward from the Lake 50 to 80 miles. This coal is also of a very superior character, and is preferred by many to the Erie coal. Like the Erie coal it is principally used for steam and fuel, and also for smithwork. Neither of these coals are as highly bituminous as the Blossburg. In 1849, 1,910,474 bushels were received at Cleveland. It is sold at the same rate as at Erie, from $2.50 to $3 per ton. Erie and Cleveland coal is worth in Buffalo, $4 per ton. From these two places, Erie and Cleveland, large shipments are made to different points around the Lakes, on both sides. It is freighted at cheap rates, by vessels which fill up their cargoes with this description of loading. Canadian ports are largely supplied from these places with this article. In 1848, the shipment of coal from Cleveland to American and Canadian ports, was 131,200 tons. Another source of supply of bituminous coals, is from beds lying on the line of the Michigan and Illinois canal, in Illinois, distant from Chicago, on Lake Michigan, 6 to 80 miles. This canal was opened in 1848, and but little coal came to market that year. In 1849, 5,150 tons reached Chicago. The upper part of these beds furnish coal highly charged with sulphur, which confines the use of it principally for household purposes. Boats and other machinery make but little use of it for steam, it being so destructive to grate-bars and boilers. But the quality of the coal is improving the deeper the beds are worked, and the prospect is that coal of equal quality will soon be raised from them, as free from this objectionable matter as the Cleveland and Erie. The production of these mines is not definitively known, but a writer in the Cleveland Herald, estimates the entire consumption of coal by the Lake region in 1853, at 300,000 tons."

NEW YORK STATE.

Statement of tolls received at three of the principal lake ports of New York, viz. Buffalo, Black Rock, and Oswego, in the years 1845 and 1846, showing an advance of fifty per cent. in two former years:

1845, - -	$677,922
1846, - -	1,013,478

Total value of property shipped from, and arriving at, Oswego, by canal, for two seasons:

	1852.	1853.
Property arriving, -	$16,415,334	$20,159,202
Property cleared, -	10,746,637	14,316,960
Total, - - - -	$27,161,971	$34,476,162

Tolls collected at Oswego:

1852.	1853.
$314,436 88	$392,730 71

The arrivals at Oswego, from Canada, the past season, 1853, amount to about 400 tons more than from American ports.*

MICHIGAN.

Value of Exports.

	Years.	Port of Detroit.	From all ports in Michigan.
	1840,		$1,305,860
	1842,	$1,108,000	
	1846,	2,495,333	4,647,608
	1847,	3,883,318	7,119,832
Foreign and coastwise,	1851,	4,076,464	

Michigan Tonnage, enrolled and licensed:

Years.	Tons.	American.	British.
1846, - - - -	26,928	4001	235
1847, including steamers,	35,145	4041	235

* Hunt's Magazine.

TOTAL COMMERCE OF THE LAKES.—CLOSE OF 1847.

By a report furnished by the Topographical Corps, through the Secretary of the Treasury, at the close of the year 1847, we have an official account of the Lake Commerce in 1846, of which the following is a summary:

Net value of the *bona fide* trade for 1846—being nearly double the amount in 1841, - - - - - -	$63,164,910
Amount of registered, enrolled and licensed tonnage on the lakes for 1846—being nearly double the amount in 1841, - - - - - - - -	tons, 106,836
Number of clearances and entries, - - - -	Number, 15,855
Goods exported and imported; the whole American lake tonnage, in 1846, [besides 30,000 tons British,]	tons, 3,681,688
Goods exported and imported in 1841, - - - -	$2,071,892
Number of passengers conveyed, in 1846, not less than	250,000
Amount of passage money paid, - - - - -	$1,250,000
Number of mariners employed, - - - - - -	6,972
Cost of shipping, in 1846, - - - - - -	$5,341,800
Population dependent upon the Lakes as the means of communicating with a market, - - - - -	2,928,925
Steam tonnage of the lakes, - - - - 60,825	Ts. 136,836
Sailing tonnage, - - - - - - 46,011	
British shipping employed in the American lake trade, - - - - - - - 30,000	
Receipts in 1846, - - - - - - - - -	$13,184,910

GENERAL COMMERCE OF THE LAKES,—IN 1848.

Aggregate value of exports and imports, - -	$186,485,267
of passenger trade, - - -	1,000,000
of American shipping, - -	7,231,247
Total, - - - - - - - - -	$194,716,514

	Tonnage.	Seamen.	Steamers.	Sailing vessels.
American,	167,137	10,907	61,791	105,346
Foreign,	35,904			
	203,041			

WESTERN RIVERS.—COMMERCE IN 1846.

From the same official source, we add the amount of the steamboat navigation of the Mississippi and its tributaries, - - - - - - -	Miles, 16,674

Steamboat tonnage on the Western Rivers in 1842,	Tons, 126,278
" " " " in 1846,	" 249,055
4000 boats of other kinds, of 75 tons average each,	" 300,000
Tonnage of flat boats, making two trips a year,	" 600,000
Merchandize transported, 1,862,750 tons, of the value of	$61,914,910
Total commerce of the Western Rivers, - - -	183,609,725
Total cost of the river craft, - - - - - -	12,942,355
Sustained at an annual expense of - - - -	20,196,242

WESTERN RIVERS.—COMMERCE—(*continued*)—IN 1848.

Net value of commerce, - - - - -	$256,233,820
Value of vessels, - - - - - -	18,661,500
	$274,895,320
Number of hands employed, - - - -	35,047

Estimate of the gross value of internal commerce on the lakes, - - - - - - -	$283,187,134
On the Western Rivers, - - - - -	512,467,640
	$795,654,774

Value of the commerce of the Western Lakes and Canada, taken from Mr. Andrews's Report to the Secretary of the Treasury:

Lake Ontario.—The value of commerce on this lake in 1851, was $30,000,000, and its licensed tonnage, 38,000 tons.

	Value.	Tonnage.
Lake Erie, - - - - - -	$209,712,520	138,852
Lakes north and west of Lake Erie:		
Huron, Michigan, Superior, -	60,000,000	30,000

Total Tonnage of the Lake Commerce.

Enrolled.		Entered.	Cleared.
Steam.	Sail.	Foreign and Coasting.	Foreign and Coasting.
Tons. 77,061	*Tons.* 138,914	*Tons.* 9,469,506	*Tons.* 9,456,346

1852.*—Value, $300,000,000; employed 80,000 tons of steam, and 140,000 tons of sailing vessels.

* State Engineer's Report of the Canals of New York.

VII. MARYLAND DIVISION

OF THE GREAT ALLEGHANY COAL-FIELD.

The superficial coal area within this State we have computed at 550 square miles.

The topographical details of the published maps differ so much, that it is quite impracticable to be precise in estimating the areas and subdivisions of the Maryland coal region. The external boundary of the entire field is sufficiently defined: we are not so certain of the interposing areas of the subordinate rocks, which divide the district into at least three portions. The geologist of the State appears to have experienced the inconveniences consequent on so imperfect a topographical survey. In page 48, report of 1836, the Frostburg coal area is stated to be 180 miles square. In the final annual report, of 1840, page 18, the area is given at 90 miles, and by another statement 135 miles. Our own admeasurement is 150 miles. These discrepancies arise, evidently, from the uncertainty of the point adopted as the southern termination of the district in question. Taking the Frostburg region at 180, (the largest admeasurement,) the middle area between Negro and Meadow mountains, at 120, and the northwest or Youghagany field, at 250, the aggregate of bituminous coal land in the State of Maryland is 550 square miles. We make this statement with some hesitation; but we conceive the entire amount will not exceed, and will possibly fall short of what we have computed.

"The statements which I shall here make in relation to the extent and location of this [the Frostburg] coal-field, are derived either from my personal observation, extending through several months, or from the statements and surveys of those intimately acquainted by a long residence with this region, and therefore may be relied on. There has been so much written by interested parties in the last few years with regard to this region,—statements so wild and extravagant put forth; maps and drawings of sections, so entirely the work of imagination of interested speculators, that in the examinations here, one has more difficulty than in a new country. These statements have been a serious injury to those holding really valuable property, inasmuch as all they could say of their property, was equalled, nay, exceeded by those who had to depend solely on ideal maps, facts and figures to sell their stocks. The whole extent of the Big Vein of workable mercantile coal does not exceed 20,000 acres. I am sure the sum added together that is claimed by the various corporations, would exceed 60,000 acres. There is therefore 40,000 acres of this vein which have no existence, contending in the market with 20,000 that have a real bona fide existence, every acre of which is worth more than the highest price ever paid for it."*

As before stated, we assigned 150† square miles for the productive area of the Cumberland or Frostburg coal-field. By reason of the basin-shaped conformation of its stratification, and by the uprising of the subordinate old red sandstone formation to the surface, this eastern area is separated from that to the westward by a belt, a few miles broad. The second coal area, situated beyond the great back-bone ridge of the Alleghany mountain, has an uncertain southern termination; being separated by another denuded belt of red sandstone, from the third coal-field, which thus fills up the remaining part of the northwest angle of Maryland.

Although the aggregate, 550 square miles, appears small when compared with some of the vast areas appertaining to other States, yet, in productive value, and in advantages of locality, we conceive that it is greatly superior to the bulk of the coal land which is situated beyond the State line, on the west, and intermediate between that boundary and the Ohio slope. As an accession to the resources of Maryland,—for the substitution of what was regarded, a quarter of a century ago, as almost a worthless appendage to the State, for that which now promises to be the most productive, may justly be deemed an accession,—it cannot fail to be appreciated for its almost immeasurable importance. After long years of expenditure, in constructing canals and railroads, to communicate between these abundant coal deposits of the mountains and the seaboard, this enterprising State has but now seen the partial completion of her principal works; and it only remains for her to reap the reward to which the perseverance of her citizens entitles her.

From amongst the various reports, public and private, of the Maryland coal region, or more properly speaking, of the Frostburg or Cumberland portion of it—for of the back country we know very little indeed—we cannot positively determine the number of workable coal beds; even in the best explored portion of the latter district.

There are, according to Dr. Ducatel, at George's Creek Valley, four workable coal seams, which have an aggregate thickness of thirty feet.‡ South of this, at Westernport, two veins are mentioned, comprising eight feet: and beyond these are four or five others, imperfectly known, but probably are continuations of the preceding, or of a part of them. At "Dug Hill," another position in George's Creek Valley, the reporter enumerates ten coal seams, which average four feet each, or forty feet in the aggregate. We are not informed how far these may be repetitions of those before mentioned.§

* Dr. James Higgins's Fourth Annual Report, Baltimore, 1854.

† [Although some maps represent this field as extending to the top and eastern side of Savage mountain, its occurrence there is denied by Dr. Higgins, whose estimate of the field is 120 square miles.—Report of 1854.—H.]

‡ State Geological Report, in 1836, pp. 50—54. The assays of these coals will be found at the conclusion of this book.

§ Annual State Report for 1840, pp. 28—33, and plates i. ii. and iii.

A published report of the George's Creek Coal and Iron Company contains a section of the excavations which have been made at this "Dug Hill," which we infer is the place, on which, more recently, has been conferred the name of Lonaconing. There are shown here ten coal beds, of which four only are workable, and of which the aggregate thickness is thirty feet. That of the six others is only ten feet.* The corrected Lonaconing section, plate iii. of the State Geologist's Report of 1840, exhibits six workable seams, which have an aggregate thickness of thirty-five feet; the other four seams amount only to six feet. Below this thirty-five feet series, viz. from Lonaconing down to Westernport, twenty-five feet of coal are known, but are chiefly made up of small seams, of which about fifteen feet are workable. By these data we make out fifty feet, as the maximum workable coal of the Frostburg region: but according to Dr. Ducatel, not more than forty-five feet can be calculated upon.

In another part of the basin, at the works of the Maryland Company, in one position, the explorations have developed three seams of coal, amounting to twelve feet thickness. At Mount Savage are six other seams, forming in their aggregate twenty-six feet of workable coal. These form part of the general group.†

At a place called Barrellville, in the Cumberland district, eight veins occur, whose average is three feet six inches each, in thickness.

Portions of the areas of the lower beds are destroyed by the erosion of the valleys. For instance, George's Creek, according to Dr. Ducatel, "has scooped out its bed through twelve hundred and fifty feet of perpendicular elevation; while Jenning's Run, he observes, has, in the short distance of six miles, cut, both longitudinally and transversely, even into the subjacent red sandstone." The lateral ravines have also subtracted largely from the area of the lower beds. It was the knowledge of these extensive denudations and removals, especially in the most mountainous portions of the coal-fields, and in those districts where the coal formation undulates, that gave rise to our previous remarks on the necessity of making large allowances for barren or inaccessible ground, when calculating coal areas. We could point out considerable districts, towards the northern termination of this Alleghany coal-field, where, certainly, not one acre in ten, and often not one acre in a thousand, contains a bed of coal in a workable condition, or even a single ton of that mineral.

In the northwest angle of Maryland, part of the coal measures are cut out by the Youghagany river; also by two parallel zones of the inferior red sandstone, along Deep Creek; and there is an extensive sweeping away of strata along the Potomac valley, which is a trough at least fifteen hundred feet in depth. When due deductions are made for these interruptions to the continuity of the coal formation, our

* George's Creek Coal and Iron Company's Report, 1836, p. 11, Map and Sections.

† Report, in 1844, by Messrs. Silliman and others.

estimate of five hundred and fifty square miles, will be found a very liberal one.*

This principle has been fairly observed in the last State Report, when applied to the Frostburg or eastern coal-field. By attending to a rule so obvious and indispensable, the geologist is compelled to reduce the area of actual coal bearing surface to 135 square miles, or 86,847 acres.

According to the foregoing data, the result gives, as the gross amount of coal in the entire basin, supposing the whole to be accessible, 6,305,137,287 tons, and the available quantity, on the ordinary mode of calculation, will be upwards of four thousand millions of tons.†

The price of Cumberland coal delivered at Tidewater, Georgetown, was, in 1838, about 20 cents per bushel; a higher price than is usually obtained in Philadelphia. The cost of mining was $1 per ton, and of transportation by canal, to tide supposed to be $2.85 per ton: to which must be added the respective profits of the land owner, the producer, and the merchant. This estimate was thought at the time, to be considerably below the mark.‡ The representations and reports of interested parties, all strenuously advocating their individual or local claims, on the attention of Congress and the public must, of course, be received with a requisite degree of caution. At this distance of time, we shall doubtless, be pardoned this remark, while necessarily reviewing the statistical merits and details of the entire coal resources of the country. We continually meet with the unreserved and unqualified assertions of these claimants, that the coal of their particular mine or district, no matter where, is the best yet discovered, for every practical use. Now, as they cannot all be the best, it follows that a good deal of exaggeration prevails, in some of these cases. It is not inappropriate to state here, that there are probably a dozen or more of coal companies, in England, Wales, and America, who announce through the press, that their particular coal has been decided by the agent of the Great Western Steamship Company or some other steamer company, to be the best generator of steam of all coals yet tried. One gentleman has also conclusively shown, that one ton of the bituminous coal of Cumberland, Va., is, in mechanical effect, equal to two tons of anthracite.

However, the test of science restores all things to their true value. The examinations of Prof. Johnson, in 1844, have dispelled many illusions; and have assigned to all the principal varieties of

* Vide the map and profiles appended to the State Report, for 1840.

† We may form some estimate, from this computation of available coal within an area so small as scarcely to be noted upon our map, of the enormous quantity in the aggregate of the American coal-fields. There needs no fear about exhaustion, at least not before the termination of a great many thousand years, according to the present rate of consumption. In regard to the Old World, also, the progress towards exhausting the numerous coal-fields is comparatively insignificant. We have computed, from five coal-fields in Great Britain, a production of 116,000,000,000 tons, or more than 5,500 years supply for consumption and exportation, on the present scale. The Lancashire coal-field will yield 8,400 millions; the Mid-Lothian district, 5,710 millions; the Newcastle, 9,000 millions; the South Wales basin, 64,000 millions of tons, available.

‡ Communications to the Committee of Congress on a national foundry, pp. 71, 147, 157.

American coals their appropriate place in the catalogue: and here the Cumberland coal takes the very highest place, in the series, in the order of evaporative power.

The analyses of the Cumberland or Frostburg coals show that the quality is, generally, of the kind denominated dry or close burning; an intermediate species between the fat, bituminous, caking coals like those of Pittsburg, for instance, and the non-bituminous varieties, like the Pennsylvania anthracites. The largest, or ten feet seam at Lonaconing, contains twenty per cent. of volatile matter, and there are some beds which do not possess more than from thirteen to fifteen per cent. of volatile matter including moisture. The carbon in these coals amounts to from sixty-eight to eighty-one per cent., which circumstance accounts for the deservedly high reputation, as generators of steam, that they have enjoyed.

Were it needed in the process of iron manufacture, there is no difficulty in making good coke, from the majority of the Cumberland coal seams. All these varieties have undergone a chemical examination by scientific experimenters. Professors Silliman, Shepard and others have shown that the main or ten feet Frostburg seam, which having been the longest worked, has conferred a character on the Cumberland coal, contains but 13.34 per cent. of bitumen, besides 1.66 of water. Such an amount as 82 per cent. of carbon which these analysis show it to possess, while at the same time it retains enough of the properties of the flaming coal, carries its own best commendation, and places it very high, if not the very highest, in the scale of American coals; a reputation which is fully sustained by the subsequent investigations of Prof. Johnson. It closely resembles some of the Stony Creek semi-bituminous coals of Pennsylvania, in all the essential particulars; except that the latter does not swell nor cement so much in burning.

We learn that a sucessful experiment in driving locomotives has lately been made on the Hudson River Railroad with coke from the Cumberland Coal Company's coal. 1854.

A special report of the president of the Chesapeake and Ohio Canal Company furnishes the details of an experiment at sea, 22d October, 1839, made on board the United States steamer Fulton.

The object was to test the Cumberland semi-bituminous coal against a highly bituminous Liverpool coal.

The chemical characters of each of these kinds might have at once suggested the exact results obtained. The Cumberland coal having more carbon, would acquire a more intense and concentrated heat; and, as it possessed less bitumen, would give out less smoke than the fat coal.

The former from the same reason, would have its fragments less changed and cemented; while the latter would be caked or agglutinated in a mass, like all caking coals. The Cumberland coal made most clinker; the Liverpool coal possesed most sulphur.*

* Special Report of the Pres. Ches. and Ohio Canal Co., 1839, pp. 38, 39.

With a knowledge of the chemical composition of various qualities of coal, it is superfluous, at the present day, to institute a series of experiments like this, between anthracite and bituminous coal, or their modified varieties, now perfectly well understood. That any fuel of the specific nature of the Frostburg coal, can readily produce all the comparative results afforded by the experiment, there cannot be the slightest doubt.

Brick.—It is stated that the experiment of using coal for the purpose of burning brick, has been successfully tried in Maryland. Each ton of coal will burn four thousand brick, with two hands to tend the furnaces. 1854.

This experiment has been in successful operation with anthracite, at Bridesburg, in Philadelphia county, for the last year.

The application of bituminous coal to the purpose of iron making after the method of the English works, has proved so successful, that between the years 1840 and 1844, five blast furnaces and two rolling mills were erected in Maryland and Pensylvania, upon this principle instead of the old charcoal furnace.*

Some of the coal of this region, within four miles of the town of Cumberland, was submitted to the examination of the late David Mushet a few years ago. He remarked that "it was the very best bituminous coal he had ever met with," and considered it well adapted to iron making. Three specimens of the varieties of iron ore of this region were, at the same time, reported upon by Mr. Mushet. The results of his analysis were as follows:

Brown fibrous hematite, of excellent quality, yielded of best cast iron, - - - - - - -	62.6
Common argillaceous iron stone of the coal measures, yielded, - - - - - - -	34.3
A very fine argillaceous iron stone, yielded of best cast iron, - - - - - - - -	41.4

These were probably selected specimens, and are above the average result.

PRODUCTIVE CAPACITY.

In the report of the Baltimore convention, December, 1834, is introduced an extract of a report of Mr. Roberts, to the directors of the Baltimore and Ohio Canal Company, to the following effect:

"As each square mile of the Great Vein alone—thirteen feet thick—would yield more than sixty millions of tons, if it could be exported at the rate of five hundred tons per day, it would require four hundred years to exhaust one square mile of the great coal vein!"†

There is surely a great miscalculation here. In the first place,

* Letter of the Committee of the Iron and Coal Trades of Pennsylvania, April, 1844.
† Report of the Baltimore Convention, 1834, p. 49.

the seam of coal contains little more than twelve, instead of sixty solid millions of tons per square mile, at the thickness named; and, according to the usual allowance in these estimates, would not yield more than from eight to ten millions of tons. Messrs. Silliman, in a subsequent report of the same locality, in 1838, state that this main or ten feet coal seam can only be worked nine feet or three yards.* This gives, for the solid cubical quantity in the ground, nine millions of tons per square mile. Making the customary allowance of one fourth for waste, for pillars, broken ground, casualties, &c., the available amount is 6,750,000 tons, instead of sixty millions. At five hundred tons mined per day, it would, therefore, with these data, require little more than forty, instead of four hundred years, to exhaust one square mile. Even on this corrected scale, the resources of this locality are demonstrated to be of very productive character; surpassed, probably, by none on the eastern margin of the Alleghany mountain range.

"The veins of coal in this region, which we have to consider as of present importance to the State, are the Big Vein, the six-foot vein, and the forty-four inch vein. Especially worthy of consideration is the big vein, as its coal is that which has given the high reputation to our Maryland coal, that which constitutes to a great extent the real capital of most of the incorporations in this county, and which must be for a long time the basis for valuable tolls on the Chesapeake and Ohio Canal. The thickness of this vein varies in different sections of the coal field, being thinner on its north-eastern border, on the extreme edge of which it is about nine feet; at Frostburg its workable thickness is about eleven feet, whilst in the middle and south-western sections fourteen are claimed by those holding property there. The average thickness of workable mercantile coal, is about eleven feet. . . .

"The veins which will for a long time to come, furnish the country with Cumberland coal, are the Big Vein, from which by far the largest quantity will come; the six-foot vein, and the forty-four inch vein. The Big Vein alone comprising 20,000 acres of good mercantile coal, of workable thickness, say eleven feet, contains in every acre 17,747 tons of coal, the whole vein therefore contains 354,933,333 tons. Deduct one-fourth for wastage of every kind, and we have 266,200,000 tons of mercantile coal. The six-foot vein contains in each acre 9,680 tons of coal; this multiplied by its number of acres, 80,000, will give 774,400,000 tons of coal for this vein. Deducting as above, and we will have it capable of furnishing 580,800,000 tons of coal. The four-foot vein contains also 80,000 acres; each acre contains 6,050 tons of coal, the whole vein therefore contains 484,000,000 tons of coal. Deducting as above, and we have 363,000,000 tons as the quantity which it can furnish. Those three veins alone then will supply the following quantities:

* Report of the Maryland Mining Company, by Messrs. Silliman, 1838, p. 15.

Big Vein, - - - -	266,200,000
Six-foot Vein, - - - -	580,800,000
Forty-four inch Vein, - - -	363,000,000
	1,210,000,000 tons."*

The statistical returns of 1840 show the production of bituminous coal in Maryland, to be 222,000 bushels: equivalent to 7,928 tons. At that time the means of transportation were very limited, and access to the mines was difficult. But the quantity is obviously under-rated. As early as 1832, 300,000 bushels were annually sent down the Potomac river. Very little of this descended lower than Harper's ferry, but the quantity increased every year.

Fig. 3.
Geological Profile of the Coal Basins of Maryland.

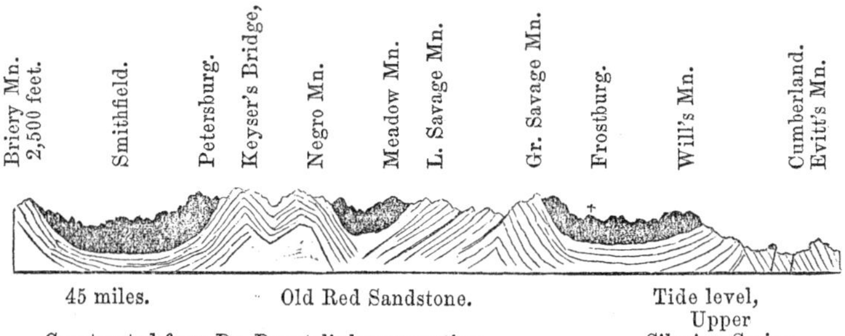

Briery Mn. 2,500 feet.
Smithfield.
Petersburg.
Keyser's Bridge,
Negro Mn.
Meadow Mn.
L. Savage Mn.
Gr. Savage Mn.
Frostburg.
Will's Mn.
Cumberland. Evitt's Mn.

45 miles. Old Red Sandstone. Tide level,
Upper
Constructed from Dr. Ducatel's larger section. Silurian Series.

As relates to the two other coal areas which lie in this state to the west of the Alleghany ridge, and of which we have been able to say so little, we possess little or no information.

We find in the Cumberland Miner's Journal the following table of the coal sent to market from that region, from 1842 to 1853, inclusive, which shows the annual increase.

Years.	Jenning's Valley.	Braddock's Valley.	Piedmont Region.	Total.	
	Tons.	*Tons.*	*Tons.*	*Tons.*	
1842	757	951		1,708	
1843	3,661	6,421		10,082	
1844	5,156	9,734		21,890	
1845	13,738	10,915		24,653	
1846	11,440	18,555		29,795	
1847	20,615	32,325		52,940	
1848	36,571	43,000		79,571	
1949	63,676	78,773		142,449	
1850	76,950	119,898		196,848	
1851	122,331	135,348		257,679	
1852	174,891	159,287		334,178	
1853	234,441	225,813	73,725	533,980	and up to Aug. 1854, 353,154 tons.
	764,027	841,020	73,725	1,678,773	

Showing an increase of 199,802 tons, during 1853.

* Dr. Higgins, 1854.

The increase of foreign coal imported into the country during the last year, was in round numbers about 30,000 tons. This would give an increase in the supply of Bituminous coal sent to the seaboard, independent of the Virginian coal trade, of 229,802 tons.

From a recent local paper we find it stated, "that the aggregate capital invested by the various companies in the Cumberland district, is in excess of $16,000,000."

COST OF TRANSPORTATION, TOLL, ETC.

These, being matters of annual revision, it would be useless to occupy space in detailing.

In 1846, the toll on the Chesapeake and Ohio canal, viz. from the Cumberland depot to Georgetown, or tidewater, was fixed at half a cent. per ton, per mile.

On the Baltimore and Ohio railroad, the rates of transportation, for the same year, were as follows:

The Cumberland road, 2 cents per ton, per mile.

To Baltimore, 1846, $3.00 per ton, and $2.50, in 1847.

To Washington city, $3.56 per ton.

1853. To Washington city, $2.10½ per ton, by Chesapeake and Ohio canal.

By a report in 1847, it is announced that the Chesapeake and Ohio canal will be opened to Cumberland in 1849, and it was estimated the coal of the Frostburg district will be delivered at the low rate of $2.50 to $3.00 per ton, at tidewater.

The current price of coal in Baltimore, in 1848, $6.00 to $6.25 per ton.

In 1848, the cost of transportation of coal over the Chesapeake and Ohio canal, from Cumberland to Georgetown, was fixed at four and a quarter mills per ton per mile.*

The whole number of vessels employed in 1852 in the transportation of Cumberland coal from Baltimore and Alexandria, was 1424.†

* Report of State Engineer on New York Canals.

† Report of C. Coal Company, p. 15.

MARYLAND.

LIGNITE.

In the Geological Report of Ann Arundel county, lignites or fossil mineralized wood, are stated to occur in great abundance. The ashes, which are derived from the spontaneous combustion of this lignite, form the principal material employed in the manufacture of alum and copperas, at the Baltimore works, on Locust Point.*

On the western shore of the Chesapeake Bay, on the banks of the Magothy, there occurs a considerable deposit of lignites, apparently in the upper secondary or green sand formation. These lignites are associated with iron pyrites and amber, the latter of which contains nests of insects converted into amber, and appears to have been formed around the smaller twigs of the wood from which the lignites have been produced.

Six miles below this locality, on the banks of the Severn, is another deposit of lignites and amber.†

* Report of Maryland, 1836, p. 30.

† Transactions of the Maryland Academy of Science and Literature, Vol. I., Part 1.

VIII. PENNSYLVANIA DIVISION

OF THE GREAT ALLEGHANY COAL-FIELD.

Estimated area 15,000 square miles, or 9,600,000 acres. Like most of the other States, Pennsylvania possesses no geological map. It is greatly to be regretted that the unfortunate pressure of the times and the imperious demand for the resources of the State for other objects, have made the postponement of the publication of the map, for which materials were collected during the geological survey, commencing in 1836, a matter of necessity.

In point of magnitude, this is the largest of the eight divisions of the Alleghany coal-field, with the exception of Virginia. As regards geographical position, it is, at the present day, more fortunately circumstanced than any other.

Some years ago, when geological investigation was in its infancy, a committee of the coal trade of Pennsylvania reported that the bituminous coal formation, within the State, covered 21,000 square miles.*

Our own computation is, that about 15,000 square miles are actually occupied by the carboniferous formation, or that which is usually denominated the coal measures. Various other statements have been made as to the size of this area. Among these is an announcement that "Pennsylvania contains more bituminous coal than all Europe. All Europe contains about 2000 square miles of bituminous coal: Pennsylvania has 10,000 square miles."† Here seems a remarkable misconception as to the area of coal in Europe; because, without passing on to the continent, the United Kingdom of Great Britain and Ireland alone contains more than 11,800 square miles of coal formation. In five out of fifty-one coal-fields within the latter area, eminent practical men have computed the available contents at 116,000 millions of tons. That of Pennsylvania, on the authority of the state geologists, contains 300,000 millions of tons.‡

For ourselves, we think that sufficient local details have not yet been acquired to enable a computation to be made of the available amount of coal within this extensive region. A superficial area of 15,000 square miles, to yield 300,000 millions of tons, would require an average thickness of twenty feet of accessible or workable coal,

* Report on the Coal Trade, in 1834. Senate Journal, Vol. II., p. 488.

† Hunt's Merchant's Magazine, August, 1845, p. 138.

‡ Hazard's United States Register, Vol. V., p. 99; Harrisburg Intelligencer; also, Mr. Biddle's Pottsville Address, 18th January, 1841; also, Mining Journal, of London, October 23d, 1841. The previous paragraph is a specimen of the absence of correct statistical information, which has characterized the publications of the day, on the comparative magnitude of coal areas. More than one of those cited here contain the announcement that "Pennsylvania contains *ten thousand times* more bituminous coal than Great Britain and Ireland."

throughout every acre of that immense district. But, judging from the annual geological reports, there are very few points where twenty feet of coal could be worked at any one locality, and the average of aggregate thickness of such veins appears to be from ten to fifteen feet, viz.: at the best exposed positions where sections could be obtained,* it is well known that a very large area towards the northern extremity of the coal measures contains but a very small fractional portion of productive coal. We have seen statements that this field contains ten workable coal seams.† It is possible that within the entire series from the conglomerate upwards, ten such seams may exist—but we have not seen a position where more than half of that number could be approached, and seldom more than two or three are available at any given locality, except in the centres of the basins.

Until the publication of the final report, and until the sections, obtained with so much care during the geological survey, furnish the means of computing with greater accuracy than do the crude and isolated details at present within reach, it seems inexpedient to pursue this investigation any further.

We have extracted from the State Geological Annual Reports, the following notes as to the greatest number and thickness of seams of bituminous coal in the Pennsylvania Division, at any one workable or available spot.

Localities.	No. of workable beds.	Total thickness.
Philipsburg—Moshannon,	2	8 feet.
Near do.—Goss',	2	12
Clearfield Creek—Wright's,	3	8
Karthaus,	4	15
Bennett's Branch—Section,	3	10½
Toby's Creek,	3	7½
Little Toby's Creek,	2	8
Portage Railroad,	4	10½
Conemaugh River,	4	19½
Laurel Hill—Western Summit,	3	11
Lockport—Conemaugh River,	4	12
Rogers's Mill— Do.	1	6
Ligonier Valley,	4	17
South of the Youghiogheny,	3	—
Elk Lick—Main Section,	4	22
Cogenhouse—Lycoming county,	2	9
Lick Run, and Queen's Run,	2	8
Tangascutack,	2	6
Ralston—Lycoming county,	2	8
Blossburg—Tioga River,	1	3½
Alleghany River,	5	16
Conemaugh,	3	10¼
Somerset county,	4	20
Frostburg—(Silliman,)	4	26
Average of 24 localities,		10½ feet.

* Rogers's Six Annual Reports of Pennsylvania.

† Geography of Pennsylvania, p. 122; Logan's Memoir on the Coal-fields of Pennsylvania and Nova Scotia; Proceedings Geological Society of London, Vol. III., Part 2.

GEOLOGICAL NOTICES.

At the period of the author's first acquaintance with the bituminous and anthracite coal-fields of Pennsylvania, nearly twenty years ago, he entered upon the investigation of American geology, imbued with the prevailing impression at that time generally advocated in Europe, and taught by nearly every geologist of eminence, that the anthracite deposits were of older origin than those of bituminous coal. In fact, the presence of anthracite was at one time thought to be conclusive evidence of a transition or grauwacké period, in contradistinction to the bituminous coal of the secondary formations.

The attendant circumstances of position, structure, mineralogical, and general characters, in the accompanying formations,—differences seemingly apparent even in several species of the coal vegetation,—all tended, at the outset, to confirm those suggestions, in a field which had previously received no scientific examination of moment.

By degrees, however, more correct views prevailed. Obscurities gradually cleared up; a host of intelligent observers almost simultaneously entered the field. At first, forming separate opinions derived from limited data or from a circumscribed range of observation, geological speculations were as numerous as were the observers. The pursuit was new to nearly all; the ground was almost untrodden. On entering upon so interesting a field, what more natural among ardent followers of the most fascinating of sciences than a variety of opinions? What more probable, at the commencement, than the advocacy of systems which more matured judgment abandoned as untenable?

As facts accumulated, and opinions were interchanged, difficulties vanished. The points of difference, at first so numerous, almost ceased to exist as new light came in. The energetic labours, applied at once over the greater part of the United States; the frank co-operation, during several successive years, of the various explorers, entitle them to all praise, and must ever render it a remarkable epoch, wherein was accomplished one of the most rapid and successful geological developments, that has occurred in the history of the science.

Let it not be inferred, however, that we consider the work as finished; it would be more proper to state that the general outlines only of American geology have now been satisfactorily traced, leaving for future observers to fill up the details.

Respecting the identity of the anthracite and the bituminous coal-fields of Pennsylvania, it is right to state that this view was entertained and advocated by Mr. Featherstonhaugh, in a course of geological lectures, delivered in Philadelphia, in 1831, illustrated by diagrams. It was then maintained that "the anthracite basin of the Wyoming valley exactly resembles the regular coal basins of Europe, of the bituminous kind."* The present writer, at that time, held an opposite opinion, as did other geologists; among whom a professor of

* Report of Mr. Featherstonhaugh's Tenth Geological Lecture, in the United States Gazette, May 8th, 1831.

high eminence, writing in 1833, remarked,—"We have in the United States three deposits of anthracite: the largest is in Pennsylvania; the next largest in Rhode Island; and the smallest in Worcester. I have examined them all, and have come to the conclusion, that all the rocks containing this coal are at least as low down, in the series, as the transition class; and I am rather of opinion that they all lie below the independent coal formation of Europe."*

Something like this view seems to have been held by Mr. Eaton, in 1830; for he stated the anthracite regions of Lehigh and Carbondale belonged to his "Third Carboniferous or Lower Secondary formation," while the bituminous district of Bradford, Tioga, and Lycoming counties, comprised his "Fourth Carboniferous or Upper Secondary."†

Even as late as 1837, in the geological report of Indiana, we meet with this passage:—"It is not likely that anthracite coal will ever be found in Indiana; because that mineral is usually found in the primitive and grauwacké formations."‡

By this time, however, the doctrine of the supposed antiquity of the Pennsylvania anthracites had been abandoned, by common consent. It seemed no longer debatable in the United States. The reconnoissances of the various geologists of Pennsylvania, appear to have sufficiently established the perfect geological identity between the formations of coal under their separate aspects; and the results of the state geological survey put the matter beyond all possible doubt.

We may, however, mention that "the anthracite formation of the United States," is even now regarded by some French geologists as "belonging to the upper portion of the transition series," and is still considered by them as older than the bituminous or true coal formation.§

The absence of fossil shells in the shales of the American anthracite beds, and their presence in those of the bituminous coal strata, both in the United States and in New Brunswick, have had their influence in leading to the opinion of a difference of geological age in these formations.

STATISTICAL NOTICES OF PENNSYLVANIA BITUMINOUS COAL.

At what precise period the mineral coal of Western Pennsylvania first came into use does not appear. By the treaty of 1758, between the Indians and the Proprietaries, as the Penn family and their coadjutors were then styled, the boundary of their lands extended eastward along the Alleghany, or Endless mountains, across Pennsylvania, so far as they range through that state. These lands had been con-

* Report on the Geology of Massachusetts, by Edw. Hitchcock, 1833.
† Eaton's Geological Text Book, 1830, pp. 39—43. Map and Sections.
‡ First Report of the Geology of the State of Indiana, by David D. Owen, 1837, p. 30.
§ Burat. Du gisement des Combustibles Fossiles, 1846, p. 49.

veyed by the Five Nations, in 1736; but from the vagueness of their definition, had been long a subject of dispute. The whole district, now known to us as the anthracite region, appears to have been confirmed by the Delawares, in 1737 and 1749.

According to Mr. Sergeant, the last purchase made by the proprietaries from the chiefs of the Six Nations, was in November, 1768. It enclosed all the area lying south of a line commencing at Owego, on the north branch of the Susquehanna; down to Towanda, and up that creek to the head waters of Pine Creek, and thence down the same and up the Susquehanna to the Indian town of Kittanning, on the Alleghany river; and down the Alleghany and Ohio rivers to the south line of the province.* It will be seen that, with the exception of that portion which lies northward of Kittanning, and which was not purchased until 1784, the proprietaries, by this purchase, came into possession of the whole bituminous coal-field of Pennsylvania, stretching from Towanda on the north-east, to the south-west angle of the state, a distance of two hundred and seventy miles, besides the northern or Wyoming anthracite region. The cost price or purchase money of these magnificent coal-fields appears to have been the sum of *ten thousand dollars* only.†

In the provincial maps, as early as 1770 and 1777, the sites of beds or "mines" of coal were marked on the shores of the Ohio. In the vicinity of Pittsburg the outcrops of coal seams were not noticed, or at least were not made use of until after these dates. In 1753, when the position was first examined by Washington, there was probably no white man living within the limits of the present site of that city; and, in 1775, we are told that there were not more than twenty-five or thirty cabins or houses standing there.‡

The value of mineral coal was well known to all who had seen or heard of its employment in Europe; but the abundance of timber in the newly acquired territory rendered the substitution of any other description of fuel quite unnecessary.

Among the first positions where land was acquired from the commonwealth, for the sake of the coal it contained, was one on the upper waters of the Susquehanna, near the present town of Clearfield, and the Indian village of Chincleclamoose, where the horizontal coal seams are very conspicuous. A tract of coal land was taken up and patented here, as early as the 1st November, 1785, by Mr. S. Boyd; but it was not until after the lapse of nineteen years, that a quantity of coal was forwarded eastward of the Alleghany mountain. The first ark load descended the Susquehanna from this place, in 1804. It was sent down by Mr. W. Boyd, and was landed at Columbia, on the Susquehanna, a distance of two hundred and sixty miles; "and it was a matter of great surprise," he observes, in an account of this experiment, "to the inhabitants of Lancaster county, to see an article with which they were wholly unacquainted, brought to their

* View of the Land Laws of Pennsylvania, by Thomas Sergeant, Esq., 1838, p. 31.

† Geography of Pennsylvania, Trego, 1843, p. 19.

‡ Ibid., p. 171.

own doors."* This movement was followed by more ark loads forwarded by other proprietors from neighbouring sites; and, subsequently, a limited trade in bituminous coal has been carried on, along the towns and iron works of the Susquehanna, partly in periods of freshets by means of arks, and partly by canal boats, during the last thirteen years. But it was not until 1828, that the first cargo of Pennsylvania bituminous coal reached Philadelphia from Karthaus, and some was also forwarded to Baltimore from the same source.

From the Congressional returns, obtained during the taking of the census of 1840, it is seen that the bituminous coal-field of this State produced 415,023 tons; employing 1798 workmen, and a capital of $300,416. It is generally supposed, that this return was below the actual production, and it is certainly most disproportionate to the amount of capital.

There will always be great difficulty in ascertaining the bituminous coal production of Pennsylvania, or indeed that of any part of the Alleghany coal-field; because the frontier is extensive, and the avenues from it are numerous, while at the same time there exists no machinery or organization for ascertaining the annual consumption of fuel, for manufacturing or domestic use; particularly in a country where every farmer is at liberty to extract coal for himself or his neighbours.

Anthracite, on the contrary, has but a few channels to market, and these are public routes on which weekly statements of the tonnage conveyed are made, and almost every ton is under supervision and record, between the mine and ship-board. We think that the quantity of bituminous coal mined is about a million of tons. It has been estimated at much more, but this is mere surmise. In 1845–6, the quantity of coal received on the Ohio at and near Pittsburg, was reported officially at under 700,000 tons; of which 200,000 tons descended the Ohio to other markets, out of the State. This seems about the proportionate increase since the returns of 1840. Various circumstances are daily contributing to enlarge the demand for coal in the valley of the Ohio; the most obvious of these are the diminution of wood in the vicinity of great rivers; the multiplication of furnaces of iron and salt-works; of steam engines and steam-boats; of manufacturing establishments, and the remarkable accession of population every year.

We have ventured to make an approximate estimate of 1,750,000 tons as the annual production of bituminous coal in the United States —which quantity is more than double the actual return in the year 1840. Of this amount, we have apportioned 1,000,000 of tons as the quantity yearly raised by Pennsylvania.† It must be borne in mind, that the bituminous coal-fields of America are still, and probably for centuries will continue, the great forest regions of the country, where mineral fuel, except in the cities, is very little resorted to for

* Journal of the Senate of Pennsylvania, 1834, Vol. II., pp. 481 and 561.
† 1851—Bituminous coal mined in Pennsylvania supposed to be 1,300,000 tons.

domestic uses, and where, at the present day, comparatively but a small amount is consumed by the iron-works, steamboats, &c.

To show, however, the extreme uncertainty, the difficulty, of estimating the actual quantity of bituminous coal raised in Pennsylvania, we may mention that one calculation, which is circulating in the newspapers, gives as the probable amount of bituminous coal mined in this State, in 1847, 10,000,000 of bushels; equivalent to 357,000 tons.

In the absence of the final Geological Report of the State, the reader may peruse with advantage the masterly description of the Appalachian coal strata, by Professor H. D. Rogers, and the paper on the physical structure of the Appalachian chain, by Professors W. B. Rogers and H. D. Rogers, in the Transactions of the Association of American Geologists and Naturalists, 1843.*

In the first communication it is shown that the coal is distributed "in a series of parallel and closely connected synclinal depressions, the direct result of the system of vast flexures, into which the whole of the Appalachian rocks have been bent, by the undulatory movements that accompanied the final elevation of the strata, and terminated the era of the coal."

EXTENT OF INDIVIDUAL COAL SEAMS.

In the article from which the last paragraph is quoted, are some very interesting facts on the great extension of certain coal seams in the Appalachian system. We have no space here, to do justice to this truly philosophic memoir. We must restrict ourselves to citing a single example. This is the great seam which is finely exposed at Pittsburg, and along the Ohio and Alleghany rivers, and nearly the whole length of the Monongahela, and is denominated "the Pittsburg seam." With the advantage of competent assistance, the author has traced this bed through Pennsylvania, Virginia, and Ohio. "The longest diameter of this great elliptical area is very nearly two hundred and twenty-five miles, and its maximum breadth is about one hundred miles. The superficial extent of the whole coal seam, as near as I can estimate it, is about fourteen thousand square miles." But these limits, he continues, though wide, fall very far within those which the bed anciently occupied, which "must have been at least thirty-four thousand square miles;—a superficial extent greater than that of Scotland or Ireland."

If, as the writer conceives is probable, and in which we entirely coincide, this seam is identical with the great bed which occurs in all the anthracite basins of Pennsylvania, "we shall then behold, in all its conditions of gradation, from anthracite to semi-bituminous, and to highly bituminous coal, a single stratum measuring at the most moderate calculation, four hundred and fifty miles in length, and two hundred miles in breadth, and covering a space of at least ninety thousand square miles."

* Pages 433 and 474, 531.

The author has, with much ability, traced the regular gradation which this remarkable bituminous coal bed experiences in size; diminishing gently from southeast to northwest:—that is to say, from twelve or fourteen feet thick on the southeastern border, to eight feet at Wheeling and Pittsburg; and, still more westward in Ohio, to five or six feet.

LOCAL STATISTICS.

It would be altogether inconsistent with the plan of the present work to enter upon the details of individual coal operations throughout this area. Many of these will be found adverted to in the annual geological reports of this district, particularly as relates to the region west of the Alleghany range.

Towards the northern and northeastern side of this range, the seams seldom attain a greater thickness than three or four feet. In the vicinity of Philipsburg, and along the valley of the Moshannon to its head waters, several good coal beds appear. Three of these are from four to four and a half feet thick each, and one main seam is, including some slaty partings, nine feet thick. A detailed section of this local district, crossing the Alleghany Mountain to the Bald Eagle valley at its base, was published by the author of the present volume, in the year 1832, in the Monthly American Journal of Geology, conducted by Mr. Featherstonhaugh. We believe this was the earliest geological section, in detail, of any portion of the Alleghany Mountain, and of the coal-field overlying its eastern escarpment.* Mr. McClure's transverse sketches, can scarcely be said to form the exception. This section was reconstructed from accurate admeasurements and actual levelling, on a greatly extended scale, in the succeeding years 1833 and 1834.

Near Karthaus, eight coal seams have been traced, amounting to twenty feet thickness; but three only of these are workable—the largest being six feet.†

Near Farrandsville and Queen's Run, two beds of three or four feet each, have been worked for some years. This coal is in good repute, but of limited area. At Ralston, and in the detached small coal basins which border the Lycoming creek, the coal measures occupy a comparatively small thickness. The two principal seams have about eight feet of workable coal. There are commonly four seams, altogether, existing within the formation in the northeast extremity of the Alleghany coal-field; but it is seldom that more than two workable beds occur in the same locality. This region has been minutely investigated by the present writer, at various periods since the year 1831. The coal is well adapted for iron making.

At Blossburg, and around the head waters of the Tioga river, from three to six seams occur, but not more than one or two have been mined, and the coals are sent by railroad into the State of New York.

* Monthly American Journal of Geology, Vol. I., p. 433.
† Report to the Clearfield Coke and Iron Company, 1839, W. R. Johnson.

One of the beds, at one point, appeared to be six feet thick, but in general the seams are about three feet each. A geological survey of this region was made by the writer, in 1832; and, at subsequent periods it has formed the subject of several communications to scientific journals. In 1831–2, the country was then in a state differing little from the primitive wilderness; but time has changed its aspect, and a large amount of business and travel is said to be done here. Few districts have been more fully illustrated, geologically, than that of Blossburg,* and in many respects it is very interesting to the geologist and the naturalist.

Fig. 4.

Transverse Section of detached Bituminous Coal basins in Pennsylvania.

Lycoming Coal Basin. Blossburg Coal Basin.

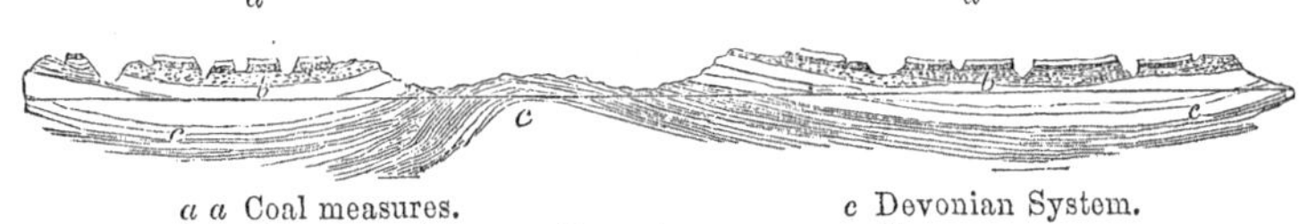

a a Coal measures.
b b Old red sandstone or Devonian.
c Devonian System. Chemung group.

Generally speaking, the bituminous coals of the northeast end of the great Alleghany coal-field, here subdivided into numerous small detached coal basins and outlying patches, consist of dry coals, yielding a considerable per centage of gray ashes; burning with but little tendency to cement or cake in the open fires, and yielding not much smoke. Sometimes rather sulphurous. We might employ a familiar mode of comparison with the fat coals of England, and perhaps those of Western Pennsylvania and Virginia, in remarking that, in their domestic use, the former never accumulate a sufficient quantity of soot to render the sweeping of chimneys necessary; they remain comparatively clean, for many years together.

The Lycoming, or Ralston basin is illustrated by the last section, which shows its relative connection with the Blossburg basin. This Lycoming basin consists of two areas, separated to the depth of one thousand feet, by the valley of Lycoming creek. The coal and iron ore beds are open here, on both sides, but the present amount of business is but small.

PRODUCTION AND CONSUMPTION OF BITUMINOUS COAL AT PITTSBURG.

Besides the main bed of workable coal at Pittsburg, which is there about six feet thick, there is another seam of less value, on account

* Mineralogical Report on the Coal Region in the environs of Blossburg, R. C. Taylor, 1832. Trans. Geol. Soc. of Pennsylvania, on the coal-field of Blossburg. R. C. T. 1835. Magazine of Natural History, Vol. VIII., p. 529; London, 1835. R. C. T. Philosophical Magazine, London, 1836. R. C. T. Penny Mag. London, do. American Journal of Science, Vol. XLI., No. 1. Section across the Blossburg and Lycoming Coal Basins. Section across the Towanda and Loyalsock Coal Basins. R. C. T. July, 1841. Trans. Assoc. of American Geologists and Naturalists, 1843, Art. by R. C. T.

of the intermixture of slate that it contains. These are considerably above water level. It has been ascertained, during the process of boring for salt water, in the vicinity of Pittsburg, on the opposite side of the Monongahela river, that four good seams, besides two small ones, lie at a considerable depth below the surface. The whole depth bored was six hundred and twenty-seven feet. The four coal beds were each about three feet and a half in thickness, and were reached at the respective depths of two hundred and eighty, four hundred and forty, four hundred and eighty, and five hundred and eighty feet. Gas was evolved from each of these veins, and continued to discharge for three or four weeks. The salt water was reached at six hundred and twenty-five feet, and rose to the height of thirty feet above the surface; discharging at the rate of seven thousand gallons in twenty-four hours.

This city and the manufacturing establishments in the vicinity, form the great focus for the consumption of bituminous coal in this State.

In the year 1825, it was estimated at one million of bushels, or 35,714 tons. In 1833, it was returned [in bushels] at 255,910 tons —there being ninety steam engines in operation.* In 1834, eighteen iron foundries, eleven rolling mills, and one hundred and twenty steam engines were at work at Pittsburg and its environs. In 1838, the consumption, by engines and the advance of manufactories had greatly increased, there being now three hundred steam engines and as many factories.† In a communication to a Congressional committee on a national foundry, in that year, it was announced that the quantity of coal consumed was seven millions of bushels per annum, and of that exported three millions of bushels; in all ten millions of bushels, or 357,140 tons, of 2,240 pounds; each bushel weighing 80 pounds.‡

It was computed, in 1842, that the consumption in Pittsburg alone had now risen to eight millions of bushels—or 285,714 tons; the aggregate production at the same time, being 420,000 tons. The number of steamboats owned in the district was eighty-nine, of an aggregate tonnage of 12,436 tons.§

In the annual message of the Governor of Pennsylvania, January, 1846, we find it announced that the consumption had reached the following amount:

	Bushels.	Tons.
Consumed in Pittsburg and its vicinity,	13,000,000	464,286
Exported from that port, down the Ohio,	6,000,000	214,286
Production, - -	19,000,000	678,572‖

* Journal of the Senate of Pennsylvania, 1833, p. 482.

† Proceedings of the Union Canal Convention at Harrisburg, in 1838, p. 12.

‡ Report of the Committee on a National Foundry, 1838, p. 60.

§ Geography of Pennsylvania, from Harris's Directory.

‖ Message of Governor Shunk, 7th January, 1846.

	Bushels.
1853. Consumed in Pittsburg and its vicinity,	22,205,000
Exported from Pittsburg to other places,	14,403,921
Total bushels, - - -	36,608,921

Averaging the price of the coal exported in 1853, at ten cents per bushel, and it gives the sum of over $1,400,000 brought into Pittsburg from abroad for coal.

The Pottsville Mining Register gives the amount of bituminous coal consumed in the Pittsburg district, as follows:

1853, - - - - - Bushels, 26,708,921

Price of Pittsburg coal. Pittsburg coal at the mines, $1.25; with duty and freights to Philadelphia, $5.25 to $6 or $7.

The freight varies so much that it is impossible to state the exact price per ton in Philadelphia.

The progress of improvement may be noted in relation to the advance of population.

1753, No white man living here.	
1813, Population of Pittsburg, - -	5,748 persons.
1840, " " - - -	40,000 "
1850, " " - - -	83,000 "

In the year 1846, were built fifty-three steamboats, which cleared from the wharves at Pittsburg, having an aggregate tonnage of 8,551 tons, and costing $684,000.

In 1847, fifty-six more steamers were built, whose tonnage was 9,954 tons.

The tonnage owned in Pittsburg, on the 1st September, 1847, was as follows:

Steam tonnage, - - - - - -	24,472
All other kinds, - - - - - - -	2,546
Total, - - - - - - - -	27,018

Thus rapidly did this city spring up in the wilderness. Her population now employs more than twenty millions of capital in these active pursuits; and communicates, by means of fifty thousand miles of steam navigation, with almost every part of the valley of the Mississippi.*

Iron Works.—These are more extensive, probably, than those of any other city in the Union.

In 1848, there were 11 rolling mills, of which 8 are capable of producing 4000 tons, each, of manufactured iron, annually, and employ about 150 hands to each mill. The pig metal, supplied from the charcoal furnaces along the river, is about 75,000 tons a year.†

1854.—There are, it is said, 17 large rolling mills; 12 principal or

* Hazard's Register of Pennsylvania, Vol. XVI., p. 22.

† De Bow's Commercial Review, Vol. V., 376.

large foundries; 20 glass manufactories; about 20 engine and machine shops; 5 large cotton factories; 4 large flouring mills, besides some smaller ones; and it is estimated that there are more than 100 steam engines in operation in the city and vicinity, besides those above named.

The Philadelphia Ledger of August 23d, 1854, estimates the number of foundries in Pittsburg at 38.

FROSTBURG BITUMINOUS COAL REGION.

EXTENSION NORTHWARD INTO PENNSYLVANIA.

We will complete our notice of this detached or frontier coal area by adverting to its small peninsular extension into Pennsylvania.

In the absence of the official geological reports, wherein the boundaries may be expected to be defined, we can only roughly estimate this at about 25 square miles, of actually available coal-land.

In the details of the coal seams there is little difference on either side of the State line. From various statements, we are apprised that about ten yards of coal may be calculated upon, within the area we have mentioned.* Assuming this to be correct, and that it extends beneath that entire surface, there are in the ground 750 millions of tons of coal on the Pennsylvania side. But the basin-form arrangement of the strata, and the reduced area occupied by the upper beds, seem to forbid this process of calculation, and to demand a considerable deduction, into the details of which it would be profitless to enter here.

The expense of mining and conveying this coal to tide, was calculated at $3.81 per ton, exclusive of the profits to the landowner, and his lessee, the merchant, &c., and interest of capital. The Special Report of the Chesapeake and Ohio Canal, Nov. 16, 1843, estimates the cost to tide, $3.17.† It is contemplated, however, independent of the large expected sale of the coal on the seaboard, to employ it on the spot in the manufacture of iron, for which, from its excellent properties as an intermediate quality between the fat coals and the anthracites, it seems to be well adapted, and is eminently entitled to consideration.

See our account of these coals in the Maryland division.

NORTH-WEST FROM FROSTBURG.

The coal region in this part of the Alleghany Region, has been detailed in Transactions of the American Geologists, &c., in 1842, in a memoir on the Physical Structure of the Appalachian Chain, by Messrs. Rogers. The following sketch is reduced from the section which illustrates that memoir.

* Report to the Alleghany Coal Company, in 1841, W. R. Johnson; and Reports of other Coal Companies in the vicinity of Frostburg. Also third annual report of H. D. Rogers.

† Special Report of the Pres. and Dirs. Ches. and Ohio Canal Co., Nov. 16th, 1843, p. 17.

Fig. 5.

Section across the eastern portion of the Alleghany Coal-Field, in Pennsylvania, showing the undulations or gentle flexures of that region.

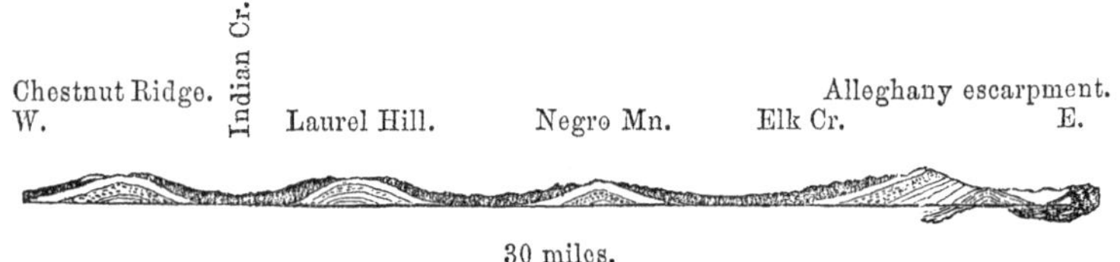

CANNEL COAL.

Near Greensburg, in Beaver county, Pennsylvania, is a bed of Cannel Coal, about eight feet thick, resting upon three feet of ordinary bituminous coal. This Cannel is light, compact, ignites with great facility, and burns with a strong bright flame.*

A similar quality of coal is found in Kentucky, Ohio, Illinois, Missouri, Indiana, and, we believe, in Tennessee. It is not commended for any purpose of iron making and manufacture, but is well approved of for steam engines.

1854. About six miles west from Greensburg, Pa., it is said, a vein of cannel coal has been discovered. Also a small vein in Cambria county, near Ebensburg.

STATISTICS OF BITUMINOUS COAL PRODUCTION.

PENNSYLVANIA COALS DESCENDING EAST.

Cumberland Bituminous Region commencing at the south-west angle of that portion of the State whose market for coal lies at the seaboard, we have little to report upon its supply heretofore; as the facilities of transportation by canal and railroads are only now brought to completion, and the amount of trade is therefore prospective, merely. The high reputation that this coal enjoys will secure it a ready sale, and even a preference, if the coal merchants can furnish it at reasonable rates. These coals will be conveyed from the south-west Pennsylvania collieries to the Atlantic, through the State improvements of Virginia and Maryland.

* Trego in the Geography of Pennsylvania, p. 179.

By the Pennsylvania State Canals, from the Alleghany Coal-field.

		1843.	1844.	1845.
		Tons.	*Tons.*	*Tons.*
JUNIATA,	Shipped and sent East from Hollidaysburg,	14,510	19,000	25,319
	Retained at Hollidaysburg, . . .		330	459
WEST BRANCH CANAL,	Dunnsbury, ...	5,448	10,475	12,415
	Williamsport, . . .	2,464	1,110	625
		22,422	30,915	38,818

There being no other outlets for the bituminous coal, to the eastward, with the exception of the small district near Cumberland, than through the State improvements, it is probable that the above statement very nearly represents the aggregate which is shipped in this direction from the Alleghany coal region. A large portion of this is deposited, on the route, at the iron works and towns of the interior.

The distribution of the remainder, for the use of the cities on the seaboard, is, for the most part, through the following channels.*

Returns in bushels and pounds reduced to the common denomination of tons of 2240 pounds.

Years.	Shipped East at Hollidaysburg.	Shipped by the W. Br. Canal.	Rec'd at Northumberland.	Sent by Union Canal.	By Tidewater Canal.	By Delaware and Chesapeake Ca'l.	Arrived in Philadelphia by water.
	Tons.	*Tons.*	*Tons.*	*Tons.*	*Tons.*	*Tons.*	*Tons.*
1843		7,912		2,079	3,923		
1844	19,000	11,585	657	1,705	7,547	6,114	
1845	25,319	13,040	853	2,650	11,963	11,291	
1846				2,400	12,438		
1847				2,702		6,334	1,058
1850				3,599			
1853	46,425						

Consumption of Coal in the Glass Manufactures of the United States.

In 1846, by a report on the subject, it is shown that in the nineteen establishments for making flint glass in the United States, there are annually consumed as follows:—

* From the annual reports of the respective companies enumerated in the table, and also from the Canal Commissioner's Reports. Also Hunt's Merchant's Magazine, August, 1845, p. 186.

Bituminous coals, chiefly Pennsylvania and Virginia, and $\frac{1}{25}$ part only foreign, - - - - - - -	44,640
Pennsylvania anthracite, - - - - - - -	5,500
Total, - - - - - - -	50,140

Besides 8,666 cords of wood.

*Importation of Foreign Bituminous Coal into Philadelphia, from Great Britain and the British Provinces.**

Years.	Bushels.	Tons.	Value.	
1833	84,510	3,018		The respective proportions of quantity are as follows: Received from Great Britain, . . . 6,733; " Nova Scotia, . . . 43,026; Tons, ... 49,759†
1834	51,286	1,831		
1835	10,056	359		
1836	145,904	5,211		
1837	140,362	5,013		
1838	274,282	9,792		
1839	406,081	14,508		
1840	36,195	1,292		
1841	244,588	8,735		
1849		4,567		
1850		7,698	$8,766	Duty of 30 per cent. not included.‡
1851		12,149	20,947	
1852		16,623	34,442	
1853		12,246	19,751	

Importations of Virginia coal into the port of Philadelphia, in 1847. During this year there were measured, 268,790 bushels = 9,600 tons.§

Quantity of bituminous coal measured at the port of Philadelphia for following years:—||

	Tons.
1850, - - - - - - -	101,395
1851, - - - - - - -	421,860
1852, - - - - - - -	61,573
1853, - - - - - - -	16,388

BITUMINOUS COAL DESCENDING WEST.

The coal business of the Monongahela Slackwater Navigation, in the year 1845, as partially shown by the amount which passed through Lock No. 1, to Pittsburg, was

1845. 2,657,488 bushels = 95,000 tons of 28 bushels each for Pittsburg.

1846. 2,575,375 bushels = 91,977 tons " for the use of the city.

" 5,206,495 bushels = 185,930 tons " for the lower trade, or for export down the Ohio.

* Chiefly British coal imported for the Philadelphia Gas Works—two months only of this year, in 16 vessels.

† Hazard's United States Register, 1839.

‡ Custom House Returns. 1854.

§ Bicknel's Reporter, January 4th, 1848.

|| Report of the Board of Trade, 1854.

RECEIVED ON THE OHIO.

Bituminous coals received on the Ohio, at and near Pittsburg, chiefly down the Alleghany, Ohio, and Monongahela rivers and their tributaries, estimated for the year 1845, at 678,572 tons.

A small but increasing amount passes down the Western Division of the Pennsylvania canal, to Pittsburg.

	Tons.
1843, - - - - - - - - -	973
1844, - - - - - - - - -	1360

BITUMINOUS COAL GOING NORTH.

From the detached coal-field of Blossburg, near the head of the Tioga river, a railroad of forty miles conveys about 30,000 tons annually to the Chemung canal, at Corning, in the adjoining State of New York. This coal-field is capable of supplying at least 100,000 tons per annum, for a long period of years.* Its business, at no time in a state of much activity, appears now to be on the increase, and its advance is greatly facilitated by the recent reduction of tolls on the State canals of New York.

These are now reduced to only 2¼ mills per ton of 2240 lbs. per mile; and for all coal conveyed for the use of the salt works, the chief destination of the Blossburg coal, toll free. This forms a striking contrast to the toll on mineral coal in Pennsylvania, which is no less than 4 mills 4 dec. per ton per mile on canals, and 6.7 mills on railroads.

From the detached coal-field near *Towanda*, a comparatively small supply passes northward, under the present condition of the public improvements. These being perfected, will soon become the channel for the conveyance of a considerable annual amount of bituminous coal into New York State.

1854.—Blossburg sends to market during the season 300 tons of coal daily by the Corning and Blossburg Railroad.

Exports of Pennsylvania Bituminous Coal from the Port of Erie, via Lake Erie.

		Tons.	Value.
	1845	8,507	$21,218
	1846	21,534	53,835
Received at the port, 70,000 tons,	1847		

Exports northward from the Blossburg Bituminous Coal Region.

The demand for this coal increases, and we are assured that its quality, as it penetrates the mountain, improves.

* Mr. R. C. Taylor's Blossburg Report, in 1832.

	Tons.
The railroad company received toll, in 1847, for -	29,110
To which may be added for domestic consumption, by the locomotives, machine shops, &c., - -	1,000
Total in 1847, - - -	30,110

1853.—The Canal Commissioner's Report states, according to a letter from H. K. Strong, in the Miner's Journal, January, 1854, "that the tolls paid on bituminous coal leaving east from Pittsburg and Hollidaysburg amount to $45,156 17. It is therefore probable that the increase in bituminous coal cannot be less in 1853 than 300,000 tons."

Rates of Toll on the State Canals and Railroads in Pennsylvania, per mile, in 1846, 1847, *besides* 13 *mills per* 1000 *lbs. on the Alleghany Portage Railroad.*

	ON CANALS. Cents.	Mills.	ON RAILROADS. Cents.	Mills.
Mineral coal, Gypsum, and Iron ore, per 1000 lbs. weight,	0	2	0	3
Which rate is equivalent to per ton of 2240 lbs.	0	4.4	0	6.7
Each 100 miles of transportation, therefore, is	44	8	67	2
On each Union Canal boat, engaged in the coal trade, per mile,	1	0		
State of New York—1846.				
Toll on the State Canal,—on Mineral Coal, per 1000 lbs. per mile,	0	1		

In a report of the President of the Chesapeake and Ohio Canal Company in 1843, the estimated prime cost per ton of the Cumberland coal, delivered in New York is $4.67.*

In 1844, Alleghany bituminous coal sold at from 16 to 18 cents per bushel =$4.50 to $5.00 per ton, in Philadelphia; brought 210 to 250 miles by canals.

During a portion of 1845, bituminous coals from the interior obtained no higher price in Philadelphia than 14 cents per bushel; but towards the close of the season the largest sales were effected in this city at 21 cents per bushel = $5.75 per ton of 2000 lbs.; the cost of the freight and toll being $2.62½ per ton, from Farrandsville and the adjacent mines, to Philadelphia.

In the winter of 1846–7, Alleghany coal, in Philadelphia, was 18 to 19 cents per bushel; Virginia coal 20 to 22 cents per bushel. This being on a very small scale is naturally very fluctuating. The current price of Alleghany coal, in Philadelphia, in the winter of

* Special Report of the President and Directors of the Chesapeake and Ohio Canal Company, November 16th, 1845. Also, Report of the probable revenue of the Chesapeake and Ohio Canal, 1834, p. 40. "To quarry this coal," says another authority, "costs about 20 cents a ton." See American Quarterly Review, March, 1829.

1848, was 25 cents per bushel, or $7.00 per ton. 1853, 20 cents per bushel, wholesale price.

The external trade, coastwise, in this coal is quite unimportant.

Boston, in 1846, received but 4900 bushels = 175 tons, from Philadelphia, and can purchase Nova Scotia coal on much lower terms.

THE ANTHRACITE REGIONS OF PENNSYLVANIA.

INTRODUCTORY REMARKS.

It has heretofore been customary to describe the Pennsylvania anthracite fields as occupying three distinct, yet corresponding areas. These are familiarly spoken of as, I. The Schuylkill or Southern coal region; II. The Middle coal region; and III. The Wyoming, Wilkesbarre, or Northern region. At the present day, they, particularly the middle district, are known to consist of a great many separate or subordinate basins. Any attempt of ours, in this place, to unravel the local intricacies of this series of coal basins, would but complicate a description which we design only to be general. Fortunately, their investigation has been committed, by the commonwealth, to able hands; and the results, we anticipate, will be of a useful character. Until these are before the public, it seems advisable to adhere to the old local classification; and in continuing for the present, the order heretofore observed, although obviously defective, we shall perhaps best consult the convenience of our readers.

During the last twenty years the anthracite deposits of Pennsylvania have acquired no small celebrity. They have attracted towards them a larger amount of capital than ever before was invested in mineral operations in the United States; and consequently, have called into exercise a corresponding amount of productive industry. One result of this state of activity, is, that able geological investigators, and writers of intelligence, have by no means been wanting to demonstrate, not alone to the proprietors, but to the commercial and manufacturing interests, and to the scientific world, the enormous value of these concentrated resources. To Pennsylvania, the almost exclusive possession of this species of combustible, within reasonable distance of the seaboard, is a boon of inestimable price, which places her in a position of enviable superiority, and baffles speculation as to the point, to which it may ultimately elevate her. The statistical details, scattered through the following pages, justify such conclusions, notwithstanding that the financial difficulties by which the country was beset, a few seasons ago, deeply affected the best interests of the state; and somewhat retarded the progress of improvement in the anthracite regions, in common with that of all others. Happily, that period of gloom and depression has at length passed by, and energy and enterprise have succeeded to apathy and despondency.

HISTORY AND PROGRESS OF ANTHRACITE PRODUCTION.

With the exception of three or four detached basins or patches, of very limited extent and value, in other states, Pennsylvania is the great depository of anthracite, on the North American continent. The entire area is made up of a numerous suite of coal basins, produced by alterations of anticlinal and synclinal axes, which range nearly parallel with the base of the Alleghany mountain. An acquaintance with the extent and number of these separate little coal troughs has only been acquired after years of investigation; and in fact is, even now, very far from complete. As in the case of the bituminous coal fields of this state, it is much to be desired that the state geological report and map should be published, and that the public should, after the lapse of twelve years from the commencement of the survey, derive some benefit from that important work.

Allusion has already been made to the geological age of the anthracite districts, and to its obvious agreement, in that respect, with the bituminous series. Respecting this point it is remarked by one of the best authorities in the science, that these deep anthracite basins, abounding in curious structural features, and containing thick seams of coal, are highly interesting by the geographical position which they occupy. "More than forty miles distant from the general denuded margin of the main or western coal-field, they nevertheless present, in the character of their strata, and of the rocks, upon which they repose, unequivocal evidence that they and the bituminous basins were once united."* But it is worthy of note, that at the north-eastern extremity of the bituminous coal area of the Alleghanies, it approaches within ten miles only of the Wyoming coal basin, which contains the hardest species of white ash anthracite. In the map which we annex, we have been able to point out this proximity, accurately.

Passing now to the historical development of resources so valuable to the interests of Pennsylvania. In 1749, the lands between Mahanoy creek on the east side of the Susquehanna river, and the Delaware north of the Blue Mountains, were obtained from the Indians; apparently, in part or the whole, the present counties of Dauphin, Schuylkill, the south parts of Northumberland, Columbia, and Luzerne; Northampton, Monroe, and Pike.† The space thus defined comprehends the lands between the Blue or Kittatinny mountain range to the south, the Susquehanna to the west, and a line drawn from the point of the mountain at the mouth of Mahanoy creek to the mouth of Lackawaxen creek, at the New York State boundary, and at the junction of that creek with the Delaware river; being one hundred and twenty-five miles long, and thirty miles average breadth. All this little territory of 3,750 square miles embraces the entire group

* Origin of the Appalachian coal strata. By Prof. H. D. Rogers—in Trans. Assoc. Amer. Geol. and Nat., 1843, p. 436.

† View of the Land Laws of Pennsylvania. By Thos. Sergeant, Esq., 1838, p. 30.

of anthracite basins, which are comprised in what are usually styled the southern and middle coal-fields, and the whole was transferred by the Indians to the proprietary government, for the sum of *five hundred pounds.* From the same territory has been acquired, within the last quarter of a century, nineteen millions of tons of coal, of the value of seventy-five millions of dollars; and in the production and disposition of which fifty thousand persons derive their support.* In the year 1847, three millions of tons were brought to tidewater, whose value there, was twelve millions of dollars.

In the purchase made by the proprietaries, in 1768, by which they acquired from the chiefs of the Six Nations the bituminous coal-fields within the State, they became also owners of the northern, or Wyoming anthracite basin; the value of which, at that time, was very little anticipated by either of the parties to the contract.

From all that we can now trace, it would seem that anthracite was first observed, and its combustible properties tested, in this northern district, in 1768. On the authority of an article in the "Memoirs of the Historical Society of Pennsylvania," the adaptation of this newly observed substance to the purposes of fuel, was discovered by certain blacksmiths, about the year 1770, two years only, after the ratification of the treaty of purchase, and three years previously to the laying out of the borough of Wilkesbarre, by the Susquehanna Land Company of Connecticut. The first cargo of this coal was sent down the Susquehanna, in boats, and reached the United States armory at Carlisle, in 1775: but it was not until 1808, that grates were constructed at Wilkesbarre, to burn it for domestic use, under the direction of Judge Fell.† It is very probable that the fine natural sections of the coal measures in this valley, occasioned by the cutting of the river Susquehanna through its margin, at two points, and the deep lateral ravines which also laid bare the thick bed of coal along its borders, were the means of displaying the carboniferous strata, and, as now, facilitated their development, at so early a period. Not less than three millions and a half or probably four millions of tons of anthracite have been sent to markets on the seaboard from hence, since the year 1829. The latter year was the commencement of the coal works at Carbondale.

In the Schuylkill division of what has been customarily called "the southern coal region," anthracite appears to have first attracted the notice of the scanty population, settled near the present site of Pottsville, about the year 1790.‡ It is extremely probable, that an obscure knowledge of its existence, and an undefined surmise of the combustible properties of this mineral substance, had existence some years earlier; especially as it had been seen, and partially tested, and had been spoken of in the Wilkesbarre district, for twenty years previously. The number of emigrants who arrived in the country,

* Speech of Mr. Ramsay in Congress, April 29th, 1844.
† Memoirs of the Historical Society, Vol. II., p. 154. Art. by Erskine Hazard, Esq.
‡ Packer's Report in Senate Journal, Vol. II., 1833–4.

and who were accustomed to the use of coal, also facilitated the knowledge of this description of fuel.

We are the more confirmed in this opinion from having seen a large map of Pennsylvania, published in 1770, from Scull's older map, in which the site of "Coal" is marked, and shown to prevail about the head waters of Schuylkill creek, and stretching thence westward, to those of the Swatara, and to "the wilderness of Saint Anthony."

Near the eastern extremity of the same southern region, on the lofty ridge which overlooks the valley of the Lehigh, anthracite was accidentally discovered in an enormous mass, open and bare to the very surface. This occurred in the year 1791. It was the first knowledge of the now celebrated Mauch Chunk mines, which, even at the present day, are worked as an open quarry. Having purchased from J. Weiss the newly found site of the "Summit Mines," the "Lehigh Coal Mine Company" was formed in 1793, for the development and working of this unproved combustible; but it was not until 1814, that the first twenty tons were conveyed down the Lehigh and the Delaware rivers, at great labour and cost, to Philadelphia, where a few wagon loads had preceded them, from the Schuylkill district, in 1812. It was as late as the year 1820 before the comparatively large quantity of 365 tons of anthracite reached their destination at Philadelphia.

The first volume of the Memoirs of the Historical Society of Pennsylvania contains "a brief account of the discovery of anthracite coal, on the Lehigh," from the pen of T. C. James, M. D., which was read on the 19th of April, 1826. The author states, that in the autumn of 1804, having, in company of a friend, crossed the Blue Mountain, they found themselves bewildered in a secluded part of the Mahoning valley, and at length obtained shelter for the night, at a solitary mill, kept by Philip Ginter. This was the individual who discovered the coal on the Mauch Chunk mountain, and who conducted Dr. James and his companion to the spot where is now the open mine, or rather quarry, of anthracite.

"At that time there were only to be seen three or four small pits which had much the appearance of rude wells, into which one of our guides descended with great ease, and threw up some pieces of coal for our examination. After which, whilst we lingered on the spot, contemplating the wildness of the scene, honest Philip amused us with a narrative of the original discovery of this most valuable of minerals, now promising, from its general diffusion, so much of wealth and comfort to a great portion of Pennsylvania. His only resource being that of a hunter in the back woods, he was, on the occasion alluded to, returning, towards evening, over the Mauch Chunk mountain, entirely unsuccessful and dispirited, having shot nothing. As he trod slowly over the ground, his foot stumbled against something, which, observing to be black, he took up. Having listened to the traditions of the country respecting coal in the vicinity, it occurred to him that this, perhaps, might be the "Stone Coal" of which he had heard. He accordingly, on the next day, carried it to Colonel

Jacob Weiss, who, being alive to the subject, brought the specimen immediately to Philadelphia."*

The result was the formation of a company to work the newly discovered coal, which, however, was neglected until 1806, when two hundred or three hundred bushels were brought down; but, not being understood, it failed to give satisfaction, and the enterprise was again suspended for several years. The writer closes his communication by stating, that he commenced burning the anthracite coal in the winter of 1804, and had continued the use of it until that time, 1826; "believing, from his own experience of its utility, that it would ultimately become the general fuel of this [Philadelphia] as well as some other cities."

The Mauch Chunk Railroad, of nine and a half miles, was begun January 12th, 1827, and was finished in May, of the same year, since which time the whole mountain has been intersected by railroads, tunnels, inclined planes, schutes, and numerous other works, and contains a large population of operatives.

The "Middle Region" is the most complicated of the three, being made up of a series of axes of elevation and depression; in the troughs of which thick bodies of coal occur, which, even now, are but imperfectly explored, but become subjects for investigation in proportion to the gradual advance in the demand for coal, and consequent increase in the value of those tracts of land which are so fortunately situated as to contain it. From its wild, mountainous, and inhospitable character, unattractive to settlers and little adapted to cultivation, this district was the last to make known its buried treasures; as, owing to the natural difficulties of approach, it was the latest to become the theatre of industrial operations. The progress of geological discovery was still farther retarded by the prevailing ignorance of the elementary principles of stratification, deposition, and extension of coal seams; each exposure of mineral coal being conceived to be, and treated as an isolated mass and a local deposit; such, for instance, as that on the summit of Mauch Chunk was long considered to be.

Yet the proprietaries were not altogether ignorant of the existence of coal within these limits, for we have seen, in the large North American Atlas, published by Faden, of London, in 1777, from an earlier map of 1770, that coal pits or mines are marked in the neighbourhood of Mahanoy Creek, above Crab Run.†

Passing from these early notices, we find that the coal trade of Pennsylvania, which had its beginning in 1820, had, in 1833, already arrived at a magnitude so much beyond anticipation, that the Senate of the Commonwealth considered it expedient to appoint a committee of inquiry, and to invest it with powers to investigate extensively into the subject.

That committee appears to have most sedulously applied itself to the duty assigned. The report produced by its chairman, Mr. S. J.

* Abbreviated from Dr. James's paper, in Memoirs, Vol. I., p. 316.

† Atlas of North America, Faden, London, 1777, from Scull's Map of Pennsylvania, 1770.

Packer, exhibits evidences of great care and labour. It concentrates a multitude of local and statistical facts, then for the first time presented to the public; and deservedly merits the eulogium which we desire to bestow upon it. The ground and the subject-matter were comparatively new, although the interests involved were considerable. The difficulty was enhanced by the uncertainty or complicated character of some of these interests, at the time. We believe it was nevertheless conceded, on all sides, that this document,—the earliest official report on the coal business of Pennsylvania, or indeed of that of any portion of the United States,—was drawn up with ability and perspicuity, and evinced much practical good sense. Superseded as it now is, or inevitably must soon be, by the mature developments of more recent times; by laborious private investigations; by highly accomplished observers; and beyond all, by the results of the State geological surveys, under the zealous superintendence, during many years, of one of the ablest geologists of the day, yet the report of Mr. Packer, in 1834, will never be thrown aside as useless. Although, strictly speaking, not scientific, but illustrated in the coarsest manner, it will always be regarded as a business-like memorial,—adapted to the times and circumstances,—and a valuable contribution to the mining and statistical information of the day. In fact, it belongs to, it is identified with, and greatly illustrates, the commercial and industrial history of Pennsylvania.

In succeeding years, detached notices, both practical and scientific, of local sections of the anthracite coal basins, have appeared in various cotemporary publications. Among these we name Silliman's American Journal of Science; the Transactions of the American Geological Society; the Monthly American Journal of Geology; the Journal of the Franklin Institute; the Transactions of the American Philosophical Society; the Journal of the Academy of Natural Sciences; the Miner's Journal, and the Anthracite Gazette, both of Pottsville; the Commercial List, of Philadelphia; the Mining Journals, of London and New York; the Mining Review; the Transactions of the Geological Society of London; the Transactions of the Association of American Geologists; the Annual Reports of the State Geologist; and several other occasional and local authorities. All these have been extensively quoted by that portion of the public press, including the Registers and Magazines, which is more especially devoted to the circulation of useful, practical, and statistical information.

Among the class of periodicals and occasional documents, not strictly scientific, yet comprising authentic communications of a business character, may be named the numerous annual reports of companies, committees, and associations, and of the State Legislature and of Congress, bearing upon the staple products of Pennsylvania; their avenues to markets; their modes of transportation, both internal and coastwise; their adaptation for domestic consumption; and, finally, the facilities they furnish to manufacturing enterprise. We advert to this temporary and commercial literature, because of its remarkable diffusion, its cheapness, its influence, and its employment, in this country, to an extent unknown in any other part of the world. It forms an economical substitute for books of a more expensive and pretending character, and may be found in every man's hand.

CANAL AND RAILROAD SYSTEM IN RELATION TO THE ANTHRACITE DISTRICTS OF PENNSYLVANIA.

Names of Railroads and Canals.	Canals.		Railroads.		Total cost.
	No.	Miles.	No.	Miles.	
					Dollars.
Lehigh Navigation, ...	1	87½			4,455,000
Lehigh and Susquehanna Railroad, . .			1	20	1,350,000
Mauch Chunk and Summit Railroads, &c. .			1	36	831,684
Delaware Division of the Pennsylvania Canal,	1	43			1,734,958
Beaver Meadow Railroad, and branch, .			1	38	360,000
Hazelton Railroad, ...			1	10	120,000
Buck Mountain Railroad, ...			1	4	40,000
Summit Railroad, ...			1	2	20,000
Delaware and Hudson Canal—partly in New Jersey, ...	1	108	1	16	3,250,000
Morris Coal Canal, in New Jersey, . .	1	102			4,000,000
The Schuylkill Navigation, . . .	1	108			5,785,000
The Reading and Pottsville Railroad, . .			1	98	11,590,000
Little Schuylkill and Tamaqua Railroad, .			1	20	500,000
Mine Hill and Schuylkill Haven and Extension, to Swatara, ...			1	55	550,000
Danville and Pottsville, 44½ m. unfinished, .			1	29½	680,000
Mount Carbon Railroad, ...			1	7	155,000
Do. and Port Carbon Railroad, . .			1	2½	120,000
Schuylkill Valley Railroad, and branches, .			1	25	300,000
Mill Creek Railroad, ...			1	6	120,000
Railroads by individuals, ...				120‡	180,000
Under-ground Railroads, ...				200‡	75,000
Lyken's Valley Railroad, ...			1	16	200,000
Wiconisco Canal, ...	1	12			370,000
Swatara Railroad, ...			1	4	20,000
North Branch Canal—division, . . .	1	73			1,491,894
Do. do extension, . . .	1	90			1,298,416
Union Canal,* and Pinegrove branch,* .	1	90			1,000,000
Dauphin Co.'s Railroad, ...			1	52	1,500,000
Baltimore and Susquehanna Railroad,* .			1	60†	1,000,000
Susquehanna Tidewater Canal,* . . .	1	45†			1,000,000
York and Cumberland Railroad,* . . .			1	26	60,000
Mine Hill Railroad, ...			1	14½	396,117
Williamsport and Elmira Railroad, . .			1	25	496,000
Corning and Blossburg Railroad, . .			1	40	600,000
Nesquehoning Railroad, ...			1	5	50,000
Room Run Railroad, ...			1	6	40,000
Lackawanna and Western Railroad, . .			1	50	
Pennsylvania Coal Co. Railroad, . .			1	52	
	10	758	25		

The foregoing table shows the principal State and private canals

* Estimated for coal purposes.
† Partly in Maryland.
‡ Bowen's Pictorial Sketch Book of Pennsylvania. Phila., 1852.

and railroads which are in direct communication with the anthracite mines of Pennsylvania, and which were constructed almost entirely for the purposes of the coal trade, since the year 1821. We believe that this statement is below the actual result, and might be materially increased, independently of the capital invested in the mines and in the coal operations of this important region.

The whole may be estimated at $45,000,000. In addition to these, the Catawissa, Williamsport, and Elmira roads of 172 miles, which are now completed, and in operation, make a continuous line of 263 miles from Philadelphia by the Reading Railroad and through Schuylkill county, to Elmira; where there is a connection with the great chain of New York improvements.

The Union canal has been widened to Pinegrove, and will be widened to the Schuylkill the present season.

The Lackawanna and Western Railroad runs from Scranton to Great Bend.

The Danville and Pottsville Railroad, to which the United States Bank contributed largely, is not in use, and will scarcely be revived, being superseded to some extent by the Catawissa road. It is unfavourably located on account of the steepness of the grades.

The North Pennsylvania Railroad is in progress. It runs through the Lehigh and Wyoming Valley coal-fields, extending up the North Branch of the Susquehanna to Waverly in New York, connecting with the improvements of that State. It promises to become one of the greatest outlets for Pennsylvania coal.

Within a year Baltimore and the intermediate region will have the following new coal route—

Baltimore to York, Pa., - - -	58 miles.
York to Bridgeport, - - -	26 "
Bridgeport to Sunbury, - - -	54 "
Sunbury to Shamokin, - - -	20 "
	158 "

The "Susquehanna Railroad" runs from Bridgeport, opposite Harrisburg, to Sunbury.

S. W. Mifflin, Esq., Civil Engineer, has recently made a survey of a railroad from Columbia along the east side of the Susquehanna, passing nearly the following places. Columbia, Safe Harbour, Mortie Village, (north of it,) Rawlinsville, Oxford, New London, Newark, (Del.,) Christiana, Newcastle, intended chiefly as an outlet for coal.

GENERAL CANAL AND RAILROAD SYSTEM OF PENNSYLVANIA.

	CANALS. Miles.	RAILROADS. Miles.	Miles.
Those of the State, (1848,) -	848	118	} = 720
Those belonging to Companies, (1848,)	432	602	
Private railroads to mines and under ground, - - - - -		320	
	1280	1040	

The gross receipts on the several lines of canal and railroad for the fiscal year ending November, 1851, amounted to $1,793,624, being an increase over 1850, of $25,417.36.

Canal and railroad tolls ending December 1st, 1852, amount to $1,938,574.

1853. From the Railway Times we extract the following:

Pennsylvania railroads, - - - - -	64
Miles, - - - - - - -	1464
Constructing, - - - - - -	897
Cost, - - - - - -	$58,494,675

PENNSYLVANIA ANTHRACITE.

COMPARATIVE ADVANTAGES OF ANTHRACITE OVER BITUMINOUS COAL, FOR DOMESTIC PURPOSES.

The author of the work on "*Fossil Fuel*," devotes a page or two, with great propriety, to the subject indicated above, and, if our space permitted, his views should be introduced, without curtailment here. He observes, that "the smoke given off, during the combustion of flaming coal in most large towns, especially the prodigious volumes of it emitted from the chimneys of manufactories, form a serious annoyance in many situations." We must quote a passage on the smoky nuisances attendant on the consumption of this fuel in London and other English cities, agreeing as we do in many, although not all, of the opinions of the author.

"A very striking contrast to the murky exterior of some of the large towns in this country, [England,] is presented by the appearance of the city of Philadelphia, over which, notwithstanding its thousands of coal fires, constantly kept up, there is no smoke. The inhabitants mostly burn the anthracite, a substance resembling the stone coal, or culm, of Wales; the carbonaceous or stony coal of Kilkenny; the glance coal of the Germans; and the blind coal of Scotland.

These coals are difficult to kindle; [no difficulty at all to a Philadelphia housemaid;] but when once thoroughly ignited they burn for a long time; [upon the admirable principles which experience has suggested, whether for open grates or for the infinite variety of stoves and furnaces common to all the houses in Philadelphia, New York, or Boston;] they make a hot glowing fire, like charcoal, without either flame or smoke."

The author goes on to account for the non-use of anthracite in Great Britain, which possesses a far larger area of anthracite than exists in America, or any other part of the world.

"It is owing to these coals commonly emitting noxious vapours that they cannot be pleasantly used in dwelling-houses in this country, though they are in considerable demand among malsters, dyers,

&c.; more especially for the furnaces of steam-engines and breweries, in those situations where smoke is a nuisance."

And what habitable place is there, among communities of men, not even we believe excepting an Esquimaux Indian's, in which smoke is not considered an intolerable nuisance, an atmosphere unfitted for living and breathing in? Let the author and his readers take the word of one who, like most Europeans, from early custom, long preferred the brightly blazing, yet sulphurous and smoke-producing bituminous coal, to the non-blazing, yet cleanly and economical anthracite: let him and them be assured, that, with the familiar modes, the ready appliances, and the improved methods, now in universal use in the Atlantic cities of America, there cannot be a reasonable apology for hesitating as to the choice of the two combustibles for domestic use. The difficulty suggested about ignition, even were it found so in practice, is deprived of all weight from the consideration, that with ordinary attention, a fire, when once kindled in the fall of the year, may be kept up for months, if needed. The supposed tendency of anthracite to emit a greater amount of noxious vapours during combustion than bituminous coal, is contradicted by the daily experience of those who employ the former in their apartments, and is much less objectionable, on that head, than bituminous coal.

Our tables of analysis at the end of this volume, will, if doubts remain, decide this matter. In fact, they show that the anthracites contain less sulphur than the blazing coal; the consequence of that obnoxious substance having almost entirely been expelled, during the conversion from bituminous coal into anthracite. The following analyses are from the Report of Prof. H. D. Rogers, in relation to Pennsylvania coals of each description:

	BITUMINOUS COALS. Sulphur per cent.	ANTHRACITES. Sulphur per cent.
The Karthaus Coal,	2.70	
Blairsville Coal,	2.60	
Lehigh Anthracite,	-	0.91
Pottsville, White Ash do.	-	0.60
Do. Red Ash,	-	0.48

We have neither soot in the chimneys, nor smoke in the atmosphere.

AREA OF ANTHRACITE FORMATIONS IN PENNSYLVANIA.

The original estimate of the superficial area of the three anthracite regions in Mr. Packer's report, was nine hundred and seventy-five square miles, or six hundred and twenty-four thousand acres.* It is customary for every writer on Pennsylvania statistics or topography, to adopt these admeasurements. Professor Rogers, however, reduces this computation materially, and offers two hundred square

* Report to the Senate, 1834, p. 438.

miles—one hundred and twenty-eight thousand acres, as the approximate area.* This result presents such a contrast with former estimates, that were we not familiar with the general accuracy of that gentleman's estimates, we might suspect some error, unless the calculation be limited to the two southerly regions, as is probable.

A different result has been published by Mr. S. B. Fisher, a district surveyor, long engaged in the anthracite regions.†

		Sq. Miles.	Acres.
I.	The Southern or Schuylkill coal-field, workable coal measures, exclusive of the external margin of conglomerate, 67,500 acres+8,450 acres on Broad Mountain,	119	75,950
II.	The Middle Coal District, comprising several basins, stretching from Shamokin to the Lehigh, over Broad Mountain, but exclusive of some of the small basins north of Hazleton. Mahanoy coal-field, 59,450a+the small basins 26,075,	133	85,525
III.	Wilkesbarre or Wyoming, or Northern Coal District—estimated by H. Colt,	120	76,805
	Total, with the foregoing exceptions,	372	238,280

Our own computation, although by no means exact, as even now the boundaries are not generally determined, is as follows:

		Sq. Miles.	Acres.
I.	The Southern or Schuylkill coal region, consisting of three principal basins, viz. 1. The great southern basin; 2. The basin north of Mine Hill; 3. The basin south of the Mahanoy creek,	164	104,960
II.	The Mahanoy and Shamokin, principal basins, including several minor basins or troughs,	75	48,000
	the eastern group of basins, at least twenty-six in number,	40	25,600
III.	The Wyoming coal-field in one area, but broken into several subordinate undulations or basins,	118	75,520
		397	254,080

* H. D. Rogers in Trans. Assoc. American Geologists, Vol. I., p. 436.

† Miners' Journal of Pottsville, March 29th, 1841.

To render our exposition of the Pennsylvania group of anthracite basins more complete, we have prepared the map hereto annexed. It will be seen that, although in a few details, hereafter to be supplied, something might still be amended, yet our draft is greatly superior in topographical minutiæ, in geological features and physical characteristics, to any other which has heretofore been published. We should have been happy to have had a precedent to follow, in a State geological map, but, as none yet exists, we have been thrown on our own exertions and resources to produce the present one.

I. THE SCHUYLKILL COAL REGION OR SOUTHERN DIVISION AND GROUP OF ANTHRACITE BASINS.

So many reports have been published relating to the property of individual coal companies, and to the general interests and characters of this district, that we cannot undertake even to enumerate them. On the more prominent of these, however, we shall bestow some cursory notices.

Without designing any injustice towards other regions, it may be expected, in the present instance, that we exercise somewhat less of brevity; seeing that for some years it has occupied an extremely prominent position in the mineral statistics of Pennsylvania. It appears entitled to this consideration at our hands, inasmuch as it lies the nearest to the ports, or places of shipment; it has employed the greatest amount of active capital; has been the most extensively worked, and the most assiduously investigated; and, moreover, has called into exercise the largest quantum of practical and scientific intelligence ever concentrated in one mineral area on the American continent.

Among the earliest original describers we may mention Professor Silliman, who, in his Journal of December, 1830, called attention to the extraordinary development of anthracite at Mauch Chunk.

We have already adverted to the excellent report by the chairman of the Pennsylvania Coal Committee, in 1834. One section of this report especially relates to the Schuylkill coal-field, as then understood, and to its mining statistics, up to that period. The summary of these details, for the year 1833, is as follows:

Coal mined and sent to market, - -	429,933 tons.
Capital invested, - - - -	$5,022,780.*

To this special report succeeded many temporary as well as annual reports of coal companies, to which we cannot advert in detail; and many incidental statements quickly followed, in connection with railroads, canals, public improvements, and private enterprise.

* Packer's Report, p. 458.

Most of these were illustrated by maps, sections, and other instructive diagrams, appertaining to the country in question. Some are on an extensive scale, and embody a large mass of information. In fact, it may be stated that few coal districts in the world have received more ample illustration, within a similar period, than has the Schuylkill coal-field, from the united exertions of the topographer and the geologist, the chemist, the operative miner, the engineer, the artist, the economist, and a variety of subordinate contributors and fellow-labourers.

As a state undertaking, the annual reports of the geologist, on the progress of his survey, have a primary claim to consideration and commendation. We await, with confidence, the final report, which will complete our knowledge of this interesting country, and of the results of many years of indefatigable research.

It is only by an elaborate survey, such as that ordered by the State—for we can scarcely contemplate a private undertaking of this magnitude—that we shall be led to a right understanding of this subject, and receive enlightenment on phenomena so irreconcilable with those displayed elsewhere.

We are greatly in need of detailed maps of all the anthracite basins.

SKETCH OF THE SCHUYLKILL OR SOUTHERN COAL-FIELD—COMMENCING EAST.

Fig. 6.

Section of the Mauch Chunk Anthracite Region, looking East.

Locust Mn. Mauch Chunk Summit. Sharp Mn.

Red Shale. Coal Measures. Red Shale.

The remarkable exposition of an enormous mass of anthracite, which was quarried to open day on the summit of the Mauch Chunk Mountain, near the eastern termination of this coal-field, excited, for a time, no small attention, not only in the United States, but in Europe.* We have already cited some historical notices attending this discovery, and the difficulties originally encountered in rendering it available. The great bed on the summit at Mauch Chunk is estimated to be from fifty-five to sixty feet thick, including the inferior seams, and also some shale beds. The open quarry here is more than thirty acres in area of excavation: in fact, there are now several of these open workings. At the neighbouring coal works of Room Run, the main bed is found to be fifty feet thick, resting, as at Mauch Chunk, upon a thick bed of under-clay, filled with stigmaria. From these two working points, have been abstracted and sent to the mar-

* Silliman's American Journal of Science, 1830.

kets of Philadelphia, New York and Boston, about three millions of tons of anthracite.

NESQUEHONING OR ROOM RUN SECTION.

Fig 7.

Section at Nesquehoning, or Room Run, looking East.

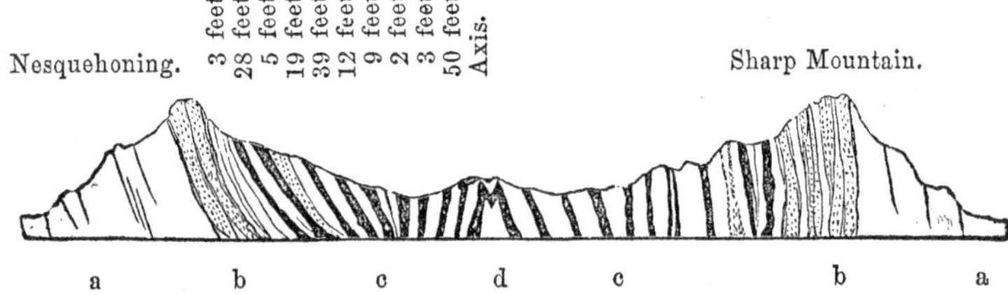

aa Red Shale and Red Sandstone. Course N. 65° E.
bb Conglomerate.
cc Coal Measures.
d Anticlinal Axis or Saddle.

The remarkable thickness of the coal at the summit appears to be the result of the doubling back of the twenty-eight feet seam upon itself, making an aggregate thickness of fifty-six feet.

There are about twelve coal seams on either side of the axis forming this basin; eight of these were in work in 1847, and had an aggregate thickness of one hundred and sixty-three feet of anthracite, of which a fair proportion consisted of merchantable coal. These were all south-dipping seams; those on the corresponding or opposite side of the axis, apparently contained at least an equal amount of good coal, probably the repetition of the same veins.

It is by no means easy to convey a true representation of the position of the coal beds on the summit at Mauch Chunk, owing to the numerous contortions of that part of the general basin. The extensive works recently put in operation, will go far to develope the intricacies of this singular district. The further advance of these undertakings in regard to production, will be detailed in a subsequent page. At present, we must make a hasty travel from end to end of this basin; stopping, for a brief space, at a few prominent points on the way.

Leaving Mauch Chunk and the Room Run mines, which have acquired some celebrity through the notices of travellers, geologists, and men of science,* and produced in 1837, 334,929 tons, the next important position is that of Tamaqua.

TAMAQUA.

The section we here introduce, represents the arrangement of the

* Journal of the Academy Natural Sciences, Philadelphia, Vol. V., p. 17. Experiments on anthracite, by L. Vanuxem.

coal seams on each side of the valley or synclinal axis; those on the Sharp Mountain or south side, being nearly vertical.

Fig. 8.

Transverse Section of the Schuylkill Coal Basin at Tamaqua, looking towards the West.

Sharp Mountain. Locust Mountain.

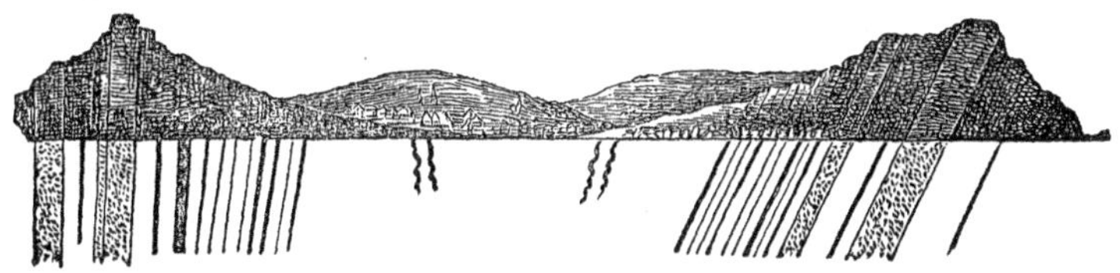

But it is especially interesting on account of its containing in the Sharp Mountain, a coal seam which is no less than seventy feet thick, nearly vertical. This valuable coal area was illustrated, in 1835, by an article in the Transactions of the Geological Society of Philadelphia, accompanied by a "petrographical map and section," by Mr. Koehler.* These illustrations exhibit a transverse section of the entire coal-field at that place, and the position of thirty-two workable coal seams, of three feet thick and upwards. It does not appear to have been suspected at the time, by the author, that this number represented duplicated seams, arranged in synclinal form; and that, consequently, the true number represented was not more than the one-half of his supposed series, which embraced a total thickness of many feet of coal, the largest seam then known, being twenty-eight feet. The anthracite of Tamaqua is less indurated than that near the Lehigh. An investigation into the nature of this coal and its ashes had been previously instituted by Professors Bache and Rogers.†

The Tamaqua Mines sent to market via Little Schuylkill.‡

Year		Tons	Year		Tons
1848,	- -	162,626	1851,	- -	310,307
1849,	- -	174,758	1852,	- -	325,099
1850,	- -	211,960	1853,	- -	384,443

Passing by Tuscarora, and by several mining villages that have sprung up within a few years, we reach Pottsville and a circle of colliery establishments in its vicinity; the focus wherein is concentrated an enormous amount of productive and manufacturing industry, such as has few parallels in the new world considering the short period since the origin and development of its resources. The magnitude of the mining business of the Pottsville district will appear from the tables which we shall furnish in another place.

Numberless are the documents, reports, maps, and statistical state-

* Trans. Geol. Soc. of Pennsylvania, Vol. I., p. 326.
† Ibid. Experiments on the ashes of anthracite, Vol. VII., pp. 158—162.
‡ Little Schuylkill Report.

24

ments, published and unpublished, that have contributed to give celebrity to this district. A transverse section of this coal-field, in the meridian of Pottsville, has been for some years in every body's hands. It appears to have resulted from the combined observation of many local explorers, practical operators, and mining engineers and surveyors. The same mistake was made here, as we have mentioned at Tamaqua, in reference to the repetition of beds by means of undulations in the strata, and by the recurrence of several alternate axes of elevation and depression. In 1836, the supposed consecutive series of coal seams numbered in Mr. Wilde's section, seventy-eight, which had an aggregate thickness of four hundred and fifty-four feet of coal. Several years afterwards, this number had been increased to one hundred and eight, and the total thickness was augmented in the like proportion.

Of late years, the accuracy of opinions so confidently expressed by merely operative persons, without the intervention of geological investigation, gradually began to be doubted, by those geologists who had acquired experience in unravelling the intricacies of districts of complicated structure, and in studying, on an extended scale, the effects produced by disturbing agencies of great power. We are unable to say, with precision, with whom these new and correct views originated. Like most subjects of this nature, they were, doubtless, the result of gradual development, as facts and illustrations accumulated. Even now, many details have to be collected, ere all the phenomena of this district, the *bouleversements* of its mineral beds, and the repetition of its groups, shall be thoroughly elucidated.

Our own views, in this matter, entirely coincide with those of the State geologist, and we perceive also, that Mr. Logan, provincial geologist of the Canadas, who has examined the Schuylkill coal-field, adopts similar opinions. All recent investigations tend to confirm these conclusions, and to show that the series of coal seams in the Pottsville section, which were formerly considered to be so numerous, and to embrace an astounding thickness of anthracite, ought to be reduced to one fifth, in certain localities, on account of the repeated flexures within the general coal area, which occasion many recurrences of the same seams.*

The necessity for revising all the old statements which may be found, regarding the local topography of the anthracite country, and which have been transferred into works of standard repute, both in America, and in Europe, will appear from a very few examples; but, at the same time, owing to the remarkable subdivision of the groups, it becomes no easy task.

A description of the mines of the southern coal region was officially published some years ago; wherein it was stated that there were at the Room Run mines fifteen beds of anthracite, whose aggregate

* Proceedings Geol. Soc. of London, Vol. III., "On the coal-fields of Pennsylvania," by M. E. Logan, March, 1842, p. 707. The author adverts to the stigmaria beds in this field, associated with the coal, and by them he was enabled to detect the inverted position of the strata.

thickness was two hundred and forty-two feet. Like most of the early reports, which were generally made by unscientific persons, this contained glaring errors and exaggerations; by reason of counting the same seams two, three, or more times over. Another report, from the same quarter, announced the total thickness of coal in nine proved beds, on the north side of the basin only, at one hundred and eleven feet, and ninety feet on the south side; being two hundred and one feet in the aggregate.* This is no otherwise incorrect than as relates to the doubling of the series.

At the commencement of operations at Beaver Meadow, in 1836, the present writer was informed by the engineer, Mr. Wilde, that he had proved three hundred feet of coal there, but it appears to be admitted, now, that about forty feet constitute the principal working series, in that quarter.

A report to the legislature of Pennsylvania, by a committee appointed to examine the Swatara coal mining district, in 1839, announced the presence of seventy veins there, of from three to more than twenty feet in thickness each, then known, and many others were supposed to remain undiscovered.† No accurate transverse section has yet been made, across the field, in the meridian of Pinegrove; but from what we already know of that region, it is probable that at least eight repetitions of the same series of coal seams occur there, and the number of actual and distinct seams is, consequently, lowered to a comparative few.

We shall quit this branch of our subject, for the present; merely observing, by the way, that a clear insight into the actual arrangement of the coal-fields, respectively, can only be completed after the construction of more authentic maps, sections, and diagrams, and by more actual admeasurements, for the purpose of geological elucidation, than are now at hand. A very large and, apparently, very valuable portion of the aggregate area, remains, still, in nearly its original obscurity.

The local maps, which have been issued in a coarse and cheap form, for the temporary requirements of parties having local interests, are somewhat numerous; but they have very little scientific merit or artistical pretension. Illustrations of a far superior character to these are now called for; and the augmented value of mineral property here, would render the cost of such a work, a comparatively light burden upon the owners.

June, 1854.

The following statement upon a subject of increasing interest in coal statistics, has been kindly furnished us by a gentleman accustomed to the scientific examination of coal deposites.

An account of recent shaftings and borings through the coal strata of the Southern Coal-Field near Pottsville, Pennsylvania.

"For some years past there has been amongst land-holders, coal

* Report of the Lehigh Company, January, 1844, p. 24.
† Report to the Legislature, by H. K. Strong, 1839, p. 29.

operators, geologists, and others interested in the development of the coal region, a growing sense of the importance of exploring the deeper lying veins, which geology taught us to believe must exist near the bottom of all the coal-basins. These explorations became the more imperative, as our mines near the surface were being exhausted. Of course the shafts or borings intended to develop the lower coal strata were generally begun upon those estates where a point could be selected near the outcrop of the lower large white ash coals. As in other situations, these veins are not brought sufficiently near the surface by the anticlinal axis or rolls of the strata, to make them particularly valuable for present operations. The first shaft was commenced by Alfred Lawton, Esq., in 1845, on the west side of Mill Creek, not far south of Mine Hill, upon the lands of Henry C. Carey, Esq. Near this point the Mammoth White Ash vein is extensively worked above water level. This shaft was sunk 72 feet, then a boring commenced and penetrated to the depth of 122 feet, into the Primrose vein, after which it was abandoned and remained thus until 1851, when E. W. McGinnis, Esq., resolved to prosecute the exploration by sinking the shaft to the Primrose, and then boring to the Mammoth vein. The success of this enterprise was so decided, that it was immediately determined to continue the shaft ($10\frac{1}{2}$ by 18 feet) down to the Mammoth vein. The following table exhibits the total thickness of various kinds of strata found during the process of boring, and indicates the depth from the surface at which some of the layers were found:

	Thickness.			Depth from Surface.		
Rock and slate, - -	119	feet.				
Primrose coal, - -	3	"		122	feet.	
Rock and slate, - -	68	"				
Holmes coal, - -	4	"	6 in.	194	"	6 in.
Rock 64 feet, slate 10 feet,	74	"				
Coal, - - - - -	1	"		296	"	6 in.
Slate, - - -	51	"				
Coal, - - - - -	1	"	8 in.	322	"	2 in.
Rock 54 feet, slate 18 feet,	72	"				
Coal, 7 feet vein, -	8	"	6 in.	402	"	8 in.
Slate, - - - - -	14	"				
Mammoth White Ash,	22	"		438	"	8 in.

But to preserve the order of time, we should observe that before continuing the enterprise of Mr. Lawton, Mr. McGinnis had prosecuted an extensive boring operation on the East Norwegian Creek, upon the lands of the Delaware Coal Company. This exploration begun in 1850, by a shaft which was sunk to the depth of 170 feet, when the search was continued by boring to a total depth of 656 feet. The following table shows the results obtained:

	Thickness.	Depth from Surface.
Slate and sandstone, -	42 feet.	
Coal and black dirt, -	18 "	
Rock and slate, with iron ore,	431 " 8½ in.	491 feet, 8½ in.
Black dirt, cut May 1st, 18	6 " 6 in.	498 " 2½ "
Slate, 4 ft. 6, and rock, 4 ft. 6,	25 "	523 " 2½ "
Coal, - - - - -	5 "	528 " 2½ "
Rock and slate, - -	127 " 9 in.	656 "

This boring terminated in slate. Being located farther south from the outcrop of the Mammoth vein, than the first mentioned shaft, it was necessary to penetrate deeper in order to reach that vein. It is, however, likely that, at no very distant period, the Delaware Company will be justified in continuing this shaft to reach the large White Ash vein.

The third deep boring was undertaken by the North American Coal Company, who employed Mr. P. W. Shaeffer, late of the State Geological Survey, to direct the operation. After careful exploration of the surface, he located upon the Company's land in the valley of Crow Hollow, just below the celebrated Primrose vein. This boring was commenced in October, 1852, and finished in 1853. The following table shows the results:

	Thickness.	Depth from Surface.
Sandstone, slate and rock,	88 feet, 1 in.	
Holmes coal, - - -	4 " 2 in.	92 feet, 3 in.
Sandstone and slate, -	57 " 5½ in.	149 " 8½ "
Dark brown slate, with iron ore and a seam of coal,	35 "	184 " 8½ "
Slate, with leader of coal,	10 " 2 in.	194 " 10½ "
Rock and some slate, -	189 " 1½ in.	384 "
Slate, - - - - -	21 "	405 " to the

Mammoth White Ash vein.

These are but the beginning of a series of enterprises for the development of our lower large coal seams, which must undoubtedly form the chief source of supply for future demands."

CAPITAL EMPLOYED IN THE PRODUCTION OF ANTHRACITE COAL.

The Statistical Return to Congress, in 1840, exhibits the	Tons.
amount of anthracite raised, - - - - - -	859,686
of men employed in mining it, - - - -	2,977
of capital invested, - - - - -	$4,334,102

Like most other returns on this branch of industry, made at this period, this is evidently below the mark. It was shown, in 1839, that in the Schuylkill mining district alone, capital to the amount of $10,360,555 was invested.*

As far back as the year 1833, when the coal trade was, comparatively, in its infancy, the capital invested in the production of anthracite, and in the means of transportation, in canals and railroads, in the purchase of coal lands, and in working capital, was $19,176,217.†

This calculation does not include the value of store-houses, wharves, landings, &c., in Philadelphia, New York, Boston, and other places; or the value of the vessels and the capital employed in shipping this coal.

In 1839, a statement, emanating from some persons well acquainted with the business affairs of this neighbourhood, showed that the capital engaged in mining and transporting anthracite, including the lateral railroads, engines, cars, wagons, boats, land and houses, amounted to the sum of $7,394,375.‡

In 1842, at a public meeting of persons engaged in the coal trade, in Schuylkill county, 31st January, a report on the coal statistics of that county was made; by which it appears that the capital invested under the foregoing heads, with the addition of the Reading Railroad, the Danville and Pottsville Railroad, and the Schuylkill Navigation, amounted to $17,526,000.

Population engaged in, or entirely dependent on the coal trade, - - - - - - - -		17,000 persons.
Number of horses employed in boating, and at the collieries, - - - - - - - - -		2,100 horses.
Agricultural products annually consumed,	$588,572	
Merchandize consumed, - - - -	918,325	
		$1,506,897

These statements are understood to comprehend only the coal production of the county of Schuylkill.

At that time, there were in use, in this county, thirty steam-engines, amounting to upwards of 800 horse power: 22 of these engines were manufactured there. 850 canal boats, 2,100 horses, 145 miles of railroads, and 3,900 railroad and drift cars, were in full employ: and all this had originated within about fifteen years.§

In 1846, the number of steam-engines employed in the collieries of Schuylkill county alone was 68, having an aggregate horse-power of 2,018; in addition there were constructed 38 others, having a power of 903 horses: in all 106 engines and 2,921 horse-power.‖

M. Chevalier, who investigated the canals, railroads, and resources

* Report to the Legislature of Pennsylvania, on the Swatara Mining District, 1839.

† Report to the Legislature of Pennsylvania, on the subject of the Coal Trade, 1834.

‡ Miner's Journal of Pottsville, January 5, 1839.

§ Miners' Journal of Pottsville, January 1, and February 5, 1842.

‖ An. Rep. Board of Trade, April, 1847.

of the United States, in 1839, remarks of Pennsylvania,—after describing the extent of her canals and railroads, both public and private,—projected and completed,—that such a result would be very remarkable on the part of an ancient people, who had, for a long period, applied themselves to the perfecting of their communications. It appears prodigious, when one remembers, that all these works, with only one or two exceptions, were not commenced,—were not even projected,—in 1825.

"Il faut faire un effort pour concevoir comment une population aussi restrainte que celle de la Pennsylvanie, a pu entreprendre, et achever à peu près, une pareille masse de travaux dans un délai aussi court."*

I. SCHUYLKILL COAL REGION.

CENTRAL AND WESTERN DIVISIONS OF THE SOUTHERN BASIN, OR GROUP OF COAL BEDS, EMBRACING SCHUYLKILL COUNTY.

The annual returns of coal production from the different coal-fields are becoming more difficult of classification, every year, and must of necessity be abandoned for a more general system, especially as relates to the districts usually known as the southern and middle coal-fields. As these regions become blended together by a net-work of railroads, the original distinctions become obsolete, or are lost. Already, for instance, some of the coal of the Wyoming or northern regions descends the Lehigh, and swells the returns made from that channel, which already receives the coal of half a dozen other basins.

So also the Pinegrove region sends part of her coal by the Union canal, and part by the Reading Railroad or the Schuylkill navigation; thus affecting the results of the yearly returns. In like manner, the anthracite of the middle region will reach market by various routes, and be blended with the contributions of all the other districts.

Annual Production of anthracite sent, by canals and railroads, to market from the mines, exclusive of the coal consumed upon the spot, or in the vicinity, from the commencement. This return from Schuylkill county embraces also that part of it, which is known as the Swatara or Pinegrove district, and also the Little Schuylkill district.

We have adopted the returns of the Pottsville Miners' Journal and Reading Railroad Reports, including in the table from 1837, the returns of the Pinegrove district. The amount of the Little Schuylkill district will be found below.

* Histoire et description des voies de communication aux Etats-Unis, tome premier, p. 542.

Years.	Tons.	Years.	Tons.	Years.	Tons.	Years.	Tons.
1825	6,500	1833	252,971	1840	476,151	1847	1,650,831
1826	16,767	1834	226,692	1841	602,545	1848	1,714,365
1827	31,360	1835	339,508	1842	573,273	1849	1,683,425
1828	47,284	1836	432,045	1843	700,200	1850	1,702,926
1829	79,973	1837*	540,152	1844	874,850	1851	2,184,240
1830	89,984	1838	446,875	1845	1,131,724	1852	2,517,493
1831	81,854	1839	465,247	1846	1,295,928	1853	2,551,603
1832	209,271						

The production of anthracite from the *eastern or Lehigh division* of the southern basins, and central group of coal beds, will be separately arranged under the returns from the Lehigh district, in a succeeding page. The return for 1847 was 643,973 tons, and the aggregate, since the opening of the trade to the 1st January, 1854, is 9,756,598 tons.†

The following table embraces the Little Schuylkill or Tamaqua coal district, having its outlet at Port Clinton, which, in 1847, produced 106,401 tons, and altogether from the opening of the district, in 1832 to 1854, inclusive, furnished 2,247,446 tons, as its quota. The arrangements, in progress at this establishment, will ensure a continued accelerated supply for the future.

Years.	Tons.	Years.	Tons.
1832,	14,000	1843,	31,000
1833,	40,000	1844,	57,345
1834,	34,000	1845,	75,998
1835,	41,000	1846,	91,007
1836,	35,000	1847,	106,401
1837,	31,000	1848,	162,626
1838,	13,000	1849,	174,758
1839,	9,000	1850,	211,960
1840,	20,000	1851,	310,307
1841,	40,000	1852,	325,099
1842,	37,000	1853,	‡384,443

We have stated that the Schuylkill coal-field supplied three species of anthracite. The environs of Pottsville furnish two of these; and we are indebted to the editor of the Miners' Journal, of that place, for a statement of the relative proportions of each kind, that were sent from thence, in the year 1846.

		Including Pinegrove.
White ash coal,	703,000 tons.	703,000 tons.
Red ash coal	539,000 "	592,928 "
	1,242,000 "	§1,295,928

The journal last named informs us that the number of colliery

* Pinegrove included in the following years.
† Reading R. R. Report, 1854.
‡ Report of Little Sch. Nav. Co., 1854.
§ Miners' Journal, January 30th, 1847.

establishments in Schuylkill county, in 1846, was 142, including ten in the Swatara region: of these 35 collieries are below the water level, viz., they are drained by steam power. There were, further, 22 collieries in preparation, of which ten are below the water level.

The following is a summary of a table contained in the columns of the Pottsville Miners' Journal, compiled by Messrs. Barnum, C. W. Peale, and J. M. Wetherell, embracing all the collieries in the Schuylkill Coal Region,* up to June, 1853, with two collieries on the Lorberry Creek Railroad. There are about two miles of underground railroad in the Lorberry Creek region, not in the table. From this chart we sum up the following information:—

Total number of collieries, - - -	113
Red Ash, do. - - -	58
White Ash, do. - - -	55
Number of operators, - - -	82
Underground railroads, miles, - -	124½
Of which through solid rock, do. - -	6¼
Steam engines employed in mining, - -	210
Aggregate horse power, - - -	7,071
Equal to man power, - - -	42,426
Power for hoisting and pumping, horses, -	3,805
For pumping only, - - do. -	1,375
For breaking and screening coal, do. -	1,891
Miners and labourers employed at collieries, -	9,792
Horses, - - - - -	468
Mules, - - - - -	569
Miners' houses out of towns, - - -	2,756
Whole capital invested in these collieries, -	$3,462,000
By individual operators, about - -	$2,600,000
Deepest slope, yards, - - -	353
Shortest do. - - - -	33
Thickest vein, worked at Hecksherville, feet,	80
Smallest, " " "	2

"All the coal lands now worked in the county are owned by six corporations and sixty individuals. About twenty-five of the owners reside in Schuylkill county, and the balance abroad.

"Not one solitary ton of coal was mined by any corporation in Schuylkill county during the year 1853—the whole product of two millions five hundred and fifty-one thousand six hundred and three tons was mined by individuals.

"The coal rent will average about thirty cents a ton. The product of 1852, in Schuylkill county, was $2,551,603 tons. This would give an income of $765,480 to the landholders, in the shape of rents, for the year."

During the agitation of the tariff, in 1846, at Washington, it was stated by Mr. Cameron, of Pennsylvania, that thirty years ago, coal

* First Anthracite coal field, whose outlets are at Mount Carbon, Port Carbon, Schuylkill Haven, Port Clinton, and shipped by Reading Rail Road and Schuylkill Canal.

was entirely unknown in this country; yet, in 1846, it gave employment to five millions of days' work, annually. It kept in movement a thousand ships of a hundred and fifty tons each, and afforded a nursery for the training of six thousand seamen, who earned three millions of dollars yearly. It gave circulation to a capital of fifty millions of dollars. It kept in activity fifteen thousand miners, and sustained a mining population of seventy thousand souls, who annually consumed upwards of two millions worth of agricultural productions, and more than three and a half millions of dollars worth of merchandise.

Schuylkill County, Pa.—This county has an area of seven hundred and fifty square miles. Elevations above tide:—Port Clinton, 400 feet; Pottsville, 610 feet; summit of Broad Mountain, 1653. Drainage, north, south, east and west. Fall of the Schuylkill river in the county, from Pottsville, 220 feet. Population of the county about 65,000.

NUMBER OF OPERATORS IN THE SCHUYLKILL COAL REGION.

We observe, in a statistical report of this district,* some details, which appear to be deserving of a passing notice. "There were only one hundred operators engaged in mining coal, in the whole Schuylkill region, during 1847. As the expenses of mining increase, the number of operators are gradually diminishing. This is apparent from the fact, that, although the number of *collieries* have increased during the last year, the number of *operators* in the region have diminished, down to about one hundred. Three years ago, they numbered about one hundred and forty. Some of the larger operators now work five or six collieries."

The Schuylkill county collieries are situated as follows:—

	1848.	1850.	1852.
Above water level,	101	120	62
Below "	42	44	49
Total, -	143	164	111

Price of Labour in the Coal district of Pottsville.

Periods.	Wages per day.		Observations.
	Miners.	Labourers.	
	Dollars.		
1831	1.00	.82	The tariff of 1828, on coal and iron, in full force.
1840	1.00	.80	The reduced tariff in operation. Wages paid in goods, making a difference of 15 to 20 per cent. against the labourer.
1842	87½	·70	Paid in traffic—one half the labourers had no employment.
1844	1.10	.76	The tariff of 1842 took full effect; all were employed, and labour was in demand.
1845	1.13	.80	Business continued as in the previous year.
1846 } 1847 }	1.25	.83	Business improving in activity; all wages paid in money, as has been the case for four years.

* Pottsville Miners' Journal, January, 1848, and 1851.

The details of the foregoing table were arranged, from the books of one of the most important coal companies in the United States, with a view to show that there had been an advance in the price of labour in the mineral districts of Pennsylvania, under the influence of the tariff of 1842; a state of things which had been assumed, somewhat theoretically, to be unfounded.*

We have ventured, towards the commencement of this volume, to suggest, for reasons assigned, that the imposition of duties on imported bituminous coal has had, and will continue to have, very little influence on the prices of anthracite.

Table of Prices of Anthracite in Philadelphia, New York, and Boston, in the following years.†

Years.	Philadelphia. Wholesale, per ton of 2,240 lbs.	New York. Retail, per ton of 2,000 lbs.	Boston. Retail, per ton of 2,000 lbs.
	Dollars.	*Dollars.*	*Dollars.*
1839	5 50	8 00	9 00 . . to . 10 00
1840	5 50	8 00	9 00 . . to . 11 00
1841	5 00	7 75	8 00 . . to . 9 00
1842	4 25	6 50	6 00 . . to . 6 50
1843	3 50	5 75	6 00 . . to . 6 50
1844	3 37	5 50	6 00 . . to . 6 50
1845	3 50	5 75	6 00 . . to . 7 00
1846	4 00	6 00	6 50 . . to . 7 00
1847	3 85 to 4 00	5 50 to 6 00	6 50 . . to . 7 00
1852	3 75	6 00 to 7 00	
1853	4 25 and 50		7 00 . . to . 8 00
1854	4 50 to 75		8 50 . . to . 9 00

In 1840, mining labour was $5.00 to $7.00 per week; in 1846, $8.00 to $10.00 per week.‡

SWATARA AND PINEGROVE DIVISION OF THE SCHUYLKILL REGION.

Having its outlet by the Union Canal.

Proceeding westward, after leaving the Schuylkill river, we arrive at the Swatara Coal District, which in 1839, and nearly for the first time, was investigated by direction of the legislature.§ Although possessing a great number of very excellent coal seams, the business enterprise has heretofore greatly languished, on account of difficulties, both local and temporary, in this quarter; especially those arising from inadequate facilities of transportation. Many of these disadvantages are already overcome, and the rest will vanish in proportion as the near prospect of profitable remuneration awakens the energy of the proprietors and adventurers.

Many associations for coal mining, canal or railroad purposes, have

* Pottsville Miners' Journal, June, 1846, and subsequently.

† Report of the Board of Trade of Schuylkill county, April, 1846, and subsequently.

‡ Speech of Mr. Cameron, in Congress, July 22d, 1846.

§ Report to the Legislature of Pennsylvania on the Swatara Mining District, 1839, p. 29.

here invested capital, and these are already busily occupied in carrying out their respective undertakings.

We insert the following statement of the quantity of anthracite which was sent down from the Pinegrove or Swatara district, by the Union Canal and Feeder, from the Pottsville Miners' Journal. We will observe that the first communication between the Swatara coal region and the Union Canal, was effected in 1833. In 1847, a railroad being completed from the district around the head waters of the Swatara and the Reading Railroad, 42,145 tons of anthracite were conveyed hence to market at Philadelphia, by the Schuylkill route, in addition to 67,437 tons via the Union Canal.

Years.	Tons.	Years.	Tons.	Years.	Tons.
1834	6,911	1841	17,653	1848	61,530
1835	13,891	1842	32,381	1849	78,299
1836	11,710	1843	22,905	1850	70,919
1837	17,000	1844	34,916	1851	
1838	13,000	1845	47,928	1852	75,000
1839	20,639	1846	58,926	1853	80,660
1840	23,860	1847	67,457		

Within the area drained by the Swatara river and its tributaries, the main coal-field separates into two forks, one striking towards the north-west, the other directed towards the south-west, and stretching almost to the Susquehanna. The north fork, which is a double basin, is eighteen miles in length, and the southern one, which is a single basin, is thirty miles; and at the termination of the former, they are more than ten miles apart.

The entire breadth of the Swatara region is occupied by a series of anticlinal and synclinal axes, by means of which the numerous coal seams are repeatedly brought to the surface; exhibiting, in the aggregate, indications of a vast amount of anthracite.

Fig. 9.

Section of the Swatara Coal Field—Pennsylvania.

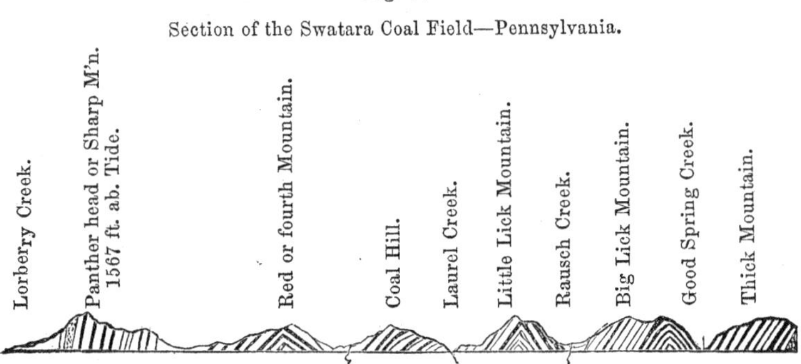

Heretofore, the wild and broken character of the surface has opposed difficulties to the complete geological development of the numerous highly inclined strata. It is only by means of the gradual progress of local improvement,—of the operative mining industry of of the country, and of a more correct system of surveys,—that we shall acquire a thorough insight into the structure of this region.

The committee on the Swatara district, in 1839, were compelled from the difficulty of the task, to abandon all expectation of obtaining a list of the coal seams. They stated, however, that about forty beds, of from three to more than twenty feet in thickness, had then been opened. The lowest number of seams in a condition to be worked, they estimated at seventy, without passing into the North or Broad Mountain. We know well at the present day, the erroneous nature of this view, and that what was then taken to be one general series or continuous group, was in fact a series of repetitions of a much smaller number, as we have already suggested.

From the very remarkable physical characters of the Swatara region, at the point or pivot from whence diverge the two forks, of which we have spoken, it would be but reasonable to expect an unusual extent of disturbance through the crushing influence of a movement so stupendous as we perceive must have there taken place. We could scarcely fail to suspect the prevalence of such phenomena as frequent lines of dislocation and cross fracture; and, accordingly, we perceive abundant evidence of the fact, in the numerous transverse ravines, of great depth, at right angles to the strike of the strata, and now forming so many outlets or channels for the drainage of the interior area. The thirty-six breaks or gaps, which in 1839 were viewed as particularly advantageous circumstances, favourable to the mining economy of the district, we are led to believe, mark, with unerring exactness, the sites of original fissures or fractures, crossing perpendicularly the longitudinal axes of the coal measures. Whether, as has been suggested, there has been any lateral movement or heaves, to disturb the linear continuity of the strata, we are at present unaware. Hitherto we have perceived no evidence of such a tendency; we are only surprised to observe so little interruption to the general range of the coal beds.

DAUPHIN COUNTY COAL DISTRICTS.

The North-West, or Bear Valley Fork, in Schuylkill and Dauphin counties, will, before long, be the theatre of very extensive mining operations. It contains a numerous suite of coal seams, two or three times repeated; some of these are of great power, and all of them are of excellent quality, especially adapted to domestic use. They are of the white or gray ash variety of anthracite, but of easy combustion; while those at the southern side of the region, towards the Sharp Mountain, consists of the red-ash and free burning kinds; a description which is highly esteemed for domestic fuel, but not so efficient for the iron furnace. An investigation of the western extremity of this district, in the fall of 1846, has been the subject of an elaborate report by the present writer, addressed to the Lyken's Valley Coal Company.* About 60,000 tons of coal were mined here prior to 1848, and sent down by the Wiconisco Railroad. Since then the amounts are as follows:†—

Year	Tons
1848	
1849	25,325
1850	37,763
1851	54,200
1852	59,857
1853	69,007

Fig. 10.

Section of the Black Spring Gap, Lebanon county, looking West.

Sharp or Third Mountain. Fourth or Red Mountain.

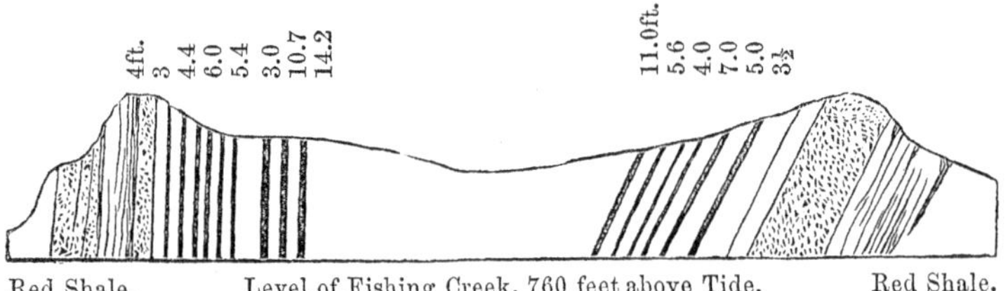

Red Shale. Level of Fishing Creek, 760 feet above Tide. Red Shale.

Scale 1600 feet to an inch, both vertical and horizontal.

The South-west Fork terminates, acutely, within two miles of the Susquehanna river, and formed the subject of two very detailed reports, in 1840, by the present writer.‡

After passing Fishing Creek Gap, and entering Lebanon county, at Black Spring Gap, fourteen beds of coal, viz. eight on the south side and six on the north side of the axis, were proved; whose aggregate thickness of freeburning red-ash coal, was ninety-one feet. At Rausch Gap, nine southern seams were proved. In number and volume the series then decreases, as we proceed through Dauphin county, and the mountain contracts in breadth, until, at eight miles from the Susquehanna, it presents a very slight prospect of advantageous working, further to westward. Finally, it contracts into a narrow insignificant ridge, terminating entirely before reaching the Susquehanna river.

The Dauphin and Susquehanna Coal Company's lands, which occupy more than one-third of the length of the whole of this coal-field has been further proved, since the publication of the first edition, in running several tunnels across the southern part of this basin (as suggested by Mr. Taylor) by which three other veins lying above those of the section of Rausch Gap, on the Company's lands, have been discovered. One of these tunnels, in Yellow Spring Gap, four

* Report to the Lyken's Valley Coal Company, by R. C. Taylor, December 16th, 1846.

† Pottsville Miners' Journal, 1854.

‡ Reports on the Mineral Lands of the Stony Creek Estate, and those of the Dauphin Coal Company, by R. C. Taylor, President of the Board of Directors, 1840.

miles west of Rausch Gap, cut through a mass of semi-bituminous coal fifty feet thick, the declination of the rocks on each side indicating it to be a synclinal axis. Of course the same number of veins, twelve or thirteen on the south side of the valley, will be penetrated on the north side by this tunnel, now driven about sixteen hundred feet in.

The railroad communication to Baltimore on the one side, and to Philadelphia on the other, is now complete, and the coal from this west end of the basin—anthracite, transition semi-bituminous—is very much liked, and commands the highest price for steaming and other purposes.

The Dauphin Coal Company's mines, now in operation, sent to market

in 1851	- - -	20,000 tons.
1852	- - -	33,639 "
1853	- - -	29,000 "

CHANGE OF CHARACTER IN THE COAL OF THE SOUTH-WEST FORK.

It is within this small district, along a distance of more than ten miles, that the coal parts with a portion of its anthracite character, and assumes that of a semi-bituminous, dry, and blazing coal; resembling the steam coal of South Wales, and having a slight tendency to adhere or cake. It possesses properties which eminently entitle it to the reputation it has already acquired, as a valuable species of fuel; the more remarkable as occurring in an anthracite region, and as being part of the same beds which consist of compact, pure anthracite, as we advance towards the Lehigh.* The gradations and modifications of this semi-bituminous Stony Creek coal, adapt it for a variety of purposes, and a selection of any required quality can here be made. Our table of Assays, in another part of this volume, shows this transition with great distinctness.

Among the early notices of the passage from the purest known anthracite to a species which is not only less dense and ponderous, but contains a gradually increasing portion of volatile matter, we find that of Mr. Packer's Report, in 1834, who adverted to the already well-known change in the condition and character of the Schuylkill coal, when traced from the eastern to the western extremities of the region. His remark, in substance, is, that " extending westward the coal becomes somewhat lighter; the specific gravity of the Mauch Chunk coal being 1.494; the Schuylkill, in the vicinity of Pottsville, 1.453; and the Pine Grove, Wiconisco and Stony Creek, about 1.400. The latter is somewhat more inflammable and easy of ignition, or, to use a prevailing idea, 'partakes more of the bituminous character.' "

This bituminous character of the last mentioned coal had been investigated, some time previously, by Professor Renwick, and was

* Report, Tables, iii., ix., x., xi., xii., and xix.

especially adverted to by him in a printed circular, in 1832; also, by Prof. W. R. Johnson, in his letter, dated April, 1838, to the National Foundry Committee.

CHANGES AND VARIATIONS IN THE MINERAL CHARACTER OF COAL SEAMS.

In an appendix to the Stony Creek Report, already alluded to, the writer, who was familiar during twenty-five years preceding with precisely similar phenomena in the basin of South Wales, endeavoured to illustrate these parallel cases by a series of analyses both of American and Enropean varieties of coal. These details were derived from well known authorities. Following up the plan, we have prepared a table of at least eleven hundred analyses of coals in all parts of the world, as a fitting and elaborate illustration of the present work. These tables will be found at the end of this volume. The authorities cited are the following:

In Europe.—Mushet, Ure, Dufrenoy, Berthier, Thomson, Schafhæutl, Robin, Karstén, Ffyfe, Richter, Kirwan, Richardson, Murchison, Lyell, Varin, Baudin, Piot, Elie de Beaumont, Flachat, Logan, Liebig, Pelouze, Regnault.

In America.—Troost, Ducatel, Jackson, Hayes, Clemson, Ellet, Johnson, Frazer, Shepard, Olmsted, Bache, Silliman, Rogers, Booth, Lea, Renwick, Chilton, Boyé, Owen.

To all these we must especially add the highly comprehensive and instructive memoir of Prof. H. D. Rogers, in the Transactions of the Association of American Geologists, on the Appalachian Coal Fields. This paper is a rich contribution to American geology; and it is due to the writer thereof to state that his views on the subject of the transition from anthracite to bituminous coal had been stated through the medium of his public lectures, in 1837, and were more fully developed in 1843.

For local statistics we have frequent acknowledgments to make to the newspaper press, among which we are bound to name the Miners' Journal, and the Anthracite Gazette, of Pottsville; the Mining Journal, of New York; and the Commercial List and other leading journals of Philadelphia. From such prolific sources as Hunt's Merchants' Magazine, Hazard's Register, the American Almanac, &c., we have been permitted to glean many valuable details. To proceed:—

That changes in the composition of the coal, from a partially bituminous state to that of anthracite, occur in other parts of the world, several instances will be found in these pages. Those of South Wales and of the Donetz coal-field in Russia, and on a smaller scale in several of the basins of France, are cases in point.

In like manner the flaming coals of the Auvergne and Bourbonnois basins, are shown to contain a complete series of dry coals with a short flame; of fat coals with short flame; of fat with long flame; and dry with long flame. Also, in the basin of Commentry, we find, at some hundred metres of distance, and upon the same margin of that basin, the anthracite of Chambled, and the dry coal with long flame of Ferrières.*

The Schuylkill coal region possesses striking points of resemblance, although upon a somewhat different scale, to the great coal basin of South Wales. The latter is ninety-four miles long, and averages about fifteen miles in breadth. Of this elongated area, the western seventy-four miles consist of anthracite beds, with the exception of some partly bituminous seams on the south side of the basin. The remaining twenty miles, on the east, consist of dry bituminous coals, semi-bituminous coals, and steam coals; but, occasionally, the same section yields beds of coal of modifications of all these varieties.

The Schuylkill region is sixty-five miles long; extremely attenuated for eight or ten miles, at its western extremity, so as to be unproductive there to any important amount. For about eighteen miles out of the sixty-five, at its western prolongation, the quality of coal is of an intermediate character, like the steam coal of Wales. It then gradually passes to free burning anthracite on entering Schuylkill county, from Lebanon; and, still further east, to the hardest and purest anthracite. We have, elsewhere pointed out a similar passage from the extremes of each quality, like those in Russia, and we add that of the coal basin of St. Gervais, in Hèrault, France.

GEOLOGICAL MODELS.

In intimate connection with the physical and economic geology of the Schuylkill coal region, the author may be here permitted to advert to a species of illustration, for the first time introduced in America, by himself. In 1840, a geological model was constructed of the western half of the Schuylkill coal district and its vicinity, upon a scale of two inches to the mile. The area so represented comprises seven hundred and twenty-three square miles, or 460,800 acres, being in breadth sixteen miles, and in length forty-five miles. This model was first exhibited to the Association of American geologists and naturalists, at their meeting in Philadelphia, in April, 1841, to illustrate an address " on the most appropriate modes of representing geological phenomena."†

Dr. C. T. Jackson, in his first general report of New Hampshire, in 1841, also takes occasion to recommend, strongly, the process of geological illustration through the aid of models. He, however, states his regret at the expensive nature of such works, and the length of time requisite for their completion.‡

* Annales des Mines, Vol. I., 1842, p. 96.

† Silliman's American Journal of Science, Vol. XLI., p. 81, 1841. Also, Transactions of the Association of American Geologists and Naturalists, Vol. I., p. 81.

‡ Report of New Hampshire, by Dr. C. T. Jackson, 1841, p. 36.

NOTE.—In Europe, geological models and models exhibiting lines of railroads, mines and harbours, and even estates, are much coming into use; although the frequency of employing so desirable a mode of representation is impeded by the heavy cost of such works. This objection applies with even stronger force in the United States, where capital is less abundant and where economy is indispensable.

We perceive by a paper read to the Institution of Civil Engineers, in London, "on the

II. MIDDLE ANTHRACITE DISTRICT.

FIRST SUBDIVISION. THE SHAMOKIN COAL-FIELD.

In proportion to its magnitude, this is the richest and most regular of all the Pennsylvania basins. The coal seams are unusually abundant. The writer ascertained the outcrops of upwards of twenty, towards the centre of the basin, in 1847, and more remain to be developed. Some of these are of large size, one being twenty-seven feet; and one, much larger, is called the Big or "Mammoth Vein." Its maximum thickness is said to be here about fifty feet.

At Bear Gap, towards the western termination of this field, is a fine natural display of outcrops of all dimensions. Like all the anthracites in their progress westward towards the Susquehanna, their specific gravity diminishes, and they become softer. "The coal of the extreme western end of the basin visibly partakes of the characteristics of bituminous coals; it crumbles to some extent during combustion, and it contains carburetted hydrogen gas; this ingredient constitutes nearly 1.11 part by weight of the coal of this locality."* A cubic yard of this coal, here, weighs one ton and seventy-three pounds. Specific gravity, 1.371.

At Shamokin, an immense quantity of coal occurs in the northern rise of the strata, which are here displayed in a double axis. The beds vary from five to eleven and a half feet in thickness, with the exception of the "Mammoth Vein," which consists, it is said, of sixty feet of coal of various qualities and degrees of purity. In general, the Shamokin coal has a white ash: it is a free-burning anthracite, of fair repute in the domestic market, and from the vicinity of Mount Carmel eastward; it is sufficiently hard for smelting iron. The large vein appears to be identical with that seen at Tamaqua, at the Swatara region, in the Wyoming coal-field, Beaver Meadow, &c. The direction of the Shamokin beds is N. 82° E.

Amount of coal sent to market from the Shamokin district, via a railroad of sixteen miles to Sunbury on the Susquehanna, for the nine years in operation, from 1839 to 1847, inclusive, 119,311 tons.

construction of geological models," by Mr. J. B. Denton, May, 1842, that the cost was stated at from 2*s.* 6*d.* to 3*s.* 6*d.* = $0.60 to 75 cents per *acre*. Of course, the expense is influenced by the scale on which the mode is projected. As Mr. Taylor's model of the Schuylkill region contains 460,800 acres, it would amount to a considerable sum, even at one-fiftieth part of the above named prices.

Again, the author referred to, states that a model, showing the line of a railroad or canal, would cost ten pounds [£10,] per mile. There are about one hundred and seventy miles of canals and lines of railroads and projected railroads shown on Mr. Taylor's model. If these rates include all the preliminary expenses of surveys and the collection of details, they are perhaps not so greatly overrated.

Since writing the foregoing paragraphs we have seen the commencement of a model of the Shamokin and Mahanoy coal basins in Pennsylvania, and rejoice at the adoption of this excellent mode of representation.

We refer our readers to an article containing much information on the comparative value of different kinds of coal for the purpose of Illumination, by Andrew Fyfe, M. D., F. R. S. Jameson's Edin. Journal, Vol. XLV., pp. 37—49. Silliman's Journal, January, 1849, pp. 77—86; March, pp. 157—167.

* Pottsville Correspondent.

This coal is of the white ash species. We have seen an analysis, which assigns 89.99 as the proportion per cent. of carbon in the harder variety. There were sent to market in

	Tons.
1848, - - - - -	19,356
1849, - - - - -	19,650
1850, - - - - -	19,921
1851, - - - - -	23,989
1852, - - - - -	25,846
1853, - - - - -	15,500

"Arrangements are making to smelt iron in the Shamokin basin with the coal and ore found there, where there are facilities for additional supplies from the rich ores of Montour ridge, for which coal could be exchanged. In fact, coal is now transported from the Shamokin basin to the iron works of Danville, distant thirty miles, the distance to the Wilkesbarre basin being fifty miles. Hitherto the latter region was depended upon for the supply of the various works at Danville, which required 200,000 tons annually."*

Until the year 1854, no extensive mining operations had been undertaken in the Shamokin coal-field. At present very large preparations are in progress for the shipment of coal from this region, and another year will probably suffice to put the collieries lately commenced, in a condition to furnish great quantities of coal to the several markets. The various improvements by which the products of this basin will be brought to market, are either finished or are rapidly approaching completion. The total amount of coal shipped from this region during the six years from 1847 to 1853, inclusive, was 134,172 tons. This coal was transported to Sunbury over the Sunbury and Shamokin rail-road, at that time known by the name of the Pottsville and Danville Rail-road, but never entirely completed. The communication between Sunbury and Baltimore, was by canal via Havre de Grace. The following companies have been formed for the development of the Shamokin coal-field; where extensive arrangements and large investments have been made for the efficient prosecution of their business.

The Locust Mountain Coal and Iron Company, owning 6250 acres of land, a portion of which lies in the Mahanoy region. This company is constructing three coal-breakers, a steam saw-mill, and one hundred miners houses. It has an outlet by the Coal-run, Mine-Hill, and Sunbury rail-roads. (See the table of analysis.)

The Coal-Run Improvement Company, (adjoining the preceding,) owns 3000 acres, of which 2200 acres is coal land. Three collieries, with breakers, and fifty miners' houses, are in preparation. The Coal-run rail-road, ten miles long, is on this tract, and it has access

* Pottsville Correspondent.

to the New York, and Catawissa, and to the Philadelphia and Sunbury rail-roads.

The Green Ridge Improvement Company, owning 2500 acres, all of which is coal land. It will connect by a seven-mile track with the Philadelphia and Sunbury rail-road. It is constructing two large collieries and coal-breakers, and 100 miners' houses.

The Philadelphia and Sunbury Rail-road Company, owning 3000 acres, all of which is coal land.

The Locust-Gap Coal Company, owning 2000 acres, of which 1000 acres are coal land.

The Summit Coal Company, owning 2000 acres, of which 1000 acres are coal land.

The Furnace-Run Coal Company, owning 2000 acres, of which 1000 acres are coal land. The lands of this company are near the town of Shamokin.

The Big Mountain Improvement Company, owning 6000 acres, of which 2000 acres are coal land. The improvements are two collieries, two coal-breakers, and 100 miners' houses.

The Coal Mountain Coal Company, owning 1000 acres, all of which is coal land. This company has a capital of $375,000, and in twelve months will be working two collieries, one of which will have a breaker. The other is a hard coal, intended for iron smelting, in which large coal is used. In their second year this company expects to mine from 150 to 200,000 tons.

The Carbon-Run Improvement Company, owning 2000 acres, of which 1800 acres are coal land. Its improvements are two collieries, one large coal-breaker, and eighty miners' houses.

The Mahanoy and Shamokin Improvement Company, owning 2064 acres, of which 1200 acres are coal land. Outlet by the Tiverton rail-road to the Susquehanna.

The Zerbe's Run and Shamokin Improvement Company, owning 2234 acres, of which 1400 acres are coal land. Outlet by the Tiverton rail-road.

We refer our readers to a Report of Professor Henry D. Rogers, on the coal lands of the Zerbe's Run and Shamokin Improvement Company, 1850.

Messrs. Longenecher, Baumgardner and Helfenstein, owning 1800 acres.

Messrs. Helfenstein and Boyd, owning 500 acres.

Hon. C. W. Hegins Dewart and others, owning 250 acres.

There have been erected and now finished, or nearly so, within the Shamokin region, during the past two years, not short of 500 houses for the operatives, who will be engaged at the various collieries—we have enumerated more than fifteen different mining operations, each of which requires at least one coal-breaker—some have several breakers, and a corresponding number of workings actually opened.

Besides these improvements there are many accessories required, such as steam saw-mills, smitheries, &c., all of which have been provided to suit the exigencies of the region.

AVENUES TO MARKET FROM THE SHAMOKIN COAL REGION, COMPLETED OR IN A FORWARD STATE IN 1854.

A rail-road twenty-seven miles in length, is finished from Mount Carmel to Sunbury, at the junction of the two main branches of the Susquehanna river. From this point the coal of the Shamokin region will find its way to Baltimore and other southern markets, by the Pennsylvania canal, which follows the course of the Susquehanna river.

As both the main branches of the Susquehanna river, as well as its tributary, the Juniata, are improved by canal navigation, a very large tract of territory which has hitherto had no communication with the coal-fields, will be opened as a market for the products of the Shamokin basin.

Williamsport, upon the west branch of the Susquehanna river, is now the terminus of Williamsport and Elmira railway. This road penetrates the centre of the State of New York, and it will soon be made continuous to the Shamokin region, by its junction with the Sunbury and Erie rail-road at Williamsport. The Susquehanna and Harrisburg rail-road is now far advanced, and when finished, it will afford a means of transport from Sunbury to Baltimore, and intermediate points. Thus Sunbury will form a centre, from which the trade will diverge in both directions, either by rail-road or canal.

Upon the completion of the Sunbury and Erie Railroad, the coal of this region will find a transport to the Lake and intermediate points. The distance from Sunbury to Erie will not exceed 270 miles. Several facts of importance have been made public by Mr. Longenecker, the president of the Philadelphia and Sunbury Railroad, in an able discussion of the business prospects of the Sunbury and Erie road. It appears that when this line shall be opened, that a large demand for coal may be expected from the Lakes and the Lake country. Already as much as 300,000 tons of coal have been sent in a single year from Philadelphia to the Western Lakes, and delivered, at a cost of more than $8 per ton, at Lake Erie. The total distance of transport could not be less than 700 miles. The impetus given to a trade which can struggle into life against such obstacles must be great when a first class railroad shall give free play to its hampered movements.

Mr. Longenecker shows that cars of the Sunbury and Erie Railroad can go out loaded, and return, not only with cargoes paying expenses, but freighted with such goods as will pay a more profitable remuneration than the coal carried westward. Comparing the actual length of the route from Mount Carmel, as a central point of the Shamokin Basin, to Erie, as a favourable coal depot upon the Lake, with the actual length of other routes from the Lackawanna region, and others to the nearest harbour upon Lake Erie, Mr. Longenecker finds the result favourable to the Shamokin region, whilst he presumes that the grades of the Sunbury and Erie Railroad, and other advantages which will appear in the working of the road, will secure to the Shamokin region a better outlet to the Lake market than is enjoyed by any rival coal-fields.

Besides the railroad from Mount Carmel, another, 13 miles in length, extends from Treverton, near the western end of the Basin, to a point on the Susquehanna River, 11 miles below the town of Sunbury. This railroad intersects the other great works just mentioned by which a northern or southern market will be reached. A bridge is nearly completed which will connect this railroad with the Pennsylvania Canal upon the opposite side of the Susquehanna.

A railroad will also be completed next summer which will extend from Mount Carmel to a point on the line of the Catawissa Railroad—a distance of 11 miles.

By this connection the Shamokin region will have an outlet to New York City.

The Mine Hill and Schuylkill Haven Railroad is already extended to Ashland, 22 miles from Schuylkill Haven. It is believed that this road will shortly be continued five miles further to enter the Shamokin coal-field. When this shall be done there will be an avenue to market at Philadelphia over the Reading Railroad.

A projected improvement which is likely to be undertaken, and which will open a desirable avenue of trade to the Shamokin region, is a railroad to connect this coal field with the Delaware River at Delaware City. This point would constitute a favourable depot for supplying steam vessels, for shipment of coal to all ports now reached from the Delaware River, and for a brisk local trade upon that stream. A reconnoissance of the ground has established the fact that such a road would possess a very advantageous descending grade from the coal region to the river, a circumstance, the value of which has been exemplified in the case of the Philadelphia and Reading Railroad.

SECOND SUBDIVISION, OR MAHANOY BASIN.

This area contains a splendid suite of anthracite beds, none of which have been yet worked, nor has there been made a railroad for conveying the coal to market; so that, at the time we write, this valuable coal basin remains entirely unproductive.

Advancing eastward, we arrive at the Girardville coal works, now temporarily abandoned for want of the means of transporting the coal, but where an enormous development is exhibited. In structure and quality, the anthracite is more solid and ponderous than at the positions mentioned westward; a cubic yard weighing one ton and four hundred and sixty pounds. Specific gravity 1.600.

This extensive coal property, upwards of fourteen miles in length, was the munificent gift to the city of Philadelphia, of the celebrated Stephen Girard, and is destined, ere long, to yield a splendid addition to the corporate revenues. Some years ago it furnished a considerable quantity of anthracite to Philadelphia, but the business has been for some time suspended.

THIRD SUBDIVISION. EASTERN GROUP OF SMALL BASINS.

Retaining, for the sake of convenience, the topographical arrangement heretofore in general use when speaking of the Pennsylvania anthracite districts, we have to remark that this is much more complex in its physical features than the southern region, which we have previously noticed. We shall convey the best idea of this middle region, by describing it as made up of a somewhat numerous series of troughs or elongated basins, and separated from each other by as many anticlinal axes or flexures, which commonly bring up to the summits of the ridges the underlying red sandstone and shales. So frequent are these undulations, that on one estate of eight miles in length which has come under our examination on the eastern portion of the region, we observed six or seven of these nearly parallel basins, whose prevailing direction is about ten degrees to the north of east. The general surface is thickly covered with diluvium, so as to conceal the outcrops of the coal seams; but in general there is an unusual agreement between the configuration of the surface and that of the basins themselves: the coal measures, which are the highest strata in geological position, occupy the bottoms of the ravines, while the red shales and sandstone, upon which the anthracite is imbedded, ascend from beneath them to the highest crests of the mountains. Thus, therefore, we learn to regulate our researches by the general rule, and seek for the superior rocks in the lowest positions, or the synclinal valleys.

Taking these local basins either as a group or individually, we find them much shallower than occurs in the Pottsville region. This fact was also inferred by Mr. Logan, from the frequency with which the conglomerate is brought to the surface.* In the present instance, too, we are struck with the circumstance of the greatly reduced thickness of the entire coal series and subjacent conglomerate, as compared with the same rocks in the southern field; even so near as Tamaqua, eight or ten miles only to the south.

The basins of which we have been speaking, are separately indicated, with his usual perspicuity, by Prof. H. D. Rogers, in his third annual report.† Details of these are doubtless reserved for the final

* Proceedings Geol. Soc. London, Vol. III., p. 710, 1842.
† Third Report of the State Geologist, p. 25.

report of that survey. In the meanwhile we may observe that this country contains many associations for the working of anthracite deposits. Most of these companies have published statistical reports of their respective localities, with maps, analysis, and other suitable illustrations. During the last five or six years prior to 1846, the development of these valuable depositories of mineral fuel proceeded with languor, owing to the financial embarrassments of the country. Of late, much of this property has changed hands: new areas have been laid open, and a spirit of activity has again manifested itself with the returning prosperity of the times.

The section below represents the relative position of the various anthracite basins which are crossed by a meridian line from Mauch Chunk mines to the Wyoming coal valley, a distance of thirty-eight miles.

Fig. 11.

Transverse Section of the Pennsylvania Anthracite Basins, in a north and south direction, looking west.

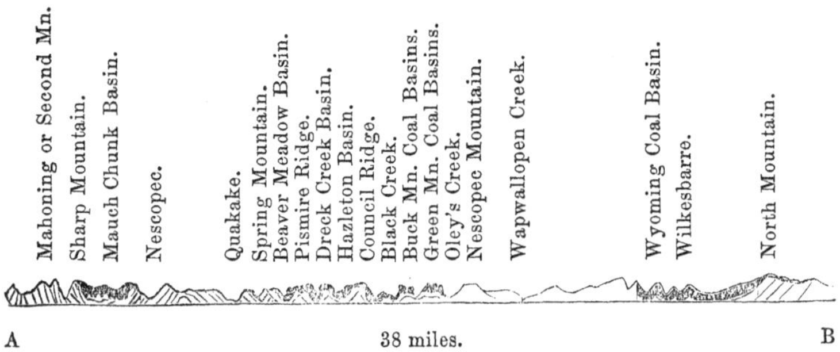

Mahoning or Second Mn.
Sharp Mountain.
Mauch Chunk Basin.
Nescopec.
Quakake.
Spring Mountain.
Beaver Meadow Basin.
Pismire Ridge.
Dreck Creek Basin.
Hazleton Basin.
Council Ridge.
Black Creek.
Buck Mn. Coal Basins.
Green Mn. Coal Basins.
Oley's Creek.
Nescopec Mountain.
Wapwallopen Creek.
Wyoming Coal Basin.
Wilkesbarre.
North Mountain.

A 38 miles. B

The eastern group of coal basins belong to various associated companies. Those which have made the greatest progress in mining coal and making rail roads, are the Summit Company, the Hazleton, the Sugar Loaf, the Beaver Meadow, and the Buck Mountain Companies. Their workings are chiefly limited to the main seam, or rather to a group of three or four seams, amounting to about twenty-eight feet, the coal of which is a remarkably pure anthracite, highly esteemed. Our notice of these establishments must be necessarily brief.

The *Summit*, now *Spring Mountain* Coal Company's operations until lately have been carried on, in great measure, as an open quarry, not possessing a rock roof. The entire mass is estimated at twenty-seven feet thickness, in four beds or divisions, and from the undulation of the strata, it is brought up two or three times to the surface within the general area. The works are recently carried on by means of a vertical shaft, and the produce is as follows:—

	Tons.
In 1846	11,868
1847	32,280
1848	65,410
1849	102,599
1850	43,793
1851	116,517
1852	139,627
1853	135,137

The *Hazelton* mines are worked by "slopes" and steam-power, on both the north and south sides of the basin; that is to say, dipping in opposite directions to the centre. About twenty feet thick of coal, in three seams, are mined at this colliery. The basin contains two or three smaller beds, as at the Summit mines. The specific gravity of the coal is 1.550.

Amount of coal sent down to market from the commencement of the colliery in 1838 to 1844, inclusive, was

	Tons.
	271,066
In the year 1845	70,659
1846	98,150
1847	105,766
1848	105,169
1849	92,480
1850	54,236
1851	113,354
1852	130,628
1853	124,250

The *Sugar Loaf* mine takes the coal of the north slope of the Hazelton basin, and from the same general bed.

The aggregate in four years, ending 1845, is 85,439 tons only. Circumstances of a financial character have interrupted the coal operations for two or three years, but they are now on the point of renewal on an enlarged scale.

	Tons.
1849	11,359
1850	12,106
1851	36,712
1852	41,763
1853	44,914

Beaver Meadow Colliery exhibits a considerable section of coal, raised by steam power. A cubic yard of this splendid anthracite weighs one ton and four hundred and sixty pounds. Its analysis, together with that of the other mines we have enumerated, appear in our general tables. The returns from the Beaver Meadow mines commenced in 1837. From that year to 1842, inclusive, a period of six years, the aggregate sent to market was

	Tons.
	231,894
In 1844 - - - -	70,379
1845 - - - -	77,227
1846 - - - -	85,648
1847 - - - -	109,363
1848 - - - -	85,681
1849 - - - -	73,961
1850 - - - -	27,571
1851 - - - -	42,263
1852 - - - -	46,280
1853 - - - -	55,997

The *Buck Mountain* contains four coal seams, the thickest being twenty-two feet. The company here sent down

In 1843 - - - -	2,844
1844 - - - -	13,749
1845 - - - -	23,914
1846 - - - -	46,103
1847 - - - -	50,847
1848 - - - -	71,101
1849 - - - -	85,819
1850 - - - -	103,937
1851 - - - -	104,456
1852 - - - -	104,202
1853 - - - -	77,457

Colerain.

1850 - - - -	2,076
1851 - - - -	39,523
1852 - - - -	37,781
1853 - - - -	58,012

Cranberry Company.

1849 - - - -	36,153
1850 - - - -	22,493
1851 - - - -	30,651
1852 - - - -	48,920
1853 - - - -	51,217

East Sugar Loaf.

1852 - - - -	12,566
1853 - - - -	30,315

There now exist facilities, by means of several railroads, and by the Lehigh Navigation on the east, by the Schuylkill Navigation on the south, and the Pennsylvania Canal on the west, for transporting to tidewater an unlimited supply of mineral fuel, unsurpassed in point of purity, probably, in the world. So late as 1834, the Coal Committee reported to the state legislature that "the whole quantity of coal mined in this middle anthracite region, was estimated at only five hundred tons, which were hauled in wagons to supply the neighbouring districts." What has been accomplished since, appears from the preceding notes.

III. THE NORTHERN OR WYOMING, WILKES-BARRE, AND LACKAWANNA, ANTHRACITE REGION.

In the western half of this elongated basin, the coal formation occupies the beautiful valley of Wyoming; the remainder extends eastward to the Carbondale works, the coals of which almost entirely go to New York, and are of first rate quality.

In geological character this is but a repetition of the first and second coal-fields below, although it has been less disrupted. Mr. Logan constructed a transverse section in 1842. Mr. Featherstonhaugh had made one in 1830. Here are several coal seams, varying from three to thirty-two feet thick; but their number is not yet fully ascertained.* Near Wilkesbarre, the principal coal mine or bed, consists of a series of layers, amounting to twenty-nine feet thickness; of which only eighteen feet are, or lately were, worked. This is mined by leaving pillars of fourteen or sixteen feet square, and the coal is extracted by blasting; commencing with the upper seams. There are several mines towards the west from this position; some of which are accessible from the Susquehanna river. They are worked by means of open galleries, twenty to twenty-four feet in height. These are generally of the denomination of red or gray ash coals; those to the eastward are commonly of the white ask kind.

It was formerly thought that the Wyoming coals were inferior in quality to those of the other districts. This evil reputation was, in great measure, derived from the impolitic method of mining, during the early years of coal operations in this valley, whereby much inferior coal was permitted to go to market. Where regard is had to a proper selection of the purest seams, or parts of seams, the coal is entitled to a character equal perhaps to that of any other. In fact, there is here, as in every part of the anthracite fields of Pennsylvania a great variety of coal, even in the same general seam.

The existence of this combustible was, apparently, known much earlier than that of the southern coal-fields; and we are informed

* Silliman's Journal, 1830. Also, Hazard's Register, Vol. X., p. 319.

that it was furnished to the United States armory at Carlisle, in 1775 and 1776; but that it had been in use since 1768, in small quantities.

Carbondale is the most important working point from the Lackawanna region, at its northeast end; from whence, in 1829, a railroad and the Delaware and Hudson Canal were opened to convey the coal to the Hudson river, and thence to New York; the amount transported the first year being 7,000 tons.

In 1834 and 1845, the capital invested in this coal undertaking, was stated to be as follows:

	1835.	Tons. 1845.
Canal and Railroad, 123 miles, -	$2,305,599	2,910,558
Colliery establishment, canal boats, lands, &c. - - -	862,501	245,971
	$3,168,100	3,156,529

*Table of the amount of Anthracite exported from the Northern Basin or Division.**

Lackawanna District.		Pennsylvania Coal Company. ‖	Wyoming or Wilkesbarre District.		Sum Total.
Years.	Tons.†		By Canal or River. §	By Lehigh Route. ¶	
			Tons.	*Tons.*	*Tons.*
1829	7,000				
1830	43,000				
1832	84,600				
1833	111,770				
1834	43,700				
1835	90,000				
1836	103,861				
1837	115,387				
1838	78,207				76,321
1839	122,300				122,300
1840	148,470				148,470
1841	192,270		32,917		225,187
1842	227,605		47,246		274,851
1843	242,000		57,740		299,740
1844	251,005		116,018		367,023
1845	269,469		178,401		451,836
1846	318,000		192,503	5,865	516,368
1847	388,203		289,898	10,246	678,801
1848	437,500		237,271	10,425	685,196
1849	454,240		259,080	19,590	732,810
1850	432,339‡	111,014	243,250	32,153	818,758
1851	472,478	316,017	336,000	25,071	1,150,166
1852	497,839	426,164	319,341	41,990	1,285,234
1853	494,209	512,700	442,511	26,235	1,475,672

The quantity thus far registered, as sent from this region between 1829 and 1846, inclusive, is 3,732,686 tons; which is evidently less than the amount exported from the basin, besides the home consumption. It is impossible to tell the amount which passed down the river.

* Report Reading Railroad, January 10th, 1853.
† Pottsville Miners' Journal, January 14th, 1854.
‡ Delaware and Hudson Canal Company.
§ Reading Railroad Report.
‖ Pennsylvania Coal Company's Report.
¶ Lehigh Company's Reports.

In 1828, there was but a solitary house on the site of Carbondale. It contained in 1833, 2000, and in 1840, 2,398 persons, chiefly employed by the company, or in transportation, &c.

There is no description of fuel for the use of the Hudson river steam vessels, in higher repute than the Lackawanna coal.

The Coals of the Lackawanna or Carbondale district are transported to New York by the Delaware and Hudson Canal, 108 miles; railroad, 18 miles; and river navigation, 91 miles. Total, 217 miles.

The coals of the Wyoming district descend the Susquehanna 194 miles, to tide at Havre; the returns are from the Canal Commissioners reports; distributing coal at numerous points along the river.

Passed the Tide water Canal in 1845	-	58,131
" " " 1846	-	67,905

The Lackawanna is considered the lightest of the white-ash coals that come to market; the usage of the trade formerly assigned thirty-three bushels to the ton; Schuylkill thirty, and Lehigh twenty-eight bushels. Our tables of specific gravities of the whole series of Pennsylvania anthracites will, however, best exemplify this.

Mr. Logan, in a communication to the Geological Society of London, in 1842, states that he had taken some pains to construct a section of the Wyoming basin, at Wilkesbarre; and furnishes the details of the formations there. The coal beds he estimates at 14 or 15 in number, with an aggregate thickness of 70 to 80 feet.

The Pennsylvania Coal Company at Pittston are preparing new openings, in addition to those worked last season. They estimate their productions for the present year at 750,000 tons, which is an increase of 237,000 tons on their last season's business. 1854.

We beg leave to make a few brief extracts from the able Report of Prof. H. D. Rogers, on "The Geological and Mining Resources of the Lackawana Basin, which includes the lands of the Delaware, Lackawanna and Western Rail-road Company," 1854; referring the reader to the Report, for more extensive information.

"It is impossible to estimate with precision, until researches now in progress are completed, the total thickness of the coal measures in the deepest parts of the Wyoming and Lackawanna basin, nor to count with accuracy the number of the available beds of coal in those localities. For my present purpose, that of a general sketch of the geology and vast mining resources of this valley, it will be sufficient to state here, that exact measurement has already disclosed, in the vicinity of Wilkesbarre, the widest and apparently the deepest portion of the coal-field, the existence of from 1000 to 1,200 or more feet of coal-bearing strata, and the presence within these of sixteen or eighteen separate beds of coal; two or three of them being compound seams of great size, and about ten or more of the whole series being permanently of ample dimensions for profitable mining. This depth of the coal measures, and the number of the contained coal seams grow less, of course, from the centre of the basin towards its two margins, and also towards its two contracting extremities.

"*On the Scranton Anthracites.*—In the immediate neighbourhood of Scranton, a portion of the Coal Basin, where the coal measures are unusually well developed by natural features in the topography, and through the researches directed by the companies, the coal rocks, counting from the upper surface of the Seral or lower conglomerate to the highest sandstones of the plateau south-west of Hyde Park Village, disclose, upon careful measurements, an aggregate thickness of about seven hundred feet; and in this depth of strata the whole number of coals, large and small, amounts to no less than twelve, not estimating as separate seams any layers which might be regarded as subdivisions of compound beds. The assembled thickness of these twelve plates of anthracite is not less than seventy-four feet, taking for some their mean, for others their minimum, dimensions; and the thickness available for market, under judicious mining, I would estimate at thirty-nine or forty feet. These aggregates, arrived at through careful personal observation and many patient measurements, exhibit certainly an unusual amount of coal in so moderate a depth of strata, being nearly eleven feet of the former to each one hundred feet of the latter; or of good saleable coal, the high proportion of six feet to every one hundred feet of rock.

"The proportion of solid carbon,—the amount of which in coals, from the best practical researches on fuel, must be accepted as very nearly the measure of their absolute heating strength,—is, in the instance of these Scranton anthracites, about eighty-seven to eighty-eight per cent. of the whole mass, a ratio only about two per cent. less than distinguishes the dryest or least gaseous varieties in the Lehigh coal fields, while the difference is amply compensated for in the gain of this amount of ignitible, inflammable gases—hydrogen and carburetted hydrogen,—which serve materially to increase the promptness of kindling, and rapidity of burning, or the total amount of heat evolved in a given time.

"With a view to exhibit more distinctly the excellences of the class of free-burning white-ash anthracites, such as those I have above described, I will conclude this essay, with a condensed survey of the principal qualities essential to a good fuel for producing steam, or for domestic uses:—

"1. It should possess great actual heating power.

"2. As far as consistent with the foregoing, it should kindle quickly, and burn fast, generating the largest amount of heat in the shortest time.

"3. Its earthy matter should be small in quantity, and difficult to fuse; it will thus make little clinker, demand but little raking of its fires, and undergo but little waste in consequence.

"4. It should contain but little sulphur.

"5. The volatile ingredients of the coal should be free inflammable gases, not bituminous matters forming smoke; and they ought to be barely abundant enough to assist rapidity of combustion, as the larger the proportion of fixed carbon, the greater seems the heating power.

"6. They should not be too tender on the fire, nor yet too refractory; a certain tendency to fall to pieces spontaneously while burning, but not an over amount of this, is a great desideratum, as it confers activity and steadiness of combustion; too much of it impedes combustion by increasing the friction of the air passing through the fire.

"7. The lower the temperature at which an anthracite will kindle and maintain itself burning, the more manageable, more active, and more economical will it prove.

"8. The better a coal unites the tenacity necessary for economical transportation, with this medium amount of frangibility on the fire, the larger the effective result of a given quantity, from the time it leaves the mine.

"9. And the greater the aggregate of positive heating power, rapidity of combustion, and compactness of stowage compatibly assembled in a coal, the nearer does it approach the ideal standard of a perfect fuel."

The preparations for sending coal to market from the northern and central regions of Pennsylvania, are going forward vigorously. From the New Jersey Central Rail-road Report for 1854, we obtain the following information relative to the connecting lines of rail-road now in the course of construction. "The Lehigh Valley Road, from Easton to Mauch Chunk and the Lehigh coal-fields, is expected to be completed during the year. A connection with Pittsburg and Ohio, over the Pennsylvania Central Rail-road, has been advanced by the completion of the Dauphin road from Harrisburg to Port Clinton, leaving only 33 miles from that place to finish a junction with the Lehigh Valley road at Allentown.

"The Warren Rail-road, diverging to the north-west from Hampton Summit, to connect with the Delaware, Lackawanna and Western Road, is now under contract, and the construction actually commenced. The two tunnels at Vass Gap and Van Ness Gap, will be the heaviest works,—the whole is expected to be out of the way and the road opened in June of next year. The construction of the southern division of the Delaware, Lackawanna and Western Rail-road from the Delaware river,—where it intersects the Warren road,—to Scranton, where the northern and finished division of the road terminates, has been some time in progress, and is far advanced. It will be finished simultaneously with the Warren road. Before these two latter works are completed, it is expected that the Oswego, Syracuse and Binghampton road will be in operation from Oswego to the Erie Rail-road, and the business of Lake Ontario be brought over that new avenue, ready on the completion of the Lackawanna and Warren roads, and the laying of a third rail on the Central road, to become a new and important element of traffic to the latter."

IV. BROAD TOP MOUNTAIN—SEMI-BITUMINOUS COAL-FIELD IN BEDFORD AND HUNTINGDON COUNTIES.

The area of this small detached coal-field we have not seen announced. Our own calculation is about forty square miles.

Of how many coal seams this basin is made up, we have not ascertained, beyond the six or seven that we have examined. They vary in thickness from three to eight feet. An article in the Transactions of the Geological Society of Pennsylvania, by the present writer, in 1835, furnishes, in conjunction with the Section, Pl. VII., a view of the principal geological features of this coal-field. The analysis, in the same paper, furnished by Mr. T. G. Clemson, shows not more than 17 per cent. of volatile matter, and 70 per cent. of carbon.

An analysis from coal of this region, in the State report, shows only 11 per cent. and 84 per cent. of carbon; while a third examiner, Johnson, finds 16 per cent. of the one and 77½ per cent. of the other.

The requirements of the neighbouring iron works, &c., occasion but a small amount to be annually mined; while the insulated position of the coal-field, in the absence of a canal or railroad, prevents a coal business of consequence being done here to advantage. There are no returns on record of the quantity produced in this small region, for the reason assigned. The peculiar character of this coal, which greatly resembles the Stony Creek coal, although from its remote position it will for many years remain in reserve, must always confer a high value upon it for furnace, foundry, and steam purposes.* The mountain, in which these coal seams occur, "is a broad irregular plateau, having several spurs, running out towards the bounding valleys."†

1853. According to a recent report upon the Huntingdon and Broad Top Railroad and Coal Company, by Messrs. W. F. Roberts and Henry K. Strong, the Broad Top coal-field has an extent of eighty square miles. The measures are generally horizontal, or nearly so, and undisturbed. See Report.

1853. The company are constructing a railroad from Huntingdon to Hopewell, and will probably extend it to Bedford, making the whole distance from Huntingdon to Bedford fifty miles. See Report.

We are aware that, by geologists, the foregoing details of the Pennsylvania anthracite regions may be thought somewhat too statistical. Yet if geology be estimated with relation to its practical usefulness, and its economical sense; if it be viewed in its tendency to benefit the community; to designate new channels for productive industry; and,—especially as regards the New World,—to almost boundless sources of remuneration, we need not terminate this portion of our subject with an apology for the abundance of that kind

* On certain coal beds near Broad Top Mountain, described under the denomination of "Bituminous Anthracite."—R. C. T. in Trans. Geolog. Soc. of Penna., p. 176, 1835.

† Geography of Pennsylvania, p. 184.

of information. It would be difficult, indeed, to select a finer field, wherein to demonstrate the practical and useful applications of the science, than the country from whence we write.

PRODUCTION OF PENNSYLVANIA ANTHRACITE.

Years.	Periodical aggregates and increase.	Sent to market in given years.	Aggregate supply at corresponding periods.	Enrolled and licensed tonnage of Philadelphia.	Coastwise arrivals at Philadelphia.
		Tons.	*Tons.*	*Tons.*	*Vessels.*
1820	First cargo of coal sent to Philadelphia. In this year there were raised at Mauch Chunk, in the Lehigh region, which continued the only source of supply for five years,	365	365	24,117	877
1825	The first returns from the Schuylkill coal district, and which, with Mauch Chunk, formed the two sources for four years,	34,893	53,935	29,421	1,195
1829	First returns from the Lackawanna district, which, added to the former, produced in the three coal regions, during five years,	112,083	355,015	27,494	2,210
1834	The Swatara district opened, forming the fourth for three years,	383,547	1,941,735	46,653	2,686
1837	A fifth region, the Beaver Meadow, came into operation,	881,026	4,131,548	58,237	7,776
1838	Another colliery, the Hazleton, in the middle region, now commenced, and, with the five others, sent to market,	739,293	4,870,841	60,161	10,860
1839	The Shamokin and the Sugar Loaf mines, now contributed to the supply, altogether furnishing	819,327	5,690,168	63,790	11,188
1840	The united exportation from all these localities was	865,414	6,555,582	67,045	9,706
1841	The Wyoming, or Wilkesbarre district, now first appears in the returns, 32,917 tons. To this may be added the amount mined, during several preceding years, by the Lykens Valley Company, estimated at 60,000 tons,	1,108,899	7,607,398	71,588	11,738
1842	General production this year,	1,018,001	8,715,399	100,641	10,457
1843	Includes the first returns from the Buck Mountain,	1,263,539	9,978,938	104,340	7,659
1844	General production,	1,631,669	11,610,607	114,894	8,016
1845	General production, including many new collieries,	2,023,052	13,633,659		11.476
1846	General production comprises a great number of mines, opened within a few years, particularly in Schuylkill Co.,	2,343,992	15,977,651		14,971
1847	General production, business increase 638,317 tons,	2,982,309	18,959,960		18,069
1848	. . . . . .	3,089,238†			24,483
1849	. . . . . .	3,242,866	25,123,781	‡188,087	25,169
1850	. . . , . .	3,356,894	28,456,420	206,497	27,555
1851	. . . . . .	4,429,458		222,428	27,060
1852	. . . . . .	4,993,471	37,872,538	229,443	31,394
1853	. . . . . .	5,195,151	43,060,491	252,451	29,456

In the foregoing summary are exhibited the periodical yearly pro-

* Ending June 30th. † Reading Railroad Report.

‡ Report of the Philadelphia Board of Trade.

duction of anthracite, and the aggregate amount of tons sent to market by various avenues, at the corresponding periods. Of the home consumption for domestic use, for manufactories, blast-furnaces, forges, rolling-mills, and the coal which is conveyed by other routes from the mining districts than the railroads and canals, we possess no returns or direct means of judging.

As the best illustration of the rapid progress of the coal trade, and of its influence on the domestic commerce of Philadelphia, since the opening of the mines, we annex to our table, in the fifth column, a statement of the enrolled and licensed tonnage of Philadelphia, and in the sixth column, the coastwise arrivals at the same port, and in the corresponding years.

There are various tables in circulation, but none are complete in all the details. In compiling the foregoing statement, we have adhered to the official returns of the various companies as far as possible, assisted by the annual summary published in the Pottsville Miners' Journal, and the Philadelphia Commercial List.

PHILADELPHIA, READING AND POTTSVILLE RAILROAD,

For the Transportation of Anthracite.

The increasing demand for the anthracite of Pennsylvania, the rapid introduction of this combustible into most of the Atlantic states, and the cities of the seaboard, for domestic purposes as an economical substitute for wood, and its enlarging consumption in manufacturing establishments, and almost every where, where steam was the moving power, accelerated, at a corresponding rate, the requirements of the owners and uses of the collieries for adequate facilities of transportation.

The project of a continuous railroad from the mines of Schuylkill county to the river Delaware, the nearest point of shipment at Philadelphia, first received the legislative sanction in 1833, and in 1835, the arrangements having been organized, the work was commenced.

On the 16th of July, 1838, the railroad between Reading and Norristown, uniting there with the Philadelphia and Norristown railroad, altogether a distance of fifty-four miles, was opened for the conveyance of passengers. In December, 1839, the line of railroad between Philadelphia and Reading was opened for transportation. Subsequently, the works were extended at its southern extremity to its terminus at Richmond, on the Delaware, near Philadelphia, and northward to Port Carbon, near Pottsville. On the first day of January, 1842, the first locomotive and train passed over the whole line, between Pottsville and Philadelphia, a distance of ninety-three miles.

The following statement exhibits the amount or quantity of coals conveyed on this railroad, and the receipts for transportation of the same, from the commencement, in 1841, for the years ending November 30th, annually, compiled from the annual reports of the Company.

Statement of Coal descending the Reading Railroad, from Pottsville to Philadelphia and intermediate points.

Years.	Tons of 2,240 pounds of coal only.	Freight and toll for coal.	Freight and toll per ton.	Maximum.
		Dollars.	*Average.*	
1841	850			
1842	49,752			
1843	218,711	278,840	$1 27	
1844	421,958	448,508	1 $06\frac{3}{10}$	
1845	814,279	886,939	1 $08\frac{9}{10}$	
1846	1,188,258	1,600,667	1 35	
1847	1,360,681	1,698,664	1 25	
1848	1,235,044	1,386,605	1 12	
1849	1,097,761	1,648,900	1 50	
1850	1,351,507	2,071,731	1 53	
1851	1,605,084	2,018,870	1 53	1.60
1852	1,650,912	2,150,677	1 42½	
1853	1,582,248	2,254,694	1 42½	

From the same official reports we perceive that the amount of running machinery, in employment in each year, on this railroad, was as follows:

Machines, Cars, &c.	1843.	1844.	1845.	1846.	1847.	1848.	1849.	1850.	1851.	1852.	1853.
Locomotive Engines, ...	31	47	54	72	77	84	86	92	89	103	103
Coal Cars, ...	1592	2456	3104	4559	4606	4606	4623	4567	4579	4579	4792
Freight Cars, ...	208	265	294	482	502	408	463	550	557	662	684
Passenger and Baggage Cars, .	19	19	21	19	22	31	30	28		30	38
Extra Cars for wood tenders, &c. .			46	34		23	24	24		25	25
" Stationary Engines, and Portable Steam Engines, .			5	15							
Horses, ...			31	50		48	37	46		49	66
Carts and Wagons, ...						23	15	23		27	33
Snow Ploughs, ...						7	7	7			7

The number of passengers transported, the receipts for their fare, and the number of miles travelled, are thus:

Years.	Number of passengers.	Miles travelled.	Receipts.
			Dollars.
1843	56,554	2,457,439	71,895
1844	66,503	3,159,909	92,362
1845	63,719	3,049,422	103,411
1846	88,641	4,154,214	141,749
1847	97,463	4,560,260	156,201
1848	105,720	5,106,698	174,958
1849	95,577	4,516,968	155,908
1850	92,726		148,378
1851	127,590	5,298,573	152,432
1852	155,164	6,401,406	168,430
1853	211,819	8,524,527	225,763

The total length of lateral railroads connecting with the Reading railroad, under other charters and corporations, but all contributing to its business, using its cars, and returning them loaded with coal, was about ninety-five miles, in 1846.*

According to the annual general account of the company for the year ending 30th November, 1846, the total cost of the railroad, locomotives, cars, real estate, depots, and materials, is $11,589,696. In November, 1853, $17,905,018.

During the year 1846, there were cleared from Port Richmond, the shipping depot of this company, on the Delaware:

	Number.				
	7,485	vessels, carrying	892,464	tons†	of Schuylkill coal.
Also	1,468	"	181,792	"	Lehigh coal.
Total	8,953		1,074,256		

In 1847, 11,439 from Port Richmond.

Arrivals of Vessels at Richmond.

1849	4,497	
1850	7,549‡	1,075,344 tons.
1851	8,623	1,211,605
1852	9,047	1,226,649
1853	7,384§	1,088,167

SCHUYLKILL NAVIGATION,

For the transportation of Anthracite—length one hundred and eight and a quarter miles, lockage six hundred and twenty feet.

This canal, which was commenced some years before the importance of the Pottsville coal-field was known, or even suspected, affords, through the annual reports of its directors, an unerring criterion of the rapid advancement of the coal trade.

* Miners' Journal, January 23d, 1847.
† Philadelphia Commercial List, January 16th, 1847.
‡ Pottsville Miners' Journal.
§ North American.

Table of the annual number of tons of Anthracite, the amount of toll received thereon, and the average rates of toll per ton on the Schuylkill Navigation, from the commencement of its coal trade in 1825; compiled from the company's annual reports.

Years.	Tons.	Toll.	Average per ton.	Years.	Tons.	Toll.	Average per ton.
		Dollars.	*Dollars.*			*Dollars.*	*Dollars.*
1825	6,500	9,700	1 49	1842	491,602	235,544	0 50
1828	47,284	46,201	0 97	1843	447,058	214,451	0 58
1830	89,984	87,192	0 97	1844	398,887	123,259	0 43
1832	209,271	199,784	0 95	1845	263,587	79,800	0 33
1834	226,692	204,490	0 90	1846	3,437	1,757	0 33
1835	339,508	310,475	0 91	1847	222,693	122,405	0 55
1836	432,045	399,472	0 91	1848	436,602	173,479	0 41
1837	523,152	484,799	0 92	1849	489,208	331,965	
1838	433,875	385,024	0 88	1850	288,030*	190,650	
1839	442,608	381,198	0 86	1851	579,156	218,660	
1840	452,291	373,400	0 82	1852	800,038	416,954	
1841	584,692	482,460	0 82	1853†	888,695	582,654	

M. Chevalier correctly remarks, that the value of this anthracite region [gisement] has literally become immense; and its workings [exploitation] have accomplished a revolution in the domestic economy of the Atlantic states.‡ This canal was commenced in 1817, but it was not until 1825 that anthracite commenced to form the principal part of its tonnage.

In 1833, the number of canal boats used on the Schuylkill Navigation, was 580.

In 1843, the number of loaded canal boats which passed down the Schuylkill Navigation, was as follows:—

Covered boats, adapted to the direct trade from Pottsville to New York, averaging sixty tons, - - - - -	278
Open canal boats, for coal, - - - - -	434
Lime boats, and miscellaneous, - - - - -	58
Registered, as passing the Fairmount locks, - -	770
Additional boats of 120 to 160 tons each, built in 1848,	65
Scow boats, 75 tons, - - - - - -	21
Coal cars, - - - - - - -	383
Cars and trucks for merchandise, - - -	10
1848.—Total number of boat loads, - - -	4,173
Tonnage descended, - - - - -	436,602
Average loads each, - - - - tons,	105
Maximum loads, - - - - - - "	178
1850.—Coal boats of the company 125 tons each, or 351 boats of 150 tons each, - - - - -	473

* Great damage by freshet.

† Coal delivered at way points short of Philadelphia, 155,750 tons.

‡ Histoire et description des Voies de communication aux Etats-Unis, par M. Chevalier, Paris, 1840.

1852.—New coal boats of the company,	56,	- tons,	160
1853.— " "	16,	- "	125

The usual boating season is thirty-five weeks, annually.

The tonnage has been annually advancing, from 32,000 tons in 1826, to 340,000 tons in 1845; 1852, 800,038. The charges on the transportation of coal have been reduced during this period.

In 1826 the tolls received amounted to somewhat less than 1½ cent per mile.

In 1843 the whole charge, including freight and toll, less than 1¼ cent. per mile.

In 1845 the whole charge, including (toll ⅓ of 1 cent) less than 1 cent per mile.

The actual toll received each year from 1825 to 1852, is shown in the preceding table.

In the latter year, the universal voice of those concerned in the coal business, and of the great body of residents and proprietors of manufactories along the line, having for some time called for further improvements of the navigation, the stockholders resolved to enlarge the works, so as to pass boats of three times the former tonnage. Accordingly, the whole of the season of 1846 was devoted to this undertaking, and the navigation was necessarily suspended during that year.

Four thousand men were employed on these works, and it was with great difficulty that an adequate supply of mechanics could be raised to carry on the work with a rapidity commensurate with the wishes of the directors.

Amidst many difficulties and embarrassments, caused by repeated freshets and the destruction of half completed works, the great object was accomplished at the close of 1846, in a satisfactory style, and with a rapidity which admits of no parallel in the history of the internal improvements of this country. The report of the president and managers of the company, 4th January, 1847, details all these circumstances, and the final result.

The capacity of the present navigation, is therein stated as being nine times as great as the canal when originally opened to the trade. It now averages more than seventy feet in width, and six feet deep. It is adapted for boats of one hundred and eighty tons burthen, and will be adequate to the convenient transit of a million and a half of tons.

The aggregate investment of the company on the 1st January, 1847, including all liabilities, as well for the construction of the work as for boats and cars, &c., amounted to $5,785,667.

Besides the canal boats specified above, the company possess three hundred cars, to run upon the lateral railroads.

Anthracite Coal Tonnage for the year 1853, *on the Schuylkill Navigation.*

The following distribution has been made:—

Total tons, - - - - - -	888,695
Delivered at points between Mount Carbon and Philadelphia, - - tons,	155,750
Carried by way of Delaware and Raritan Canal, to New York and its vicinity, -	474,105
Shipped Coastwise in sailing vessels from the Schuylkill, - - - - -	85,000
Delivered at various points on the Delaware river and bay, - - - -	22,800
Delivered in the city of Philadelphia and vicinity, for families, manufactories, &c.	151,040
Total tons, - - - -	888,695

The state of the affairs of the Schuylkill Navigation Company on the 23rd December, 1853, was

Assets, - - - - - - -	$265,426 45
Doubtful debts, - - - - -	16,366 53
The amount for the cost of the works, docks and landing, and real estate of the company, and all expenses for repairs, and discount and interest on loans and losses beyond the net income, - - - - - -	10,292,427 20
Total, - -	$10,574,220 18

This statement does not include the boat and car loan trust, which amounts to $236,337 85.*

Quantity of Schuylkill Coal annually sold in Philadelphia for home consumption, which descended by the Schuylkill Navigation.

	Years.	Tons.	Years.	Tons.
Delivered at Philadelphia.	1836	61,944	1841	89,000
	1837	71,916	1842	88,000
	1838	98,707	1844	97,600
	1839	100,694	1847	226,610
	1840	90,000	1853	151,040

The registration of the consumption in Philadelphia, appears to have been discontinued. The quantity which was received by Philadelphia in the year 1847, by railroad, was 203,540 tons.

* Schuylkill Navigation Report.

Freights of Schuylkill Coal annually from Pottsville in the month of October, in each year, per ton.

		1844.	1845.	1846.	1847.
		Dollars.	*Dollars.*	*Dollars.*	*Dollars.*
Freight and toll from Pottsville.	To Philadelphia by Canal,	0 85	1 00		2 00
	To Philadelphia by Railroad,	1 25	1 40	1 70	
	To Richmond,			1 60	2 00
	To New York,	2 25	2 50	2 70	
Freights from the depot at Richmond on the Delaware.	To New York,	1 00	1 10	1 00	1 30
	To Albany and Troy,	1 50	1 50	1 45	
	To Hartford, Conn.	1 65	1 65		
	To Salem,				2 37½
	To Boston,	1 70	1 70	1 50	2 50
	To Fall River and Providence,	1 35	1 40	1 30	1 80
	To Baltimore,			75	
	To Washington,			80	

Rates of Toll and Transportation by Railroad to July 1st, 1854.

From	Mount Carbon to Richmond,	$1 70
"	Schuylkill Haven,	1 65
"	Port Clinton,	1 45
"	Mount Carbon to Philadelphia,	1 60
"	Schuylkill Haven,	1 55
"	Port Clinton,	1 35

Rates of Toll by Canal to July 1st.

From	Port Carbon to Philadelphia,	$00 70
"	Mount Carbon,	00 69
"	Schuylkill Haven,	00 67
"	Port Clinton,	00 55

1854.—The freights by canal fixed by the boatmen, $1 80 to New York, 80 cents to Philadelphia from Pottsville and Port Carbon, and five cents less from Schuylkill Haven.

On the enlargement of the Schuylkill Navigation, in 1846, it was estimated by the directors that the actual cost of freight from Pottsville to New York, with an adequate supply of large boats, would be about $1 35 per ton. The freight in August, 1847, was, however, $2 00 to Richmond only.

The direct exportation of coal from the Schuylkill region, which descended the Schuylkill Navigation and the Delaware and Raritan Canal, to New York.*

Years.	Loaded Canal Boats.	Tons.
1839		27,000
1840		64,388
1841	1,354	78,296
1842	2,243	126,554

* Annual Reports.

Years.	Loaded Canal Boats.	Tons.
1843	2,045	119,972
1844		111,521
1845		116,610
1846	Navig. suspended.	
1853		474,105

1852. Freight and tolls of the Schuylkill coal to tide-water at Richmond have averaged the past year about thirty cents per ton more than by the latter to the wharves on Schuylkill. Freight per railway, $1 50 to $1 80; its cost at tide-water from $2 75 to $3 30.

We have seen a statement, purporting to exhibit the importance to Pennsylvania of the trade in anthracite: that in the year 1845, independent of the quantity consumed in the State, there were shipped to other states, as much coal as amounted at the average value [$4 00 per ton,] at the place of shipment, to the sum of $5,000,000.* In 1847, the value of the anthracite shipped at tide-water for other states, was not less than $10,000,000.

COMPARATIVE VALUE OF THE ANTHRACITE OF PENNSYLVANIA.

With the exception of iron, in the mining and manufacture of which, according to the census returns of 1841, more than twenty millions of capital were employed in the United States,—nearly eight millions of which appertained to Pennsylvania,—the coal production of this State then gave employment to more workmen, and more capital than all the other minerals in the Union combined, or more than 2½ times; as is shown in the following:

	United States in 1840.			Pennsylvania in 1840.		
	Lead.	Gold.	All other metals.	Anthracite.	Bitumi's Coal.	Total Coal.
Men employed in mining,	1,017	1,046	728	2,997	1,798	4,795
Capital invested, . . .	$1,346,756	$234,325	$238,980	$4,334,102	$300,416	$4,634,518
Exclusive of the canals, railroads, Persons,	2,791	Sum tot.	$1,820,061			

In 1847, the capital invested in the canals and railroads, communicating with the anthracite region of Pennsylvania alone, amounted to more than forty millions of dollars. This is wholly independent of the capital employed in the coal regions, and the trade consequent upon it.

* Anthracite Gazette, February 7th, 1846.

RAPID AUGMENTATION OF PRODUCTION AND CONSUMPTION.

With the exception of the two distressing years, from 1841 to 1843, when every species of property was alarmingly depreciated, and all business appeared to be paralized, the anthracite trade in its various departments, although not yielding enormous profits, steadily and rapidly acquired importance, from the period of its commencement. Each succeeding year saw new fields explored; new deposits discovered; more enterprise exerted in opening approaches from the seaboard to the coal beds, and more avenues for the transportation of the mineral fuel through the wilderness; more capital invested in this fruitful branch of industry and commerce. Each year added to our acquaintance with the extent, the limits, the existence, of these vast carboniferous masses; and advanced our progress in the science of industrial and economic geology. As in the building of the mighty Egyptian pyramid, each year saw arise, with augmented bulk, and still increasing magnitude, that immense fabric which had commenced from nothing.

The time has not yet arrived when an elaborate description of the Pennsylvania coal-fields can be satisfactorily presented. That task will, no doubt, ere long be executed by able hands, under all the advantages which the influence and the resources of the State government can confer. In the mean while, a selection from the notes which have accumulated under the author's hands,—although derived from the sources common to all,—in the absence of data more scientific, and of statistics more official, will be the best substitute he can offer.

The following table in the diagram form, so far as extends down to 1842, is compiled by adding to the amount sent to market in each year, the quantity of coal on hand at its commencement, and deducting the surplus remaining at its expiration. The succeding years represent the production, annually. From 1842 to 1848, the returns are of the amount which reached tide.

Adhering to the metaphor which he has employed, the writer presents his readers with a diagram of this statistical pyramid;—built, not with granite, nor sienite, nor ponderous marble, but with Pennsylvania anthracite, and reared by her industrious citizens and free labourers.

Which of the two fabrics,—that "of the olden time," or this, in the building of which most of us have had a hand,—is most conducive to "the good use of man?" let the economists say.

Mr. Gliddon tells us, that the Egyptians, in days of yore, builded their pyramids "from the top downwards"—and so, too, have we constructed *our* pyramid. We are, therefore, not without good precedents, as venerable as they are substantial for such a practice.

Fig. 12.

STATISTICAL PYRAMID OF THE ANNUAL CONSUMPTION OF AMERICAN ANTHRACITE.*

Years.	Tons.
1820	365
1821	1,073
1822	2,240
1823	5,823
1824	9,541
1825	33,699
1826	48,115
1827	58,567
1828	73,403
1829	104,403
1830	142,734
1831	216,820
1832	318,871
1833	395,365
1834	451,636
1835	601,935
1836	626,526
1837	724,539
1838	773,836
1839	835,553
1840	876,049
1841	919,793
1842	1,158,001
1843	1,263,539
1844	1,631,669
1845	2,023,052
1846	2,343,992
1847	2,982,309
1848	3,089,238
1849	3,242,866
1850	3,356,899
1851	4,429,458
1852	4,993,471
†1853	5,195,151

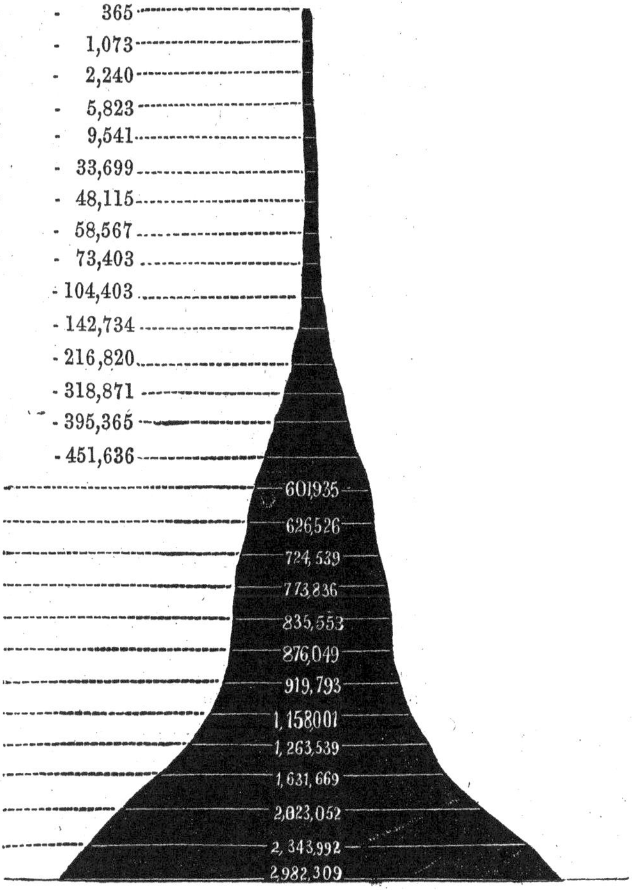

* It will be observed, that the top of Mr. Taylor's pyramid does not enlarge in strict proportional ratio; but as this ocular mode is among the best to give a proper idea on the subject, we have made no alteration from the cut as in the former edition. We have reluctantly rejected our first intention of continuing the pyramid so as to exhibit its actual proportions, but finding the outline for the succeeding years to 1853 would extend far beyond the margin of the page, we have simply adopted the plan of inserting the tons for each year, leaving to the imagination of the reader to give the vast dimensions for the base of a structure destined to increase year after year beyond all limits.

† Pottsville Miners' Journal, and Reading Railroad Reports, 1853.

And now let us compare our work with theirs.

We are told that it employed 100,000 men, during twenty years, to construct the pyramid of Cheops, or Shoopho, and that ten preceeding years were requisite merely to prepare the materials, and to convey them from the quarry, a distance of twenty miles. Now this, the largest of all the pyramids, contained an area of 3,300,000 cubical yards. Allowing the ordinary computation, that a cubic yard of anthracite is equivalent to one ton in weight, and as we have brought up from the bowels of the earth about nineteen millions and a half of cubic yards or tons, and have conveyed them an average distance of at least one hundred and fifty miles, we have, in twenty-five years acquired materials about sufficient for the erection of six such structures, united in one. Such are the colossal proportions of our Pennsylvania Pyramid.

To be sure, the outline is not symmetrically beautiful, nor in that strict geometrical proportion which the eye delights to dwell on, or which the architect loves to contemplate. True, its upper portion is somewhat too attenuated; but then we make it up in the base. At the beginning of the work, we admit, the outline had more of the spire than of the pyramid about it; but then we were short-handed; and, moreover, as soon as we discovered our error, we lost very little time in correcting that, as the books show; and we soon adopted more substantial proportions.

The base is now so broad and so firmly planted, and the structure has withstood such heavy storms of late years, that no fears for its permanence and future increase can now be entertained.

Success, then to our great Pennsylvania Pyramid! May its proportions increase and overshadow the land!—may it ensure protection, security, and prosperity to all who seek its shelter, or who labour around its base!—to all who contribute towards its enlargement!

1854. The coal consumption of the present year is estimated at 6,325,000 tons, allowing an increase of 15 per cent. on the consumption of last year. The Cumberland Journal, from which we gather this fact, regards it as tolerably certain that the supply of coal for the present year will fall short of this demand, and that the presumed deficiency will be large.

ANTHRACITE.—STATISTICS OF PRODUCTION.

Lehigh Company's Mines. So rapid was the increase in the demand for anthracite, after the export trade commenced in the year 1825, [although the home market originated in 1821,] that in ten years, as appears from the official returns in 1835, the number of coasting vessels that received freights of the coal which had descended the Delaware, was one thousand and sixty-nine.

Amount of Anthracite sent down from the Company's mines at Mauch Chunk and Room Run.

	Years.	Tons from the Company's mines.	Years.	Tons from the Company's mines.
11 years, previously to	1831	207,887	1844	219,245
	1831	44,683	1845	257,740
	1833	123,441	1846	274,623
	1837	200,000	1847	334,929
	1838	154,693	1848	336,570
	1839	142,507	1849	379,285
	1840	102,264	1850	424,258
	1841	78,164	1851	480,824
	1842	163,762	1852	510,268
	1843	138,826	1853	476,976*

Including Room Run Mines.

General statement of the Coal conveyed on the Lehigh Navigation.

Until so late as 1837, the only coal sent down to market was from the Lehigh Company's own mines, in the Southern Region.

Years.	Periodical increase.‡	Tons of coal transported.	Tolls chiefly arising from coal.
			Dollars.
†1836	The annual quantity had increased from 365 tons in 1820 to	148,211	110,905
1837	By the addition of the Beaver Meadow coal from the 2d District	224,095	149,266
1838	The Hazleton Company's mines came into operation, and the whole	214,211	125,411
1839	The Sugar Loaf mines commenced in addition, making this year	221,850	141,300
1840	Additional mines were put in work in the 2d Region, in this and following years.	225,585	143,335
1841	The reduction is owing to the damage to the canal by a freshet.	143,038	65,792
1842	Quantity sent down by the Lehigh Navigation,	272,553	157,844
1843	" " " " (some delay on account of strike of boatmen,)	267,826	173,660
1844	A considerable improvement in the trade of this year.	377,094	170,759
1845	A very favourable year for transportation.	429,492	203,405
1846	do. do. do.	522,297	
1847	do. do. do.	643,612	
1848	. . . .	680,746	
1849	. . . .	801,246	
1850	. . . .	722,622	353,130
1851	. . . .	989,650	428,566
1852	. . . .	1,114,231	486,555
1853§	. . . .	1,080,544	550,054

* Including 83,721 tons from the Room Run.
† History of the Lehigh Coal and Navigation Company, 1840, p. 61.
‡ Lehigh Company's Report for the several years.
§ Annual reports of the Managers. The separate amounts of toll on the coal are not distinguished in the published returns.

Shipments of Lehigh Coal from Bristol to Philadelphia, and ports on the Delaware.

Years.	Tons.	Years.	Tons.	Years.	Tons.	
1821	525	1834	80,000	1842	50,780	
1825	11,245	1835	85,000	1843	77,840	Lehigh Co. Others no separate account kept since 1844.
1830	12,601	1839	42,000	1844	35,972	
1833	25,000					

LEHIGH COAL STATISTICS.

The following table from the Pennsylvania Canal Commissioners' annual reports, shows the amount of anthracite received on the Delaware division of the State Canal, at the undermentioned points.

	1844.	1845.	1846.	1847.	1849.	1850.	1851.	1852.	1853.
	Tons.	*Tons.*	*Tons.*	*Tons.*	*Tons.*	*Tons.*	*Tons.*	*Tons.*	*Tons.*
Sent southward from Easton, .	301,956	392,821	522,990					774,460	635,137*
Rec'd at Bristol,	290,105	335,199	387,786	450,178	434,475	363,323	534,532	566,742	440,500
Coal ship'd from Bristol by the Del. Division of Penna. Canal,	. .	. .	. .	. .	. .	. .	1,594	481	1,167†

Shipments and clearances from Philadelphia and Bristol, laden with Lehigh coal, in 1846, 1468 vessels, exclusive of boats, carrying 181,792 tons.‡

Aggregate of the Production of Anthracite,

From the commencement of the trade in 1820 to the end of 1853, taken from the Report of the Reading Rail Road, January 9th, 1854.

		Tons.
The total of coal brought down by the Lehigh Navigation, including that from the Lehigh Company's mines, Beaver Meadow, Sugar-Loaf, and other collieries, - - - - - -		9,756,598
The Schuylkill region has sent to market by the Reading Rail-road and Schuylkill Navigation, during the same time, including Little Schuylkill, - - - - -	2,247,446	22,295,396
Pinegrove, - - - - - - -		707,616
Lackawanna, - - - - - -		7,024,640
Wilkesbarre, - - - - - -		2,713,586

* Canal Commissioners Report, 1854.
† Commercial List.
‡ Philadelphia Commercial List, January 16th, 1847.

Shamokin, - - - - - -	234,633
Lyken's Valley, - - - - - -	245,232
Dauphin Company, - - - - -	82,639
Total,	43,060,340

The various Reports differ as to the aggregate amount of coal; the following is as correct as can be obtained, by including the Little Schuylkill Company in the Schuylkill Region. The table is taken from Reading Rail-road Report and Pottsville Miner's Journal, January 9th, 1854.

The Report of the Board of Trade gives the aggregate at 43,629,889 tons, February 6th, 1854.

Lehigh Coal Trade in 1852 and in 1853.

	1852	1853
Mines worked by the Company.		
Summit Mines, - -	429,786	393,255
Room-Run, - -	80,487	83,721
Beaver Meadow, - -	46,280	55,997
Spring Mountain, - -	139,627	135,137
Colerain, - - -	37,848	58,012
Sugar-Loaf, - -	41,763	44,914
East Sugar-Loaf, - -	12,566	30,315
Hazleton, - - -	130,628	124,250
Buck Mountain, - -	104,202	77,457
Cranberry, - -	48,920	51,217
Wyoming, - -	41,990	26,235
	1,114,231	1,080,544

In the annexed diagram, bearing the title of "*the Stream of the Lehigh coal trade*," we have exhibited, under one view, the whole extent, past and present, of the trade, from its origin. We represent, in this sketch, the main stream and the lateral channels through which the anthracite finds its way to market. True it is, the stream has heretofore flowed somewhat irregularly; sometimes embarrassed by natural obstacles and unexpected calamities; sometimes interrupted by temporary causes; yet has it ever pursued its accelerated course, and onwards advanced in an accumulated volume. Notwithstanding every check, always has it surmounted all barriers; always maintained its progressive character. Always has it given promise, in no distant time, of a mighty flood; prolific—fertilizing—reproductive. Success, then, to the *Stream of the Lehigh coal trade!* Onward may it flow; swelling in its volume; bearing on its surface and in its inmost depths, the elements of prosperity to all who embark upon its waters! For ourselves, and in our day, we perceive but the beginning. We approach but the fountain head;—the margin of a stream to whose capacity we can suggest no ultimate limit. We see, but darkly, the outline of that magnificent future, to which all things are tending, when its projectors shall cease to exist.

Fig. 13.

The Stream of the Lehigh Coal Trade.

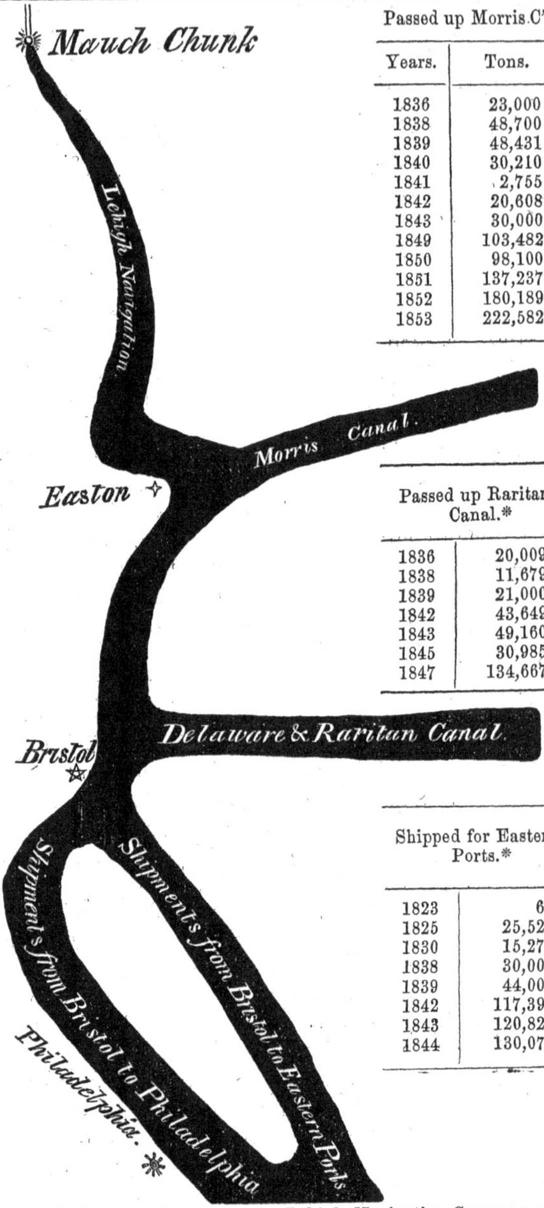

Mauch Chunk to Easton.

Years.	Tons.
1820	365
1825	28,393
1830	41,750
1835	131,250
1838	214,211
1840	225,585
1842	272,553
1843	267,826
1844	377,094
1845	429,453
1846	522,297
1847	643,971
1848	680,746
1849	801,246
1850	722,622
1851	989,650
1852	1,114,231
1853	1,080,546

Easton to Bristol.

Years.	Tons.
1836	112,082
1838	151,788
1840	171,210
1842	211,817
1843	197,000
1844	290,105
1845	335,199
1846	387,786
1847	450,178
1849	580,934
1850	507,323
1851	691,810
1852	782,388
1853	659,909†

Shipped to Philadelphia.*

Years.	Tons.
1821	525
1825	11,245
1830	12,601
1842	50,780
1843	77,840
1846	181,792

Passed up Morris C'l.

Years.	Tons.
1836	23,000
1838	48,700
1839	48,431
1840	30,210
1841	2,755
1842	20,608
1843	30,000
1849	103,482
1850	98,100
1851	137,237
1852	180,189
1853	222,582

Passed up Raritan Canal.*

Years.	Tons.
1836	20,009
1838	11,679
1839	21,000
1842	43,649
1843	49,160
1845	30,985
1847	134,667

Shipped for Eastern Ports.*

Years.	Tons.
1823	62
1825	25,528
1830	15,271
1838	30,000
1839	44,000
1842	117,397
1843	120,826
1844	130,074

*. No returns have been made for several years by the Lehigh Navigation Company relating to the following Tables.
Passed up the Raritan Canal.
Shipped to Philadelphia.
Shipped for Eastern Ports.
† From the Company's Reports.

EXPORTS OF ANTHRACITE FROM THE SCHUYLKILL AND DELAWARE.

Statement of the number of tons of anthracite, which had descended from the Lehigh and Schuylkill navigations, and which were shipped for exportation *coastwise*, together with the number of vessels of all sorts, [brigs, schooners and sloops,] freighted therewith at Philadelphia or Bristol.

This statement is exclusive of the coal which passed through the Delaware and Raritan, and the Morris canals.*

Years.	Lehigh Coal, shipped coastwise from Bristol and Philadelphia. Vessels cleared.	Lehigh Coal, shipped coastwise from Bristol and Philadelphia. Tons.	Schuylkill Coal, shipped coastwise from the Delaware, at Philadelphia and Port Richmond. Vessels cleared.	Schuylkill Coal, shipped coastwise from the Delaware, at Philadelphia and Port Richmond. Tons.	
	No.		*No.*		
1822			4	181	
1825			190	19,378	
1830			644	63,137	
1833	322	30,753	2,010	198,168	
1835	1,069	70,194	2,361	267,139	
1836	261	30,076	3,225	344,812	
1837	51	6,549	3,070	328,304	
1838	96	38,977	2,695	278,268	
1839	158	44,000	2,561	286,990	
1841			3,065	367,812	
1842	From Bristol.	117,397	2,134	256,080	
1844	1,127	130,074			
1846	1,468	181,792	7,485	892,464	From Port Richmond only.
1847			11,439	1,375,000	From Port Richmond only.
1849		116,830			From Port Richmond only.
1850			7,549	1,075,344	From Port Richmond only.
1851			8,623	1,211,405	From Port Richmond only.
1852			9,047	1,226,488	From Port Richmond only.
1853	†379	63,180	‡7,384	1,088,167	From Port Richmond only.

There have been no returns of the number of vessels at Bristol for several years by the Lehigh Company.

PRICES OF ANTHRACITE IN PHILADELPHIA.

Average Retail Prices for unbroken Coal, delivered in Philadelphia, per ton of 2,240 pounds, chiefly derived from the Commercial List, and from Bicknell's Reporter, and the Pennsylvania Inquirer.

Years.	Lehigh White Ash.	Schuylkill Red Ash.	Years.	Lehigh White Ash.	Schuylkill Red Ash.	
	Dollars.	*Dollars.*		*Dollars.*	*Dollars.*	
1828	6 50	7 00	1844		3 50	See also the diagram of Lehigh prices of anthracite, [Fig. 14.] These prices are, of course, only approximate.
1830	6 50	6 50	1845		3 50 to 4 00	
1832	6 00	6 00	1846	4 50	4 25 to 4 50	
1834	5 00	5 25	1847	5 00	4 50 to 4 75	
1836	6 00	7 50	1848			
1838	5 50	6 00	1851	4 50 to 4 75		
1840	5 50	5 50	1852	4 50	3 75 to 4 00	
1842		4 25	1853	5 to 6	5 50	

* Partly taken from the Commercial List of Philadelphia, January 9th, 1847.
† Philadelphia Commercial List, December 31st, 1853.
‡ North American.

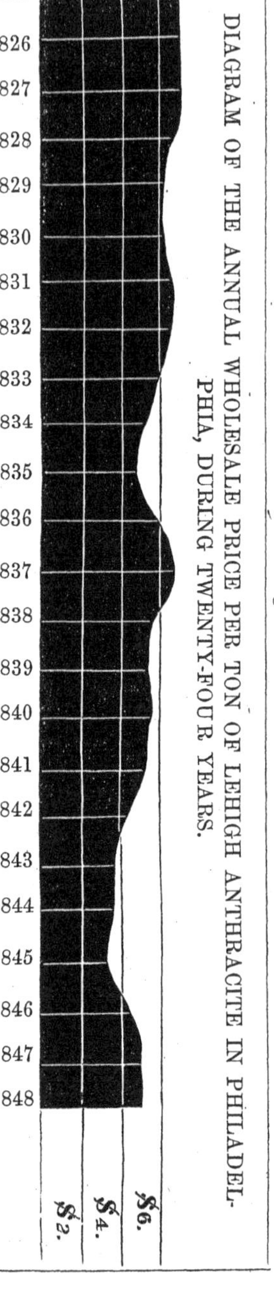

Fig. 14.

DIAGRAM OF THE ANNUAL WHOLESALE PRICE PER TON OF LEHIGH ANTHRACITE IN PHILADELPHIA, DURING TWENTY-FOUR YEARS.

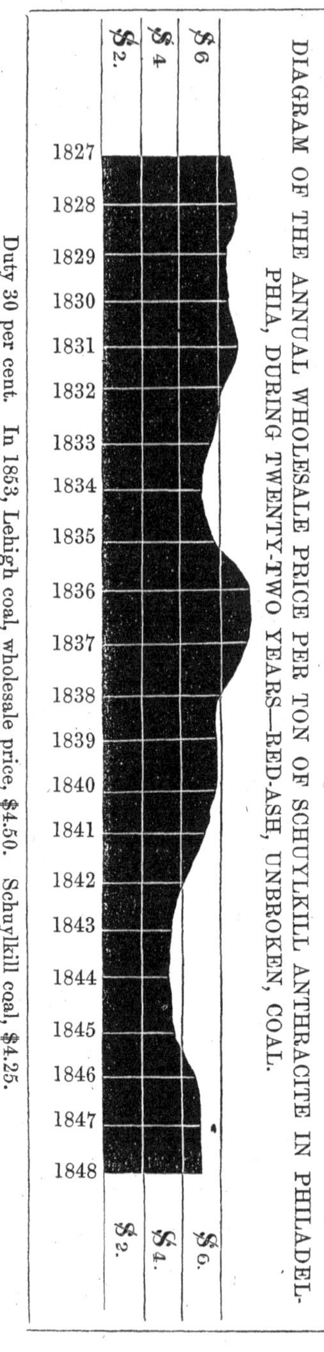

Fig. 15.

DIAGRAM OF THE ANNUAL WHOLESALE PRICE PER TON OF SCHUYLKILL ANTHRACITE IN PHILADELPHIA, DURING TWENTY-TWO YEARS—RED-ASH, UNBROKEN, COAL.

Duty 30 per cent. In 1853, Lehigh coal, wholesale price, $4.50. Schuylkill coal, $4.25.

DELAWARE AND RARITAN CANAL.

Pennsylvania coals [Schuylkill and Lehigh] which passed through the Delaware and Raritan Canal to New York.

Years.	Tons.	
1842	171,754	
1844	267,496	
1845	372,072	
1846	339,923	of which 134,667 tons were Lehigh coal.
1847	540,200	" 405,533 " Schuylkill coal.
1850	568,403	Lehigh and Schuylkill coal.
1851	769,602	Lehigh and Schuylkill coal.
1853	474,105	carried by Delaware and Raritan canal to New York.

Number of Canal Boats which cleared from Bristol, on the Delaware Canal.

Years.	Boats.
1842	4603
1845	7361
1846	7785
1847	9208

Total number of clearances from the port of Philadelphia, in 1846, of coals of all descriptions, 8,953 vessels, averaging 120 tons, and containing 1,065,228 tons, in addition to that shipped in boats from the Lehigh.* In the year 1847, the number of clearances of vessels laden with coal from Port Richmond, near Philadelphia was increased to 11,439.

Rates of Commission as regards coal, adopted by the Philadelphia Board of Trade.

Commission on Sales,	5 per cent.
Receiving Commission,	10 cents per ton.

Coal down the Susquehanna and the Tide Water Canal.

	1851.	1852.	1853.	
Anthracite,	129,276	135,316	185,501	tons
Bituminous,	20,673	18,992	21,601	"

DOMESTIC COMMERCE OF PHILADELPHIA.

The following tables show the progressive increase, in periods of five years, of the *enrolled and licensed tonnage*, engaged in the trade of Philadelphia, and of the total *registered*, enrolled, and licensed tonnage of that port; which increase is, in great measure, attributable to the coal trade of Pennsylvania, within the last twenty years.†

* Philadelphia Commercial List, January 16th, 1847.

† Statistical Annals of the United States. Adam Seybert, M. D., Phila., 1818, and other authorities.

Years.	Coastwise. Licensed and enrolled Tonnage.	Total Tonnage registered, enrolled and licensed.	Years.	Coastwise. Licensed and enrolled Tonnage.	Total Tonnage registered, enrolled and licensed.
1791	3,222	57,520	1841	71,588	105,805
1795	7,325	90,946	1842	57,749	100,641
1800	8,032	103,011	1843		104,340
1805	11,000	87,254	1844		114,894
1810	15,803	125,258	1845		End of June 30th.
1815	22,360	87,254	1849		188,087
1820	24,117	83,575	1850		206,497
1825	29,421	95,011	1851		222,428
1830	24,236	72,215	1852		229,443
1835	49,860	*86,445	1853		*252,451
1840	67,045	103,944			

There are a few unavoidable, but not very material, discrepancies in some of these returns, owing to the different sources from whence the data have been obtained.

FOREIGN COMMERCE.

As regards the *foreign commerce of Philadelphia*, our returns exhibit a great falling off, whilst that of New York and Boston has considerably augmented. The following abstract is sufficient to show the relative proportions of the foreign trade enjoyed by these three principal ports.

Value of Exports, domestic and foreign, from the Custom-House returns, from the Ports of Philadelphia, New York, and Boston.

Years.	Philadelphia.	New York.†	Boston.
	Dollars.	*Dollars.*	*Dollars.*
1843	3,059,171		
1844	3,664,696	28,526,739	
1845	3,916,833	30,422,672	9,370,857
1846	5,118,054	35,607,367	8,575,384
1847	8,589,265		
1849		39,736,969	
1850		60,119,248	9,141,652
1851		87,653,849	†10,198,180
1852	5,828,571	71,023,509	
1853	5,680,759	‡93,909,998	†21,000,000

* Report of the Board of Trade.
† Hunt's Magazine.
‡ New York Journal of Commerce. Foreign exports $91,000,000.

Table of the Foreign Arrivals and Departures, and of the aggregate Value of the Exports and Imports.

	Philadelphia. Foreign Commerce.			New York, City and State.			Boston and Massachusetts.		
Years.	Vessels entered and cleared.	Tonnage entered and cleared.	‖ Value of Imports and Exports.	Vessels.	Tonnage entered and cleared.	¶ Value of Imports and Exports.	Vessels entered and cleared.	Tonnage entered and cleared.	Value of Imports and Exports.
			Dollars.			*Dollars.*			*Dollars.*
1821		147,918	15,550,689		326,435	48,144,657			
1825		*	25,485,545		543,610	87,057,252			
1830			12,993,915		558,000				23,528,791
1832		†	14,194,324			79,214,347			25,868,280
1835			16,129,112		834,054	118,536,569			26,596,146
1842	893	173,581	11,152,515	9,166	1,921,810	85,453,000	4,204	662,887	27,794,000
1844	901	168,810‡	7,453,006	12,012	2,895,176	91,088,693	4,865	712,787	27,667,843
1845	787		8,603,912			81,154,479			25,375,658
1846	917	173,000	8,829,925	11,568	3,176,981	111,289,696	5,830	796,721	34,507,073
1847	1,255					147,915,531	4,088		
1848							5,981	**	
1849	1,082					137,395,220	6,296		
1850	975		16,575,440			198,453,390	5,509		‡‡
1851	1,092		19,524,797			219,015,427	5,686	1,006,280	45,008,009
1852	1,118		22,284,278	‖‖		‖‖ 201,373,227	5,927		50,051,288
1853	1,050		26,845,785	7,163	3,239,000	287,987,650	6,145		

Value of the General Commerce, foreign and domestic.§

Imports and Exports.

Years.	Philadelphia.‖	New York.¶	Boston.‡‡
	Dollars.	*Dollars.*	*Dollars.*
1845	11,733,590		30,962,734
1846	13,426,669	111,289,696	31,202,050
1847	20,725,202	147,915,531	
1848			
1849		137,395,220	
1850	16,576,440	198,453,390	34,797,815
1851	19,524,797	219,015,427	40,106,319
1852	22,284,278	201,373,227	
1853	26,845,785	††287,987,650	§§64,317,379

* Hazard's Regis. of Penna., 1828.

† Dictionary of Commerce,

‡ Geography of Pennsylvania, Trego, p. 145. Commercial List and Philadelphia Prices Current. American Almanac, 1845–7. Hunt's Merchant's Magazine. Niles's Register. M'Culloch's Gazetteer. Emigrants' Directory, 1820. Commerce and Navigation of the United States, 1844.

§ Custom-house returns.—Commercial List, 1854.

‖ Philadelphia North American.

¶ Hunt's Magazine.

** De Bow's Review.

†† New York Journal of Commerce estimates foreign exports and imports at $281,000,000.

‡‡ American Almanac.

§§ Boston Post.

‖‖ New York Journal of Commerce.

*Table of Importations into the Port of Philadelphia.**

	Value.	Duties.
1847	$12,145,937	$2,904,749
1848	10,700,865	2,762,093
1849	10,160,479	2,714,965
1850	13,381,759	3,361,112
1851	14,871,992	3,673,123
1852	16,455,708	4,033,909
1853	21,165,020	5,278,083

Official Statement of Duties on Imports into Philadelphia.

1849 - - - - -	$2,714,965
1850 - - - - -	3,361,112
1851 - - - - -	3,673,123
1852 - - - - -	4,033,909
1853 - - - - -	5,278,083

In the subjoined statement we have shown the amount of *tonnage, owned, registered, and enrolled,* of three of the principal commercial ports of the United States, at stated periods, whereby the contemporaneous advance of their trade is made apparent—compiled from official returns.

Ports.‡	1810.†	1831.	1834.	1839.	1843.	1844.	1845.	1846.
	Tonn'ge.	Tonn'ge.	Tonn'ge.	Tonn'ge.	Tonn'ge.	Tonn'ge.	Tonn'ge.	Tonn'ge.
New York,	268,548	286,438	359,222	430,000	496,965	525,162	625,875	655,695
Boston,	149,121	138,174	212,536	203,615	201,323	210,885	227,994	241,520
Philadelphia,	125,258	79,968	85,520	96,862	104,340	114,894	147,812	148,058

Ports.	1847.	1848.	1849.	1850.	1851.	1852.	1853.
	Tonn'ge.	Tonn'ge.	Tonn'ge.	Tonn'ge.	Tonn'ge.	Tonn'ge.	Tonn'ge.
New York, . . .	646,043	733,077	796,491	835,867	931,193		1,149,133
Boston, . . .				313,192	342,936		450,492
Philadelphia, . .			188,087	206,497	222,428	229,443	252,451

For the more complete illustration of the relative commercial importance of these ports, we have added a table of their *foreign arrivals and coastwise arrivals,* respectively, during the years subsequent to 1810.§ The later years are from the Philadelphia Commercial List.

* Philadelphia North American, January, 1854.
† Seybert's Statistical Annals, p. 308.
‡ Hunt's Magazine, and Report of Board of Trade for Philadelphia.
§ Hazard's United States Register; Philadelphia Commercial List, and other sources. Commerce of Boston, Hunt's Merchant's Magazine, Vol. X., 1844.

Years.	Foreign Arrivals. No. of American & Foreign Vessels. New York.	Boston.	Philadelphia.	Coastwise Arrivals. New York, exclusive of sloops.	Boston, including sloops.	§Philadelphia, including sloops, &c.	Coal Trade of Philadelphia.* Clearances of vessels of all descriptions with coal.	Tons of coal shipped from the port.
1810			405			1,477		
1825			484			1,195	190	19,378
1830	1,510	642	374		2,398	3,287	644	63,137
1832	1,808	1,064	428	4,500	3,538	2,849	1,592	158,442
1834	1,932	1,156	430		3,527	2,686	1,575	156,154
1835	2,094	1,302	421		3,879	3,573	2,361	267,139
1836	2,293	1,452	421		3,944	3,764	3,225	344,812
1838	1,790	1,553	464		4,018	10,860	2,791	317,245
1840	1,953	1,628			4,251	11,188 }		
1841	2,118	1,791			4,446	11,738 }		
1842	1,962	1,719	†	3,803	3,862	10,457 }	includ'g the coal trade.	
1843	1,832	1,688		4,734	4,964	7,659 }	exclusive of the coal trade.	
1844	2,208	2,174	472	5,360	5,009	8,016 }		
1845	2,043	2,304	387	5,770		8,029 }	Port Richmond.	
1846	2,289	2,113	459	4,663	6,683	6,018 }	8,953	1,065,228
1847	3,147	2,734	657	4,864		17,083	11,439	1,375,000
1848	3,060	3,101	542		6,118	23,941		
							P. Rich'd in coal trade.	
1849	3,237	3,183	585		6,199	24,584	4,497	
1850	3,487	2,838	518		5,978	26,737	7,549	
1851	3,888	2,838	579	7,139	6,334	26,428	8,623	
1852	3,822	2,864	679		6,286	30,715	9,047	
1853	‖3,927		‡566		9,000	29,456	7,384	

It is becoming more and more difficult to obtain a correct exhibit of the coastwise commerce for the different ports.

From this view, it is seen that in the greatest increase in the number of foreign arrivals, Boston stands the first, while New York is the second; and on that of coastwise arrivals, if we include the coal trade, Philadelphia considerably outnumbers those of the two other ports. The apparent diminution in the coastwise arrivals at Philadelphia, from 1843 to 1846, is owing to the omission of all the small craft which it had been customary to include. So great are the discrepancies among these statements, that it is impossible to know which to select. There seems no rule observed, by which the actual state can be known through the returns, which can be increased or diminished at pleasure, according to the number of the smaller vessels incorporated therein.

As far as the port of Philadelphia is concerned, the annual returns in the table are exclusive of all ships, barques, brigs, and schooners,

* Commercial List, January 16th 1847.

† Commerce of Philadelphia.—Custom-house returns.

‡ Including steamers, 580.

§ Report of the Philadelphia Board of Trade and Commercial List, 1854. We annex the coastwise arrivals during the year 1853, from the Philadelphia Board of Trade. Ships, 96; barques, 109; brigs, 529; schrs, 6,325; sloops, 3,709; steamers, 1,136; barges, 6,525; boats, 11,098; total, 29,456.

‖ New York Journal of Commerce.

in the service of the U. S. government. They are also, with the exceptions marked, entirely independent of the enormous amount of coastwise shipping engaged in the coal trade.* These we have given, where we possess the data. The vessels at Port Richmond are chiefly employed in the coal trade.

Statement of the *enrolled and registered tonnage* of New York, Boston, and Philadelphia—employed in the foreign and coasting trade, including temporary registers, and exclusive of the fisheries:† omitting fractions, distinguishing the foreign from the domestic tonnage.

Years.	New York. City and State. Foreign Tons registered.	New York. City and State. Coasting tons enrolled and licensed.	Boston and Massachusetts. Foreign Tons registered.	Boston and Massachusetts. Coasting Tons enrolled and licensed.	Philadelphia. Foreign Tons registered.	Philadelphia. Coasting Tons enrolled and licensed.
1793	45,355	13,986	135,599	51,402	60,924	4,579
1800	97,791	51,553	213,197	75,080	95,631	7,380
1805	121,614	67,812	285,689	86,413	77,238	10,016
1810	188,556	83,536	352,806	107,260	109,628	14,255
1815	180,664	100,960	299,298	115,327	77,199	19,875
1820	115,632				59,458	24,117
1825	159,327				65,590	29,421
1830	110,163				47,979	24,236
1835	200,780				51,588	34,857
1840	202,370				52,268	51,676
			Boston only.			
1842	226,072	233,401	157,116	36,385	42,891	57,749
1843	237,240	259,725	165,482	37,116	39,445	64,894
1844	253,888	271,273	175,330	35,564		
1845			191,853	42,146		

	New York City.‡ Registered, enrolled, and licensed tons.	Registered, enrolled, and licensed tons.	Philadelphia. Registered, enrolled, and licensed tons.
1848	733,077		
1849	796,491		188,087
1850	835,867	313,192	206,497
1851	931,193	342,936	222,428
1852	1,016,599	381,088	229,443
1853	1,149,133	450,492	‡252,451

1846—tonnage owned by New York, 655,695; by Massachusetts, 541,520; by Pennsylvania, 148,058. 1853—New York, 1,150,000.

Note.—In commercial navigation, the *registration or enrolment* of ships at the custom-house, is designed to entitle them to be classed among national shipping, and to enjoy the privileges of the country and port, to which they belong, and in which they have been built.

Licenses are granted under certain regulations; among which are

* The returns for these years in the table are from a statistical statement in the United States Gazette, February 17th, 1847. In Bicknell's Reporter they are thus stated—in 1845, 4,620; in 1846, 7,046.

† Seybert's Statistical Annals, p. 321—324, and subsequent sources.

‡ Hunt's Magazine.

their limiting the vessels to certain maximum proportions, and not to be square-rigged vessels, or propelled by steam. These licenses contain an accurate description and admeasurement of the vessel, the names of which may not be changed, and their owners must give security by bond as to the employment of the vessels, which are restricted to the uses assigned.*

The entrances are not enumerated, because they correspond so nearly with the Table of Clearances.

Number of Clearances of American and Foreign Vessels engaged in General Commerce from the principal Ports of the United States, for the year ending June 30th.†

	1847-8.			1849-50.			1851.		
Ports.	No. Ves'ls.	Tonnage.	Crew.	No. Ves'ls.	Tonnage.	Crew.	No. Vesl's.	Tonnage.	Crew.
1 New York, .	2,401	758,745	30,247	7,303	2,149,098				50,000
2 Boston, . .	2,060	281,874	14,412	3,906	546,752		3,103		
3 New Orleans,				843	369,937				
4 Baltimore, .				521	126,819				
5 Philadelphia,	583	143,143	6,155	479	111,618		556		
6 California, .				623	180,128				

1853‡—No. of vessels cleared in New York, 3,236; tonnage, 1,484,000. In Philadelphia, 580.

Steam Engines employed in the Coal business in Schuylkill County§ *in* 1846.

	No.	Horse-power.
In Pottsville, previous to 1846,	68	2018
In Pottsville, built in 1846,	23	636
In Minersville and Port Carbon,	15	267
Employed in 1846, - -	106	2921
" 1847, - -	167	4465
" 1850, - -	169	4818
" 1852, - -	205	

Steam Engines employed in Schuylkill County in 1852.

39 engines engaged in hoisting coal, with an aggregate power—horses, - - - - -	1,040
64 engines engaged in pumping and hoisting, with an aggregate power of—horses, - - -	2,724
102 engines engaged in breaking and screening coal, with an aggregate power of—horses, - -	2,745
205 engines.	
Aggregate steam power engaged in the coal trade, -	6,509

* McCulloch's Commercial Dictionary. † Registry of the Treasury.

‡ New York Journal of Commerce.

§ From the Pottsville Miners' Journal, January 23d, 1847, and 1851. Also annual reports of the Pottsville Board of Trade.

Which may be put down as equal to the labour of 32,545 men.

1852.—In Schuylkill county there are engaged in the coal trade,	179
In Lehigh Region,	64
In the Lackawanna,	55
Engines,	298

In the valley of the Schuylkill were, in 1847, 324 miles of railroad and 108 miles of canal; in constructing of which have been expended upwards of $19,000,000, while the improvements in railroads and canals, in connection with the transportation of anthracite in the Lehigh valley, is ascertained at $7,045,000; in other avenues $8,000,000; and in the whole more than $37,000,000.

In justice to individual enterprise, at an early period of the employment of an almost untried combustible, we are bound to note that in January, 1825, Messrs. Jonah and G. Thompson, of Philadelphia, completed for their Phœnix Nail-works, on French Creek, a steam engine in which anthracite was employed. We understand this was the first successful application of this fuel to the generation of steam.†

EMPLOYMENT OF ANTHRACITE IN IRON MAKING.

In the "Revue Generale de l'Architecture," M. Michael Chevalier published in 1840 an account of the anthracite basins of Pennsylvania. His statements contained nothing particularly remarkable, save that they brought down the condition of the operations in coal to a later period than that of Mr. Packer's report in 1833, upon which they are obviously based. He remarked, that the Americans have found out the means of making anthracite available,—not only for manufacturers,—but what was equally novel, for domestic purposes: so that it has not only almost superseded the use of wood in eastern Pennsylvania, but in most towns and cities along the Atlantic shore. New applications of anthracite are discovered, and the Pennsylvania iron masters, in imitation of Mr. Crane, have successfully applied this combustible in their furnaces.

For domestic purposes its use has been greatly aided by the employment of stoves, the adaptation of which to this species of fuel has been advancing, from year to year, in a continued series of improvements, until there is little left to amend or desire.

After enumerating some of the difficulties attendant on its first introduction, and on the acquiring a knowledge of its properties, which were made apparent almost by accident, M. Chevalier adds,—"Mr. Wetherill, one of the principal manufacturers in Philadelphia, showed me in 1835 the place where, twenty years before, he had dug a hole to bury the anthracite, then looked upon as incombustible refuse."‡

* Pottsville Miners' Journal.
† Monthly American Journal of Geology, G. W. Featherstonhaugh, Vol. I. p. 72, 1831.
‡ Revue Generale de l'Architecture, 1840.

In relation to the present estimation in which anthracite is held, we may trace its growing importance in exact proportion as, year after year, new methods of application were, almost involuntary, invented, and as one difficulty after another was surmounted.

Long within the experience of the present writer, large areas of Welsh anthracite land heretofore neglected and commanding only insignificant prices, have acquired a value wholly unexpected. It is no farther back than 1828 that we find Mr. Bakewell, a geologist of no slight eminence in his day, lamenting that the quality of the Welsh coal was "so inferior," and, in fact, so impracticable as to be of little comparative use.*

Let us hear what is now said of this formerly despised combustible, by an intelligent authority writing from the same region:—"Anthracite may be termed a native mineral, containing ninety-four to ninety-six per cent. of carbon; burning without smoke or clinker in the grate, and almost wholly free from sulphur. One hundred tons of this anthracite are equal in effect to a hundred and forty-four tons of bituminous coal. Therefore, it enables steam vessels to carry, in the same space, nearly twice the quantity of effective fuel; while the use of anthracite in these vessels lessens the cost of stoking five-sixths. The wear and tear of bars, boilers and furnaces, owing to the absence of sulphur, is less. Furnaces of the same dimensions yield, on the average, forty per cent. more iron with anthracite, without any additional cost for labour. Anthracite pig iron is found to possess greater strength and tenacity than any other. In re-melting, the iron runs more fluid, and is very strong—a union of qualities most desirable, but seldom met with; and, owing to the intense and continued heat of anthracite, some of the richest iron ores, not fusible with bituminous coal, are now easily smelted."†

The rapid progress made in the manufacture of iron in America, by means of Pennsylvania anthracite, since the commencement of the process in 1840, and even during the subsequent years of unexampled prostration in every department of business in this country—especially unpropitious to the introduction of a new branch of manufacturing industry—attest the growing importance of this description of fuel. In this State no less than thirty-six furnaces have been erected during this interval, and several others are reported to be in progress. Those completed yielded in 1845–6, at the rate of 107,200 tons per annum of anthracite iron: being one-third of the entire production of pig iron in the United States heretofore. In 1846, the production of forty-three anthracite furnaces was estimated at 119,437 tons. To this statement must be added a corresponding proportion of refined, puddling, rolling mills for bar and railroad iron, and other works, in which this fuel is now solely used.‡

The Board of Trade of Schuylkill county published the following statement of the number of furnaces and rolling mills in Pennsylva-

* Bakewell's Introduction to Geology, 3d edition, p. 181.

† Mining Journal of London, Vol. X., p. 189, 1840.

‡ Letter of the Committee of the Iron and Coal Trade of Pennsylvania.

nia and New Jersey which employed anthracite as a fuel, and were in operation previously to April, 1846: premising that there were only four anthracite furnaces in activity prior to 1842:

	Tons.	Tons.	
42 furnaces, producing of pig-iron per week, - - - - -	2360 or	122,720	annually.
27 rolling mills, manufacturing annually,		114,500	"*

It has been somewhere maintained that coal which yields a *red ash* never works well in blast furnaces, in consequence of the sulphur it contains. It is urged that this sulphur can never be effectually gotten rid of, except by the complete combustion of the coal and of the sulphuret of iron which prevails—the process of coking, whether in ovens or pits, only reducing the *per* to the *pro*-sulphuret. *White-ash* coals, it is therefore suggested, should always be selected for blast furnace work, whenever practicable.† We conceive that these observations were intended to have reference to the varieties of Welsh bituminous coal in the iron districts, and does not apply to anthracite here. If so, no comment is needed.

In Pennsylvania, the subject of the comparative values of red-ash and white-ash anthracites used in blast furnaces, has been discussed by practical persons, most of whom conceive that the one is equally advantageous with the other. In one respect the preference is given to the white-ash variety, on account of the greater density and compactness of its structure than the red-ash coals, which are softer, and are supposed to make a less strong fire. But with respect to the theory that red-ash coals contain more sulphur than the white, it has yet to be proved that in an equal average weight of each there is any appreciable difference, taking one coal seam with another.

It is well known that the white-ash coals of Pennsylvania contain a larger amount of carbon than the red-ash species, and that their specific gravity or density is correspondingly greater. The excess of carbon in the one being balanced by an increase of earthy and ferruginous matter in the other. Yet this excess of earthy matter, containing among other materials a small amount of sulphuret of iron, is very insignificant, and would scarcely produce any perceptible difference in the iron produced by the agency of that variety of coal.

To ascertain the respective amounts of ashes in these two classes of anthracite, we have consulted a variety of tables of analysis of Pennsylvania coals—the results are as follows:

	Per cent.
Twenty-three analyses of different white-ash coals give an average of ashes, - - - - - -	4.62
Twenty-one analyses of red-ash coals in Schuylkill region,	7.29

* Fourteenth Annual Report to the Coal Mining Association, Pottsville, April 1846, p. 9.

† Data for the use of blast furnaces, by S. B. Rogers, Nant-y-glo.

The red-ash has, therefore, only about two and a half per cent. more of earthy matter,—of which portion only eight or ten per cent. consists of iron—than the white-ash.

But what is more directly to the point, is the summary of results of Professor Rogers's analysis of both kinds of anthracite from the Pottsville district, wherein the red-ash had absolutely less sulphur than the white.

The introduction of anthracite and the hot blast in the iron-making districts of South Wales, has materially changed the relative proportion of the materials. Formerly, when the bituminous variety was employed in the coke furnaces of Monmouthshire, the materials necessary to make one ton of iron was, as stated by Mr. Rogers, of Nant-y-glo, as follows:

Coke, containing 4032 lbs. of carbon,	2 tons or	4480 lbs.
Calcined iron-ore, containing 2240 lbs. of iron,	$2\frac{1}{4}$ "	5040 "
Limestone, - - 2105 "	$1\frac{1}{4}$ "	2240 "
Atmospheric air, 360,000 cubical feet about	$1\frac{1}{4}$ "	2700 "
To make one ton of iron,		14,460 "*

In the United States, the employment of either species of coal and the make of iron from each, will be partly governed by local circumstances, and particularly by the proximity to the main deposits of fuel; while large quantities of iron will, for a long period, be made through the agency of wood. The most rapid advance in iron making, of late years, is in the vicinity of the anthracite districts of Pennsylvania, where, as has been already stated, thirty-six blast furnaces, employing anthracite alone, have been put in operation since the year 1840.

We will endeavour to present an epitome of the iron making of Pennsylvania,—premising that many of these returns are extremely defective.

Years.	Blast furnaces.	Forges & Rolling Mills.	Pig iron made tons.
1828	44	78	24,822
1830	45	84	131,056
1842	213	169	151,885
1843		"	190,000
1844		"	246,000
1846	317	"	368,056
1847	317		389,350
1849			253,370
1850	304	200	198,813
1853			368,056

* Mining Journal of London, 1840, 1841, and subsequently.

By a pamphlet on this subject published in 1847, by C. G. Childs, there were in the year 1840, by the census returns,—

			Tons.	
Of furnaces, - -	213	producing	98,395	cast-iron.
From the bloomeries, forges, and rolling mills of the State,	169	"	87,244	bar iron.
Fuel consumed in the process, *chiefly charcoal*, - -		"	355,903	
Number of workmen employed in all these operations, including mining fuel and ore, - - -			11,522	persons.

Pennsylvania State is emphatically the great store-house of the Union, in the way of coal and iron. In the ten following counties are located the principal iron-works:

	Iron Works.
Berks county has - - - - -	44
Lancaster - - - - - -	30
Clarion - - - - - -	30
Huntingdon - - - - - -	28
Blair - - - - - - -	27
Chester - - - - - -	25
Venango - - - - - -	21
Columbia and Montour - - - -	20
Centre - - - - - -	20
Armstrong - - - - - -	18

PRODUCTION.

The Committee of the Coal and Iron Association of Pennsylvania reported in July, 1846, an estimate of the iron manufacture of this State, from which we derive the following summary:

	1847.		1850.	
	No.	Tons of iron annnually made.	Furnaces in blast.	Tons.
Furnaces operating by the use of charcoal, .	274	248,569		
Furnaces employing anthracite, . . .	43	119,487		
	317	368,056	349	198,813
Increase in the number and production in four years, being at the rate of 142 per cent. in that period,	104	216,171		

Decrease from 1847 to 1850, 190,537 tons, or 49 per cent. in three years.

Capital employed in the production of this amount of Pig Metal.

		1842.	1846.	1850.
	Dollars.	*Dollars.*	*Dollars.*	*Dollars.*
Charcoal furnaces, $47 capital to every ton of pig metal manufactured. Anthracite furnaces, $25 capital to each ton of pig metal manufactured.		8,560,418	14,669,818	12,921,576
Increase of capital invested in four years.	6,109,400			
Capital required in the conversion of the pig metal.				
One half the aggregate made, converted into bars, hoops, sheet iron, nails, &c., at $20 the ton,	3,680,560		5,520,740	
The other half, into castings, at $10 per ton,	1,840,280			
			20,190,658	

Population employed in this branch of industry.

	1846.	1850.
In mining the anthracite and ore, -	4,978	30,103
In making the charcoal, - -	12,428	
Persons dependent on these for their subsistence, - - - -	69,624	11,513
Population connected with the production of iron, and persons deriving support by the labour in the *conversion* of the iron, estimated at a similar number, - - - -	87,030	238,000
Total,	174,060	269,616

Without reckoning those who are connected with the manufactories of iron, machinery, &c., or in the transportation and sale of coal and iron, or in the business of railways, canals, &c.

In the collection of "Documents relating to the Manufacture of Iron in Pennsylvania," published on behalf of the convention of Iron Masters, in 1850, is an appendix by Charles E. Smith, Esq., Chairman of the Committee on Statistics. This article contains fifteen very valuable illustrative tabular statements, some of which we have taken the liberty of condensing in our pages.

Blast furnaces 298, and 6 bloomeries, one-fifth using anthracite, 304.

Present capacity, tons, 550,959.

	1847.	1849.	1850.
Made in tons,	389,350	253,370	198,813

According to the statement of Governor Johnson, the condition of the Iron interest in Pennsylvania, 1st January, 1851,* was,

Total number of Iron Works of all descriptions,	504
of men employed in the process,	41,616
of horses employed, -	13,562
Capital invested in land, buildings, machinery, &c.,	$20,502,076

The amount of toll collected upon Iron at all the offices on the Pennsylvania State Improvements, for 1853, $122,383.†

We take from a local paper‡ the following list of 16 iron establishments which, during this year, 1854, will furnish 160,000 tons of railroad bars, representing in production, $5,500,000 for labour, and 1,826,000 tons of raw materials, as follows:

Pig iron, required	1¼ ton per ton of rails,	213,333	tons.
Coal used,	5¼ tons per ton of rails,	840,000	"
Iron ore,	3½ tons per ton of rails,	560,000	"
Limestone,	1⅓ tons per ton of rails,	213,333	"
Total number of tons raw material,		1,826,666	

The capital employed in these establishments is reckoned at $10,000,000. They support a population of 92,500 persons. We give the list of the mills, and their production of rails, in 1854:

	Tons.
Montour Iron Works, Danville, Pa., - -	18,000
Rough and Ready, " " - -	4,000
Lackawanna, Scranton, " - -	16,000
Phœnix Iron Works, Phœnixville, Pa., - -	20,000
Safe Harbour, Safe Harbour, " - -	15,000
Great Western, Brady's Bend, " - -	12,000
New Works, Pittsburg, " - -	5,000
Pottsville Iron Works, Pottsville, " - -	3,000
Cambria Iron Works, Cambria, " - -	5,000
Trenton Works, Trenton, N. J., - -	15,000
Massachusetts Iron Works, Boston, Mass., -	15,000
Mount Savage Iron Works, - - -	12,000
Richmond Mill, Richmond Va., - - -	5,000
Washington Rolling Mill, Wheeling, Va., -	5,000
Crescent Works, Wheeling, Va., - -	5,000
New Mill, Portsmouth, Ohio, - - -	5,000
Tons,	100,000

* Speech of Governor Johnson, 1851.
† Report of Canal Commissioners.
‡ Philadelphia Bulletin.

Statement of the supplies of Pennsylvania Iron received at Philadelphia from the interior of the State, by canals and railways, during the year 1846.*

	Tons of 2240 lbs.	
Pig iron and castings,	66,568	
Wrought iron, - -	15,588	
Blooms, - - -	6,278	
Nails and spikes, - -	4,745	
Total,	93,179	Value, $3,944,055
In 1847, - - - -		" was 5,687,670

Pennsylvania now produces 368,056 tons of pig iron, valued, when worked into bars, castings, &c., at $23,921,960.†

Statement of the importation of Foreign Iron into the Port of Philadelphia.

		Aver. of 8 years, 1832 to 1839.	1842.	1846.
		Tons.		
Iron, chiefly railroad iron	76,132			
Steel, ...	2,276	79,738	3,714	3,575
Manufactured iron, . .	1,330			

The recent reduction is attributable to the vast extension of the domestic production.

Amount of Iron imported into Philadelphia during year 1853, *comparing with previous years, from Annual Report of the Board of Trade.*

	1850.	1851.	1852.	1853.
Iron, tons, ...	4,844	10,966	13,322	24,644
Iron, bars, ...	359,722	297,007	233,081	375,077
Iron, bundles, ...	136,423	153,859	120,028	194,338

Imports at Philadelphia.

Foreign iron imported into the District of Philadelphia, during the year 1853:

	Tons.	cwt.	qrs.	lbs.
Railroad Iron, - - -	337	4	2	19
Rolled Iron, - - -	14,870	3	3	9
Hammered and sheet Iron, -	1,988	6	1	9
Pig Iron, - - -	20,083	11	0	22
Old and Scrap Iron, - -	105	4	0	6
Castings, - - -	57	2	1	20

* Abstract of a table published in the North American, January 2d, 1847.
† See Childs on Coal and Iron.

	Tons.	cwt.	qrs.	lbs.
Chains, &c., - - - -	464	14	3	25
Steel, - - - -	1,037	13	3	7
Anvils, &c., - - -	122	8	0	10
Nails, &c., - - -	32	3	1	5
Hammers, - - -	11	1	3	21
Iron Wire, - - -	13	15	1	0

Within a year or two, and extending to the present time, the manufacture of iron has greatly increased; and the construction of new works on the largest scale, and the enlargement of old ones, are pushed with such activity that it would be impossible to enumerate them, were it necessary, in a work like this. The attention of the traveller is attracted by the appearance of new works in unexpected places, and announcements of the more prominent ones are constantly made in the newspapers. Of these, we condense from the Public Ledger the following in relation to the Cambria Iron Company's Works.

"These works are situated at the confluence of the Little and Great Conemaugh rivers, immediately below the thriving borough of Johnstown, near the western slope of the Alleghany mountain. The property embraces about twenty-five thousand acres. The lands, with four charcoal furnaces, and other improvements, were purchased a few years ago, for the sum of three hundred thousand dollars. They are accessible by the western division of the Pennsylvania Canal, the Portage Railroads of the State, and by the Pennsylvania Railroad, which traverse them east and west. The country has been here upheaved by some internal convulsion, and with it on both sides of a narrow valley, the richest deposits of iron ores, bituminous coal, hydraulic cement, fire brick clay, and limestone, in strata contiguous to each other. The principal vein, adjoining the furnace and rolling mill, (carbonate of iron,) lies over the coal measures, about two hundred feet above the bed of the Conemaugh, and sixty feet above the tunnel head of the furnaces. It covers an area of one hundred and sixty acres, and contains an inexhaustible deposit of ore, which smelts very freely, requiring but little limestone, as a flux. It makes a superior quality of pig, either for forge or foundry. There are five coal veins besides, four above the level of the canal, and seven feet vein below, which the company will have to work with a shaft fifty or sixty feet deep, if ever they should need it. The second vein lies upon a bed of hydraulic cement, which is five feet thick and of good quality, and the third upon a bed of limestone, two feet thick, which is used for fluxing the ores in the furnaces. The great advantage of the estate is, that it presents one of the most eligible sites for furnaces, rolling mills, foundries, and other iron working establishments in the world, nature having provided an exuberant store of wealth, and left it to the art of man to take it out easily and roll it down to the tunnel head of the furnaces, a labour-saving machine, worked by the power of gravitation. The Cambria Iron Company has a capital of a million of dollars, fully paid in. Besides the four blast furnaces, which have lately been enlarged and improved, and are now yielding a product of two hundred tons of pigs per week, the company have nearly completed four other blast furnaces, for smelting with coke the iron ores taken from the face of the hill, and calculated to produce each five thousand tons of pigs per annum, making an aggregate of about twenty-eight thousand tons per annum. They have also finished a large rolling mill, six hundred by three hundred and fifty feet in extent, in the shape of a cross, with sixty puddling stacks, twelve heating furnaces, four scrap furnaces, &c., and which, when in full operation, will produce more than one hundred tons of railway iron per day, or thirty thousand tons per annum. This, we believe, is the largest single mill in existence. The engine for the four blast furnaces is four hundred horse power. To drive it, there are eight boilers to each furnace, thirty-two boilers in all, sixteen of them forty feet long and thirty-six inches in diameter, and sixteen thirty-four feet long and thirty inches in diameter. The steam cylinders are two of twenty-eight inch bore and eight feet stroke, and the blast cylinders are seventy-two inches bore and six feet stroke, making at full speed seventy thousand cubic feet of blast per minute. The fuel for making steam and heating the hot blast pipes, is furnished by the gas extracted from the furnace stack flues below the tunnel head, and ignited under the boilers. The aggregate horse power of the mill is four hundred and forty, driving four engines. The engine for the squeezers is forty-two inch stroke, eighteen inch bore and eighty horse power; for the roughing rolls, it is six feet stroke, twenty-six inch bore, and one hundred and fifty horse power; for the rail mill it is four and a half feet stroke, twenty-six inch bore, and one hundred and fifty horse power; and for driving blowers, shears, pumps, saws, &c., the engine is forty-two inch stroke and sixteen inch bore. There are sixteen boilers of thirty-six inches diameter, and forty feet long."

Gas, as an auxiliary in smelting iron, in Pennsylvania.

The quantity of coal [anthracite] usually required in the iron works, has of late been reduced by the process of heating the blast by the gas from the top; and the steam engine is worked by heat derived from the same source—the boiler being at the top of the furnace.

The following account of the construction of blast furnaces for smelting iron with anthracite, was prepared by the editor of this edition, for the American Journal of Science, in which it appeared in 1848. The details are those of the Chikiswalungo furnace, near Columbia, Pennsylvania, of which the editor is senior proprietor.

"The anthracite which we have found best adapted for smelting iron, is that of the Big Vein of Wilkesbarre, furnishing the coal of the 'Baltimore Company;' the 'Diamond,' of M. C. Mordecai; and the 'Black Diamond,' of Robarts, Walton & Co.; all of which are from the same strata and of the same quality. These coals command about twenty-five cents a ton more in the general market, than the others in this part of the coal region. The Pittston coal is of a good quality, but being softer, it makes more waste in handling, and being more distant, transportation downwards is somewhat higher, and boatmen will not go there for freight, if they can get it lower down. The North Branch canal extension being completed, this coal will in future have its natural outlet northwards.

"The Pine-Grove region supplies an excellent coal for domestic purposes and the generation of steam. It is not hard enough for smelting iron, although it can be used.

"Some of the Shamokin beds, or those of the Coal Mountain Company, we judge to be well adapted for smelting iron. We have not been successful in using Plymouth coal.

"Passing eastward to the Schuylkill and Lehigh region, we again meet with good iron smelting coal. Most of the Lehigh coal will answer this purpose, except that of Buck Mountain, which is a free burning white-ash, well adapted for generating steam. The Broad Mountain white-ash coal is among the best for iron smelting. Red-ash coal generally burns too freely for this purpose, although well adapted for steam purposes.

"The occurrence of inexhaustible strata of anthracite coal in Pennsylvania, has attracted the attention of miners and practical men generally, to its use in smelting iron. With charcoal, this process requires a peculiar location, and a large capital, to be invested in extensive woodland tracts, which are generally mountainous, and consequently cheaper, being unfitted for agriculture. This renders the construction of the necessary roads difficult, and transportation expensive. The number of workmen employed in wood-cutting, coaling, and hauling, is large, and the expense of horses and wagons, forms a considerable item. Charcoal being a soft, porous material, much of it is wasted in transportation and handling, and large sheds are required to store and keep it dry. These various contingencies require the general manager to have industry, judgment, and good business habits. In using anthracite, the exact expense of the fuel is known, the transportation being by railways or canals extending to most of the mines, and if the furnace is placed near such public works, there will be but little waste of coal in its final transportation. There is but little waste in the transportation of ore, which is of course common to both kinds of fuel.

"The earlier attempts at smelting iron with anthracite in the ordinary furnace, failed so completely, that it was by some deemed impossible to accomplish it; while others, looking to a different construction for a solution of the problem, devised various structures, more remarkable for ingenuity than utility; later experiments having proved that no such modifications are necessary, except perhaps a higher inclination of the bosh and a less contracted tunnel-head.

"Incandescent anthracite has the peculiarity of being rapidly extinguished when struck with a blast of cold air, the loss of heat from this source exceeding that resulting from combustion; and although this phenomenon does not take place when the temperature exceeds a certain point, the vast accession of cold air in a blast furnace, may be sufficient under slightly unfavourable conditions, to produce it at any time. Hence a hot blast, which is economical when charcoal is used, becomes an essential element of success with anthracite; and its temperature should not be less than is sufficient to smelt a slip of lead opposed to a jet of it near the twiers. Anthracite being a very dense and concentrated fuel, the amount of air thrown in must be much greater than when charcoal is used. Success, therefore, depends upon the quantity as well as upon the temperature of the blast.

"The necessary amount of oxygen can be secured only by means of the proper machinery, and a certain velocity of the blast; and in consequence of this fact, the false opinion that the effect depends merely upon the velocity or sharpness of the jet, is universally maintained. In consequence of this view, the exit pipe is reduced to a small size, and the quantity secured by increased velocity under a high pressure; which causes much of the blast to be lost, as among the multitude of joints to be made air-tight, it is impossible to secure them all. Besides this, the machinery is liable to injury from the great and unnecessary strain upon it.

"The stack or main structure of a blast furnace, is a quadrangular pyramid, the lower portion of which has an arched passage through the middle of each side, leaving four large piers of masonry. Three of these passages (A, fig. 1) are named twier-arches. The junction of these arches forms an open square about one-third the diameter of the stack, in which the hearth, (which requires renewal from time to time,) is built up with large cut stones of siliceous conglomerate or sandstone. Near the top of the hearth, the inner portion of the four arches is closed by forming a square with eight large sows or iron beams, four of which are shown in section at S in fig. 2, their position nearly corresponding with the dotted square in fig. 1. The dotted circle in fig. 1, indicates the internal face of the fire-brick lining (*l*, fig. 2) at its widest part, and also the top of the bosh,* (*b*, fig. 1, 2.) The lining being circular and the lower portion a square, the former is supported upon four plates (*q*) of such a form as to close the angles of the latter, and at the head of the furnace it is continued in an ordinary brick-work chimney, (*z*, fig. 3,) leaving one or two large vacant spaces for the purpose of filling. The chief use of this chimney is to protect the workmen from the heat.

"The two posterior piers have a circular passage (fig. 1, 2, *g*) for the admission of the blast pipe, *p*, which descends from the hot-oven (*o*, fig. 3) at the head of the stack. This passage is sometimes continued through the front piers, which renders the front or working arch, cooler, and gives more ready access to the twiers. The blast pipe is carried by appropriate branches to the posterior and lateral

* This word is from the German word böschung, a *slope*.

twiers, *t*, *t*, *t*, fig. 1, the former being seen in longitudinal—and one of the latter in transverse section, at *t* in fig. 2.

"To prevent the twier from being destroyed by contact with the fire of the hearth, it is made of an interior and exterior cone of wrought iron, with a stream of water circulating between them. The twier receives the nozzle, and this the belly-pipe, which is attached to the terminal upright portion of the blast-pipe, called the ball-and-socket joint, from having a connexion of this kind. Behind this, at *k*, there is a small aperture, to insert an iron rod to detach any slag that may cling to the twier. As the smelted materials within the hearth frequently rise above the twiers, it is evident that, in case the blast should be accidentally checked, they would flow into the blast-pipe. To prevent this, a valve (*v*, fig. 2,) is interposed, which is kept open by the blast, but falls the moment the pressure is removed.

"The cavity of the hearth, (H*h*, fig. 1, 2) where the metal is reduced and retained, bears a very small proportion to the size of the stack. In the figures, which are drawn to a scale of one-eighth of an inch

Fig. 1.

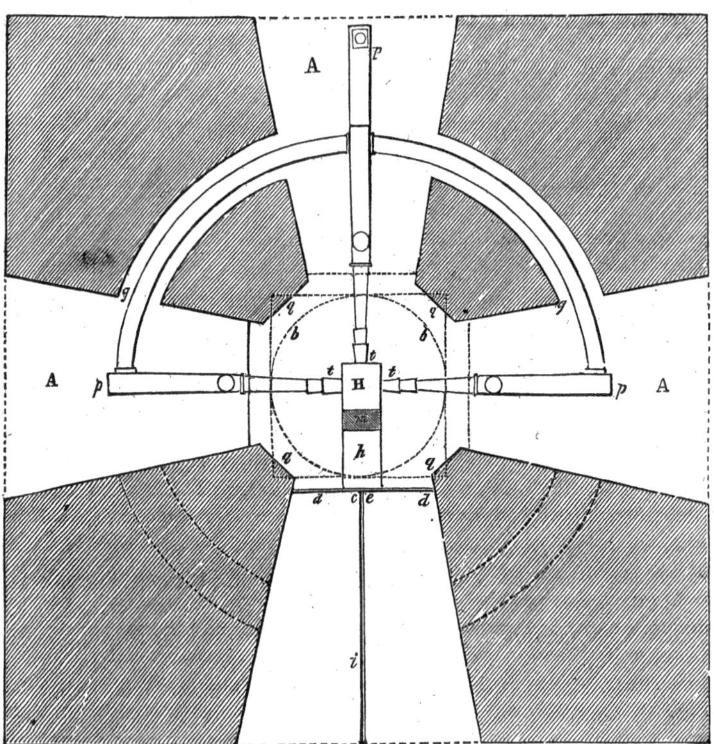

to a foot, it is two feet wide, and five and a half feet long, but enlarges upwards in a slight degree. The back and front of the hearth are separated by a partial partition called the temp, (*m*, fig. 1, 2,) made of fire-bricks, and which does not extend to the bottom of the hearth. The horizontal dotted line, (*h*, fig. 2,) shows the level to which the smelted materials can rise before running over the dam-stone (*d*) at *c*, where the cinder will escape first, being the lightest, whilst the smelted iron occupies the bottom. To prevent the fluid matter from being forced out by the blast, clay is rammed beneath the temp, around the twiers, and upon the surface, at *h*, where it is retained by heavy iron plates, which are raised every few hours, to allow the cinder to run off along the level of the top of the dust-plate *c, i*, whilst the metal is run off every twelve hours, at the lower level of *d*, through an aperture at the bottom of the dam-stone. The dam-stone is defended in front by a large iron dam-plate (*de*, fig. 1) against which the dust-plate *c, i*, rests. The lower edge of the latter rests upon the ground, which is raised to about the level of the bottom of the hearth *d, e*.

Fig. 2.

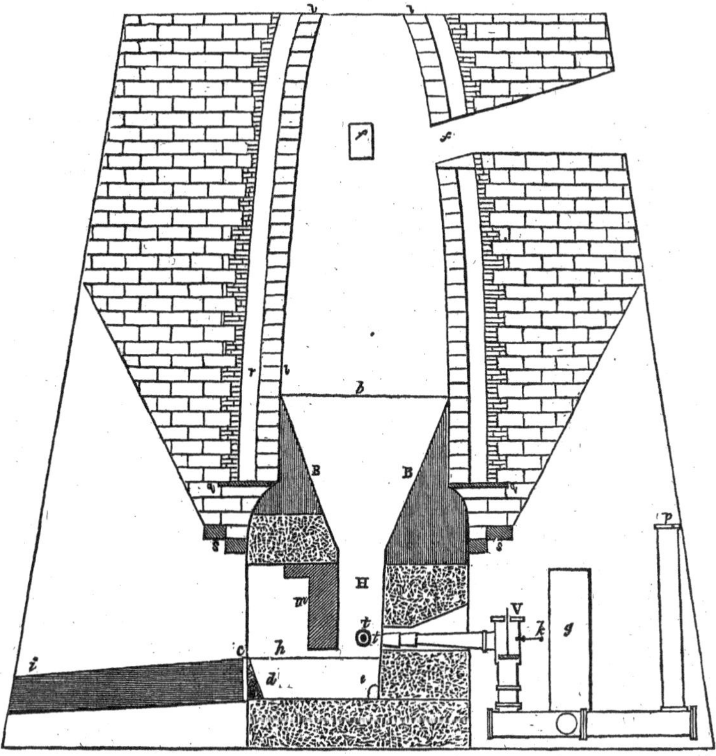

Explanation of Figs. 1, 2.

A, twier arches.	*e*, tapping place.	*k*, twier key.	*q*, square of the hearth.
B, bosh,	*f*, flues for boilers.	*l*, lining.	*r*, space for loam.
b, greatest diameter.	*g*, passage.	*m*, temp.	*s*, sows.
c, cinder run,	*Hh*, hearth.	*n*, sconsh'n.	*t*, twiers.
d, dam-stone & plate.	*i*, dust-plate.	*p*, blast-pipe.	*v*, valve in blast-pipe.

"Fig. 2 would represent a transverse section of the stack, if the left half were symmetrical with the right. In this case the temp *m*, and the open space in front of it would be filled with stone to the bottom of the hearth, and *e* would represent the place of exit for the iron.

"The bosh B B, fig. 2, (shaded vertically,) resembles a large funnel except that its termination at H, fig. 1, 2, is square. It is built of fire bricks, except its lower portion in front, where, in consequence of the open temp arch (*m*), a large stone (*n*) called the sconsh'n, is laid across the front portion of the hearth. The inclination of the bosh is seen to be at a higher angle than when charcoal is used; but it may vary considerably without affecting the result. When in blast a few months, the bosh increases in steepness from the abrasion of its surface, and the hearth partakes of this enlargement; so that instead of being a parallelogram, it assumes an oval form. The enlargement of the hearth continues until its walls become so thin, that the radiation of the heat will prevent the inner portion from melting away any further; and in case the temperature diminishes, the inside will be protected by a coating of slag. I have known a furnace to be in successful action, when the hearth had been so much enlarged as to have the middle portion of the inmost back sow (*s*, fig. 2) melted away, permitting the blast to escape until the aperture was closed with tenacious clay. In this case the under surface of the sow was in contact with the brick wall usually built beneath, as an additional barrier to the escape of the heat.

"Towards the head of the furnace, there are three equidistant apertures (*f*, fig. 2) to admit the waste flame, first under the boilers, then through a return flue in them into a hot oven, which is placed in part upon the top of the stack posteriorily and laterally. When a separate engine is employed, the oven is placed upon the front side of the top, and the flame passes into it by a single aperture.

"The boilers are in this case three in number, twenty-six feet long, forty-five inches in diameter, with a return flue eighteen inches in diameter. They are represented as *w w*, in fig. 3, where the course of the flame is represented by the arrows leaving the outlet of the flues in the stack, and passing beneath and through the boilers into the hot oven *o*, which has one or two high chimneys to secure a proper draft. For the purpose of exhibiting the position of the boilers, a part only of the brickwork which supports and encloses them is represented in the figure and the minor details of construction are omitted. Figure (3) is an elevation and partial section of the right side of fig. 1, 2, showing a twier arch, with the aperture for the admission of the blast, the parapet upon the top, and the chimney (*z*) around the tunnel-head. The engine* is placed upon the ground on this side, the boilers extending to the bank against which the structure stands. When the convenience of a bank or hill cannot be had, it is evident that both the boilers and oven might be placed on

* The engine is of 100 horse power, and the diameter of the furnace 10 feet. It is capable of smelting 90 tons a week.

or near the ground, if the chimney were sufficiently high, (not less than seventy feet,) and the walls built so as to be free from crevices.

Fig. 3.

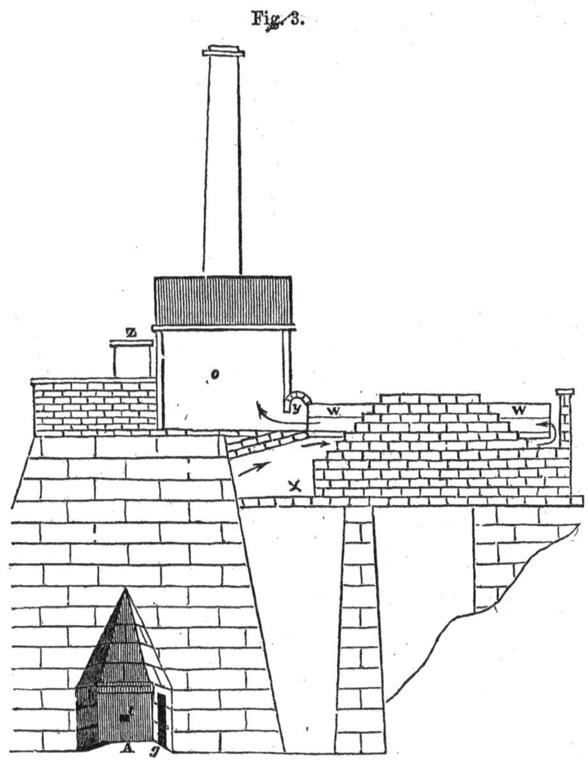

"The boilers are supported on large iron beams, (partly shown below the bricks at *x*,) between which arches are turned in their longitudinal direction. There are several doors in the position of *x*, to allow the flues to be cleared from ashes, &c. These doors open into the arches beneath, and there are others along the sides for the same purpose. The boilers are usually placed in contact with the oven; but the passage *y* (which extends to the chimneys) is proposed to be left to turn more or less of the flame into the chimneys, which will place the relative distribution of the heat to the oven and boilers under control, a point which seems not to have been hitherto attained. This might also be accomplished by separate chimneys to the oven and boilers. In either case, the chimneys must be supplied with a damper, which is best placed upon the top.

"If four simple boilers were used, the flame might be passed under one pair and return under the other; or the oven might be placed upon the bank, which would afford a good foundation for it and its chimneys; but the distance which the heated blast would be required to travel before reaching the twiers, would be an objection.

"The hot-oven is built and arched over with brick, and strongly bound externally with iron, the heat being sufficient to destroy supports passing through it. It is sufficiently large to contain a small forest of upright flattened pipes about ten feet high, with an internal cavity of about four by seven or eight inches, the thickness of the metal being about an inch. These are maintained at a red heat, the blast through them preventing their destruction. They stand upon two large pipes or cylinders about a foot in diameter, and from twelve to fifteen feet long, with a single row of apertures, (*a*, *d*, fig. 4,) and one or more (*b*, *c*) large enough to admit of a double row of apertures. Over the neck of each aperture a detached collar is placed, into which a pipe is firmly cemented, and the heads of two pipes on adjoining cylinders are similarly connected by an auxiliary pipe forming a semicircular or gothic arch, as represented in section in the upper part of the figure. The blast entering the first cylinder at *a*, meets a partition near the middle, and has to pass through the seven openings and pipes across the arched heads into *b*, and so on to *d d*, when it passes in the opposite direction to its place of exit at *g*, whence it descends to the twiers. The partitions are not in the middle of the cylinders, because by the time the air has passed half through them, it requires more room on account of its expansion." H.

Fig. 4.

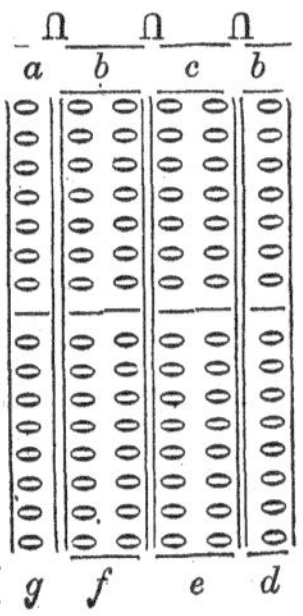

a b c b

g f e d

WIRE CABLES.

Wire ropes or cables adopted in mining shafts and on inclined planes, in preference to hempen ropes.

It does not appear that the use of wire cables in the hard anthracite collieries of Schuylkill county, has been altogether successful as a substitute for hempen rope in the "slopes." The difficulty arises from the chafing of the cables, and the clogging of the rollers by fragments of anthracite, which injure the wire cables much more rapidly than it does the hempen ropes.

We annex some valuable information on the subject of Iron Wire cables, taken from the Foreign portion of the first edition.

Employment of Iron Wire Cables in the coal pits where steam engines are used.

M. le Bergmeister Klotz has published a note in the "*Archiv für Mineralogie*," which article has been translated by M. Ch. Combes.

The translator states that twisted iron wire cables have been introduced in the shafts of extraction in the mines of the Hartz. The economy in the costs of extraction resulting from these wire cables, compared with those formerly made of hemp, soon determined the engineers and proprietors of the mines of Saxony, of Prussia, and ılmost all Germany, to adopt them. In France they have been slow o follow this example. The economical details, very circumstan-ially presented in M. Klotz's notice, appear to call for the attention of proprietors of mines, and of manufacturers of wire, in all coun-ries; and we present the readers of this volume with some valuable practical results, obtained under the supervision of the Prussian min-ng engineers, which are, probably, perfectly new to them.

Hempen Cables.—Towards the close of the year 1833, the provid-ng and maintenance of the "cables of extraction," in several of he collieries in the district of Essen and Werden, in Prussia, were iven to the enterprise and competition of the manufacturers of hempen ropes; their remuneration consisting of a price agreed on for each 100 scheffel [= 147.9 English bushels] of coal extracted from he mine, or shaft.

The names of those collieries, their depth of shaft, and the prices paid for the service of those hempen cables, are as follows:

That of Saelzer, whose vertical shaft is 216 Engl. feet deep, and hat of Neue Aack, which is 308 feet, Engl. deep, paying fr. 0.3608 = 3½*d.* (three pence halfpenny) = $0.07 per 100 scheffel [= 147.9 Engl. bushels] = about five tons. These terms are equivalent to $\frac{7}{10}$ of one penny Engl. or 0.01\frac{4}{10}$ American currency, per ton, as cost r service of the hempen cables.

The colliery of Wische, for a vertical shaft of 81 lachters = 152.604 metres [= 500 English feet,] paying fr. 0.4638 [= 4.494*d.* = $0.09] per each hundred scheffels. This is equivalent to $\frac{9}{10}$ths of one penny, or 0.01\frac{8}{10}$ per ton, for use of the ropes.

That of the Kuntswerk, whose shaft is 46 lachter = 152.6 metres = 283 Engl. feet,] paying 2 silbergros 6 pfennigs = fr. 0.3092 = 3 pence Engl. = $0.06 American,] per 100 scheffels. This harge is equivalent to $\frac{6}{10}$ of a penny or 0.01\frac{2}{10}$ per ton.

Subsequently, the price was lowered, in this case, = 2$\frac{2}{10}$*d.* Eng-ish, = 0.04\frac{4}{10}$ per 100 scheffels. Therefore the payment for the use of the hempen ropes was reduced to $\frac{44}{100}$ of a penny, or $0.00.88 per ton. But, it is added, at this last price, the contractor of cables suffered a loss, which he was able to prove.

The colliery of *Braut in Küpers wiese* has had for several years a contract which still subsists, by which it pays for the extraction of each 100 scheffels, by an inclined shaft, of 39 lachter [= 240 feet,] 3.596*d.* Engl. = 7$\frac{2}{10}$ cents. This sum is equal to 0.719*d.* = 0.01\frac{44}{100}$ per ton.

Without following M. Klotz through the details of the compara-ive value, weight, durability, cost, and power of the wire cables, at all the shafts, we will select the first only, the colliery of

Salzer and Neue Aack.

1. *Results with the hempen cables.*—In 1833, two cables were placed in these shafts, which lasted six months and fourteen days.

They cost 917 fr. [= £37,] and raised 19,645 tons of coal. Thus the expense was $\frac{45}{100}$ of a penny = \$0.00$\frac{9}{10}$ cent per ton.

At the end of that time, in 1834, two other new hempen cables were placed in the shaft, which lasted six months and seventeen days, and the charges, amount of work, and expense per ton, were about the same as the preceding.

A third and last set of hempen ropes lasted eight months and a half; and, calculating from the data furnished by the author, the expenses amounted to 0.54 penny = 1.08 cent per ton, of coal raised.

Results with the annealed wire cables.—In 1835, the first two cables of this description were adopted at this colliery. The two, together, were 931 feet in length: weighing 898 livres [= 988 lbs.] one of them lasted one year three months and twenty-four days. The expense per each 100 scheffel amounted to fr. 0.0655 = 0.126*d.* = \$0.00$\frac{25}{100}$ per ton. The other lasted fourteen months, and the charges per ton were 0.16*d.* = 0.32 cent.

Results with un-annealed wire ropes.—In 1838, two cables of this kind were fixed, which lasted eleven months; their cost and expenses formed a charge on the coal produced, of only penny 0.089, = ct. 0.178 per ton.

On comparing these four results, their proportionate expense on the extraction of the coal is as follows:

		Penny. Ct.	
1. Contract with the hempen rope manufacturers,		0.70=1.40	per ton.
2. Expense when using hempen cables,	1st and 2d set,	0.45=0.90	per ton.
	3d set,	0.54=1.08	
3. Expense when using annealed wire cables,	1st cable,	0.126=0.25	per ton.
	2d cable,	0.160=0.32	
4. Expense when using unannealed wire cables,	2 ropes,	0.089=0.178	per ton.

Colliery of Ver. Henriette.

Result with hempen cables.—In 1835–6, two hempen cables lasted six months and a half. The cost on the amount of coal raised in that time was fr. 0.4329 per 100 scheffel = *d.*0.84 = ct. 1.68 per ton.

Result with annealed wire cables.—In June, 1836, two wire cables were fixed at these mines; which, after having broken five times, were laid aside at the end of four months, eleven days, of work.

The extraction of coal cost, during this time, paying first cost, fr. 0.2061 per 100 scheffel; which is equivalent to *d.*0.4, = ct. 0.8 per ton.

Two other wire cables were next introduced, and endured six months and three weeks, and were removed after having broken eleven times. The expenses attending the raising the coal, during this second period, including first cost, was remarkably low, being only fr. 0.1277 per 147 bushels, which is equal to *d*.0.247, = ct. 0.494 per ton, on the quantity raised.

Result with un-annealed wire cables.—In May, 1837, cables of this kind, from the Hartz, were next adjusted; which lasted nearly unto the end of 1838, or eighteen months. The expenses, per ton, of coal raised were now reduced to *d*.0.049, = ct.0.098.

Recapitulation.	Penny.	Ct.	
2. Cost of raising coal with hemp cables,	0.84=	1.68	per ton.
3. " with annealed wire cables,	0.40=	0.80	"
	0.247=	0.494	
4. " with un-annealed wire cables,	0.049=	0.098	"

The average proportionate results of 2, 3 and 4, of these and other mines of the district, are represented by the figures or numbers, 100: 38: 13.

Wire Ropes in the Mining District of the Hartz.

In an address to the Royal Cornwall Polytechnic Society, by Mr. John Taylor, communicating some things that came under his notice during a recent visit to the mining district of the Hartz, the subject of wire ropes is particularly adverted to. The opinion of so experienced an authority being of no slight value, we give it in the following extracts:

"In the mines of the Hartz nothing engaged my attention more than the universal employment of wire ropes, for drawing the ores and waste from underground. This appears to be one of the most important improvements in the economy of mines that has for some time been made; and as it is but now beginning to make progress in this country, I am induced to notice it, in the hope that what experience may have been gained in Cornwall may be gathered at the next meeting of the Society: that the matter may be discussed, and the results made more generally known.

"The merit of this invention is due to M. Albert, the able and enlightened principal officer of the mining administration at Clausthal, who gave his zealous attention to this subject; and after overcoming many difficulties, succeeded in bringing them to their present perfect state. The first information respecting the use of wire ropes, afforded to the English miner, was by Count Brenner, *Oberberg Hauptman* (the title of the chief director of the mines of a country) of Hungary, by a paper which he communicated to the British Association at Newcastle, in the year 1838, at the meeting of which he was present, and did me the honour to ask me to read it for him. The subject did not appear to attract the attention it deserved; and it was not until the return of Professor Gordon from Germany, that any attempt was made to avail ourselves of the improvement.

"Of late, several persons have engaged in the manufacture, and great rivalry seems to exist as to claims to patents, and to superiority of quality. Some are in use in Cornwall, and there may now have been time enough to have gained a certain degree of knowledge of the value of this investigation.

"In the Hartz the diameter of the wires appears so small, when compared with what one has been used to look at, as to wear a very remarkable appearance. The saving of expense, when compared with the use of hempen ropes, is stated to be very great. Those I saw were made of twelve wires; but, for great depths, the upper part is somewhat stronger. The pulleys over the shaft are seven or eight feet in diameter; and some stress is laid on this, and I do not consider it a fair trial of these ropes to work them over pulleys of much smaller diameter. I was, however, surprised when I visited the iron works at Ilsenberg, at Rothehütte, and at Konigshütte, to find wire ropes used on cranes in the foundries, and in machines for raising iron ores perpendicularly in train wagons to the top of high furnaces. In these cases, the barrels on which the ropes wound, and the pulleys over which they worked, were necessarily very small in diameter. They seemed as pliant as they need be, and to have sustained no injury in use. I was informed that for such purposes the ropes were formed of a greater number of smaller wires, by which it was found that they endured the bending to a more acute angle, without injury. I notice this, to show that they may be adapted to almost every use to which cordage is applied."

The author suggests that oxidation of the wire ropes might probably be prevented by a process now much used in France, which is by coating iron with zinc, in the same way as it is covered with tin in the manufacture of tin plate. This is termed galvanising the iron, and is very successful in protecting it from rust. It is commonly applied to wire-work that is exposed to the weather, such as trellis work for gardens, &c.; and is performed after the wire is woven into the forms required, and at a very cheap rate. It unites or solders the joints or crossings; gives the whole a very pleasing appearance, and is effectual in preserving it for a great length of time. This is more especially applicable for wire ropes that are extended in a rigid state; such, for instance, as standing rigging.

Wire Ropes in the English Collieries.

Since the foregoing communication was made, a great deal of discussion has taken place on the employment of these cables as a substitute for hemp in the English mines. Much information has been elicited by means of various correspondents in the Mining Journal and other works devoted to the occasional consideration of such matters.

An objection to their use seems to have pervaded the pitmen, and much opposition has been made to their adoption, without, as it appears, adequate grounds, except on account of their novelty.

This has been fully shown in a trial at law, 29th of July, 1844, at Durham. The question between the parties at issue was chiefly whether the wire-rope used at the Wingate colliery, and by which the men were lowered to their work, was fit and proper for that purpose or not.

It appeared that the use of wire-rope in the English collieries dated only since the end of the year 1842, when they were introduced in various parts of the northern coal-field; both in working inclines and in raising coals from the pits. Those used on the inclines were chiefly round: while those of the collieries were made *flat*, for the greater facility of rolling them on the drum.

The rope, used at the Wingate colliery, complained of, consisted of 96 wires, and was worked in consequence with another rope of 144 wires.

On account of the giving way of some of the wires of the first-mentioned rope, the workmen refused to descend to their work. It was, however, submitted to a test of ten tons weight, and a portion of it to a strain of upwards of nineteen tons; although the usual weight required was only three tons and a half. It was on the scale of one cwt. to four fathoms = four and two-thirds lb. to each foot, and was considered capable, when new, of sustaining a weight of twenty-seven tons. Evidence was adduced that nine wire-ropes had given way in the Coxhoe and Jarrow collieries; while, on the other hand, it was shown that wire-ropes were used at various collieries throughout the country, and that they were generally looked upon as much safer than hemp; that they were better, and that any symptom of weakness was sooner perceived.

The cohesion in the wires in the twisted ropes is such, that the fracture of a wire originally continuous does not essentially weaken the strength of the rope, on the same principle that the fibres of which the several strands of a hempen rope are composed, do not consist of continuous threads throughout, but are made up of a multitude of pieces, which vary from a few inches to a few feet in length each.

The decision of the jury, in this case, and the valuable evidence produced, appear to demonstrate the superiority of the wire rope over that manufactured of hemp.*

Wire ropes, both flat and round, are now in use in the English collieries, of the kinds patented by Smith and by Newall & Co. At the beginning of 1844, at the Gosforth colliery, Newcastle, two flat wire ropes made by Newall & Co., were in daily use: their weight, in proportion to hemp, being as 21 cwt. to 47 cwt., and the power of drawing increased by about 30 tons a day. One at Ince colliery, Wigan, had been in use eight months without deterioration. Those at Rainton colliery, after seventeen months work, remained in excellent condition.

Several other collieries had been using flat wire ropes for more than a year, without depreciation of their capability. Others are mentioned which have been in constant use during fourteen, sixteen, and eighteen months, uninjured. Their cheapness and durability are

* Mining Journal of London, August 3d, 1844.

asserted, by practical managers of mines, to exceed greatly those of hemp. On the Durham and Sunderland railway were thirty-seven miles of wire ropes; some of which according to the engineer, had been at work two years,—three times the duration of hemp ropes,—and are still in use and apparently very good.

Wire ropes first came into notice about the year 1836, since which time both round and flat cables are in common use in the English collieries, both for shafts and inclined planes; for the latter of which they are especially adapted.*

We perceive, however, that this subject is still open to discussion, and by no means settled. It has been charged against the iron pit ropes that frequent accidents and loss of life have occurred from their breaking; and it is even said that the flat hempen ropes are coming into use again. The colliers observe that "the hempen ropes give notice before they break but the iron ones do not."†

Iron Wire Cables in France.

The utility of this invention has been carefully investigated in France. Among other articles on this head, during the year 1845, in Annales des Mines, is a useful one on the fabrication and employment of these cables by M. Cacarrié, mining engineer,‡ and another memoir on the same subject by M. Pernollet.§

Wire Ropes and Cables in America.

These have been introduced, in some cases, sucessfully, in the United States. In July, 1839, a wire rope, three-fourths of an inch, was applied to one of the hoisting machines of the Philadelphia tobacco warehouses.

The American Railroad Journal contains an account, dated September, 1843, by J. A. Rœbling, C. E., of his introducing wire ropes on the *Inclined Planes* in Pennsylvania.

Three wire ropes, measuring in the aggregate 3400 feet, 4½ inches in circumference, were put in operation on the inclined plane No. III., of the Alleghany portage railroad in 1842; and were used a considerable portion of that season and the whole of the year 1843. The hempen ropes, heretofore used were 8½ inches circumference, made of the best Russian or Italian hemp; and could not be trusted in safety, longer than one season. Another wire rope, 5100 feet long, in four pieces, was about to be laid down on Plane No. X.

The first wire rope placed on this line, or in connection with it, was 600 feet long, 3½ inches in circumference, and had been already in operation two years or seasons, at Johnstown, January 1, 1844.

Two more iron wire ropes were put in work in 1842, one at Holli-

* Dunn's History of the Coal Trade, 1844, p. 61.
† Mining Journal of London, January 3d, 1846.
‡ Annales des Mines, 1844, Liv. III., p. 495 to 504.
§ Ibid., Liv. IV., p. 133.

daysburg, the other at Columbia, in Pennsylvania. The engineer asserts that wire cables deserve all the preference usually assigned to them over hempen ones, where the former are placed in exposed situations, and where great strength and durability are required. In the ropes in question the individual wires, as well as the strands and ropes, were separately coated with varnish during the manufacture.

The saving by wire, instead of hempen ropes, in 1844, is reported* to be $1,465 per annum for each plane, or $14,650 annually for the ten planes of the Alleghany portage railway.†

Wire Tiller ropes in the United States.—By an article quoted from the "Cincinnati Republican" in 1840, it appeared that, at the beginning of that year, 264 steamboats on the western rivers had adopted the use of wire tiller ropes; although it was only eighteen months since their first introduction. The wire tiller rope of the steamer "Commerce," which rope had been heated red hot for at least an hour, came out of the fiery ordeal without the least injury.‡

Wire ropes for the shafts and inclined planes in the collieries around Pottsville, are in use; but opinion is divided on their utility.

For further information on the subject we refer to some valuable communications in the Mining Journal of London, American Railroad Journal, and in the Journal of the Franklin Institute, Philadelphia.

* American Railroad Journal, December, 1843. Journal of the Franklin Institute of Philadelphia, January 1st, 1844. Mining Journal of London, April 27th, 1844.

† Hunt's Merchant's Magazine, Art. II., August, 1845, p. 132.

‡ Hazard's United States Register, April, 1840.

DELAWARE.

Lignite or Fossil wood occurs in the cretaceous group and green sand formation, and was especially exhibited in the lower mass of strata, in the deep cutting of the Delaware and Chesapeake Canal. Dr. S. G. Morton has adverted to this deposit in his Synopsis of the Organic Remains of the Cretaceous Group. The author observes that, "Lignite at one period was considered to be indicative of tertiary formations, but it is now frequently recognized in the green sand of Europe," and proceeds to point out its existence in similar strata on the Atlantic border of the United States.

Like the strata of this period in various parts of Europe, the lignites here are accompanied by *Amber*, [Succinite.]*

The vast beds of lignite which occupy so much space between the Missouri and the Rocky Mountains, may probably be referred to the super-cretaceous period; corresponding with the green sand formation of Delaware, rather than to the tertiary.

* Synopsis, by Dr. S. G. Morton, 1834.

NEW YORK.

PETROLEUM AND SPRINGS OF CARBURETTED HYDROGEN GAS.

THIS bituminous substance is commercially known under the name of "Seneca or Genessee Oil," in the town of Cuba, in this state. Mr. Vanuxem described this "oil spring" in his state geological report, in 1837. It is a dirty circular pool, about eighteen feet in diameter, furnishing but an inconsiderable supply of petroleum. The oil is much used by farmers, and has a ready sale.

The reporter observes that, "There is no necessary connection between oil springs and beds of coal; the presence, merely, of bituminous matter disseminated in the rock, accompanied by decomposing pyrites, suffices to account for its presence; or a depth at some former period, sufficient to give the required temperature necessary to disengage the petroleum from bituminous matter."*

On Cayuga Lake also, and at Cataraugus county, in this state, petroleum is found; and in several other localities in York state. Carburetted hydrogen also rises from the water courses in many places. Both of these matters are, in this state, connected with beds of marine shells, and with salt water. So constant, Mr. Vanuxem adds, is the accompaniment of carburetted hydrogen with salt water on the borders of the upper part of the Ohio river, that the presence of this gas is considered a sure indication of the vicinity of salt water.

The details of the number of these "gas springs," are to be found in the chemical reports of Dr. L. C. Beck, to the Governor of New York, in 1838.

Fredonia Natural Gas Lights.—Among the most interesting of the cases described by Dr. Beck, is that of Fredonia, in Chautauque county. The gas is collected by means of a shaft, sunk in bituminous slate. It is conveyed by a tube to a gasometer, and from thence, for the purpose of illumination, to different parts of the village. This gasometer had a capacity of about two hundred and twenty cubic feet, and was usually filled in about fifteen hours, affording a sufficient supply of gas for seventy or eighty lights.† For interesting details as to the employment of natural jets of gas, see under the head of Virginia, at Kanawha; also near Pittsburg.

Coal.—The intelligent mineralogist above quoted, reports that, in this state, throughout almost the whole series of its transition rocks, both anthracite and bituminous coal have numerous localities; but

* Mr. Vanuxem's First Annual Geological Report of New York, 1837, p. 195.

† Dr. L. C. Beck's Report in 1838.

invariably in quantities too small for useful or economical purposes. In Europe, this most important fossil substance gradually acquires a maximum, then diminishes to a minimum; passing from plumbago to anthracite, thence to the bituminous, through its various varieties, to the acetous bituminous; thence to fossil wood and peat; and finally terminates in the perfect vegetable.*

August, 1854. We find it stated in a Philadelphia Journal, that coal had been discovered in Steuben County, New York, and that the owner of the land had received the state bounty of $5000, as the first discoverer of coal in that state.

Cupriferous Lignites of the Catskill Mountain series.—Vegetable casts, replaced by grey sulphuret and carbonates of copper, occur in the same series of red and grey shales, as we have seen in various parts of Pennsylvania.†

The iron works in this state are of considerable importance. They rank in point of number and value as second in the Union.

PEAT OR TURF.

Professor W. W. Mather has especially directed attention during the progress of the geological survey, to this substance, both as a manure and a fuel. It is very common, and at some points has been in use for a great many years. A vast number of localities are pointed out in the State Reports, and estimates are made of the amount of peat therein.

As an instance of the value of peat, the case of a bog of 40 acres is cited, which furnishes a supply to the city of New York—where it is sold for $4.50 per cord. The peat being six feet deep, the produce of the sale per acre is $4500, a little more than a third of this being expenses.‡

In Mr. Vanuxem's district some valuable deposits of peat are also mentioned.

Dr. Emmons has, in like manner, furnished particulars of peat bogs within his geological district, and pointed out the high economical value which must, sooner or later, attach to this combustible, especially in those countries where coal is absent, or expensive.

"Perhaps it would be saying too much to assert that peat is more valuable than coal; but when we consider that for creating heat it is not very inferior to bituminous coal; that it contains a gaseous matter equal in illuminating power to oil or coal gas; that its production is equally cheap; and in addition to this, that it is a valuable manure if properly prepared, its real and intrinsic worth cannot fall far short of the poorer kinds of coal."§ ||

* Ibid., p. 196.

† Mather's Fourth Report, p. 229.

‡ Annual Reports on the Geology of New York, 1838, 1839, 1840. Mather and Vanuxem.

§ Emmons' Report on the Geology of New York, 1839, p. 216. See notes on Peat, in various parts of this volume.

|| We insert the following important information relating to the various uses to which Peat

*Consumption of Fuel in the city of New York (exclusive of foreign coal.)**

Years.	Wood in Loads.	Anthracite and Bituminous Coal.	Charcoal in Tubs.	Value of Fuel.
		Tons.		*Dollars.*
1830	297,586	Anthracite, 23,605½ / Virginia, 15,293		814,817
1832	265,192	63,417	347,792	1,327,507
1833				1,127,430
1836				1,100,480
1839	257,676	By canals, 96,431	303,284	
1840	242,944		335,895	2,500,000

Retail annual average prices of Schuylkill Anthracite, broken, per ton of 2000 lbs.

Years.	Average Price.	Years.	Average Price.
1838	$8.70	1844	$5.56
1839	8.58	1845	6.50
1840	8.00	1846	7.00
1841	8.45	1847–8	6.50
1842	7.16	1852	5.50 to 6.00
1843	5.96	1853	6.00
		1854	6.50 to 7.00

Wholesale prices by the cargo of Anthracite and Bituminous Coals in New York,† the duty $1 75 per ton from 1832 to 1847.

	1844.		1845.	1846.	1847.	‡1852.	‡1853
Quality or Locality.	Chaldron.	Per ton of 2,240 lbs.	Chaldrons exclusive of duty.			Chaldron.	
	Dolls.	*Dolls.*	*Dolls.*	*Dolls.*	*Dolls.*	*Dolls.*	*Dolls.*
Liverpool coal per chaldron, ...	8 50 to 9 25	6 60 to 7 20	7 to 10	7 to 8 00	8 50 to 8 75	7 75	9 50 to 10
Newcastle do.	7 00 to 8 00	5 40 to 6 20	6 to 9	6 to 7 00	8 00 to 8 25		
Scotch do.	6 50 to 6 75	4 90 to 4 20	6 to 8	6 to 6 50		5 50	—
Pictou and Sydney do. .	6 75 to 7 00	4 20 to 5 40	6 to 7	6 to 6 50	6 00 to 6 50	6 00	6 75
American Anthracite, per ton, 2000 pounds. . .		5 00 to 6 12	5 to 6	6 to 7 00	5 50 to 6 00	4 75 to 5 50	5 50 to 6
Virginia Bituminous, per chaldron, ...				6 to 6 50	Ton.		

The retail prices in the years quoted are from $1.00 to $1.25 higher, exclusive of the duty. The chaldron is rated at 36 bushels, and the ton is 2000 lbs. weight.

We understand the price of anthracite coal in New York is from the boats $7,50, and at the yards $8, delivered. Some of the New

is now applied, which correctly should have been added to the article on Peat in Ireland; see Introduction; but that portion of the work being already in press, we will give the extract here from London Mining Journal, July 8th, 1854. A company has been formed and now in operation, called the Irish Peat Company, having a factory in Kilberry, in Ireland, to produce from Peat, Tar, Paraffine, Oil, Naptha, Sulphate of Ammonia, and to manufacture iron with Peat—the furnaces being heated with the gas manufactured. The inefficiency of the machinery in the first instance, has been the cause of some delay, but the experiments made under the superintendence of Dr. Sullivan, have been quite satisfactory. Candles are also to be manufactured from the Paraffine obtained from the Peat in Ireland.

* Hazard's Register, 1833 to 1841. This table is incomplete. Also Lehigh Co. Reports.

† New York Commercial Advertiser, August, 1844.

‡ New York Shipping and Commercial List. Sept. 8, 1852. Sept. 3, 1853.

York papers insist upon the removal of the 30 per cent. duty on foreign coal, as the only sure preventive to the high price of this essential article.*

From the New York Shipping List, the wholesale prices of coal were January, 1854, per ton: Liverpool Orrel, from $12 to $12.50; Newcastle, $10.50; Scotch, at $8.50 and $9; Pictou and Sidney, per carge, $8.50; anthracite, $5.50 per ton, to $6. Cannel coal varies from $12 to $16 a chaldron, and even higher.

Population of the city of New York in		1840,	312,000
"	"	1845,	365,000
"	"	1850,	650,000
"	"	1853,	700,000

Emigrants and passengers arrived:

Years.	Vessels.	Passengers.	Passengers from California.
1841	2,118	57,337	
1846	2,280	115,230	
1850	3,487	226,287	
1851	3,888	299,081	18,207
1852	3,822	310,335	12,158
1853	4,107	299,425	15,517

	1850.	1851.	1852.	1853.
Emigrants arrived,	212,796	239,255	300,992	289,765

Cost of the State canals of New York,†				$30,987,335
"	"	"	" 1850,	38,986,857

Value of Imports and Exports from Port of New York.

	1846.†	1847.†	1849.	1850.‡	1851.§	1852.	1853.
Imports,	70,269,811	96,447,104	97,658,251	138,334,642	131,361,578	129,849,619	194,097,652
Exports,	36,423,762	53,421,986	39,736,969	60,119,248	43,910,640	46,427,354	67,136,642
Specie,					43,743,209	25,096,255	26,753,356
	106,693,573	149,869,090	137,395,220	198,453,890	219,015,427	201,373,228	287,987,650
Cash duties received.	17,220,635	20,128,052	21,718,624	28,047,239	31,081,263	31,332,737	43,088,225

The annexed is an official statement of the tolls collected on all the New York State canals, in each of the following years.‖

					Total to Nov. 22.
1846	-	-	-	-	$2,842,214
1847	-	-	-	-	3,635,330
1848	-	-	-	-	3,215,841
1849	-	-	-	-	3,196,412
1850	-	-	-	-	3,195,054
1851	-	-	-	-	3,291,486
1852	-	-	-	-	3,118,244
1853	-	-	-	-	3,189,740

* Sept. 7, 1854. Phil. Ledger.
† Controller's Reports.
‡ New York Journal of Commerce and New York Tribune.
§ Hunt's Magazine, February, 1854, p. 206.
‖ New York Herald.

The average number of days in which the New York canals were navigable, (accidents not included,) in the twenty-three years previously to 1847, viz. from 1824 to 1846, inclusive, was 231.*

There are about 1,100 miles of canal and inland navigation in the State.

Capital invested in the rail-roads of New York, in 1846, $12,-750,000.

Aggregate length of twenty-one rail-roads, in 1847–8, 758 miles then in operation, besides others in progress. Net income received, about 7 per cent.

Number of locomotives,	107	
" of passenger cars,	212	893†
" of freight cars,	542	
" of mail cars,	139	

Abstract of the statistics of thirty Rail-roads in the State of New York in 1852.‡

Total cost of rail-roads and equipments,	$84,034,456
Average cost per mile, - -	36,701
Aggregate amount of capital, - -	53,963,550
Aggregate length laid, - -	1901¾ miles.
Annual rate of interest generally 7 per cent.	
Number of locomotive engines, is - -	446
Passenger cars, first class, - - -	477
Freight cars, - - - - - -	4,695
Baggage, mail, emigrant and second class,	272

Railways State of New York, 1853.‡

	Miles in operation.		Cost.
31	2,432	- -	$94,361,262

The total value of all the property which cleared from and came to the Hudson River, on all the canals, in the following years:§

	Boats.	Value of Cargoes.	Tons.
Arrived and cleared in 1844	19,393	$87,782,849	
" " in 1845	20,040	100,906,298	1,428,956
" " in 1846		115,732,780	1,601,535
" " in 1848		125,827,357	1,777,466
" " in 1849		127,098,569	1,885,416
" " in 1850		140,658,009	2,475,018
" " in 1851		143,145,297	2,452,486

* Hunt's Merchants' Magazine, November, 1847, also January, 1848.
† American Railroad Journal, January 1st, 1847.
‡ State Engineer's Report Canals New York. § Hunt's Mag.

The value of the entire movement of property, in 1846, from and to the Hudson, is greater by $7,297,845, than the value of all the goods imported into the United States during the fiscal year, ending July, 1844; and exceeds by $9,039,207 the aggregate value of the imports and exports of the port of New York, in 1846; 1849, $144,732,285; 1850, $156,397,929.

*Quantities transported on the New York canals, paying tolls.**

Years.	Tons.	Value.	Tolls.
1849	2,894,732	$144,732,285	
1850	3,076,617	156,397,929	
1851	3,582,733	159,981,801	
1852	3,863,441	196,603,517	$3,118,244
1853	3,052,251	187,483,866	

Statement of the *tonnage*, *toll* and *value*, of articles of all denominations, which passed through the State canals of New York, eastward, to tide water at Albany and Troy.

Years.	Tons.	Tolls.	Value at Albany and Troy.	
		Dollars.	*Dollars.*	
1834	553,596		13,405,022	The total value of the exports from the port of New York, in the year 1847, was $7,071,795.
1841	774,334	2,034,882	27,225,322	
1842	666,226	1,749,197	22,751,013	
1843	834,283	2,081,590	28,376,599	
1844	1,019,094	2,446,374	34,183,167	
1845	1,204,943	2,646,453	45,452,301	
1846	1,362,317	2,842,214	51,105,256	
1847	1,744,288	3,635,330	73,092,414	
1848	1,447,905		50,883,907	
1849	1,579,946		52,375,521	
1850	2,033,863		55,474,637	
1851	1,977,151		53,927,508	

The total amount of tolls upon the New York State canals, received in the twenty-four years, from 1824 to 1847, inclusive, was $34,534,356.

Statement of the number and tonnage of canal boats of every class, which passed upon the canals of the State of New York, in the years 1843 and 1846, showing the comparative increase in the latter, both in number, and capacity, and amount conveyed.

Years.	Number of Boats registered.	Average.	Aggregate tonnage.	Aggregate of tons conveyed.	Av'ge conveyed by each boat.
					Tons.
1843	2,126	55	117,553	1,513,439	711
1846	2,725	62	168,287	2,268,662	832
1849	4,863	63	329,353	2,394,732	

* Engineer's Report, 1852.

Statement of the quantity and value of mineral coal, chiefly the bituminous coal of Pennsylvania and Ohio, which was transported on the New York and Erie canal, eastward.*

Years.	Shipped at Buffalo from the Lake.	Reached Tide at Albany or Troy, by canal.	Value.	Blossburg.
	Tons of 2,240 lbs.	*Tons of 2,240 lbs.*	*Dollars.*	*Tons.*
1841		8,045	15,586	
1842		8,816	18,101	
1843		6,528	32,588	
1844		8,250	55,993	
1845	873	21,339	119,496	
1846	1,461	8,414	47,116	
1847		14,055	84,000	30,110

The rate of toll on mineral coal was reduced to one mill per 1000 lbs. weight per mile, on the New York and Erie canal, on the 1st August, 1845. As the old rates amount nearly to a prohibition, this reduction secured a revenue to the New York canals, from a source which had previously yielded little or nothing.† The new and old rates are as follows:

	1845	1846
	C. M.	C. M.
Mineral coal, per 1000 lbs. weight per mile, equal to 2¼ miles per ton of 2240 lbs. per mile, -	0.4½	0.1
All coal used as fuel in the manufacture of salt, -		Free.

The cost of all expenses of transportation on the Erie canal in 1852, was $2\frac{9}{10}$ mills per ton per mile. Mr. Seymour the late State Engineer estimates the whole cost at 3⅛ mills per ton per mile. The charges for transportation on the Erie canal in 1851 and 1852 (except late in the season) have averaged $2.50 per ton for down, and $2.35 for up freights, exclusive of the charge for State tolls, being at the rate of 6.9 and 6.5 mills per ton per mile.

The earnings of the New York and Erie Railroad for the year ending September 30, 1853, was $4,318,962. Report of the Directors of the New York and Erie Railroad, November 30, 1854.

	Tons.	Value.
Tonnage of the New York and Erie canal in 1852, was, - - - -	2,234,822	$66,893,102
Tonnage of 1853 is not published but the value was - - -		74,000,000

* Annual Report of the Commissioners of the Canal Fund, January 5th, 1846.
† Hunt's Merchants' Magazine, February and June, 1846, and other sources.

RHODE ISLAND.

LOCALITIES OF COAL.

The report of the Geological Survey of this State, appeared in 1840, from the pen of Dr. C. T. Jackson. This duty, of course, embraced the examination of the coal formation, the collection of characteristic specimens for analysis, and of organic or vegetable fossil remains; wood engravings of which illustrate the report. Like all the public papers proceeding from this gentleman, it is characterized by special attention to whatever tends to those practical and useful results which constitute essentially the aim and object of his labours.

The two positions where anthracite is found are, 1st, in Cumberland county, north of Providence; and 2d, at Portsmouth, in Rhode Island, 23 miles to the south.

Cumberland.—Only a single bed of anthracite is mentioned here; dipping to the south. All attempts to mine this appear to be abandoned for the present.

Mr. Hitchcock has traced anthracite also to Middleborough, at West Bridgewater, and at Wrexham.

Bristol Neck.—Slate rocks, the grauwacke of the reporter, containing an abundance of fossil plants of the coal period—corresponding with the coal formation on the opposite shore of Portsmouth—occur here, overlaying granite; but no coal seam has yet been noticed.

Pappoose.—*Squaw's Neck*, near Bristol, contains the same series of slate and compact rock, termed grauwacke; and similar vegetable remains as at Portsmouth; but no regular bed of coal is observed.

Cranston.—On *Sockanosset Hill*—strata of similar character, with graphite and impure anthracite, show the extention of the coal formation in this direction.

Warwick Neck.—Similar carbonaceous grauwacke, which is more promising for coal than at most localities, and the local situation is very favourable for mining.

Providence.—From excavations near the Court-house some anthracite was obtained underlaying tertiary clay.*

Newport.—Anthracite beds, a few inches thick, occur in the south part of Newport, in the slate commonly denominated grauwacke.†

The whole area in Rhode Island State, where rocks of this age appear, covers about 150 square miles. The central part of this is

* Jackson, 1840.

† Hitchcock's Geology, pp. 262, 275.

overlaid by horizontal clay and sandy beds, apparently of the tertiary period, but without shells or other fossils.

PORTSMOUTH COAL MEASURES.

Anthracite.—Some instructive details respecting the beds of anthracite and their contiguous rocks, occur in Mr. Hitchcock's Geological report of Massachusetts.

We do not know the date of the first working the coal here, but the operations failed between the years 1809 and 1816, and were subsequently resumed.

Dr. Meade, in Bruce's Mineralogical Journal, January, 1820, says that the main seam of anthracite was then in work, and was 14 feet wide; yielding from 10 to 20 chaldrons a day, with the labour of only fifteen workmen. These works were soon after abandoned.

Previously to 1827, the mine was again put in operation, and the quantity of coal raised in that year, by 20 men and 5 boys, was 2200 tons, and an equal quantity of slack or small coal. The former sold at the mine for $4.50 per ton, and the latter for one dollar per ton. The agent, as is usual, represented these beds as capable, in the much hackneyed phrase, of furnishing "an inexhaustible supply;" yet we find that, in the succeeding year 1828, the mines were again abandoned as unprofitable.

Prof. Hitchcock states that "six beds of anthracite are exposed, and more than thirty are said to exist in that part of Rhode Island." The six main seams are probably reducible to three, on each outcrop of the basin; but of the "thirty," we confess ourselves somewhat sceptical.*

It appears by the reply to a circular addressed by the Secretary of the Treasury for statistical information, that the quantity of anthracite mined in Rhode Island in 1844 was only 2800 tons. The price at the mine has been uniformly $3 per ton; but the mine is once more abandoned; never having been profitable.†

In 1846 Rhode Island coal imported into Boston, 165 tons.

Mr. Vanuxem was one of the earliest investigators into the quality of the Rhode Island anthracite—in a series of experiments, published in the Journal of the Acadamy of Natural Sciences in 1825. The result will be found in another part of this work.‡

PORTSMOUTH ANTHRACITE FORMATION.

Proceeding now with the State geologist's description,—"The main formation consists of slate rock, which is here and there charged with beds of anthracite. Several small seams of coal have been found as far south as Newport. On Quaker Hill a small coal bed was struck

* Geology of Massachusetts, 1833, p. 277.

† Report December 3d, 1845, p. 338.

‡ Journal of the Acad. Nat. Sci. Phil., Vol. V., p. 17. Experiments on Anthracite, Plumbago, &c., by Lardner Vanuxem, March 15th, 1825.

many years ago, but was not wrought. Lawton's valley exhibits the clay slate without any coal beds. Butts's hill, in Portsmouth, presents a mass of stratified rocks, alternating with and overlaying the slates of the coal measures. The strata dip towards each other from each side of this eminence, from Case's coal mine, on the east, to the old Portsmouth mine on the west. In Portsmouth township several seams of anthracite occur, included between beds of carbonaceous slate, subordinate to the fine *grauwacke* rocks. These mines were abandoned by the proprietors, about the year 1825.

The coal beds are stated to be three in number, varying from 2 to 12 feet in thickness, and quite irregular.

The main Portsmouth coal mine is included between walls of slate, and is stated to measure three feet in thickness.

The analysis of the specimens obtained from hence, gave the following results:

Specimens.	First.	Second.	Third.
Carbon, ...	84.50	77.00	85.84
Water and volatile matter, . .	10.00	7.00	10.50
Ashes, ...	5.50	16.00	5.66

Hence, it is evident that this coal will burn freely.

The anthracite is obtained by blasting, in large masses, and its compactness ensures transportation without waste by fracture.

The best coal is that which is impregnated at its natural joints with per-oxide of iron and manganese, (rusty coal,) while the glassy and greyish black masses are more charged with argillaceous (and siliceous) matter, and decrepitate violently when thrown on the fire. Owing to the presence of so large a proportion of water, the coal burns with a flame; the water of composition is decomposed during combustion, and carburetted hydrogen and carbonic oxide are produced. The former gas burns with a yellow, and the latter with a blue flame.

It is evident, from their composition, that the ashes of all these coals will form slags or clinkers.*

The reporter arrived at the conclusion that there is a valuable supply of coal that can be economically obtained at the Portsmouth mines. The opinions formerly entertained as to the difficulty of burning this coal have now no value—for the people of New England did not then understand the art of burning anthracite. A calculation, founded on the supposition that there are three workable seams of three feet each, within the basin of Portsmouth, showed an amount of 37,800,000 tons of coal, available.

The northern and north-western portion of the island of Rhode Island, occupying a space of not less than eight or ten miles long by two wide, essentially form the coal basin of the island; if we limit

* Rhode Island Report, 1840.

the definition to the area which possesses that form of arrangement in its stratification. On the outer or south-eastern border of this basin, commencing in the centre of the island, the strata arching over from beneath the coal bearing formation, dip eastwardly. A cursory examination of this basin with some diagrams were made by the present writer in 1841.

GEOLOGICAL CHARACTERS.

One of the earliest impressions made on the traveller, who, in visiting this region, brings recollections of ordinary coal fields, is the primitive, or rather the metamorphic and disturbed geological character of the entire rock series, much of which is probably new to him under its changed aspect. In fact very few persons, in passing through this region, would conceive themselves in the midst of a coal formation at all.

That we might the sooner attain a correct understanding regarding these novel appearances, some transverse sections were constructed, and also a profile following the east shore of the island. The west coast, being flat and without cliffs, did not well admit of such a mode of illustration.

The first transverse section, crossing from Narraganset Bay on the west by Butts's Hill Fort, to Mount Hope Bay on the east, exhibits an uninterrupted basin-form arrangement of stratification, having coal beds cropping out on opposite margins. Pursuing his inquiry, the geological observer will no longer doubt but the whole group of strata, many hundred feet thick, constitutes an actual coal formation, although its separate members seem to have little resemblance to such rocks as usually comprise our coal fields.

Fig. 16.

Transverse Section of the Portsmouth Anthracite Basin, Rhode Island, looking North.

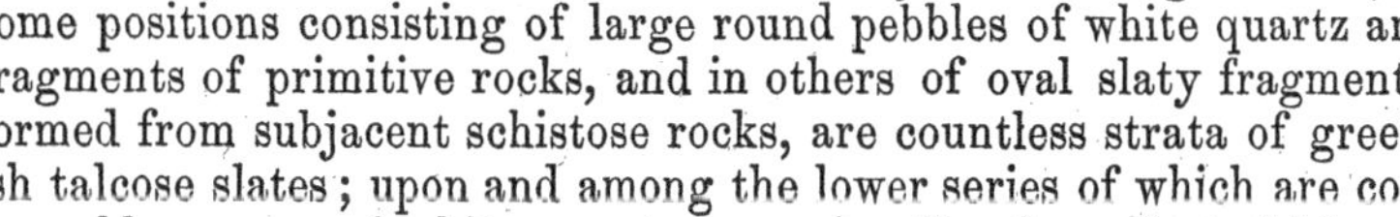

Based upon more than one mass of very coarse conglomerate, in some positions consisting of large round pebbles of white quartz and fragments of primitive rocks, and in others of oval slaty fragments, formed from subjacent schistose rocks, are countless strata of greenish talcose slates; upon and among the lower series of which are conformable seams of white quartz, occasionally three feet thick; or again a net-work of smaller quartz veins from a few inches to a foot thick, traversing both the conglomerate and the talcose slates. Among these slates occur darker laminæ, and these contain distinct impressions of the usual coal plants. Passing from these, the predominant mass consists of talcose schist, among the divisions of which may often be obscurely traced magnificent casts in relief, of ferns, pecopteris, &c. But for these intelligible characters, one might imagine the schists were of much greater geological antiquity.

Following southward along the eastern shore, the cliffs, although not lofty, are sufficiently so, with the aid of the transverse ravines, to develope the structure of the adjacent country. In the course of two miles of the cliffs of the east coast, the conglomerate beds are six times thrown up, and as often descend below the tide level. Then occur a numerous suite of twisted and contorted schists, of grey laminated slates whose surfaces singularly resemble the grain of birds-eye maple; and again another series of green, talcose, contorted schists, crowded with crystals of iron pyrites; crossed in every direction by innumerable veins of white quartz, and succeeded by compacter beds, which almost possess the qualities of sandstone. Perpendicular upthrows and heaves, and again the reverse movements, divide the whole series into large and separate sections, rising above or sinking below, water level. The inclination of the respective masses is continually changing. To the rocks we have enumerated succeeded a *melange* of metamorphic slates, of gray fissile beds, of conglomerates, quartz veins, and black shales; of veins and filons of asbestos, and of talcose laminated strata; undulating, fractured, contorted, inverted—in short, disposed with such absence of order and arrangement, as to defy the pen and pencil of the geologist to delineate.

Leaving the coast line at Clark's Mill and Creek, our *second transverse section*, of 3½ miles, crosses over to the opposite or western shore. During half this space, the metamorphic rocks alone, to which we have alluded, and which are named *Grauwacke* in the state report, appear on the surface and dip to the eastward. The schists and coarse slates, the carboniferous shales, and the quartz veins, which here seem to be appropriate to the coal-bearing series, are again seen arranged in the basin form, stretching to the coast near Lawton's Valley. We only observed one bed of anthracite, whose immediate out-crop exhibited about 1¼ foot thick, increasing as it descended. How far to the south this trough extends we did not attempt to trace; but as thin seams of coal are seen among the modified rocks, on the coast east of Newport, it is probable this arrangement continues through the entire length of the island.

Returning to our first transverse section of the coal basin, near the parallel of the Portsmouth mines. Certainly, there are many features here presenting themselves that have no parallel in our ordinary secondary coal-fields. Among these are the vast assemblage of talcose, waving slates; the veins and seams of asbestos, abundant even among the coal shales, and occasionally penetrating the anthracite itself; the quartz veins also in the coal; the unusual appearance of vegetable remains on these greenish-grey, schistose laminæ; the traversing veins of white crystalline quartz, and the plumbaginous nature of nearly all the out-crops of coal. All these characters might readily lead geologists to ally the series to the transition or primary rocks.

Details of the Coal Beds.—Yet perhaps these are entirely due to the metamorphic influence to which the whole group, in common with all others in the surrounding country, has been subjected. There are three coal seams proved on the western side, occurring at the distance of ninety feet from each other, and dipping, at an angle of 38° to the centre, but probably flattening in that direction. Towards their out-crops all the strata evince the effects of great pressure and squeezing; producing corresponding irregularities in the thickness of the coal beds, such as will probably always render the working or productive results uncertain.

In some particulars there appear to be analogies between the talcose schists and accompanying beds of anthracite in Rhode Island, and the anthracite seams associated with strata of gneiss and talcose schists, formerly considered as primitive, at L'Oisans, in Dauphiny, and the Alps. M. A. Brongniart has declared that the coal vegetation of this formation is identical with that of the true coal measures. M. S. Gras confirms this view, and states further that these were sedimentary rocks, modified by subterranean emanations. In the Alps there are many proofs of the transformation of sedimentary into crystalline rocks, as high up as the coal measures.

The old Portsmouth Mines, towards the close of 1841, had been re-opened, and several new shafts had been commenced in their vicinity. As regards facilities of transportation, no position can be more convenient; for sloops and schooners can approach within one or two hundred yards of the mine. The quality of the coal is excellent; the demand for it increases every year, and it can readily be sold, as fast as it is possible to mine it. In 1842, the price for the large coal was $5.00 per ton, and half that sum for the finest or pea coal: terms which can be commanded nowhere else at the pit's mouth, in the United States. As a proof of the value even the smallest had acquired, the owners were screening over the refuse heaps, abandoned 20 years before; and were selling the coarsest at $2 to $3 per ton; the next size at $1.25, and mere dust 75 to 50 cts. The quality of the coal improves as the depth increases. As may be inferred from the geological condition of this region, the great drawback on the prospective value of coal undertakings any where within its limits, and on the confidence so essential to such operations, arises from the irregularity of the ground; making the thickness and the continuity of any one coal seam, a matter of extreme uncertainty, even for the space of a few feet. The roof of the main worked bed is tolerably regular, and consists of a good hard slate; but the floor undulates considerably, and of course affects the thickness of coal to a corresponding extent. At one point here there were only eighteen inches of coal, between roof and floor; yet on advancing but a short distance we observed a thickness of fifteen or sixteen feet. Under these circumstances, it is difficult to assume an average. Dr. Jackson's estimate of three feet workable coal to each seam, through the entire basin, may be a safe one, but we would not like to be the purchasers on the basis of that calculation. In 1842, the slope or inclined plane of the main gangway down the crop of the vein, was three hundred feet. Lateral drifts, following the coal seam, showed about three yards thickness; but we subsequently learned that it had again contracted.

The seam lying above this had been commenced by other owners, as a colliery, in 1841. Its thickness was then six feet; both roof and floor were good and promising, being of clay slate, dark, tough, and regular. Many coal plants occur in the roof.

The plumbaginous character of the carbonaceous deposits throughout the entire range from Mansfield, in Massachusetts, to Newport, in Rhode Island,* is not devoid of interest, either to the miner or the mineralogist. At Wrentham, in Massachusetts, are several seams of highly plumbaginous coal.† At Mansfield also, Dr. Jackson mentions a bed of coal which "*was found to have been altered, and was like graphite or plumbago.*" In Rhode Island, the presence of graphite is not adverted to by the state geologist, further than to remark in his analysis, No. 2, of the Portsmouth coal, that it was *not* plumbaginous. At some new trial openings in the latter neighbourhood, on more than one out-crop, we observed that the mineral appeared to consist almost wholly of graphite. It is remarkably light, spongy, or cellular—and is collected and forms an article of sale at a good price, under the name of "British Lustre," for the usual purposes of plumbago or black lead. Asbestos occurs abundantly, running through the slates which adjoin the coal or graphite bed. Like those of Massachusetts, they are also traversed, and even the coal itself, occasionally, by numerous veins of quartz. All these circumstances combine to satisfy the most skeptical, of the modified or metamorphic character of the coal-field of Rhode Island.

Viewing it in this light, there seems to be no assignable reason why this formation in which we trace the fossil flora of the regular coal measures, should be considered any older than the secondary anthracites and bituminous coals of Pennsylvania; and we are sustained in this opinion by the declaration of Dr. Maculloch, that, "the coal of secondary origin, containing vegetable remains, is converted into plumbago by the influence of trap, as coal is, daily, in the iron furnaces."‡

It is well known, that, until within a comparatively recent period, the beds in which numerous seams of anthracite occur, in Ireland, were confidently termed by distinguished geologists, "transition clay slate, intermixed with considerable beds of quartz." Yet it is now universally admitted that the entire coal field is of no older date than the regular bituminous coal fields elsewhere.

Even the Western part of the South Wales coal basin, was at one time called the "Grauwacke series."

* Geological Survey of Massachusetts, p. 162, 1833.
† Hitchcock's Geology of Massachusetts, p. 46.
‡ System of Geology, Vol. II., p. 297.

Since the foregoing notes were collected, we have been favoured by Dr. Emmons with his volume on the so-called Taconic system, as exemplified in the northern states of the Union; including the Rhode Island coal fields. The Taconic system, according to the views of the author, and in opposition to those of many eminent geologists, embraces a series of rocks which are supposed to be older than the New York lowest series, and are characterized by a separate class of organic remains. It rests unconformably upon primary schists.

We believe that the author does not comprise within this system the coal formations of Rhode Island, although they repose upon it, and have many lithological characters in common, on account of the proximity of the schistose Taconic rocks which have furnished the greater part of the materials; and consequently they appear to possess a character of much greater antiquity than the coal and subjacent rocks usually exhibit elsewhere.

"That it is possible for a sedimentary rock to retain or assume the characters of the parent rock, is rendered highly probable by the characters of the rocks or slates connected with the Rhode Island coal beds. Here, in connection with the conglomerate, probably of the old red sandstone, there is much material which is a talcose slate, differing but slightly from the talcose slate of the Taconic system. The beds of conglomerate with which these slate beds are in connection, do not appear to be metamorphic; and the whole seems to be merely indurated or hardened slate, the original particles being talc and mica, with some fine quartz. The rock, when complete, is merely an ordinary talcose slate."

Dr. Emmons is somewhat indisposed to admit the metamorphic character of the Rhode Island coal, inasmuch as the slates and conglomerate bear no marks of the action of heat; the fossils are similar in texture to those of other coal fields, and are free from all traces of fusion; and because if sufficient heat had been applied to volatilize the bitumen of the coal, then ought the slate also to exhibit marks of having been burnt. These reasons, however, do not appear sufficiently conclusive nor do they apply to the anthracite of Pennsylvania, which exhibit no traces of fusion, neither in the coal, the slates, nor the organic remains. I believed the word "*baked*" has been frequently applied to this process, and with apparent propriety. Dr. Emmons admits that this coal of Rhode Island is traversed by veins of quartz, which might have been deposited from hot water or aqueous vapour holding silex in solution. The changing of the coal into graphite still remains to be accounted for.

As relates to the economical value of this anthracite.—On the whole, we see no reason for dissenting from the prediction of Mr. Hitchcock, "that, ere long, the anthracite of Rhode Island, and even that of Worcester, will be considered by posterity, if not by the present generation, as a treasure of great value."

Portsmouth, Rhode Island, coal imported into Boston:

	Tons.
1850 - - - - - -	1053

It is objected to this coal, that it will not succeed in an open grate, and that the cold air chills the fire in that position; but that it answers well when consumed in cylinder stoves. The only objection urged in that case, is, that it occasionally forms too much clinker. If we look at the numerous analyses of this coal, and see the small amount of foreign matter which it contains, besides pure carbon, we certainly should not expect such a result.

Quantity of Pennsylvania anthracite imported into Providence:

	Tons.
1844 - - - - - -	51,848
1845 - - - - - -	67,638

An annually increasing quantity of Nova Scotia coal is also brought into Providence.

ANTHRACITE NEAR PAWTUXET.

An announcement was made in 1846, of a new locality of anthracite in this state near the Valley Falls. Subsequently, a mining company has been carrying on some operations here, and report speaks favourably, thus far, of the success of the undertaking, and of the quality of the coal, except in regard to hardness. A depth of one hundred and twenty feet has been sunk and about five hundred tons of the anthracite were raised in the summer of 1847. The mine is about six miles south of Providence, two from Pawtuxet, and not more than two miles from tide water.

Prof. Hitchcock, in his Geological Report, described a large tract in Bristol county, a part of Plymouth county, the whole of the island of Rhode Island, and a strip on the west side of Narraganset Bay, as underlaid by grauwacke, a rock older than the coal formation, and equivalent to the Silurian and Cambrian strata of late geologists. He adds, that while publishing his final report, in 1840, "I became satisfied that a part of this region was a true coal formation, and so marked it on the map. I now advance a step further, and maintain that the whole of this tract, embracing not less than 500 square miles, is a genuine coal field that has experienced more than usual metamorphic action. The metamorphic action to which this deposit and the coal have been subject is two-fold, viz., first, mechanical; second, chemical.

"The mechanical force seems to have operated upon the strata containing the coal in a lateral direction, so as not only to raise them into a highly inclined position, but also to produce plaits or folds such as would be formed if several sheets of paper lying upon one another were taken into a man's hand, and by pressure on the opposite edges were crumpled so as to form ridges and hollows.

"The chemical metamorphosis which these rocks have experienced consists mainly in such effects as heat would produce."

He concludes his remarks on the subject thus:—"The evidence seems very strong on which I base the conclusion that the Bristol and Rhode Island deposits, with vegetable remains, possess the age and character of a true coal field, as the carboniferous period of the geologists.

I. In the first place the general outline of the surface over this field corresponds with that of a regular coal field or basin.

II. The rocks correspond essentially to those of the coal measures.

III. The number, position, strike, dip, and general character of the beds of coal already discovered in the district under consideration render it probable that it is all one coal field, or essentially one.

IV. The character of the vegetable remains found in connexion with these coal beds make it almost certain that they belong to the coal measures of the carboniferous system."

PEAT.

Block Island.—The most southern appendage of this state, and included in Dr. Jackson's geological and agricultural survey, in 1840—from whence we obtain the substance of the following details.

"*There are no trees upon Block Island*, and since wood fuel is too expensive for general use, it most fortunately happens that nature has amply provided the inhabitants with a great and almost inexhaustible supply of *peat*, or *tug*, as it is there called. Thus, almost every family owns a peat bog, which is their depository of fuel, from which they draw an ample allowance, yearly.

"Attached to every dwelling we find a '*tug-house*,' in which is stored up the winter's fuel; and each family burns from twenty-five to thirty-five cords of peat per annum. The mode of preparing it is—in case it is a first cutting—to split out cakes of it, about six inches square, which are laid upon the bank to dry in part; after which it is turned; and subsequently it is piled up, in open stacks, through which the air circulates and completes the process.

"In case an old bog is dug over the second time, *the peat is made by the hands into balls*, as large as a twelve pound cannon shot; and these are laid on the ground partially dried, and then stacked like piles of cannon-balls. They become firm, and burn very well; giving out a large and clear flame, and making a good coal. The fire-places are all arranged with peat grates, or frames made of bar-iron, large enough to fill a kitchen fire-place. On this they lay the peat, and it proves to be an excellent fuel; giving a good clear fire, suitable for all kinds of cooking, and for the warming of apartments.

"I think that most persons would give up their prejudices against peat, if they should spend a few weeks among the people of this island."

The opinion prevails that when *tug-bogs* have been entirely cut out, by throwing back the loose turf, the peat grows again in forty years, so as to fill the bogs.

Every little valley on Block Island contains a few acres of turf bog; and its depth varies from four to ten feet. There is evidently enough left, even if it does not grow, to supply the inhabitants with fuel for ages.

At a great number of points in this State, peat bogs prevail. Besides as fuel, it is used as a valuable ingredient in composts or manures. By analysis, it was found to yield 88 per cent. of vegetable matter, and 12 per cent. of ashes.

Some peat from another part of the State gave 88.6 vegetable matter, 9.5 of silex, and 1.9 of various substances.

Another contained 95.5 per cent. of vegetable matter. To these are added in the appendix, the analysis of twenty specimens of peat from various parts of the State.

This substance is held in high estimation in New England. In the Farm Reports of Rhode Island, Mr. Phinney, addressing the State Geologist, says, "I know of no way in which you could render a more essential service to the public, more especially to farmers, than by enabling them to convert their unproductive and unsightly bogs into sources of wealth. I consider my peat grounds by far the most valuable part of my farm; *more valuable than my wood lots for fuel*, and more than double the value of an equal number of acres of my uplands, for the purposes of cultivation. In the first place, they are valuable as fuel. I have for twenty years resorted to my peat meadows for fuel. It gives a summer-like atmosphere, and lights a room better than a wood fire. The smoke from peat has no irritating effect upon the eyes; it does not, in the least degree, obstruct respiration, like the smoke of wood; and it has none of that drying, unpleasant effect of a coal fire. Peat, taken from land which has been many years drained, when dried, *is nearly as heavy as oak wood, and bears about the same price in the market.*"*

* Rhode Island Report, p. 247.

MASSACHUSETTS.

MANSFIELD ANTHRACITE MINES, FIFTEEN MILES NORTH-EAST OF PROVIDENCE.

Dr. Jackson examined this coal for the sake of comparing it with that of Portsmouth.

The result of an analysis of a fair sample, was

Carbon,	87.40
Water, and volatile matter,	6.20
Ashes,	6.40

It appears on this authority, that a small seam of anthracite was accidentally discovered in Mansfield, in 1835, in the process of sinking a well. In consequence of this, numerous persons obtained leases with the right of mining, on farms in this vicinity. The first vein was only eighteen inches thick, and was not worked, on account of being so small. Subsequently, the researches conducted under the direction of Dr. Jackson, determined the existence of five beds of anthracite, the maximum thickness being five feet; and their apparent linear extent was not less than one mile.

These beds abounded in coal plants, of which many species were collected by the geologist.

Several analyses were made to determine the character of this anthracite, which seems to differ from that of Portsmouth, chiefly in having a greater amount of carbon, viz., 87, 90, 92, 96, and 98 per cent., with a reduced quantity of ashes.

From want of experience in mining, the adventurers failed to derive an adequate benefit, and the mines were abandoned in 1838.

They were re-examined, in 1839, by Dr. Jackson. One seam was proved to be seven feet thick, and its quality was good,—yielding but six per cent. of ashes.

At another mine, where a shaft had been sunk to the depth of one hundred feet, a bed of coal was encountered, but thought to be inferior.

The nature of these seams seems to correspond, as regards irregularity, with most of those known in Rhode Island. Some days the miners raised ten tons of coal, while on others they obtained but little;—yet the geologist considers that the Mansfield coal mines are still capable of being worked to good profit, if pursued with skill and judgment.*

* Dr. Jackson, on the Mansfield Coal Mines, in his Rhode Island Report, in 1840.

The usual coal plants occur imbedded in the shale and slate of the Mansfield district: but, Mr. Lyell observes, no traces of shells or corals have been discovered.

"In like manner, we find an absence of all fossils, except vegetable remains, in the anthracite coal district of Pennsylvania, and no fossils of any kind in the subjacent conglomerates and red sandstones."* In the bituminous coal shales of Pennsylvania, we, however, find several genera and species of fossil shells, and the remains of fishes and shells in the sandstone beneath. At Blossburg, numerous shells occur in the coal and iron shales.†

Import of Pennsylvania anthracite, and of American and Foreign bituminous coals, reduced to the common denomination of tons, from chaldrons, tons, and bushels, into the port of Boston. Note—there are considerable variations in the published statements of coal imported and consumed in Boston, which cannot be readily accounted for.

	American Coals.		Foreign imported Bituminous.			
Years.	Anthracite tons of 2,240 lbs.—chiefly from Philadelphia.	Richmond and Pennsylvania bituminous coal. Tons reduced from bushels of 80 lbs. and from chaldrons of 36 bushels.	Scotch and English in tons from chaldrons of 36 bushels or 2,800 lbs. reduced to tons of 2,240 lbs.	Nova Scotia and Cape Breton from chaldr's of 1¼ tons or 2,800 lbs.	Total of British and Colonial tons.	Total consumption of all kinds in tons.
1834	76,180	4,504			25,513	106,197
1835	75,722	7,575	4,287	17,650	21,937	105,234
1836	67,186	7,165	9,146	30,453	39,599	113,950
1837	80,557	3,903	11,873	37,114	48,987	138,447
1838	71,364	3,843	10,344	33,262	43,606	128,813
1839	90,485	5,160	5,880	47,482	53,362	149,907
1840	73,847	3,299	12,410	37,587	49,997	127,143
1841	110,938	4,430	14,245	37,536	51,781	167,149
1842	90,276	4,350	12,718	27,374	40,092	134,718
1843	117,451	5,354	6,862	25,230	32,092	154,897
1844	139,566	6,103	9,098	25,417	34,515	180,184
1845	165,422	10,160	15,195	42,035	57,230	232,812
1846	168,001	6,179	13,188	26,851	40,039	214,219
1847	258,093	5,671	4,256		65,203	328,967
1848	274,902	2,098	5,952		55,768	332,768
1849	262,632	743	12,800	34,531	57,971	321,346
1850‡	294,675			40,082	42,532	339,472
1851	356,758		7,394	38,027	45,421	399,708
1852	432,061		9,748	48,828	58,576	486,889
1853	364,938		4,999	57,111	62,110	426,998

Anthracite and bituminous coal are included in the first column in the latter years, taken from the report of the Engineer and Surveyor on the canals of New York.

The retail chaldron of Boston is from 2500 lbs. to 2700 lbs. weight.

Virginia coal is sold in Boston, chiefly for glass-houses; and, other-

* Quarterly Journal Geol. Soc., London, No. 1, p. 201.

† Also abundant shells in the coal shale at the Portage Railroad. See Trans. Geol. Soc. Penna., p. 255.

‡ Annual Report of the State Engineer and Surveyor on the Canals of New York, 1852.

wise, the quantity does not affect the retail trade. This coal was purchased, in 1845–8, for from $7.50 to $10.50 per chaldron.

Annual Importation of Virginia Coal into Boston.

Years.	Bushels.	Years.	Bushels.	Years.	Bushels.
1835	212,105	1841	124,041	1849	
1836	200,635	1843	149,996	1850	63,415
1837	109,275	1846	183,052	1851	78,280
1838	107,625	1847		1852	14,000
1839	144,475	1848		1853	4,600

Statement of the quantities and value, of American and foreign coal consumed in manufacturing, in the State of Massachusetts, during the year ending April, 1845.

		Tons.	Value.
	Anthracite, -	79,749	$453,411
Bituminous coal	American, -	21,948	147,917
	Foreign, -	29,578	190,405
		131,275	$791,733

Average retail prices of Anthracite and Bituminous Coal in Boston.

American Anthracite, per ton of 2,000 pounds.						Foreign Bituminous Coals, per chaldron.			
			Schuylkill.						
Years.	Lehigh.	Lackawanna.	White Ash.	Red Ash.	Bituminous Richmond, per chaldron.	Cannel.	Liverpool and Scotch and Newcastle.	Pictou.	Sydney.
	Dolls.	*Doll.*	*Dolls.*	*Dolls.*	*Dolls.*	*Dolls.*	*Dolls.*	*Dolls.*	*Dolls.*
1841	8 87	8 75	8 75	9 21		12 42	10 92	9 25	9 25
1842	7 21	6 96	6 96	7 58		10 75	9 41	8 33	9 25
1843	5 75	5 75	5 75	7 08		12 50	10 08	8 58	8 58
1845	6 00	6 00	6 25	6 25	7 50 to 10 50	9 50	7 50	6 75	7 25
1846							8 00 to 10 00	7 50 to 9 00	8 50 to 9 50
1847	6 75	6 75	6 75	7 00	7 to 10	11 00 to 12 00	10 00	8 25 to 8 50	8 50
1848	6 50 to 6 75		6 75	7 00	6 75 to 7	11 00 to 11 50		7 25 to 7 75	7 75
1853	8 to 9							8 00 to 9 00	8 00 to 9 00
1854*	8 50 to 9	8 50	8 50	9 00	10 to 12	10 00	15 00	10 00	12 00

Newcastle is quoted at Boston, wholesale, at $12.00 per chaldron.†
Pictou coal - - - - 9.50 " "
Anthracite - - - - 8.50 " "

Plumbago. In all the cases which we have cited, respecting the anthracites of Rhode Island and Massachusetts, it would appear, in accordance with the views of Professor Hitchcock, that there exists, in these regions, a gradual passage from anthracite to plumbago, or graphite. Whether this approach to graphite arises, according to the view of the learned professor, from the *age* of the enclosing rocks, or whether it be not rather the result of the obvious modifying influence of igneous operations, as we have suggested, we will not now discuss.

* April 16th, 1854—average price less. Boston Courier. † Miners' Journal, Sept. 11, 1854.

Wrentham Coal explorations, five miles from Mansfield.

These were visited by Dr. Jackson, in 1839 and 1840. Although some seams of anthracite existed here, they were adjudged to be too thin to warrant the expenditure attending their mining.

They are also noticed by Mr. Hitchcock, who states that the coal bears a resemblance to the anthracite of Rhode Island.

From the carbonaceous and pyritiferous slates of this impure coal, Mr. Lyell collected numerous impressions of the most common coal plants. Like those of Rhode Island, the slates and micaceous sandstones forming the roof of this anthracite, contain layers and veins of quartz.

Worcester.

Plumbaginous anthracite, according to the state geologist, occurs in an imperfect mica slate, or transition mica slate. This coal bed is seven feet thick, dipping moderately to the north-east; but the works were suspended in 1833.

In this coal, he observes, the metallic aspect is even much more distinct than in the Rhode Island coal; and the quantity of the plumbago is much greater. Several tons of this substance have been ground, and sold for plumbago.*

The Rhode Island coal is heavier than the purest Pennsylvania anthracite, and that of Worcester is heavier than the former, as approaching nearer to graphite.

We subjoin the specific gravities of these, determined by well known authorities; remarking, however, that its great weight seems mainly to be the result of a superabundance of earthy matter, as shown by Dr. Percy's analysis.

Purest anthracite of Lehigh, broad Mountain, &c., (Bache, Johnson and others,)	1.600 to 1.650
Massachusetts. Mansfield plumbaginous anthracite, (Dr. Jackson,)	1.710
Portsmouth, R. I. Plumbaginous anthracite, (Dr. Jackson,)	1.770 to 1.850
Worcester. Plumbaginous anthracite, (Bull,)	2.104 but doubtful
Graphite. (Dr. Ure and Beudant,)	2.080 to 2.450

This plumbaginous anthracite has lately been submitted to the examination of Dr. John Percy, of Birmingham, England. The fol-

* Hitchcock's Geology of Massachusetts, p. 279.

lowing result is published in the proceedings of the Geological Society of London.*

Carbon,	28.350
Hydrogen,	0.926
Oxygen, Nitrogen, }	2.155
Ashes,	68.569
	100.000

Mr. Lyell is of opinion that the stratified rocks, containing the plumbaginous anthracite of Worcester, consisted originally of sedimentary strata, which have been so altered by heat and other plutonic causes, as to assume a crystalline and metamorphic texture, by which the grits and shales of the coal have been turned into quartzite, clay-slate, and mica-schist; and the anthracite into that state of carbon which is called plumbago or graphite.†

The quantity of anthracite annually produced in Massachusetts is but small. We see by a late report, that the value of the mineral coal and iron ore obtained in 1845, was together, but of the value of $21,669.‡

Dr. C. U. Shepard remarks, that the discovery of anthracite at Worcester, in this State, unattended with any secondary or recomposed rocks and vegetable remains, is an apparent exception to the general rule that, "Good workable coal has never been found either in the oldest crystalline rocks, or in the newest formations of the secondary." We think this anomaly is explainable on the ground of the great change in structure and appearance of the regular coal series, and the local effect of intense heat, and consequent modification of character to which the mass has evidently been subjected.

It may even be questioned, however, he adds, whether this seam of plumbaginous mica slate, deserves the name of coal slate.

Production. 1844-5, mineral coal and iron ore mined, 21,669 tons—78 hands employed.§

Bituminous Coal in new red sandstone.

The State geologist observed "thin veins and irregular nodules" of coal in the new red sandstone of the Connecticut valley; but as it is acknowledged that "in almost every instance it appears to be the result of the carbonization of a single plant, whose form can be distinctly traced," we have small expectation that workable beds of coal will ever be met with here, any more than in other parts of the globe, in the same geological formation. It is also to be noted that

* Quarterly Journal of the Geol. Soc., No. 2, p. 205.
† Lyell on Massachusetts Anthracite, Ibid., p. 199, May, 1845.
‡ Statistics of Massachusetts, by John G. Palfrey, Secretary of the Commonwealth, 1846.
§ Hunt's Merchants' Magazine, Vol. XIV., p. 285.

in some parts of the range, these coal traces have been deprived of their bitumen, and have the character of anthracites. This is explained by the author, by the accidental presence or vicinity of trap rocks.*

Railroads completed in 1846, six principal roads—381 miles.
Aggregate length of railroads in 1853, 3,297½ "
Population in 1850, - - - 994,751

CONNECTICUT.

COAL has not been found in this State in sufficient abundance to be ranked with its valuable mineral productions. Dr. C. U. Shepard, reporting on the Geological Survey of Connecticut, announces this fact. Some effort had been made to obtain coal in a highly glazed, plumbaginous mica slate at Sandy Hook near Newton; but the result was unsuccessful and the geologist recommended the abandonment of the enterprise, without delay.†

The State geologist of Massachusetts mention some thin veins of bituminous coal in the new red sandstone of the Connecticut valley.‡ So late as November, 1847, it was announced that a bed of coal had been discovered in the town of Ridgefield, and that measures were in progress for working it.

* Geology of Massachusetts, 1833, p. 230.
† Geological Survey of Connecticut, Report 1837.
‡ Geol. Report of Massachusetts.

NEW HAMPSHIRE.

No traces of coal rewarded the researches of the State geologist in 1840 and 1841, and it appears, from his report, that there is little probability of any being found.

He points out the peculiar applicability of the science of *modelling*, to this interesting and picturesque area.*

Peat.—This useful substance is beginning to be understood and appreciated in New Hampshire; a recent notice, [1844,] informs us, that in the vicinity of Piscataguog village, a piece of swampy land which was thought of but little value, was purchased for a small consideration. The proprietor afterwards discovered that it was covered with *Peat*, to the depth of from three to six feet, and contained one thousand cords to the acre, which are valued at about two dollars per cord.†

The analysis of New Hampshire Peat, by Messrs. Whitney and Williams is as follows:

Locality.	Vegetable matter.	Silica, Alumina, Iron and Lime.
Meredith,	94.90	5.10
Canterbury,	93.80	6.20
Franconia,	73.70	26.30

Peat is employed as an important ingredient in forming a compost for agricultural purposes, in New Hampshire.

* Geological Report of New Hampshire, by C. T. Jackson, 1841, pp. 54, 161.
† Manchester Memorial, August, 1844.

MAINE.

Anthracite and Bituminous Coal.—Although in his "Report on the Geology of the Public Lands" in this state, Dr. Jackson, we believe, did not obtain actual evidence of beds of coal, yet he announces the presence, at various places, of the formation in which coal is usually found, elsewhere, and we have good reason to hope that, when explorations for that object can be more leisurely undertaken, we shall be enabled to add Maine to the other coal producing States of America.

All along the south side of the Aroostook, and stretching southward of Mars Hill, over an area of 120 to 150 square miles; and north of the Aroostook, over an undefined area as far as Temiscouata Lake, are seen, according to the Geologists, "all the marked characteristics of the regular anthracite coal formation."* The rocks certainly belong to that formation, and are frequently glazed with carbon, but no bed of coal was discoverable. This is not by any means a decisive or even discouraging circumstance; when it is remembered how unfavourable to such discoveries is the position of the surveyor whilst passing, amidst numberless privations and difficulties, through an almost impracticable and uninhabited wilderness. Accident will probably bring to light, much that must inevitably escape the scrutiny of the most practised observer.

As the "regular coal series" is mentioned, in conjunction with "the old red sandstone, resting upon the grauwacke," we should infer that the coal, if any, belongs precisely to the same age as the great central coal fields of the United States, whether bituminous or modified in the form of anthracite.

Shales, containing vegetable impressions such as usually characterize the coal measures, have been observed at Waterville and certain points in this state; but the geologist hesitates to vouch for the absolute existence of coal.† These supposed impressions, indicative of the coal measures, have been decided by Prof. Hubbard and admitted by Dr. Jackson, to be not vegetables, but true annellides, such as characterize the slates which Murchison has included in the Cambrian series.‡

Peat.—Dr. Jackson, than whom there can be no better authority on agricultural geology, directed his attention to the numerous valuable deposits or rather accumulations of peat in this state. At the localities which he designates, this substitute may be most advantageously wrought for fuel; it is applicable to the burning of lime,

* Geology of Maine.
† First Report on the Geology of Maine, p. 106.
‡ Proceedings of Assoc. of American Geologists, 1841.

and various domestic uses, as well as convertible into a powerful manure, admirably adapted for loosening and enriching clayey soils.

By an extract from a report ascribed to this author, and quoted in Silliman's Journal, we learn that in Maine, peat is found at the depth of three feet from the surface, amid the remains of rotten logs and beaver sticks; showing that it belongs to the recent epoch. The peat is twenty feet deep (thick) and rests upon white siliceous sand. This recent coal was found while digging a ditch to drain a portion of the bog.

"On examination I found that it was formed from the bark of some tree allied to the American Fir, the structure of which may be readily discovered by polishing sections of the coal, so that they may be examined by the microscope."

Dr. Jacksons analysis, shows that it contains in 100 grains,

Bitumen, - - - - -	72
Carbon, - - - - -	21
Oxide of Iron, - - - -	4
Silica, - - - - -	1
Ox. Manganese, - - - -	2
	100

"This substance is, therefore, a true bituminous coal, remarkable indeed for containing more bitumen than is found in any other coal known. I suppose it to have been formed by the chemical changes, supervening upon fir balsam, during its long immersion in the humid peat." This is a very interesting discovery; and the same substance appears to exist in other peat bogs of the State.* See notices on peat in other parts of this volume.

In his "Report on the Geology of Maine," in 1837, the author last quoted notices the accumulation of Peat at Quoddy Head, near the south-east angle of the State. This mass is 15 feet thick, and it is suggested to deprive it of its water, by means of pressure, when it will form a valuable fuel. When it is remembered that, according to the analysis of Sir Humphrey Davy, peat contains 60 per cent. of carbon, its use should by no means be discarded. It is a valuable fuel for domestic purposes and for many manufactories.

The above remarks will apply to many other localities in Maine, and the time will arrive when, wood becoming scarce, our neglected peat bogs will be in requisition.† In Germany, peat is dried in kilns, heated with the small fragments and the refuse parts of the same substance.

See under the head of Peat and Turf, in various parts of this work. We have in our introductory portion of this work, explained the mode of working the turbaries or peat bogs of France.

* Silliman's American Journal of Science.

† First Report on the Geology of the State of Maine, by Dr. C. T. Jackson, p. 32.

MICHIGAN.

BITUMINOUS CENTRAL COAL BASIN, BETWEEN LAKES HURON AND MICHIGAN.

To the series of reports made annually by Dr. Douglas Houghton and his assistants, we are indebted for the earliest notice of this coal-field.

This able geologist was appointed to commence the survey of the State in 1837, and, in the execution of duties unusually arduous and fatiguing, exhibited much zeal and perseverance.* His reports, however, up to the present time, have not received the elucidation they so much need, from geological maps of the State.

In a region which almost everywhere is thickly covered with beds of diluvium, sand and tertiary clay, which by far the greater part remains even now in its original state of primeval forest, and which, except as relates to the superficial value of the soil and timber, continues uninvestigated, it would be unreasonable to expect from the geologist much exactness of detail, as regards this important coal-field. We are yet in uncertainty as to its limits, for no part of it, strictly speaking, can be said to be satisfactorily defined. The few coal strata which exist have so little inclination, and are so completely buried beneath the more recent deposits referred to, that positions where outcrops can be observed are of rare occurrence, and these are only found along the beds of the rivers, with which this beautiful country is thickly intersected. The northern termination of the formation, of which the coal seams form an important member, is left entirely conjectural, but is supposed to reach the head waters of the Tittabawassee and Maskego rivers. This part of the State was, at the time of the survey, entirely an uninhabited region.

Dr. Houghton came to the conclusion that the coal-bearing sandstones, or, strictly speaking, the *coal basin*, occupies an extent of surface nearly oval in form, whose centre very nearly corresponds with the true centre of the peninsula.

He estimated this area at 150 miles in length, from south to north, and upwards of 100 in extreme breadth—covering an area of about 11,000 square miles. The general outline, as sketched by this geologist, is probably approximately accurate: fully as much so as the nature of the country permitted in 1840.

How much of this area may be considered an actual coal-field, is matter of conjecture, of course. From a consideration of the map

* Since penning the above passage, the unfortunate death of this talented man has come to our knowledge. He was drowned during a sudden snow storm, on Lake Superior, while engaged in his professional duty.

of the State, and from a partial reconnoissance of the district, by the present writer, in 1847, we think it a safe calculation to estimate the productive coal-bearing area at 100 miles in length, by an average breadth of 50 miles. This may be computed at 3,000,000 acres, or somewhat less than 5000 square miles.

The coal seams are thin and few in number. What are usually denominated the coal measures are comprised within a very limited thickness. The entire coal-field is evidently a shallow one, having suffered little from disturbance.

Dr. Houghton's section exhibits the coal series, as consisting of two beds of coal and bituminous shale, separated by a bed of undetermined, but inconsiderable, thickness of sandstone; the whole group resting upon a gray fossiliferous limestone, which does not anywhere exceed 14 feet in thickness.

UPPER COAL FORMATION.

Maximum thickness 30 feet,—of comparatively small area, in the central part of the coal basin. The coal is comprised in several layers, not exceeding in thickness from one to two feet each, accompanied by thin beds of argillaceous iron ore. In point of quality it appears to be inferior to the lower coal.

LOWER COAL FORMATION.

But two continuous beds of workable coal are ascertained to exist in the State of Michigan. The lowest of these lies at a small distance only above the limestone stratum, and is associated with a bed of shale, which is also sufficiently bituminous to answer the purpose of an inferior coal. The State geologist estimates the maximum thickness of this lower series at twenty feet. At Corunna, the county seat of Shiawassee county, on the border of the river of that name, is the best development of the lower coal bed that at present has been observed. It is the only locality in the State where coal has been raised for economical use, and even here the work is upon a small scale. It consists of a few shallow open pits sunk on the margin of the river and down to the level of its waters. This close proximity to the water renders the situation selected very unfavourable for mining. The coal bed is 3½ to 4 feet thick. In structure it is finely laminated, like the Ohio coal: the laminæ appear as if cemented together with bitumen. In quality it is excellent, highly bituminous, and brightly blazing; it is said to produce a very hot fire, and to be well adapted to the purposes of the blacksmith. The coal is here covered by some thin beds of black bituminous laminated slate, by a course of nodules of argillaceous iron ore and by seams of black shale and clay. Then occurs a drab-coloured gypseous clay, and over that are the gray and yellow sandstones with impressions of coal plants; from some portions of this series grindstones have been made. The fossiliferous limestone, underlying the coal, crops out in the vicinity.

However excellent this coal appears at these small openings, its quality, and even its thickness, can scarcely be said to be now fairly tested; for, as at this precise spot there is no rock covering, it can only be considered as surface or crop coal.

We have been more minute in this notice because it is one of the very few localities which have been explored for the purpose of actually mining the coal over an area of several thousand square miles.

We have, during the year 1847, examined the outcrop of coal at several other points on the Shiawassee river, but the attempts at its development are, as yet, of a very feeble nature. The coal at all these positions was accompanied by overlying courses of excellent argillaceous carbonate of iron, and by beds of gypsiferous shales.

Dr. Houghton closes his annual report for 1841, by recording his belief that bituminous coal will be found in abundance, for all the wants of the State, and that it may be fairly inferred from the facts already determined, as to the range of the coal-bearing rocks, that coal will be found at numerous other points than those now ascertained, and also in several counties where it is not now positively *known* to exist, and where no attempts have even been made to pursue its traces.

Whenever these developments are completed, or even when they are but very partially extended, and when the area, quality and amount of productive coal beds in the situations favourable for transportation are determined satisfactorily, the geographical importance of such a vast district of bituminous coal must be apparent. The influence which it will exercise over the future prosperity of this new State can scarcely now be appreciated. Centrally situated, accessible to all the upper lakes and rivers, around whose borders not a trace of a coal formation exists, the Michigan coal-field could have no rival. Should coal operations be conducted here on an adequate scale, and the means of transportation to the lakes be facilitated, the present price of mineral fuel will be greatly reduced. Anthracite in September, 1847, obtained from $13.00 to $15.00 per ton at Detroit.

That we are not alone in our estimation of a coal-field so favourably circumstanced, in this portion of the American continent, we might quote a passage in the report to his Excellency, the Governor-General of British North America, on the geological survey of Canada, by W. E. Logan, Esq., 1st May, 1847.

"The great expense attendant upon the transport of copper ore to a distant smelting locality, naturally turns the attention of those whose minds are directed to the subject of mining it, to the aid to be derived, in its reduction, from such coal deposits as are most nearly situated to the region in which it exists. The geological structure of Canada appears to promise little in regard to this useful mineral; but in the States of the neighbouring Union, there are two localities on the great chain of lakes to which the mineral region of Lake Superior belongs; one at Cleveland, on Lake Erie, the other at Chicago, on Lake Michigan; within forty and sixty miles of which, respectively, coal might probably be made available. But in the heart of the southern peninsula of Michigan, which is still nearer the metalliferous region, a third great coal-field is spread out; and in this instance the waters of Lake Huron appear to make a deep incision into the deposit in [near?] Saginaw Bay.

Saginaw Bay, therefore, appears to be the position naturally destined for the reduction of such copper ore as may result from the mineral region of Lake Superior. These ores, combined with the sulphurets reported to have been discovered on Lake Huron, seem to be sufficiently varied to give a favourable smelting mixture. The coal is of the bituminous description, and beds of fire-clay will be found supporting the seams. Unless some great change should be effected in the system of smelting copper ores, there is little doubt the produce of the Michigan mines will ultimately centre in this locality; and it can only be the operation of fiscal laws that will prevent the Canadian ores from finally reaching the same destination."

PEAT.

The geologist dwells, with much earnestness, on the abundance and value of the peat deposits of Michigan. The substance, he states, is of considerable importance, not only as a combustible, but as a manure. These peat beds are, like the coal seams, comparatively shallow, seldom exceeding four feet in thickness. It is of the fibrous kind, and by no means so compact as to form a good fuel. It most commonly overlies beds of calcareous marl, which has accumulated in the innumerable low meadows, beaver swamps, and wet prairies of the country. Michigan has been aptly designated by the Indians as "the land of lakes," and the State geologist has reported on the existence of not less than three thousand lakes within the limits of the peninsula. *Lignites* have not been met with in this country.

CENTRAL BITUMINOUS COAL-FIELD

OF THE

PLAIN OF THE MISSISSIPPI, USUALLY STYLED THE ILLINOIS COAL-FIELD.

We have separated this vast area into four divisions, corresponding with the respective States into which they extend, as follows:—

I. Indiana division.	III. Kentucky division.
II. Illinois division.	IV. Iowa division.

In making this classification, we have included no portion of the coal area of Missouri, from which there is, however, no other separation from that of Illinois than the Valley of the Mississippi. The Missouri region will be best treated on, separately, in the present instance.

I. INDIANA DIVISION.

To Dr. D. D. Owen's Geological Reconnoissance of this state in 1837–8, we are chiefly indebted for what we are able to communicate respecting this imperfectly known region.

He states that the entire western portion of Indiana proves to be rich in coal, and although wood is extensively employed, as the cheapest or most convenient fuel at present, the axe is busily at work in the primitive forests, and the rapid increase of steam-power, calling incessantly for fuel, is thinning them out from year to year.*

The reporter was unable, from the result of the reconnoissance of a single season, (1837,) to form any exact estimate of the area of this coal-field. He simply records the results of local examinations, while traversing the various counties. Limited, therefore, as is the information obtained in an area which covers several thousand square miles, we receive this contribution to American geology, with all the confidence which the known intelligence of the reporter deserves at our hands.

Bituminous coal, of the ordinary kind, abounds here, as well as cannel coal. On White river, the seams are upwards of six feet thick. Others are four and three feet thick, near Terre-haute.

* Owen's First Report of a Geological Reconnoissance of Indiana, in 1837.

If we were to reason from geological, or rather from mineralogical conditions, on the future position of Indiana, we should say that her western counties are destined to become, one day, the chief centre of manufacturing industry in that parallel.

Near the mouth of Coal creek, on the Wabash, in Fountain county, no less than six beds of coal are exposed, interstratified with argillaceous iron ore and carbonate of iron. These coal seams vary from one foot six inches to four feet six inches thick, each.

At the Sugar creek foundry, in Parke county, is a three feet bed of cannel coal; or a coal nearly approaching to that variety.

An eight feet seam of good coal occurs in one locality; another in Clay county, four or five feet thick. On Honey creek, Vigo county, a four feet coal vein; as also on Lick Fork of Busseron creek. On the Patoka, in Pike county, is a bed said to contain nine feet of coal.*

The reporter is unable to state any facts to establish the identity and continuity of these coal seams, or how far they may be repetitions of the same series. He has, however, made analyses of many of these, which will be found in our tables at the end of this work. We conceive, that they are all approximate results merely; inasmuch as they are derived from outcrops, rather than from perfect coal of the mines, which, at the time of the survey, were not in existence.

Dr. Owen concludes his second report of Indiana with a summary of its geology. We have only now to do with that part which refers to the coal formation. This area is a part of a great coal-field, which includes nearly the whole of Lower Illinois, and eight or ten counties in the south-western part of Kentucky.

In Indiana, this bituminous coal formation occupies an area of about 7,700 square miles.

The coal exhibits its vegetable origin very distinctly. Layers of charcoal, from which the woody fibre can be readily detached, are frequent in the superior coal beds. The dip of the beds is westward, and they are of the same geological age as those of the Cumberland Mountains in Tennessee.

Peat occurs in this state.†

Fig. 17.

Transverse Section of the Illinois and Indiana Coal-Field.

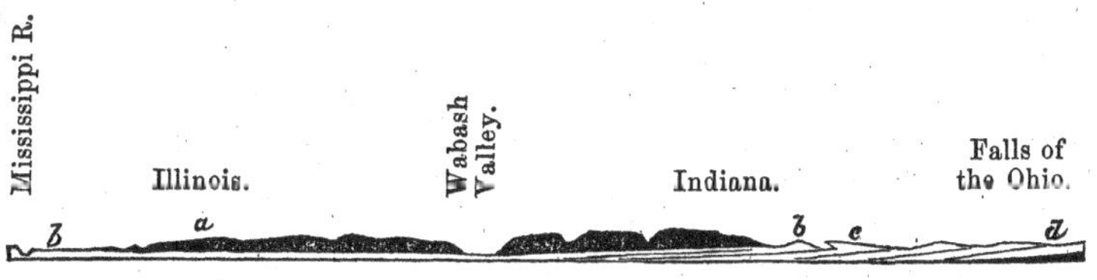

a Coal measures. bb Carboniferous limestone.
c Old Red Sandstone—Portage and Chemung Groups—Waverley Sandstone.
d Marcellus Shale—Helderberg series.

* Owen's Second Report of a Geological Survey, made in 1838.

† Ibid., p. 47.

The foregoing diagram shows a transverse section of the lower part of the Illinois coal-field, from the large section of Mr. Hall.*

The coal area of Indiana is only a portion of an immense western region, which a writer has asserted to be "fifteen hundred miles long by six hundred miles broad, and would cover half Europe.†"

We hesitate before adopting his statistics, notwithstanding that we are, on the whole, disposed to agree with those who admit the possibility of a period when one single coal formation covered that part of the continent of North America, which extends from the Lehigh to the Missouri. But as we have only to do with the existing—not the theoretical, or antediluvian areas of these coal formations—we are compelled to state that the present maximum length of the great Illinois coal region is about three hundred and thirty miles, and the greatest breadth is two hundred miles. The actual area, containing coal, in the entire space, which comprehends portions of Indiana, Kentucky, and Illinois, is 56,200 square miles.

In Dr. D. D. Owen's paper read to the Geological Society of London, in November, 1842, but not published *in extenso*, until November, 1846, we observe that the dimensions assigned by that gentleman differ very little from the preceding, except in the total area, which, he states, "equals the entire island of Great Britain." This is no doubt an accidental oversight, because that area is 83,828 square miles, or 88,052 including the Scottish Isles.

We may repeat here, that the other great bituminous coal field, which is usually named the Alleghany coal region, is in the extreme length along its centre, seven hundred and fifty miles, and in average breadth eighty-five miles, containing 65,300 square miles.

The Illinois coal field contains the caking variety, some splint, and some cannel coal.

A canal was opened in 1839, at Evansville, one hundred and eighty-eight miles above the mouth of the Ohio, which communicates with the extensive coal region in the interior of this state.

Length of Wabash and Erie canal in 1847, three hundred and seventy-four miles; cost, $5,585,000; receipts, $500,000.

BITUMINOUS COAL OF CANNELTON.

We have been furnished with details from several parties, in reference to the cannel coal of this locality. It is situated in that part of the coal field which is intersected by the Ohio river, at about 120 miles below Louisville by water, but scarcely more than half that distance in a straight line. The seam varies from three to four feet in thickness and occasionally expands nearly to five feet.‡ It is entered, for the purpose of mining, by means of an adit-level, a quarter of a mile from the Ohio, at an elevation of seventy feet above the bank of that river. In regard to geographical position, therefore, the site is

* Trans. Assoc. Amer. Geol., Vol. I., p. 267.

† American Quarterly Review, March, 1839. Also Darby's View of the United States.

‡ Letter to Prof. Silliman, by B. Lawrence, Louisville, November 20th, 1847.

unusually advantageous, and the coal can be furnished at a corresponding low price.

The coal, for an inch or two of the upper part of the seam, contains less bitumen than the rest, bearing a slight resemblance to cannel coal, and decrepitating when burning. Two or three inches of the lower part consists of a highly bituminous shale. The remainder of the mass is of the finest quality, coming out in large blocks of a foot or eighteen inches in diameter, exhibiting vegetable traces. It burns freely, yields a pleasant flame, and affords a light sufficiently strong to enable a person to read by it.*

This coal was experimented upon by Prof. W. R. Johnson, and the results are detailed in his invaluable report to the Navy Department of the United States in 1844. He observes that the fracture is often conchoidal, and the lustre dull, like that of Scotch cannel coal. The surfaces are frequently covered with films of sulphuret of iron. Specific gravity was 1,247 and 1,297, the mean of which gives 79½ lbs. per cubic foot. The mean result of two specimens gave the following proportions:

Of fixed carbon, - - -	59.40	100
Moisture and other volatile matter, -	34.90	
Earthy matter, (3.49 to 8.16) - -	5.70	

Samples of the coal have contained a much smaller amount of earthy matter—in one case not much exceeding 2 per cent.

The observations of Prof. Johnson agree with our own experience:—"From its flaky texture it speedily disintegrates into flat masses, burning with little intumescence, and scarcely any tendency to agglutination. This property allows a free passage to the air, favours rapid combustion, and causes the exhibition of an exceedingly brilliant light. Its prompt and rapid action appears to adapt it, in a remarkable manner, to the purposes of western steamboats. It seems to bear transportation better than any other sample of bituminous coal which came under notice. It was the only really available sample forwarded for trial from the great coal fields of the west."†

It is submitted to us, by the proprietors of this coal, that the specimens sent to Washington were intended to represent a perfect cross section of the seam of coal, including the inferior portions. Had the combustible experimented upon consisted solely of the better and main portion of the vein, they conceive that the result of the investigation would have probably been yet more favourable.

We may be permitted here to remark that we have, in the progress towards the completion of this work, received numberless communications from companies and individual proprietors of coal mines, in all parts of the Union. We have endeavoured to do strict justice to all, so far as our means and space permitted us to act discriminately,

* Letter of Prof. Frederick Hall, published in the National Intelligencer of Washington, July, 1843.

† Report to the Navy Department of the United States, by W. R. Johnson, 1844.

and without subjecting ourselves to the imputation of advocating particular interests. Whilst wishing success to all, we have never knowingly permitted the suggestion of interested parties to divert us from that course.

Prices of the Indiana cannel coal in 1847–8.—We are informed by a proprietor that from 2000 to 3000 bushels of this fuel are now sold daily, at the mine, to steamboats at 7 cents per bushel, $1.96 per ton. He adds that it can be afforded to manufactories near the mine at 4 cents per bushel, $1.12 per ton, and that it is, considering the *quality*, the cheapest coal in the world, at a position convenient for extensive manufactories.*

This field occurs, like most others, in the basin form: the bed of cannel coal at this place being the lowest in the series, which consists of two workable beds only, as we learn from a pamphlet recently published.† The lowest coal, as we have stated, occurs at Cannelton and at Trade Water in Kentucky; the upper seam is worked at Bon Harbour, and at various other places. At different points each of these seams varies from three to ten feet in thickness. The same bed of cannel appears also in Kentucky, Illinois and Missouri.

We observe in the pamphlet alluded to, that a ton of this coal is equivalent only to 26½ bushels on the Ohio. The work terminates with the emphatic statement, in relation to this coal position, that, having the cheapest power, the cheapest materials, and the cheapest food, it possesses the means of manufacturing the cheapest goods in the United States.

The bituminous coal of Car's Run, 160 miles below Wheeling, is put on board the steamboats for 6¼ cents per bushel, or $1.75 per ton. It is a lighter and drier coal than that of Wheeling, and less bituminous, but is considered to be better adapted for the steamboats of the Ohio.

* Hamilton Smith, Esq., January, 1848.

† "The relative cost of steam and water power, the Illinois coal-field, and the advantages offered by the West, particularly on the lower Ohio, for manufacturing;" Louisville, 1848.

II. ILLINOIS DIVISION.

So much of the coal formation as is comprised within the State is, at the utmost, 44,000 square miles; but if due deductions are made for unproductive portions, for large spaces divided or removed by rivers and valleys, the real productive area would not, probably, be found to exceed thirty thousand square miles.

There is no coal on the Ohio river nearer to its junction with the Mississippi than Saline, near Shawneetown, 116 miles above the mouth of the first named river. On the Mississippi it is rather a shorter distance, being sixty miles to Muddy creek, and thence twenty-five miles up that creek to the first coal-bed there, or twelve miles by land. Some coal operations commenced here some few years ago, having in view the supply of the towns along the Mississippi, as far even as New Orleans. The present supplies of coal to the lower country are obtained from a vast distance up the Cumberland and Tennessee rivers, but especially from Wheeling, Pittsburg, and the intermediate points, 900 miles further from the market than the Illinois coal of Muddy creek. The estimated expense of delivering this coal at New Orleans, by arks, is about $2.25 per ton: while the minimum price of coal there is 25 cents a bushel, or $7.50 per ton. In winter time from 50 to 62½ cents per bushel, or $12 to $15 per ton, have been occasionally the retail price there. This Muddy creek coal seam is a horizontal bed six or seven feet thick, above which is another vein, not heretofore worked. Coal can be thrown from the mouth of the drift into a boat. Its quality is most excellent, igniting readily, and caking together perfectly, without making much clinker. It has been used for fifty years by the old French settlers, to make edge tools, which have borne a high reputation.

What is termed St. Louis coal, supplied to the steamers, burns with a good flame, and cements like that of Pittsburg; ashes dark gray, in small quantity, and consumes with little waste. It is often mixed with yellow sulphuret of iron in flakes occurring on each face of the sectional fracture; and consequently is not, we understand, in so good repute for the purposes of iron manufacturing.

At the present day, it is impossible to state precisely how much of Illinois and the contiguous States is occupied by productive beds of coal. The true coal producing area is probably but a fraction of the space comprehended within the geological limits of the carboniferous formations in those countries. This remark especially applies to Missouri and Illinois; in the former State, the coal prevails rather in detached patches, than continuously spread over the entire space. When persons speak of the vast area occupied by the coal formations in the western country, we must understand them as referring to the external limits of these areas, for we have no detailed surveys to show the extent occupied by the workable coal beds. We believe, in saying thus much, we but express the sentiments of every local observer in that quarter.

Towards the north-west boundary of this district, several coal seams are seen in the tongue of land which lies between the Mississippi and Rock rivers. One of these beds is from five to six feet thick: its quality is fair, and evidently improves as the workings proceed.

To the south of Rock river are several good coal seams which are capable of supplying almost any required quantity of this fuel. Their local position and advantages render them of very great value to the country lying north of this.*

An article on the geology of Upper Illinois, appeared some time since, [1838] from the pen of Dr. C. U. Shepard,† in which are notices of the coal in that quarter. The outcrop of a six feet vein is described as occurring in the valley of the Illinois river. The same bed exists at Vermillionville, and is the most important of any in the district. In quality, it is a fat, bituminous coal, having equal proportions of carbon and of volatile matter.

The exact boundary of the coal-field has not been traced here on account of the great thickness of alluvial or drifted matter, which, as in Michigan, sometimes covers the coal measures to the depth of more than a hundred feet.

Dr. D. D. Owen's valuable paper was published in the Journal of the Geological Society of London, in 1846. He has ascertained that, with the exception of some slight shades of specific difference, there is a striking analogy between the fossil flora of the American fields and that of the equivalent strata in Europe. Among many others, he mentions that *Palms* are not uncommon, and some remains of Coniferæ (?) have been found. He also obtained remarkable specimens of the stumps of fossil trees (apparently palms) found standing erect with the roots attached, imbedded in slaty clay; and slender leaves have been found, in great abundance, in the near vicinity of the stumps, imbedded in the clay.‡

The author adds, that valuable beds of argillaceous iron ore exists in this coal-field, but that hitherto few furnaces have been in operation. He conceives that this ore must ultimately become an important source of income to the State, or rather to the proprietors residing in it.§

Stated production of coal in Illinois,	in 1839,	13,427 tons.‖
" " " "	in 1840,	15,417 "

* Owen's Report to Congress on the Mineral Lands. Doc. No. 239, p. 44.

† American Journal of Science, Vol. XXXIV.

‡ Models of these stems, illustrating their appearance in their original site, were exhibited by Dr. Owen, at a meeting of the Society of American Geologists and Naturalists.

§ Proceedings of the Geol. Soc., Vol. IV. Also Quarterly Journal of the same Society, Vol. II., p. 433.

‖ Hunt's Merchants' Magazine, Vol. V., p. 434, and subsequently.

Employing, according to the return to Congress, 154 workmen, and $120,076 capital; probably much underrated.

A Chicago paper states, "There are, or will be, in operation on the 1st of July, 1855, in the State of Illinois, 3,715 miles of railroad. There are now in operation, leading into the City of Chicago, 1,626 miles of railroad."

CANALS.

This State, besides the vast extent of its navigable streams, possesses 374 miles of finished canals, which cost $5,585,000, and yielded in 1845 a revenue of $500,000.

III. KENTUCKY DIVISION.

THE south-eastern extremity of this vast coal region, stretches across the Ohio river into Kentucky, and occupies eight or ten counties in the north-western part of this state,* probably about 4,500 square miles.

At Hawsville, on the left bank of the Ohio, 120 miles below Louisville, is a coal bed four feet thick. The upper eighteen inches of this bed consists of Cannel coal; the remainder is common bituminous coal, two and a half feet.

Its analysis by Dr. Jackson, is	Carbon,	48.40
	Bitumen, &c.,	48.80
	Ashes,	2.80
		100.00

The price of this coal at New Orleans, was sixty-two and a half cents to one dollar per barrel, of two and a half bushels. It is in request there for the use of the tow-boat companies.†

Hawsville is about 258 miles above the mouth of the Ohio. The coal seam is nearly horizontal—appearing on both sides the river, in a position remarkably favourable for loading into vessels lying in the Ohio. It is a compact, largely conchoidal, coal, producing a bright flame; does not cement or adhere together in burning, but on the contrary, falls into a profuse white ash, much larger in amount, practically, than the foregoing analysis exhibits.

* Owen's Report of Indiana, p. 39.
† Hazard's Register.

Although 700 miles in advance of Pittsburg, it has been hitherto, we are told, unable to compete with that coal, which is floated down the Ohio in arks, and, it is said, can be mined cheaper. The boats and arks, in which the coals are conveyed down the stream, can also be built cheaper above; and moreover, the Pittsburg and Wheeling coal is estimated better for blacksmiths' use, &c. Still the Hawsville Cannel is especially liked for steam engines. For domestic use we think it is objectionable, on account of the great quantity of very white ashes which are left after combustion, filling up the grates, &c., to an unusual degree; at least such was the case in the supplies furnished on three or four occasions when we passed up and down the Ohio in steamboats. Altogether, it is greatly inferior to Lancashire Cannel coal.

It is specifically lighter than common bituminous coal, yet heavier than Cannel coal of Lancashire and Yorkshire.

Kentucky or Hawsville Cannel, spec. grav., -	1.250
What is called the Cannel coal of Jackson county, Ohio,	1.410
Lancashire Cannel, - - - - - -	1.199
Ingolton Cannel coal in Yorkshire, - - - -	1.195

Caseyville Cannel coal, similar to that of Cannelton, in Indiana, reported on by Professor W. Johnson.*

Spec. grav. 1.39. The result of his analysis gave of

Volatile matter, - - - -	31.80
Fixed carbon, - - - -	44.50
Earthy matter, - - - -	23.70
	100.00

It will be observed that the proportion of earthy matter is unusually large, amounting almost to one-fourth, and agreeing with the practical remarks made by ourselves and others, on board the Ohio steamboats, on several occasions.

Petroleum.—Springs of this substance occur at several points.

An account is given in Silliman's Journal of a *Petroleum Oil well*, near Burksville, Kentucky. This was discovered some years ago, whilst boring for salt water, and after penetrating solid rock for two hundred feet, a fountain of pure oil was struck, which was thrown up more than twelve feet above the surface of the earth. Since that time it does not appear that the supply is sufficiently regular to furnish an important amount of this oil.†

Professor Mather reports that many hundred, perhaps many thousand, barrels, might be annually collected at the different petroleum springs in Kentucky; and that it commands a high price in the eastern markets.‡

* Coal Report to the Secretary of the Navy, 1844.
† Silliman's Journal of Science.
‡ Mather's Reconnoissance of Kentucky, in 1838, p. 28.

Gas Springs, evolving carburetted hydrogen, are not uncommon. They burn with a white flame, and are capable of useful applications, such as lighting and warming houses, boiling salt, &c.

A similar occurrence takes place at the Albion Mines, Nova Scotia; and also, on a magnificent scale, in China.

BRECKINRIDGE CANNEL COAL.

Professor B. Silliman, Jr. read before the American Scientific Convention at Washington, May, 1854, a paper on the Breckinridge Coal-field of Kentucky, of which the following is a brief summary:

"*Breckinridge County Cannel Coal.*—The coal measures in this county reach an elevation of about 500 feet above mean high-water of the Ohio river, at Cloverport. The strata lie with remarkable uniformity, and nearly horizontal, their dip being south-west, at the rate of about four inches in one hundred feet, or less than twenty feet in a mile. The surface of the country is deeply cut down by the wear and tear of the water-courses, and of atmospheric agencies. The bluffs on the Ohio, at Cloverport, as elsewhere in that region, are composed of a fine-grained granular sandstone, loosely aggregated, thickly bedded, of a light gray colour, nearly free from anything but silica, and easily reduced to a sharp, clean white sand. No traces of organic remains could be found in it; but, from its position, it must be below the coal. It will be observed that *three* coal-seams are shown. The two lower are common bituminous coal, and need not be particularly mentioned in this connection, except to say that the lower bed (from four to five feet in thickness) is believed to be the same which is worked at Hawsville. The coal which now occupies our attention is the uppermost member of the series, occurring at an elevation of nearly three hundred feet above the level of the Ohio, and capped by a thick overlying mass of sandstone and shales. This bed of cannel coal is about three feet in thickness, with a bituminous shale of some ten feet in addition, on which it immediately rests. As a consequence of its position, it covers a less area of country than the lower beds, having been removed in many places by denuding causes. It is to be looked for in the regions where I have seen it—in the uplands and hill-tops, and only there, where the general level of the country reaches a certain elevation.

"The northern margin of the coal beds in the place referred to, lies about nine miles from the Ohio river, to which point a railway has been built. Here a marked change in the topography of the country is visible. The hills become more sweeping, and the valleys less precipitous, while the contour lines wind in graceful and gentle curves around the base of the hills. The upper members of the series at this point furnish a soil of enduring strength, as is tested by the size and abundance of the forest trees. Numerous small rivulets, the head-waters of Clover Creek, afford ample drainage.

"The Cannel Coal of Breckinridge County is characterized by the following peculiarities, viz.:

"1. *Its density*, which is 1.14 to 1.16. Common bituminous coal varies from 1.27 to 1.35, and the anthracite from 1.50 to 1.85. The only coal lighter than this, so far as is known, is the so-called 'Albert Coal,'* of New Brunswick, whose density is 1.13. The cause of this low density will be sought, chiefly, in the very large amount of volatile matter.

"2. *Its Tenacity and Elasticity.*—Coals are usually brittle and inelastic. This is tough, and resists powerful and repeated blows, and rebounds the hammer like wood. The splints of this coal may be sensibly bent by pressure, and regain their original form again. A fissure in it may be sprung open by a wedge, and will close again on withdrawing it. The writer has never seen any other coal with this peculiarity.

"3. *Its Electrical Power.*—The Breckinridge coal becomes powerfully excited by friction, with resinous electricity. This peculiarity may be demonstrated very easily, and has never been before noticed in any other coal, so far as the writer has been able to learn, except in the 'Albert Coal,' of New Brunswick, before named. It is not easy to understand why other very highly bituminous coals should not have this property, but such is the fact with a large number that have been tried.

"4. *Its Chemical Constitution.*—This has been determined in the usual way, by destructive distillation, with the following results, viz.: in 100 parts we have,

	I.	II.
Volatile at redness, - -	60.27	63.520
Fixed Carbon, - - -	31.05	27.160
Ash, - - - - -	8.66	8.470
Hygroscopic moisture, - -		.777
Sulphur, - - - -	a trace	a trace
	99.98	99.927
Coke, - - - - -	39.71	36.68

"A comparison of these analyses with those of other highly bituminous coals, will show that there are very few examples recorded of so high an amount of volatile matter. For example, we find among American coals that the 'Albert Coal,' of New Brunswick, yields 61.74; the coal of Chippenville, Penn., 49.80; that of Kanawha, 41.85; that of Pittsburg, 32.95 per cent.; while the mean of the fat, caking coals of Liverpool is 37.60 per cent. The Lowmoor Scotch Cannel, and the Boghead, also a Scotch coal, are the only ones giving a higher proportion of volatile matter. In fact, the ordinary proportions of volatile and fixed ingredients in bituminous coals are completely reversed in the Breckinridge Cannel."

* Asphaltum. (See Taylor's deposition before the Supreme Court, Halifax, on the Hillsboro' Asphaltum Mine, Albert City, N. B., and other competent authority.)

We regret our limits prevent our making larger extracts on the subject.

IV. IOWA DIVISION.

THIS great bituminous coal-field has its north-western termination in Iowa Territory; occupying some ten or twelve townships, or about four hundred square miles, which we have included in our estimate of the Illinois field.

Not much was known as to the details of the coal seams here, at the period of Mr. Owen's survey, in 1840. Only one seam is adverted to, as cropping out west of the Mississippi, the coal of which was said to be of indifferent quality. Its analysis appears in our tables. There are several good coal seams on the eastern borders of the river in the Illinois portion.*

According to official returns to Congress, in 1840, there were 10,000 bushels, or 321 tons of coal raised in Iowa.

1854. With respect to the immense coal fields in Iowa, we take the liberty of quoting largely, by permission, from the valuable work of Dr. D. D. Owen, in his last Geological Survey and Report of Wisconsin, Iowa and Minnesota, published in 1852. "The country, which has been carefully examined, is the most extensive ever reported by any geologist or geological corps in this country, including, as it does, four times as much territory as the State of New York, and being about twice and a half as large as the island of Great Britain. The average width of the territory laid down in the geological map, is about 270 miles, its area exceeds 200,000 square miles." "Throughout this vast district, all the principal streams which water it, have been explored, to the number of ninety-one, and more than a fourth of these have been navigated from their mouth almost to their source in bark canoes."

"Coal and iron in abundance have been found, and other valuable minerals. The coal measures of Iowa are shallow, much more so than those of the Illinois coal field. They seem attenuated as towards the margin of an ancient carboniferous sea, not averaging more than fifty fathoms in thickness. Of these the productive coal measures are less than a hundred feet thick. The thickest vein of coal detected in Iowa, does not exceed from four to five feet, while in Missouri some reach the thickness of twenty feet and upwards. In

* Owen's Report on the Mineral Lands of Iowa and Wisconsin, 1840, p. 44. Also in revised edition of 1844, p. 53.

quality the coal is, on the whole, inferior to the seams of the Ohio valley. To this, however, some very fair beds form exceptions."

"Lignite was found on the Mankato river and its branches, which approached the cannel coal in its character. Although search was made, no regular bed of lignite was found.

"The carboniferous limestone in Iowa, if we commence where it crosses the Des Moines, between the mouth of that river and the Missouri line, ranges north as far as the confluence of the Iowa and English rivers, then for about 40 miles it is lost to surface view, reappearing on the Iowa river, in Tama county, and ranging thence in a north-westerly direction, towards the head of the stream; there, however, it is, to a considerable extent, covered up from the view by the drift, showing itself in such cases only in the cuts of streams.

"This zone of limestone has an average width of 25 miles; it circumscribes, with a short interval, the great coal field which occupies the whole of southwestern Iowa, extending north to latitude 42° 30′, and separates it from the Illinois coal field by a calcareous belt varying in width from 25 to 50 miles.

"Of this coal field (in Iowa alone, not including its extension south into Missouri,) the dimensions are as follows:—Its average width from east to west, is less than 200 miles; its greatest length from north to south, about 140 miles; its contents about 25,000 square miles. It extends, measured in a direct line, nearly 200 miles in a north-westerly direction up the valley of the Des Moines.

"The fossil corals" says Dr. Owen, "are found embedded in the sub-carboniferous limestones, and near the top of the series, always under the true coal bearing beds, never above these or included in them, and nowhere else. This geological fact holds good not only in Iowa, but throughout the entire range of the sub-carboniferous limestones in Indiana, Illinois, Kentucky, and Tennessee. In not a single instance from the range of the Cumberland Mountains in the east, to the interior on the west, has a workable bed of coal been discovered in a position beneath the strata of limestone containing these corals. In these organic remains, then, we find the surest, the most unerring guide in the search after this invaluable article of commerce, that warms our houses; that drives our steam engines, by which we navigate our rivers, lakes and oceans; that propels the machinery by which we weave our fabrics; that reduces our iron, by which we cultivate our soil, and carry on every considerable mechanical operation; that refines our metals, that contributes to the production of both the necessaries and luxuries of life, and by which we transmit intelligence, with the swiftness of lightning, to stations the most remote. Without the knowledge of this fact, millions of dollars might be expended,—have been expended—in fruitless, hopeless mining operations after geological incompatibilities."

"My judgment is, that the carboniferous limestone of Iowa contains few mineral veins—no productive ones. The workable seams of coal yet discovered in this formation, do not exceed four or five in number. Nor as our measurements indicate, is it likely that the number will be much increased, even when the coal-fields come to be fully known; since in the British coal-fields in the same depth of strata, (less than a hundred feet) a much greater number of seams than the above is rarely found."*

WISCONSIN.

Bituminous Shale and Limestone—In the lead bearing magnesian limestone of Wisconsin, are occasionally observed thin seams or lamina of a buff coloured shale, which, on being placed on a fire, burns for awhile with a moderate flame, after which the residue presents a preponderance of earthy ashes. This asphaltic shale is calcareous, and frequently fossiliferous. It has been, in the absence of other fuel, economically employed in lime burning, as it contains inflammable matter in sufficient quantity to calcine the limestone without additional combustibles.

Peat is very abundant throughout the valleys of the Mineral Region; and in a district where vegetable and mineral fuel is so scarce, seems highly probable, that it will at some future day be resorted to as an extremely valuable substitute for coal and wood. The valleys, of which we speak present a very peculiar character, in one respect: which is in the singularly level planes which are maintained in their entire breadth. They appear as if they had once been filed to a uniform level, in the manner of a dam, from bank to bank, or like artificial reservoirs from which the waters have escaped. These level bottoms consist of peat beds, to an unknown depth; and small streams meander through them, having muddy bottoms and frequently expanding in swamps. It would seem that these Wisconsin valleys, have acquired this peculiar uniformity of plane surfaces from the depositions of earthy matter in the first instance, succeeded by the growth and decay of that class of coarse aquatic vegetables, which prevail under such circumstances.

In using the term "level" we apply it only in relation to the breadth of the valleys, and not to their length; for their inclination is often considerable, that is to say, from fifty to a hundred feet per mile.

* For further information, see Dr. D. D. Owen's Report on the Mineral Lands of Iowa, Wisconsin and Minnesota, 1852.

MISSOURI STATE.

We are in possession, comparatively speaking, of but scanty geological information respecting this extremely important mineral state. We possess, as yet no geological map of this vast region,* and we have no authentic data whereby to fix the area of its coal formation. In the absence of these we have made an approximate estimate, whereby we think it very probable that at least one eighth of all Missouri is overlaid with coal measures. Every year, however, as cultivation advances, and the country becomes settled, new localities of bituminous coal are determined. We conceive that we make no exaggerated estimate in assigning 6000 square miles as the amount of coal land in Missouri; being one tenth of its entire area.

Among the earliest notices of the existence of fine seams of coal, far up the Osage river, are those of Captain Pike, in 1806. More recently, other localities of excellent coal, have been discovered nearer to the mouth of the same river, and it seems not improbable but the entire valley of the Osage river is a continuation of the same general Missouri coal-field. For the analysis of the Osage coal, see the tables at the end of this volume.†

The St. Louis limestone and the coal formation reposing upon it, have been described by Dr. G. Englemann, in the American Journal of Science and Arts, January 1847.

The thickness of this upper carboniferous or mountain limestone, is between 200 and 300 feet.

The coal-bearing strata overlie it; below, directly upon the limestone, is a sandy, and above an unctuous clay or shale; the whole about forty feet thick. On this shale rests a coal bed of three to five feet—the only workable one in this neighbourhood—covered by a thin stratum of clay, which itself is overlaid by 10 or 15 feet of a blue or brown limestone, the uppermost plæozoic stratum in the region.

Beneath the St. Louis limestone is a sandstone 50 to 100 feet thick, and this is succeeded by the lower carboniferous or pentremital limestone, which is probably 1000 feet in thickness.

Dr. H. King, who has seen this formation at the south-west parts

* A very fine Geological map of this country, Iowa, Wisconsin and Minnesota, has lately been published by Dr. David D. Owen, United States Geologist, 1852, with a Survey and Report of this vast region. We have gladly availed ourselves of the permission of the author to annex part of the map, showing the coal formations, on a reduced scale. Our copy has been shaded black to correspond with the other maps.

† Also Mr. Johnson's Coal Report to Congress, p. 539.

of Missouri, thinks that on the Osage river, this lower limestone dwindles very much, but that the sandstone and the coal stratum above it, are much more developed; and that the fine coal mines worked there, sometimes not far above, and distinct from the lead-bearing magnesian strata, are in this same lowest coal bed.

Bituminous coal, approaching to the quality of the European cannel coal, forms an important, but at present, not well defined bed in this state.

A very fair cannel coal is found at several points in Callaway county, north of the Missouri river, and also on each side of that river, 130 miles above St. Louis. Extraordinary statements have been put forth in relation to the immense thickness of the Callaway county bed of cannel coal; at one point 24 feet and at another 46 feet, in thickness. In Coal county, a few miles from *Cote-sans-dessein*, it is affirmed that a shaft has been sunk 32 feet into the coal, without getting through the stratum; [?] probably an oblique section of the seam.

Cannel coal has also been discovered, in 1848, within eight miles of St. Louis, forming a bed of remarkable thickness.

In regard to the Callaway or *Cote-sans-dessein* cannel coal before spoken of, we have seen testimonials as to its qualities from several well known scientific persons, and the results are in some measure to be inferred from their analysis. There is no doubt but the coal from this remarkable depository, is well adapted for steam purposes and for making gas, and by iron-masters it is considered to be well suited for the manufacture of iron. It cakes very readily, without much changing its form; producing a porous cake amounting to 59.95 per cent of the coal. It has but a very slight trace of sulphur in its composition; a circumstance of some importance in relation to iron works, as many of the Missouri and some of the Illinois coals contain too much sulphur for those purposes. It is lighter than ordinary bituminous coal.*

The great variations in the thickness of this mass of cannel coal, at different points—for we presume that it is, in fact, but a single bed—have led to the supposition by some, that the general arrangement is that of enormous lenticular masses, rather than disposed in a continous flat seam. We are as yet without sufficient evidence of such a fact, as it seems scarcely consistent with the uniformity of stratification prevailing in this part of the country.

Should such a disposition really exist, we might find several parallel cases in other parts of the world. For instance, the remarkable coal seam in the basin of the Basse-Loire, in France, which is distributed in lenticular spaces, instead of in sheets, the average of which masses are stated to be forty-nine English feet in thickness.

Such also is the character of the anthracite of the departments of Mayenne and Sarthe, which combustible occurs in irregular masses of various sizes, but which are never much prolonged. The same

* Notices by Messrs. Booth, Boyé, Johnson, Chilton, and others, contained in a report of the Callaway Mining and Manufacturing Company, New York, 1847.

features, in connection with the conformation of the bituminous coal, are observable in the basin of Haute Dordogne, or Champagnac, in France. Here also the coal is in lenticular masses, sometimes one hundred and twenty feet long and thirteen feet thick; but commonly the coal occurs in rognons, balls or spheres, from sixteen to thirty-three feet in diameter.

Production.—We can entertain but vague ideas as to the annual quantity of bituminous coal at present raised in this State. According to a congressional report, the amount which was mined in 1840 was only 249,302 bushels, or 8,903 tons. From much later returns, made at St. Louis, we are informed that the quantity of coals weighed at the city scale, in the year 1846, was about 1,700,000 bushels, and the estimated amount, in 1847, was 2,000,000 bushels, or 71,428 tons for that city alone. To these are added the Pittsburg coals and some anthracite, making the aggregate, in 1847, about 100,000 tons. A large portion of this advance is ascribed to the great increase in the number and business of the foundries and factories of St. Louis.

Dr. D. D. Owen, in his recent admirable report says: " It is between Rockport and the mouth of the Osage, on both sides of the Missouri river, that those immense beds of coal are found, which attain a thickness of twenty, perhaps forty feet. The Osage coal is remarkable not alone for its extraordinary thickness, but also for the peculiar character and structure of the coal itself, together with the mineral insinuations which invade it. The lightness of this combustible is such, that before imbibing water, it will float upon that fluid, indicating a specific gravity actually less than 1. In its structure, fracture and lustre, it has an appearance intermediate between cannel coal and the dull varieties of asphaltum, but it contains 31 per cent. less volatile gases than pure bitumen, and from 5 to 10 per cent. more volatile matter than the ordinary varieties of the bituminous coal of the western coal-fields.

"At the pit west of Marion this coal assumes a cuboidal, and even a sub-column or structure, somewhat analogous in miniature to basaltic trap, while at the same time a net-work of pyritiferous ores of zinc and iron have ramified its joints and fissures, and appear often in brilliant chrystalline forms—the whole bearing evidence of great local disturbance, igneous action, and gradual consolidation under pressure. It appears, indeed, altogether probable, from the peculiar character of the coal, its structure and great local thickness, that it has been subjected to a sufficient degree of heat to have fused or semi-fused the mass, under a pressure that prevented the escape of the volatile gases, transferring it, at the same time, either in this condition or by sublimation, from its original bed into some wide adjacent fissure formed by disruption of the strata, where it has then very gradually passed into the solid state. Its uniform occurrence in close proximity to an abrupt change in the geological formation of the adjacent country, and the sudden elevation of primitive rocks, together with the highly inclined position of the coal itself, furnishes abundant proof that it has been implicated in the remarkable disturbances which have convulsed the whole of the surrounding country subsequent to the carboniferous era."

Dr. Owen continues: "The first workable bed of coal which I encountered in my descent of the Missouri river was Wellington. It is from 12 to 14 inches thick, and lies a few feet above the bed of the river. At Camden, on the opposite side of the river, nearly east of Wellington, a bed of coal has been exposed 15 feet above the river, corresponding probably to the Wellington bed. It is also found at several places on the Snei, south of the Missouri."

ARKANSAS.

ANNOUNCEMENTS of discoveries of bituminous coal have, from time to time, been made in this State, particularly in the vicinity of the Arkansas valley. At Spaldries' Bluff, in Johnson county, on the north bank of the Arkansas river, above Little Rock, coal was worked a few years ago, and we presume is continued at the present time.

At the request of the writer, Mr. J. F. Frazer kindly furnished him with the following result of his examination of this coal:

Carbon, - - - - -	62.60
Volatile matter, including sulphur,	26.90
Hygrometric water, - - - -	2.00
Ashes, - - - - -	8.50
	100.00

Specific gravity, 1.396

Coal traces have been mentioned by Mr. Nuttall, as occurring near the western boundary of the State, towards Fort Smith, and extending many miles westward, at least as far as the Falls of the Canadian river, in the Indian territory, and southward to the borders of the Red river, in the country of the Choctaws.* Northward, in the Cherokee country, [Nebraska,] the coal formation has been determined as far as Grand and Verdigris rivers; and 120 miles still further north, in the Osage country, coal abounds on the Little Osage river,† and is evidently an extension of the great Illinois coal-field, which stretches from thence almost uninterruptedly for 450 miles, across the Missouri and Mississippi rivers, and thence beyond the

* Nuttall's Travels in Arkansas, in 1819. Long's Narrative, in 1819, 1820.

† Pike's Narrative, in 1805-6-7. Bradbury's Travels, in 1809, 1810, 1811. Sibley's Journal, 1817.

Illinois, Wabash, and Ohio rivers, in the north-western angle of Kentucky.*

The quantity of bituminous coal returned to Congress in 1840, was only 200 tons, evidently an incomplete account.

SOUTHERN MISSOURI.

PORTION ADJOINING TEXAS, NOW CALLED KANZAS TERRITORY.

"ABOUT one hundred and fifty miles west from the confluence of the Arkansas and the Canadian rivers, in W. long. 97°, is the western limit of the great limestone and coal formation. The coal-beds in this region are of great thickness, and are apparently extensive and numerous. This formation," says the narrator of Major Long's first expedition to the Rocky Mountains, 1820, "appears to be unconnected with the great [tertiary] coal formation along the base of the Rocky Mountains, and the sandstones of the two districts are remarkably dissimilar."†

The same coal formation, also traced in this direction by Capt. Pike in 1806, and by Mr. Nuttall in 1819, is the evident prolongation of the great coal region which traverses Illinois, Missouri, and part of Arkansas, in the direction of Texas. Beyond it, to the westward, is the great plain composed of red saliferous sandstones with gypsum; and beyond that, towards Taos and Santa Fé, the mountainous range contains a bituminous coal region.

"The geological constitution of the Prairies is exceedingly diversified. Along the eastern border, especially towards the north, there is an abundance of limestone, interspersed with sandstone, slate, and many extensive beds of bituminous coal. The coal is particularly abundant in some of the regions bordering the Neosho river, where there are also said to be a few singular bituminous or 'tar springs,' as they are sometimes called by the hunters. There are also many other mineral, and particularly sulphur springs, to be met with."

Further westward, the sandstone prevails, but some of the table plains are based upon strata of a sort of friable calcareous rock, which has been denominated "rotten limestone;" yet along the borders of the mountains the base of the plains seems generally to be of trap and greenstone.‡ But much of the middle portion of this enormous prairie region exhibits no rocky traces whatever, so much so that

* Owen's Indiana Reports; and Mather's Kentucky Report.
† Long's First Expedition, Vol. II., p. 408.
‡ Gregg's Commerce of the Prairies, Vol. II., p. 185, 1845.

"we sometimes travel for days in succession without seeing even as much as a pebble."*

Towards the head of the Osage river, coal strata prevail, and with various other localities, form detached or outlying areas, evidently portions of the coal-fields of Missouri, Illinois and Indiana. The same series are seen on the Arkansas river, near Fort Smith, and at the Canadian Fork. We have also learned the interesting fact of the existence of large beds of coal at the head of the Canadian river, and in the Raton Mountains, between Santa Fé and the Arkansas river.

To the northward, according to Mr. Nicollet, alternate areas of the cliff limestone and coal measures present themselves, on either side of the Missouri river, from its mouth to the junction of the Platte river, in 41° N. lat. To what extent the coal formation stretches to the east and west of the Missouri river within the parallel, we have no certain information, further than that the limestone ceases to appear on the surface beyond about the 97th degree of west longitude, and is there covered by sand, gravel, and erratic deposits.

WOOD COAL AND BROWN COAL.

In a communication to the Association of American Geologists, in 1845, Lieut. Johnston describes an examination made by him of a "Bluff" at Mount Waneus, on Red river. This bluff presents an escarpment of fifty feet high, in which are various seams of wood and wood-coal or lignite, intermingled with iron pyrites, and on the surface of the bluff alum crystallizes in considerably quantities. Permanent springs flow from the base, and taste strongly of alum. This formation, a sand passing into stone, was traced five miles back from the river, at the same general elevation. The seams of wood and sand alternate, and the author described them as of recent or post-diluvian origin; but it is near one hundred feet above the present low-water mark.†

The calcareous strata in the vicinity of Fort Washita contain decided cretaceous fossils.

BITUMEN AND PETROLEUM.

On the False Washita river, towards the Wishetaw Mountains, Lieut. Johnston met with a dark sandstone with a vertical dip; out of which, throughout its course, a great quantity of bitumen has flowed. A specimen of the liquid bitumen has the consistence and appearance of common tar. It occurs as mineral oil on the surface of a spring near that place. We have no information as to the age of the rock, which is in the vicinity of granite.

* Gregg's Commerce of the Prairies, Vol. II. p. 185, 1845.

† Proceedings of the Association of American Geologists and Naturalists, April, 1845, pp. 74, 76.

MISSOURI TERRITORY.

BROWN COAL OR LIGNITE FORMATION OF THE UPPER MISSOURI VALLEY.

LIMITING our description in this place to that portion of this enormous area of brown coal, which lies within the United States territory, south of 49° of north latitude, and east of the Rocky Mountains, we will place before our readers such information respecting this extraordinary region as we have been able to acquire.

From the mouth of the Missouri river upwards to the Platte river, the carboniferous and cliff limestone of the American geologists, with occasional shallow basins of the true coal formation occur, and the carboniferous limestones extend still further to the mouth of the Sioux river, lat. 42° 30′. Here commences the interesting formation first described by Mr. Nicollet, belonging to the cretaceous group, with calcareous marls containing microscopic multilocular shells, resembling those discovered by Ehrenberg, in the chalk, and other beautiful fossils of larger size corresponding with those in the chalk, the gault, and the green sand formations of Europe. Mr. Nicollet traced this group up the Missouri river for four hundred miles, but it is known to extend as far as beyond the Mandan village, to a point between Beaver river and Grand river, at about N. lat. 47° 30′; thus occupying, in that direction, 5 degrees of latitude.

At this point, and overlying the cretaceous series, commences the vast tertiary area, composed of horizontal strata of variously coloured sand, clay, shale, sandstone and coal, irregularly alternating; extending at least twenty degrees to the south-west and south, and northward, apparently, to the arctic ocean.

The first notice that we have met with of this formation is in the narrative of Lewis and Clarke's expedition to the Rocky Mountains, in 1804. The coal or lignite was first observed at twenty miles above the Mandan villages. The bluffs on each side of the Missouri are upwards of one hundred feet high, composed of sand and clay, with many horizontal strata of carbonated wood, resembling pit-coal, from one to five feet each in thickness, and occurring at various elevations above the river.

At fifty miles above the villages, similar coal seams were noted; but here they were observed to be on fire, emitting quantities of smoke and a strong sulphurous smell. This point was 1652 miles above the Mississippi. Further on, the same sulphurous coal continued for eighty miles more; strata of coal, frequently in a state of combustion, appearing in all the exposed faces of the bluffs. The quantity of this coal improved as the party advanced near the mouth of the White Earth river, eighty-five miles further, affording a hot and lasting fire, but emitting very little smoke or flame. Thence forty-seven miles to the Yellowstone river, and at a bluff, eight miles up that stream, were seen several strata of coal.

The narrator observes, that for fifty miles above this junction, there were greater appearances of coal than had yet been seen, the seams being in some places six feet thick ; and there were also strata of burnt earth, which were always on the same level with those of coal.

The explorers had thus far traced this coal formation along the banks of the Missouri, for a distance of three hundred and thirty miles. The horizontal formations of clay, loam and sand, with fragments of coal in the drift of the river, extended three hundred miles more, to Muscle-Shell river, or six hundred and twenty miles from the Mandan villages. Even above this point, washed coal continually appeared on the shores of the river; and at Elk Rapids, eight hundred miles from Fort Mandan, the high bordering bluffs were still composed of horizontal beds of clay, brown and white sand, soft, yellowish white sandstone, harder dark-brown freestone, and large round, or kidney-shaped nodules of clay iron ore. Coal, or carbonated wood, similar to that previously observed, was also seen, and was accompanied with burnt earth—probably the result of the spontaneous combustion of the coal, as was noticed for hundreds of miles below. Precisely the same phenomena were recorded, at a subsequent period, by Captains Back and Franklin, and by Dr. Richardson and others, and by Hearre, in 1769, and Mackensie, in 1789—extending, apparently continuously, and in the same parallel, full eighteen hundred miles, northward.

Returning to the narrative of Lewis and Clarke.

After reaching the grand forks of the Missouri, and ascending two or three days journey up Maria's river, northward, it was remarked that precisely the same geological character and coal strata prevailed for more than sixty miles. So far, therefore, the exploring party had been travelling through or over a ligneous deposit, of singularly uniform character, for no less than nine hundred and eighty miles, following the windings of the river. Pursuing the south fork, towards the great falls of the Missouri, coal was still observed, in bluffs of dark and yellow clay, at a distance of 2454 miles up this mighty river, and it was not until near the base of the Rocky Mountains, and after one thousand miles of travelling across it, that this great region of coal beds and lignites was passed.

On his return, Captain Clarke descended the Yellowstone, from about latitude 45° to its mouth, in latitude 48° 20′, and every where found the same series of coal and variously coloured clays, sands, and soft sandstones, as was traversed in ascending the Missouri.

The fossil bones of a supposed fish, probably a saurian, were also observed above the Big Horn river. Large quantities of brown coal were seen in the cliffs below the junction of this river, and all the highlands adjacent appeared to be composed of earthy beds of different colours, abounding in coal or carbonated wood, of an impure quality. Below the Big Horn is a large stream falling in from the south, whose Indian name implies "the coal creek," from the great quantity of that mineral upon its margin. The same coal series continued to the confluence of the Missouri, exhibiting uninterruptedly, for seven hundred miles, in addition to the thousand previously traversed, the vast persistence of this formation. The enormous area of similar strata is further shown by the discoloration of all the tributaries that enter the Missouri, both from the south and the north, from the forty-second to the forty-ninth degrees of north latitude.

On the authority of M. Sublette, these lignite beds prevail along the whole of the country watered by the Padouca [Powder] river, in beds of from three to nine feet thick; and also on the Batsoah, or Cherry river, and the south fork of the Platte river; thus bringing the formation southward to latitude 40°.

It appears probable, from Capt. Fremont's narrative, July, 1843, that the sandy and clay beds which he crossed at the head waters of the Platte and Arkansas rivers, are southern continuations of the same formation to latitude 39° and 38° where the underlying yellowish and gray limestone, containing cretaceous fossils, first made its appearance in that direction, and is traceable eastward down the Smoky Hill fork, nearly to its junction with the Republican fork of the Kansas river. Colonel Long, in 1820, descended the Canadian or South fork of the Arkansas river, in which the prevailing rock is red sandstone, with salt and gypsum. It would seem, therefore, that our tertiary lignite formation ceases before reaching the latitude of 35°. But coal of some kind has been noticed by Col. Emory, in latitude 36° 30′, at the head of the Canadian river.

Seams of lignite and wood coal are, however, observed on supposed tertiary strata on the borders of the Red river, and limestones of the cretaceous period likewise occur in the same country, in the vicinity of Fort Washita.

It was announced, in 1841, that bituminous coal, probably brown coal, had been discovered on the St. Peter's river, in the Missouri territory.

UPPER MISSOURI.

NOW NEBRASKA TERRITORY.

In 1843, Messrs. Audubon and Harris ascended the Missouri to the mouth of the Yellowstone river. The latter gentleman has furnished some account of this tertiary lignite region.* The whole series of strata, for many hundred miles prior to reaching this formation, is described as perfectly horizontal; the upper part of each bed of rock being successively intersected by the angle of descent of the river. The tertiary group is indicated by the remarkable strata which form the picturesque hills noticed by travellers, and called the Mauvaises Terres by the trappers and voyagers. Mr. Harris counted, in one place, eight seams of coal, between the river bank and the top of the bluff; varying from six inches to four feet in thickness. This coal, he observes, is very light, and ignites with difficulty, emitting a very unpleasant odour while burning. Fossilized wood is very abundant; occasionally much flattened by the pressure of overlying strata. Mr. Bell was the only one of the party who had an opportunity of witnessing the burning of the cliffs, about thirty miles above the Yellowstone, on the northern bank of the Missouri; and all agree in attributing this burning to the spontaneous combustion of the coal. We observe, that Mr. Harris states that the coal seams commence in the upper part of Nicollet's great cretaceous clay bed; and further, that there occurred in the same formation, "a substance like petroleum in colour and consistence, but without odour."*

To the foregoing brief abstract we may add, that, from the specimens brought home by the last named traveller from the vicinity of Fort Union, near the confluence of the Yellowstone and Missouri rivers, we derive incontestible proofs of a fresh-water formation. Among other strata exposed in a cliff near the fort, are thin beds of clay and argillaceous rock, both containing three or four species of fresh-water univalve shells. There is, besides, a rock, twenty or thirty feet thick, which also contains proofs of fresh-water origin, in bivale shells, leaves of deciduous trees, and bones, apparently, of a mammiferous animal.

The Upper Missouri Valley has yet to receive examination from the scientific geologist, and there can be no doubt but highly interesting results would follow from investigations in a field so rich and extensive. The committee, to whom Mr. Harris's paper was referred, close their report with the remark, that "the proofs thus afforded of a probably widely diffused fresh-water formation in the region of the

* Proceedings of the Academy of Nat. Sciences, Philadelphia, May, 1845.

Upper Missouri, reposing upon the cretaceous strata, and imbedding remains of a manifestly tertiary age, are, just at this time, invested with considerable interest, from their according with the discoveries recently made by Captain Fremont, of the presence of other and probably extensive fresh-water tertiary strata in the Oregon territory."

We have, in that portion of this volume which is appropriated to British America, supplied many additional facts respecting the central and northern parts of the great tertiary range, whose southern area we have been considering above. From the united testimony of highly competent observers, there now remains very little doubt, that a continuous tertiary coal formation stretches from the Missouri and the Yellowstone, and even from near the sources of the Platte, and some branches of the Arkansas, and the borders of New Mexico, to the far distant shores of the Artic ocean.

STATE OF OREGON.

SOUTH OF NORTH LATITUDE FORTY-NINE DEGREES.

Cascades of Columbia River.—Tertiary Coal.—At about 122° west longitude and 45½° north latitude, Captain Fremont, near the foot of the cascades, discovered in a bluff on the river, "a stratum of coal and forest trees, imbedded between strata of altered clay containing the remains of vegetables, the leaves of which indicate that the plants were dicotyledonous. Among these the stems of the ferns are not mineralized, but merely charred, retaining still their vegetable structure and substance; and in this condition a portion also of the trees remain. The indurated appearance and compactness of the strata, as well, perhaps, as the mineralized condition of the coal, are probable due to igneous action. Some portions of the coal precisely resemble in aspect the *cannel coal* of England, and, with the accompanying fossils, have been referred to the tertiary period."*

These strata appear to rest upon a mass of conglomerate rock. The vegetable fossils collected here we submitted to the examination of Mr. James Hall, who refers them to the tertiary series, and even to a very modern epoch of that deposit.†

In the cabinet of specimens collected by the United States Exploring Expedition, at Washington, are some specimens of lignite or bituminous wood, from Oregon, also of coal vegetation, apparently

* Captain Fremont's Report of the Exploring Expedition to the Rocky Mountains, p. 192, 1 843-4.

† Ibid.—Plate III., Figs. 14, 15.

of the oolite age, or perhaps yet more recent. Among these plants are leaves of dicotyledons, resembling the birch or beech, and with these occur some species of ferns.

Coal has been discovered and worked in Wallamette or Willamette Valley, nearly a hundred miles above Oregon City.

Another locality of an imperfect coal is at twenty or thirty miles up the Cowlitz river, a tributary to the Columbia river, on the northern side. We have been assured that this was the true anthracite, but Captain Wilkes only regarded it as tertiary lignite. Sir George Simpson observed large quantities of this coal on the surface, bordering this river.*

Fossil Copal, or Highgate Resin—Has been found at the falls of the Wallamette or Wilhamet, a tributary of the Columbia river, Oregon; and on the shores of the Pacific, north of the mouth of the Columbia river.†

Puget Sound.—The discovery of mineral coal, of sufficient strength for sea steamers on the Pacific coast, within the settled limits of the United States, appears to have been recently accomplished.

Coal in great quantity is announced in about the parallel of 48° north, and within easy navigable facilities of the waters of Puget Sound.

Its principal known locality is on the Strila-guamish river, about twenty miles above its mouth, where it empties into Puget Sound, [or more properly, Admiralty Inlet,] opposite to Whidley's Island. The latter is more than 50 miles in length, and is open to the Ocean direct, by the strait of Fuça.‡

Specimens of this fuel were sent to the Secretary of the Navy, and Professor Johnson was directed by him to analyze some of them, with a view to their applicability for steamers. The results are announced as follows:

Specific gravity, 1.315; weight on the merchantable state, 51 to 55 lbs. per cubic foot, according to the size of the lumps. Will require on board a steamer, about 42¼ cubic feet of space to stow one gross ton; lustre brilliant; wholly free from liability to soil.

It is composed of fixed carbon, - - -	56.84
Volatile matter, - - - - -	40.36
Earthy matter, - - - - -	2.80
	100

Scarcely increases in bulk, in the process of coking; has no tendency to agglutinate; and consequently preserves an open fire.

Nearly free from sulphur. Under a well-constructed boiler, it ought to produce from 7½ to 8½ lbs. of steam, from 212 degrees, to each pound of coal burned. Resembles the flat bituminous coal of Mercer couuty, Pennsylvania.

* Overland Journey round the World, by Sir George Simpson, Phila., 1847, p. 107.
† Alger's, Phillip's Mineralogy.
‡ Communicated from Oregon, December 3d, 1850.

STATE OF CALIFORNIA.

IN THE OCCUPATION OF THE UNITED STATES ARMY, IN 1848.—AREA 72,000 SQUARE MILES.

North Fork of the Platte River.—Coal Field.—North latitude 41½° and west longitude 107½°.—In the precipitous bluffs bordering this river, Captain Fremont observed a series of strata containing fossil vegetable remains and several beds of coal. The position of this coal formation is in the centre of the Rocky Mountain chain, and its elevation is six thousand eight hundred and twenty feet above the sea. In some of the coal seams the coal did not appear to be perfectly mineralized, and in others it was compact and remarkably lustrous. The rock above the third bed of coal, in the lower hill, is a siliceous clay slate, having a saline taste, and there were also noticed thin layers of very fine white salts, in powder.*

There being no specimens brought home from this place, it does not appear what is the actual character of the formation, but it probably belongs either to the tertiary or the oolite period. The number of beds, their thickness, and apparent extent are not mentioned: but the circumstances under which these original observations were made; the impossibility of giving more than a casual and hasty glance at the geology of the country through which the expedition passed, whilst in a state of continual danger and privation, rendered more exact details almost impracticable. This basin or deposit appears to be surrounded by granite.

Green river.—Black's Fork, Muddy Fork, and other tributaries.—North latitude 41½°, extending from 110° to 111° west longitude.

The strata near Green river were observed by Captain Fremont to contain handsome and very distinct vegetable fossils, overlying an impure or argillaceous limestone. Further westward, conglomerate rocks were seen; and near them, at Muddy Fork, occurred strata of fossiliferous rock, having an oolitic structure, and characterized by fossils apparently of that formation or age. Advancing up the stream, alternating strata of coal and clay, with distinct and beautiful vegetable remains were discovered. Coal also appeared occasionally in the hills as the party advanced, and was displayed in rabbit burrows, in a gap through which they passed over some high hills. A portion of the region thus traversed was seven to eight thousand feet above the sea.

* Captain Fremont's Report, pp. 126 and 296, 1843–4.

The section of coal strata remarked by Captain Fremont, consisted of two beds of coal of fifteen inches each, and three others which are separated by an equal number of clay beds. There is an intermediate bed twenty feet thick, which consists of indurated clay, resembling fire clay, with vegetable remains, chiefly of fossil ferns. Mr. Hall has described and figured these in the appendix to the Report. Having previously compared these fossil ferns with a large collection from the coal measures of Pennsylvania and Ohio, it became quite evident that this formation could not be of the same age. Several specimens were referred to the oolitic coal vegetation of England, and the general character of the other species, and the absence of the large stems so common in the coal period, led to the conclusion that they also belonged to the oolitic period, although the evidence is not entirely positive. For ourselves, we think that the shells would indicate a later origin. One thing appears certain, that the coal plants must be regarded as mostly of new species; and, in this respect, they form a very important addition to the flora of the more modern geological periods.*

Nearly in the same parallel of longitude, but at the distance of one hundred and fifteen miles to the south, strata of bituminous limestone, highly fossiliferous, were discovered by Captain Fremont on the return of his exploring party. The genera of fossils, Mr. Hall thinks, may possibly belong to rocks of the age of those in the vicinity of the coal above mentioned, but the species are all new. No coal was remarked here, in the hurried passage of the travellers, and the intermediate ground was not visited.

All the circumstances which have so far been brought to light, are of an exceedingly interesting character, and lead us to desire a further and more elaborate investigation.

Coal.—It is said that another coal-field has lately been discovered, which, if true, will greatly facilitate the introduction of steam navigation in the Pacific, and be the means of making California one of the most important commercial positions on the west coast of America; particularly, if ever a communication should be opened by means of a canal across the Isthmus of Panama.†

In the spring of 1847, a new coal mine was discovered near San Luis Obisco, north latitude 35°.—There are now three mines within three hundred miles of Monterey; yet coal was sold, in 1846, from an American whaler, at $5 00 a bushel.

Asphaltum and Petroleum occur abundantly in western California.

1851. Dr. Le Conte has lately been employed to investigate a coal deposit in the vicinity, viz. 12 miles north of San Diego, on the sea shore. This coal is stated to occur towards the lower part of a cliff, about 200 feet high, which forms a portion of the great *tertiary* formation that extends along the coast of Upper California. The seam of coal observed was four feet thick. The tooth of a saurian

* Appendix to Fremont, p. 297.

† Life in California, by an American, 1846, p. 224.

animal was observed in these tertiary beds, and amber occurs occasionally in them.

This tertiary coal, although inferior to the true or older coal, may subserve many useful purposes in this region, where fuel is so much needed. As usual, much pyrites accompanies this combustible.

The California brown coal bears some resemblance to jet, and leaves a light, abundant, white ash. Coal of this same quality exists between San Diego and San Luis Rey.

1853. The following statement is from the London Mining Journal. "Mr. Benham, of San Francisco, has been in Washington Territory examining coal mines. In Bellingham Bay, Mr. B. informed us there is a mine of 16 feet deep, with a dip of 45 degrees. The vein is of solid coal, with the exception of 2 strata of clay, one of 4 the other of 5 inches. The bed of the coal is 40 feet above high water mark. Over 100 tons of this coal have been placed in San Francisco market and commanded within one dollar of the highest price.

STATE OF TEXAS.

Now admitted into the North American Union. Superficial extent claimed, 397,319 square miles; but as defined by statute of first Texan Congress, 324,013 square miles. This boundary is not yet settled, next Mexico.*

Pitch Lake.—An announcement has been made of the existence, in Texas, within 100 miles from Houston,† of a small lake that closely resembles the Pitch Lake of Trinidad. It is filled with bitumen or asphaltum, and is about a quarter of a mile in circumference. During the cool months of winter its surface is hard, and is capable of sustaining a person. From November to March it is generally covered with water, which is acid to the taste; from which cause it has been commonly called the "Sour Pond." In the summer months a spring occurs, near the centre of the lake, from which an oily liquid, (probably petroleum,) continually boils up, from the bottom. This liquid gradually hardens, on exposure to the air, and forms a black pitchy substance, similar to that which forms the sides of the lake. It is said to resemble, precisely, the bitumen of Trinidad; and the Texans conceive that, at some future day, it will be valuable for the production of gas for their cities. It burns with a very clear bright light, but gives out a pungent odour.‡

* Map of Texas, published by the United States War Department, 1844.

† Near the Pond, between Liberty and Beaumont, and about twenty miles from the latter village.—Houston Telegraph.

‡ "Bitumen in several places." McCulloch.

Coal is now well known to exist in Texas, although the country has not been geologically examined. There is no doubt but coal prevails at intervals entirely across the country, in a north-east and south-west direction. Its general position is about two hundred miles from the coast.

On the Trinity river, two hundred miles above Galveston, the coal region there was investigated in 1846, and found to be more extensive than was anticipated. A company, under the title of the "Trinity Coal and Mining Company," was incorporated by an act of the Texan Congress in 1840. Both anthracite and semi-bituminous coal, somewhat like the cannel, in appearance, occur here.*

Mineral coal, in great abundance, prevails not far from the Mustang Prairie. It is also found, accompanied with excellent iron ore, in the vicinity of Nacogdoches. According to report, this coal is abundant, rich, and of a fine appearance.†

Mr. Kennedy, who has taken pains to collect information relative to the resources of Texas, although not an original investigator, says, in a work published in 1844, that "in addition to iron, the utilitarian sovereign of metals, Texas possesses coal—the grand auxiliary of the arts which tend to enrich and civilize the world. Coal, both anthracite and bituminous, abounds from the Trinity river to the Rio Grande. The coal on the latter river above Dolores, has been represented, by the agents of the "*Texas and New Ireland Land Company*," [an association broken up by the revolution in 1836,] as of excellent bituminous quality.‡

Formations of secondary limestone, with others of carboniferous sandstones, shales, argillaceous iron ore, and bituminous coal beds, are said to occupy a large portion of the interior of Texas. Westward of these occur the inferior and Silurian strata, trilobite limestones, and transition slates. Beyond all, basaltic and primary rocks of the Rocky Mountains arise; while northward is the great salt lake of the Brazos, and the vast red saliferous region traversed by the exploring expeditions of Captain Pike and Major Long, and since made more familiar to us by Mr. Gregg and other travellers.

A bed of coal extends across the Brazos river towards the Little Brazos and the San Andres, down which stream it may without difficulty be transported at high water.

Near the city of Austin, on the eastern border of the Colorado, is a peak, called Mount Bonnell, overlooking Austin, and having a fall of seven hundred feet perpendicular to the bed of the Colorado. This and other hills, although not scientifically examined, are known to contain beds of anthracite coal.

On the Rio Grande, south-west of Bexar, is a great abundance of bituminous coal. The navigation of this river is reported to be free for eight months in the year.§

* New Orleans Picayune.

† Notes sur le Texas. Documens sur le Commerce extérieur. Juillet, 1842.

‡ Texas, its Geography, Natural History, &c., by W. Kennedy, 1844.

§ Report, in 1834, to the "Rio Grande Land Company."

In many parts of the rolling prairie region, coal, of fair quality, and iron ore have been found; and it is supposed that beds of these valuable minerals extend over a great part of the country.*

We have received some recent information of the character of the country bordering upon the Rio Grande, as far up as the Presidio de Rio Grande, from the notes of Lieutenant B. P. Tilden.†

On approaching Loredo, within forty or fifty miles, by the course of the river, and extending north of that town, a coal formation is traversed during that distance. Beds of coal are frequently to be seen, as are deposits of nitre and sulphur, and also thick beds of good fire-clay, at the bases of the bluffs. These strata, and the accompanying sandstone rocks, are supposed to be a prolongation of similar strata at Guerrara, on the Rio Salado, to the south-west; as they agree in their range and dip. The writer, who apparently is not very familiar with geological phenomena, does not furnish any further details.

TERRITORY OF NEW MEXICO.

AREA 219,477 SQUARE MILES.‡

Coal is said to occur in the Sierra Verde. Perhaps it is a continuation of the great zone of lignite which stretches parallel with the Rocky Mountains, even to the borders of the Arctic Ocean, and the most northern limits of the American Continent. Or it may be a continuation of the carboniferous formation which has been noticed by Col. Long and others towards the head of the principal rivers bordering the plains.

Don Manuel Alvarez, in a letter dated May 4th, 1847, at Santa Fé, and published in a St. Louis newspaper, whilst describing the minerals of New Mexico, says,—"Coal is found in abundance and of good quality, between the Placers, in the Ratons mountains, and in many other places."

Since then, we have received the narrative of the military exploration, from the Pacific to the Missouri, by Lieut. Col. Emory. He describes the occurrence of coal, between Bent's Fort on the Arkansas river, and Santa Fé, to the north and south of the Raton Pass. That seen to the northward, at Capt. Summer's camp, is described as an immense field, the seam which cropped out being thirty feet thick.

* McCulloch, art. Texas; and Iken's Texas.

† Notes on the Upper Rio Grande, by Lieutenant B. P. Tilden; Philadelphia, 1847.

‡ United States Gazetteer, published by Lippincott, Grambo & Co.

That noticed by Col. Emory was on the banks, and near the head waters of the Canadian river, at about north lat. 36° 50′, on the 7th August, 1847. At present we have no knowledge whether this be true coal or only brown-coal, but are inclined to think it must be the true coal formation. If so, it is an extremely interesting geological fact.

UNITED STATES OF MEXICO.

AREA, EXCLUSIVE OF TEXAS, 1,650,000 SQUARE MILES.*

Currency.†

1 onza (ounce) of gold = 16 pesos (or dollars), but as gold is usually more in demand, the value varies to 17 or 18 dollars. Half, quarter, eighth, and sixteenth ounce pieces are coined.

1 silver peso = 1 dollar.
½ peso; ¼ peso, or peseta.
1 real = ⅛ of a peso, (12½ cent piece United States currency.)
½ real, or 1 Medio (6¼ cent piece United States currency.)
¼ real, or 1 quartillo, a copper coin.
⅛ real, or 1 flaco, " "

Weights.

1 monton, about Mexico = 32 quintales, but about Zacatecas, Fresnillo, &c., it is only 20 quintales.

1 carga = 3 quintales.
1 quintal = 4 arrobas.
1 arroba = 25 libras, (pounds.)
1 libra = 2 marcos.
1 marco = 8 onzas, (ounces.)
1 onza = 8 ochavos, (eighths.)

Measures of Length.

1 legua of 26.63 to a degree = 5000 varas.
1 vara = 375.9 Paris lines.
= 2.784 English feet = 135.130 of a Paris line.
The vara is divided into 4 cuartas and 48 dedos (inches.)

Official estimate of the population in 1842, 7,015,509 persons; of which only one million are whites.

We have met with no detailed geological description of coal on the Mexican Isthmus, nor on the main land; yet there is abundant

* Mexico, by Brantz Mayer, Secretary United States Legation, 1844.
† Burkart's Aufenthalt und Reisen in Mexico, Stuttgart, 1836.

reason to believe that brown coal, at least, prevails on the east flank of the central mountain range, as well as true coal near the eastern frontier.

M. Humboldt affirms that coal, and also fossil wood or lignite, are frequently found in different parts of New Spain.*

We know that a bituminous coal region crosses the Rio Grande, above Dolores, into Mexico, after traversing the greater part of Texas, and pursuing the same general range, of south-west and north-east, as the central coal-fields of the United States.

Brown coal traverses entirely the whole breadth of the Isthmus of Panama in a north and south direction, in 8° to 10° north latitude.

On the 30th April, 1842, and 5th October, 1843, decrees of the President of the Mexican Republic were issued fixing the tariff of maritime and frontier customs. These decrees fix the value in the currency of the Republic, of foreign money, as follows:†

	Mexican Currency.		French Currency.	
	Piastres.	*C.*	*Francs.*	*C.*
The £1 Pound Sterling, (20 shil. of 12 pence each,)	5	00	25	00
1 Franc, (20 sous or 100 centimes,) . .		20	1	00
1 Marc banco, (16 shil. of 12 pfennings each,)		37½	1	88
1 Réal de veillon, (34 maroédis,) . .		05	0	25
One Piastre, ...			5	00
Réal, ...				62½
Centieme, ...				05

Bituminous Coal on Salado River.—An extensive bed of excellent coal exists at Guerrera or Reveilla, a Mexican town of 4,000 inhabitants, situated on the left bank of the Salado river, one hundred and twenty-five miles above Camargo. It is now (1848) worked by an American company, and promises to be of vast importance, as it removes the principal obstacle to steamboat navigation, the want of fuel on the Rio Grande, into which the Salado empties its waters, at the distance of twelve miles from Guerrera, and eight miles by land. Both these rivers are navigable for steamboats drawing six feet of water. The existence of this bed of coal was made known to Lieut. Tilden, in a recent expedition to Loredo, and a few tons were placed on board the steamboat. It is described as "a hard bituminous coal of first rate quality," imbedded in sandstone. Silver and other minerals occur in the vicinity.

A coal formation fifty miles in breadth, probably a continuation or contemporary of that of the Rio Salado, crosses the Rio Grande from Texas, into Mexico at Loredo.

* The precious metals were at all times the principal source of attraction in Mexico. At the period of M. Humboldt's residence here, there were three thousand (3,000) mines in operation, raising annually twenty-one millions of dollars ($21,000,000) in silver, and two millions (2,000,000) in gold. Of copper, there was coined at the Mint, from 1833 to 1837, $4,712,000.

† Documens sur le commerce exterieur, Mexique, Lègislation commerciàle, January, 1844.

A very short distance above Loredo, on the Mexican shore, and within two hundred yards of the Rio Grande, a remarkably fine coal vein, eight feet thick, occurs. It is affirmed to be good in quality, and free from sulphur; burning readily, and applicable to smiths' uses.

In a country where fuel is so very costly, these coal mines must eventually be invaluable. All these mines will, probably, be worked by American industry.

Lieut. Tilden states that at twenty-five miles below Loredo, in a reddish bluff one hundred feet high, are numerous "petrifactions of roots;" from whence we might infer that it was a lignite deposit, except for the circumstance of their being within the limits of the coal formation, above described.

He speaks, also, of a great abundance of a substance commonly called red chalk or keel, in the vicinity of the eight-feet coal vein, opposite Loredo.

Whatever the geological age of the coal deposits to the southward, it seems at least now fully settled that good bituminous coal prevails as low as 27° north latitude.

The existence of coal in the mining provinces of Mexico was for some time very doubtful. Mr. Bullock, however, brought specimens from the vicinity of Real del Monte, of coal analagous to jet. It was analysed by Dr. Trail. Mean specific gravity = 1.2248: forms a coke of about 50 per cent.; the volatile portion affords a very pure coal gas. It becomes considerably electric by friction, in which circumstance it is analogous to jet, and differs from cannel coal, which scarcely shows any symptoms of electricity by friction.*

Province of Oajuca or Oaxaca.—Peninsula south of the Gulf of Mexico. Coal is stated to be very abundant in this province, which is celebrated for its mineral wealth. We are not informed as to the geological age of this coal. It has often been proposed to form a ship canal or railroad across the Isthmus of Tehuantepec, by which means the minerals of the country will be rendered accessible. The unsettled condition of Mexican affairs will prevent, for some time perhaps, the accomplishment of so important an undertaking.

Province of San Luis Potosi.—In the intermediate neighbourhood of Tampico, abundance of coal was announced, in 1847.

Province of Vera Cruz.—District of Acayuceam.—Here are several coal beds, it is reported, but none of them have been worked. They are, no doubt, continuations of those in the adjoining province of Oajuca.†

In the villages of Sayultepec and Moloacan, are fountains of petroleum.

Asphaltum or Chapapote.—In the interior of Mexico, according to a late traveller, "are Lakes of fresh water, where the Chapapote is found, bubbling up to the surface. When washed upon the borders,

* Philosophical Magazine, Vol. LXV., 1825.
† Mining Journal, February 14th, 1846.

it is gathered and used as a varnish for the bottoms of canoes. It has a pungent smell, like that of liquid asphaltum, and possesses, I think, some of its qualities.*

* Hunt's Merchant's Magazine, August, 1845, p. 164.

BRITISH AMERICA.

Area and Population of the British American Provinces, and Territories on the North American Continent, in 1846.

Provinces.	Square miles.	Population.	Rough estimate of area of coal land, square miles.	Remarks.
Lower Canada, . .	194,815	693,649		None ascertained.
Upper Canada, . .	147,000	506,055		" "
New Brunswick, . .	27,700	130,000	8,000	Dr. Gesner's report.
Nova Scotia and Cape Breton,	17,500	199,870	2,500	Nova Scotia.
			250	I. C. Breton, Sydney distr.
			104	I. of Boularderie, Cape B.
			180	II. S. coal-field, "
			Not defined.	III. W. " "
			"	Chiefly a coal formation at least.
Prince Edward's Island, .	2,134	34,666	5,000	
Newfoundland, . .	35,913	81,517	Not defined.	Magdelene Islands, coal,
	425,062	1,645,757	Say 16,000	as the mininum.
British Territory, up to 70° N. latitude and 140° W. longitude, deducting lakes and bays,	2,574,938	Unknown.		
	3,000,000			
British Honduras, . :	62,740	3,958		
Total British possessions, in North America,	3,062,740			

Years.		Population.	Sq. miles.
1852.	Upper Canada,* - -	953,239	147,832
"	Lower Canada, - -	890,261	201,989
1851.	New Brunswick, - -	193,800	27,700
"	Nova Scotia, - -	276,117	18,746
1848.	Prince Edward's Island, -	62,678	2,134
1851.	Newfoundland, - -	101,600	57,000
"	Hudson's Bay Territory, -	180,000	2,500,000
"	Labrador, - - -	5,000	170,000

* From the Office of the Chief Superintendent of Education.

Money.—Canada Currency.

1 English shilling = 1*s*. 1*d*., Halifax currency.
1 shilling currency = 10 pence, English.
20 shillings = one pound = 16*s*. 8*d*.
£1 sterling = 8 per cent. premium.
The American and Spanish dollar is 5 shillings, Canada currency.
1 Pistareen = 1 shilling, Halifax currency.
1 French five franc piece = 4*s*. 8*d*., Halifax currency.

To change Halifax currency [4 dollars = £1 currency] into British sterling, deduct one-tenth.

To change British sterling into Halifax currency, add one-ninth.

IMPORTATION.

By act of the Colonial Legislature, dated July 5th, 1843, all coals are allowed to enter the British American Colonies, free of duty.

Importation of Coal and Culm from Great Britain into British North America, from the Parliamentary Returns.

Years.	Tons.	Years.	Tons.
1831,	31,134	1840,	52,175
1832,	47,506	1841,	55,177
1836,	44,302	1844,	58,928
1837,	49,754	1845,	79,359.

INLAND COAL TRADE.—IMPORTATIONS FROM THE UNITED STATES.

From Ohio.—Bituminous Coal Imported from Cleveland.

Years.	Tons.	Years.	Tons.
1837,	6,605	1842,	2,020
1838,	2,639	1844,	1,240
1841,	1,559		

Of the coal exported into Canada from the Port of Erie we possess no details.

American coal received at Toronto in 1846, 1143 tons. Importation of American coal is diminishing annually.

There is very little reciprocity in the trade between Ohio and Canada, as may be seen by the following official statement for the year 1844:—*

* Report of the Secretary of the Treasury of the United States, January, 1845.

	Value.	No. of vessels.	Tonnage.
Exports from the port of Cleveland to Canada, - - - - -	$618,837	210	21,544
Imports from Canada to Cleveland, -	10,738	101	12,534

Value of imports of coal from the United States in 1851, $38,672 value.*

We possess no recent British returns of the amount of American bituminous coal which passed through the Welland canal; but it is understood that three-fourths of the property which now passes this canal, is conveyed in American vessels on American account.†

Tonnage, of all descriptions, on the Welland Canal.‡

Years.	Tons.	Tolls.	No. of schooners.	No. of scows.
1834,	37,917			
1837,	80,697			
1838,	95,397			
1840,	202,282	£18,037	1863	700
1841,	247,911	£18,583	1895	972
1848,	372,854		Boats of all kinds.	
1850,	399,600		4,761	
1851,	691,657		4,916	
1853,	1,050,000			

Rates of toll in 1845, on American coal on the Welland canal, for passing through the whole line, 2*s.* 6*d.* per ton. Between St. Catherine's and Port Dalhousie, 4*d.* per ton. Sea coal free of toll.

Through the Welland canal the navigation of the lakes is uninterrupted for the distance of 844 miles, from east to west, and the extreme distance from south to north is 347 miles.

The British trade on the *upper* lakes, in 1845, was only about one-tenth the value of the American lake trade, as appears from the following statement:§

American trade, valued at $1,517,132, employing 550 sailing vessels, and 49 steamers.

British trade on upper lakes, $150,000.

1853. The tolls received by State of New York from Canadian produce and property passing through her canals, is estimated at over $300,000 annually.

By a report, furnished in 1847, of the Secretary of the Treasury of the United States, it appears that in 1846 there were 30,000 tons of British shipping employed in transporting American goods on the

* State Engineer's Report Canals New York, 1852.
† Report of Lieut. Col. Kearney, U. S. T. E.
‡ State Engineer's Report of the Canals, New York.
§ Hunt's Merchants' Magazine, 1842.

lakes generally. The *bona fide* value of the American Lake trade, in the same year, is returned at $61,914,910.*

Colonel Abert, of the United States Topographical Engineers, reported in 1847, that the existing tonnage on the upper lakes, in a military point of view, is sufficient for 100,000 men. The British tonnage is small on the upper lakes, only 4,500 tons; propellers 2,500 tons.

On Lake Ontario the British have the advantage in the number of steamers, description of vessels, and number of mariners.

Lake Champlain is exclusively American.

	Tons.
American tonnage, 1846, - -	106,836
British tonnage, - - -	46,575

Tonnage of vessels employed in the Inland Trade between Canada and the United States in 1851.

VESSELS INWARD TO CANADA PORTS.				VESSELS OUTWARD FROM CANADA PORTS.			
American.		British.		American.		British.	
Steam.	Sail.	Steam.	Sail.	Steam.	Sail.	Steam.	Sail.
Tons. 1,224,523	*Tons.* 139,867	*Tons.* 845,589	*Tons.* 202,039	*Tons.* 753,318	*Tons.* 153,670	*Tons.* 564,089	*Tons.* 206,361

Inward and outward to and from Canada ports:

Steam, American, -	1,977,841	
" British, - -	1,409,678	
		3,387,519
Sail, American, - -	293,537	
" British, - -	408,400	
		701,937
Total inward and outward tonnage,		4,089,456

Trade between United States and Canada, in 1851.

Value of imports into Canada from United States,	$8,936,256
exports from Canada to United States,	4,939,280
Total imports and exports,* - -	$13,875,536

* See further details of the American Lake Commerce.

† North Penn. R. R. Co.'s Reports, 1853.

The total value of the exports during the year 1852, from Canada to Great Britain, amounted to about	£1,689,244
To the North American Colonies, -	203,036
British West Indies, - - -	3,460
United States, - - - -	1,571,130
Other foreign countries, - - - -	48,123
Total,	£3,514,993

Value of imports during same period, was	£2,667,783
North American Colonies, - -	120,238
British West Indies, - - -	1,278
United States, - - - -	2,119,424
Other foreign countries, - - -	152,899
Total,	£5,071,623

The gross amount of duties collected in 1852, was £739,263.

The above statement is taken from a volume entitled "Tables of the Trade and Navigation of the Province of Canada, for the year 1852," published by Government.*

Exportations of Coal to the United States from the British Colonies of North America [Nova Scotia and Cape Breton:] from the United States official Returns.

Years.	Tons.	Value.	American Tariff.	
			Per heaped bushel.	Per Ton.
		Dollars.		*Dollars.*
1802	233	588	Duty, 5 cents, from 7th June, 1794.	1 40
1804	388	978		
1832	41,934		6 cents, from May 2, 1824, per bushel.	1 68
1834	51,777			
1836	78,212			
1838	71,908			
1840	85,951			
1842	73,114		From August 30th, 1842, per ton.	1 75
1843	64,186			
1844	57,241			
1845‡				
1846	95,330	195,452	From Dec. 1st, 1846.	30 per cent. *ad valorem* duty.

In 1849, there were exported to the United States, 35,527 chaldrons of coal from Nova Scotia, value, £29,528.

* Hunt's Mag., June, 1853.
† Boston alone, in 1845, 42,035 tons.

There are no duties, either of exportation or importation of coals, from or to British America. A drawback is allowed by the United States on foreign coal re-exported, as in the instance of the depots of Pictou coal for the use of the British steamers. The law passed in January, 1840.

Importation of Iron from Great Britain to the British North American Colonies.

Year.	Bar.	Pig.
1844,	11,029 tons.	2991 tons.

November 5th, 1853. Importation of iron and steel, wrought and unwrought, from United Kingdom to British North America, value, - - - -	£479,220
Do. of Hardware and cutlery, - - -	138,630
Do. of Machinery, - - - - - -	6,150
Total,	£624,000

PROVINCES OF CANADA.

The area of East and West Canada is 341,815 square miles. Population in 1846, 1,199,704. Canada West in 1849, 791,000 persons. Population of the two Provinces in 1852, 1,842,265.

The result of the geological survey of Canada, as reported on by Mr. Logan,* sets at rest the question as to the existence of workable beds of coal within these provinces. None such have been traced; although there is, at Gaspé, a set of rocks overlying a series which corresponds with the old red sandstone, and the Chemung and Portage groups of New York, which rocks undoubtedly belong to the carboniferous series, though the part resting in Canada appears to be too low down to be associated with the profitable seams of coal.

Mr. Logan, in tracing the conglomerates and sandstones of this series round the Chaleur Bay, in Canada East, to Bathurst, in New Brunswick, has determined their relation to the nearest coal seams of the latter province with a considerable degree of certainty. The general dip of the Canadian part of the carboniferous deposit accords with this relation. Its slope towards the Chaleur Bay would carry it beneath the coal-bearing strata observable on the south, or New Brunswick side; while no rock of a similar quality is there seen to overlie the coal measures.

* The progress of the geological survey is marked by a preliminary report, dated December 6th, 1842, by Mr. W. E. Logan. A report by the same geologist, dated April 28th, 1844. A report from Mr. A. Murray, assistant geologist, March 14th, 1844. Report from Mr. Logan, May 1st, 1845; and from Mr. Murray, April 20th, 1845.

The reporter concludes with the observation that the conglomerate rocks with which fossilized coal plants (rarely sufficiently abundant to constitute even a very thin coal seam) are associated, within the limits of his survey, appear to be the very base of the coal series, in so far as Gaspé is concerned, and their distribution in Canada is just sufficient to show that a very narrow margin, on the north shore of the Bay Chaleur, may be considered the limit, in that direction, of the great eastern coal-field of North America.*

Black bituminous shales in the Gaspé District.—In the lowest of these were observed nodules occasionally resembling septariæ, in which the divisions or veins hold a mineral undistinguishable, in its general appearance and combustible nature, from good sea coal.

The whole group was determined by Mr. Logan to be about 1140 feet thick, and is apparently the equivalent of a part of the Hudson river group of the New York geologists. Its position is, therefore, a very considerable distance below that of the true workable coal-bearing measures, and we are not warranted in expecting coal seams to exist in it.

Here, and in the vicinity of Quebec, as in New York, erroneous expectations have been formed, and consequent disappointments have ensued, that these black bituminous shales indicate the proximity of workable coal.

Some of them hold a sufficient quantity of bitumen to yield a bright flame when subjected to a strong heat.

Carbonaceous shale and coal plants in the Gaspé sandstones.—This group, which Mr. Logan's detailed section shows to be 7,036 feet thick, appears to comprise what in the New York succession is termed the Chemung and Ithaca groups, with perhaps a portion of the old red sandstone or Devonian series. Towards the lower part are beds containing abundance of fucoid-like plants; while near the base is a small seam of coal and carbonaceous shale, together measuring three inches, which appears to hold a regular course, having a bed of clay beneath it. The middle portion of the group contains seams of argillaceous shale and sandstone, in which are balls or nodules of argillaceous ironstone.

LOWER CANADA.

Petroleum Springs.—According to the report of the Provincial Geologist, there are two petroleum springs in the neighbourhood of Gaspé Bay.

The first is situated on the south side of the St. John's river, about a mile and a half above Douglastown. The bituminous liquid oozes from the mud and shingle of the beach, at intervals, for about three-quarters of a mile.

The position of the other petroleum spring is on a small fork of the Silver brook, a tributary of the south-west arm. The liquid collects on the surface of the water, in the form of a thick dark green

* Geological Report of Progress, Montreal, May 1st, 1845.

scum, which can be taken up with a spoon. The odour could be distinguished for one hundred yards around.

Bituminous Trap Dyke, Gaspé Bay.—In some parts of the Dyke, the petroleum druces are so numerous, that there is scarcely a fragment the size of the hand that does not contain several of them, and the tar-like smell of the mineral is perceived in walking by the Dyke, at the distance of fifty yards. In some of the cavities the liquid is hardened into a resinous pitch-like condition.*

UPPER CANADA.—WESTERN DISTRICT.

Naphtha and Petroleum in the corniferous limestone—cliff limestone of Ohio. This rock, which is the highest in the geological series described by Mr. Logan as existing in West Canada, and a member of the Onondaga limestone group, contains, in the township of Cayuga, north of Lake Erie, much bitumen. When struck with the hammer, this rock gives out a peculiar odour, denoting the presence of naphtha. This substance is frequently seen occupying small cells, from which a sufficient quantity can be collected to determine its character. Near London, the naphtha or petroleum is found floating on the surface of the etangs, or stagnant waters of the Thames, and which is frequently collected by means of a piece of cloth.†

Peat.—In the vicinity of Port Daniel, in the Gaspé district, peat is extensively spread.

The Eastern Provinces of British America have for some years had the benefit of examination by several resident gentlemen, as well as travellers, highly advanced in science. We shall frequently have occasion to quote from the reports of these writers, details of a very interesting character in relation to the carboniferous formations that occupy so large a portion of the area of those countries.

* Report of Progress of the Geological Survey of Canada, for the year 1844, p. 41.

† Rapports sur une Exploration Géologique de la Province de Canada, January 27th, 1845, p. 92.

PROVINCE OF NEW BRUNSWICK.

THE area of this province is 27,700 square miles.

Bituminous Coal-field.—The entire area of coal measures within the province, is locally subdivided into several districts, of which the following are the principal:—

I. The great northern coal-field.

II. The Westmoreland, or south-eastern.

III. The Sunbury and Queen's county, or south-western.

The aggregate area of these was estimated, in 1840, at five thousand square miles. Dr. A. Gesner, in a communication to the Geological Society of London, in 1843, stated that the area of the coal-field in New Brunswick had been recently determined to be seven thousand five hundred square miles; or ten thousand square miles, including Nova Scotia, but exclusive of Cape Breton.* These coal measures are described as usually lying in long parallel troughs, or in oval basins. Since the first report of Dr. Gesner, he has explored the whole of this vast region. The result of this geological survey is, that the coal formation is found to occupy, in New Brunswick, no less than eight thousand square miles. Here the most productive coal beds prevail in the interior, while those of Nova Scotia occur on the shores of her bays and rivers, where they offer every advantage for mining operations. The coal-fields of the two provinces are united at the boundary line, and belong to one carboniferous period.† The developments of almost every season illustrate more clearly the magnitude of these coal areas, which extend from Newfoundland, by Cape Breton, Prince Edward's Island, and Nova Scotia, and across a large portion of New Brunswick, into the State of Maine.

Sir I. E. Alexander officially reported, January 5th, 1846, that the great field of New Brunswick and Nova Scotia covers a surface of upwards of nine thousand square miles; but Dr. Gesner, the provincial geologist, much exceeds that estimate, as we have seen above.

I. *The Great Northern Coal Field.*—Mr. Henwood, a geologist of high standing, observes, that "the beauty and extent of these coal measures it is impossible to describe. In fact, we pass over nothing else, from Fredericton, on the St. John's river, to Miramichi, and thence to Bathurst, a distance of at least a hundred and fifty miles. They consist of various beds of sandstone, shale, and conglomerate, with numerous thin seams of coal, few of which are more than a foot or two in thickness. The whole of this district is particularly rich

* Proceedings Geol. Society, Vol. IV., p. 182; reprinted in Journal Franklin Institute of Philadelphia, June, 1844.

† Mineral Wealth of Nova Scotia,—Gesner. Also Mining Journal, July 19th, 1845.

in fossil flora."* The coal measures, whose lowest members are the conglomerate beds, are perfectly horizontal in the banks of the Nepisiguit, near Bathurst, and these repose upon granite.

An interesting geological phenomenon has been observed here. In one of the thick beds of blue shale overlying the granite, and containing ferns and other fossil plants, occur lignites which are impregnated in their laminæ, as well as in their cross fracture, by *rich, vitreous, copper ore*, and coated with green carbonate of copper. Other lignites, also containing vitreous copper ore, occur in Nova Scotia, in the neighbourhood of Pictou, in considerable quantities, under precisely similar circumstances, within the coal formation.†

Something like this is of not uncommon occurrence in the United States, in the cupreous lignites of the red and blue shales at the base of the old red sandstone, or Devonian system. We have observed them at numerous points in Pennsylvania.

These lignites occur as casts of reeds, canes or flags; generally obscure, and the impressions of leaves seem in some degree to resemble those of the coal series above. The copper is in form of rich gray sulphuret, the surfaces of the lignites being coated with green carbonates. In more than one or two instances a good deal of expense has been incurred in exploring this ore, but we have never seen it in sufficient quantity to repay the cost. Copper seems invariably to accompany this bed of lignite; at least it is seldom unaccompanied by lignites. The latter are sometimes bitumenized. Professor Del Rio has mentioned a similar occurrence. Mr. Murchison also states that in the great copper district, which flanks the west side of the Oural mountains, the copper is wholly in the form of vegetable casts.

Near the Victoria coal mines, which are situated on the left bank of the Nepisiguit, previously spoken of, the vegetable remains occur, partly converted into coal, and partly replaced by gray sulphuret of copper. The same state of things occurs in the rocks at the Joggins, on the Bay of Fundy, within the Nova Scotia coal basin. Mr. Logan states that on the Nepisiguit an attempt was made by the Gloucester mining company, to work the deposit as a copper mine; but the irregular distribution of the organic remains rendered their operations uncertain, and induced the abandonment of them. The bed averages about two feet thick, but in one direction it appears to thin off, from four feet to nothing.‡

Coal is mentioned by Captain Bayfield, as occurring at Percé, near the entrance of Chaleur Bay.§

In Mr. Logan's very detailed section of the coal measures which are displayed in the cliffs of the New Brunswick coast, on the south side of the Bay of Chaleur, it is noted that on many of the coal plants, a very minute convoluted shell is seen, and in the shale is

* Transactions of the Royal Geological Society of Cornwall, 1840.

† Mr. Dawson, on the Geology of Nova Scotia, February, 1845—proceedings, p. 35.

‡ Geological Report of Canada, May 1st, 1845, pp. 63, 79.

§ Captain Bayfield, in Trans. Geol. Soc. of London, Vol. V., p. 87.

a small bivale. *Stigmaria ficoides* occurs in abundance and of large size.

II.—*Westmoreland or South-eastern Coal-field.*—Dr. Gesner's second report, in 1840, shows that this coal area is seventy miles in its longest diameter, and that it averages seventeen miles in breadth. "It is by no means certain that coal is contained in every part of the area; but as the outcropping of the bituminous strata has been discovered in a number of situations, it is evident that it embraces vast quantities of coal, and is of the highest importance to the province."

The Westmoreland coal measures, we are told, rest directly upon granite.

III.—*South-western Coal-field of Sunbury and Queen's Counties.*—By Dr. Gesner's first report of his geological survey, we learn that the coal measures repose upon the mountain limestone, and cross to the west of the river St. John.

On the shores of Grand Lake, in Queen's county, a company has been incorporated more than 35 years, with a capital of £30,000, to work the coal-beds which here lie horizontally a few feet above the level of the water.

An excellent coal mine has been opened on the banks of the Salmon river, which coal is said to be superior to that of the Grand Lake.

The quantity exported is very small, compared with the enormous magnitude of the coal area, as may be seen by the following table.*

In the year 1828, 66 chaldrons; 1830, 70 chald.; 1833, 138 chald.; 1834, 687 chald.; 1835, 3,537 chald.; 1839, 2,143 chald,

No later returns have reached us.

A mine of asphaltum has been discovered by Dr. Gessner at Frederick's Brook, in the parish of Hillsborough, in the county of Albert, and province of New Brunswick. The following facts exhibited at the New Brunswick mine, are physical characteristics of a true asphaltum vein.†

1. The absence of lamination in the mass.
2. Its brilliant conchoidal fracture and occasional tendency to assume a columnar structure.
3. The character and configuration of its surface-markings.
4. Its small specific gravity; not equalling nor exceeding many of the resins.
5. The general prevailing uniformity in the entire substance or contents of the vein.
6. Its aspect, fracture, divisions, purity, and especially its almost entire freedom from foreign and earthy matters.
7. The absence of all vegetable traces in connection with the material of the vein.
8. The absence of all apparent organization in its composition.

* Statistics of the Colonies of the British Empire, p. 244.

† See Taylor's deposition before the Supreme Court at Halifax, N. S., respecting the Asphaltum mine at Hillsborough. Phil. 1851.

9. Its apparent fused and liquid state originally, and its subsequent consolidation after cooling.

10. The practical restoration of its characteristic surface-markings, and its peculiar conchoidal fracture, after being once more melted and rendered soluble, and again cooled and consolidated.

11. Its not soiling the fingers, in the manner of coal.

12. Its being strongly electric.

The specific gravities of the New Brunswick and Cuba asphaltes, determined chiefly by R. C. Taylor, are as follows:

That of the Casualidad mine, mean of three observations, R. C. T. - - - - - -	1.176
That of a mine near Matanzas, mean of observations, T. - - - - - -	1.160
That of the mine at Hillsboro', New Brunswick, first observation, T. - - - - - -	1.095
Do. do. second observation, T.	1.096
Do. do. third observation, Buck,	1.097
In Glascow, by Professor Penny, - - -	1.097

PROVINCE OF NOVA SCOTIA.

"THE General Mining Association," as tenants of the Crown and of his late Royal Highness the Duke of York, are lessees "of all the mines and minerals, of every description, in the province of Nova Scotia proper, and in the island and county of Cape Breton."

These coal mines are leased for sixty years, from 1827, at the fixed rent of £3,000 sterling, $14,500 per annum. This fixed rent of £3,000 sterling or £3,333 currency, per annum, conditions for a maximum annual raising of 20,000 Newcastle chaldrons, and fixes a royalty of two shillings currency, for every chaldron beyond that quantity.*

By an arrangement made with the association in 1845, they were allowed to take out 26,000 Newcastle, or 52 London chaldrons, or 65,000 tons, instead of the twenty thousand chaldrons, stipulated for; the fixed rent remaining the same.

The company is generally known under the title of "The Nova Scotia and Cape Breton Mining Company."

The operations of the company commenced in 1827, and have hitherto been confined to the working of *coal* mines and the discovery of *iron ore*. The collieries now open and at work are four in number, viz.: the Pictou and Albion in Nova Scotia, and the Sydney and Bridgeport mines in Cape Breton.

The capital of the "General Mining Association" is £400,000—$1,936,000; and they possess fourteen thousand acres of land; besides the right to all minerals and mines within the province of Nova Scotia and Cape Breton.†

In reciting these details we, as well as our readers, cannot omit to remark the injurious magnitude of such gigantic monopolies as the one before us. In this case it covers an extent of more than *twelve millions of acres*, or three times the size of Wales. It is scarcely necessary to say that its tendency is to impoverish the people; to destroy all energy in cultivating the abundant natural resources of a fine country; to prevent all fair and wholesome competition, to narrow the scope of active and productive industry, and to discourage all individual and general enterprise. On the continuance of such a deplorable system, the rival coal proprietors of the United States, may well found their calculations of a remunerative internal trade in coal at home, with even greater safety and certainty than on the influence of tariffs and the restrictions of international regulations.

In 1834, another large company was incorporated, under the name of "The New Brunswick and Nova Scotia Land Company," having

* This extra rent charge of two shillings per chaldron, has frequently been mistaken by American writers for a tariff; or as a writer styled it, in 1846, "an excise duty of 20 cents per ton, for the support of the local government".—U. S. Gazette.

† Statistics of the Colonies of the British Empire, Martin, p. 230.

purchased of the Crown five hundred thousand acres of land lying in the centre of the province of New Brunswick, at the price of 2*s.* 6*d.* sterling—$0.60 cents an acre. The company, are by act of Parliament, to have the privilege of purchasing lands in New Brunswick, Nova Scotia, Cape Breton, and Prince Edward's Island. They are thereby authorized to mine and work, copper, tin, lead, iron, and all other minerals, except gold and silver, and *except coal and culm;* the two latter being already granted by lease to the general mining association,* in all the above mentioned places except New Brunswick.

Geology.—A small Geological map of Nova Scotia was published in 1841, by Messrs. Jackson and Alger. In 1842, Dr. Gesner's first geological map of Nova Scotia and Cape Breton was completed, and issued the same year. In 1845 another map to illustrate the geological papers of Messrs. Gesner, Dawson and Brown, appeared in the Quarterly Journal of the Geological Society of London.

On comparing the first named of this series of illustrations with the subsequent ones, an unusual discrepancy is apparent between them. According to the former the coal of Nova Scotia is restricted to a couple of spots, embracing an area so minute, as scarcely to be discernible upon the map. Estimating these coal areas by the scale of the plan, they only cover an aggregate area of thirty-three or thirty-four square miles; whereas Dr. Gesner's statement exhibits an area of carboniferous formations of two thousand five hundred square miles; while Messrs. Dawson, Logan and Brown greatly exceed even that area. However, since the provincial surveys have been completed, there can be no longer a doubt on this point, and all original errors arising from early and defective investigations are now fully adjusted.

Dr. Gesner, the first in date states, that from Pictou Harbour, in Northumberland Strait, a *central belt* of coal measure, about six miles broad, runs in a westerly direction across the isthmus, and passes between the southern flank of the Cobequid mountains and the southern coast of the isthmus along the basin of mines; and thence further westward to Advocate Harbour. The length of this central belt is about one hundred miles, and the coal is supposed to rest unconformably on old red sandstone.

These carboniferous beds lap round the eastern extremity and pass along the northern flank of the before mentioned Cobequid mountain range and the Annan Hills, whence they again pass nearly due west to Chignecto Bay in the Bay of Fundy. All the isthmus north of the line thus designated, consists of carboniferous strata, forming the northern Cumberland coal field. The Nova Scotia coast of Chignecto Bay, runs nearly at right angles to the direction of the strata, and presents an admirable section of them, nearly thirty-five miles in length. In making a careful examination of this thirty-five miles of coast, only one unimportant fault was observed. By measuring the horizontal distances between the strata, and making allowance for

* Appendix to Statistics of the Colonies of the British Empire, No. III., p. 76.

their inclination, at a number of places, Dr. Gesner estimated the total thickness of the coal measures along this sectional line, at not less than three miles, [between four and five, Lyell;—2¾ miles and 50 feet, Logan.] This is a remarkable fact; unlike, for its magnitude, any other group of coal strata on the North American continent.

The carboniferous series is, thus far, exhibited in the cliffs called the South Joggins; and Dr. Gesner mentions the existence here of nineteen small coal seams, within the horizontal distance of three quarters of a mile. They occur in an aggregate thickness of 1800 feet of strata, and vary in dimensions from six inches to four feet.

Mr. J. W. Dawson, himself a resident of the Pictou coal mines, states that the coast section on the north-eastern side of Nova Scotia, cuts at an acute angle, across two great coal troughs; the one beginning at Pictou on the east coast, and thence stretching to the west along the northern shore of the Basin of mines, [Bay of Fundy;] the other beginning at Antigonish [St. George's Bay,] and thence extending westward to the Shubenacadie river and the southern shore of the Basin of mines, [Bay of Fundy.] These two troughs are separated by a hilly range, composed of igneous rocks, and of disturbed lower carboniferous and silurian, or palæozoic strata.*

We ascertain, therefore, from the foregoing outlines, that the coal formation of Nova Scotia occupies three distinct areas, viz.: I. The Northern or Cumberland Region; II. The Pictou or Central Basin; III. The Antigonish or Southern Basin. Dr. A. Gesner adds a fourth district, which probably is an extension of the third. He states that on the south border of the Basin of mines there is an area near Falmouth and Windsor, of seventy square miles, in which, though coal has not yet been discovered, the ferns, stigmariæ, and other fossil coal plants, which the sandstones and shales of that area contain, sufficiently establish the point that it belongs to the coal measures.†

Large as is the amount of these united areas, comprising between two and three thousand square miles, it nevertheless forms but a fraction of that immense coal formation which occupies large portions of New Brunswick, Nova Scotia, Cape Breton, and Prince Edward's Island.

I. *Northern or Cumberland County Coal Region.*—This embraces the triangular area, which extends from the Cobequid mountain range and the Annan Hills, to the extreme northern boundary of the Nova Scotia peninsula; having the Northumberland Strait to the east and Chignecto Bay to the west.

In 1845, in a geological report to the provincial government of Canada, Mr. W. E. Logan published a section of the carboniferous strata, within this region, as developed at the Joggins, a continuous cliff, eighty to one hundred feet high, on the south shore of the Bay of Chignecto, Bay of Fundy.

* Quarterly Jour. Geol. Soc. London, February 1st, 1845, J. W. Dawson, p. 26.

† Transactions of the Geological Society, London, 1843. Proceedings of the Geological Society, London, 1843, Vol. IV., pp. 178—190.

This section is one of the most remarkable ever accomplished, and may be quoted as a model of close investigation, and extraordinary accuracy in developing an enormous series of beds. It comprises the vast group of coal measures which are displayed along the cliffs, locally named the Joggins, of the sea shore of Chignecto Bay. This locality, so singularly favourable for taking the strict admeasurement, and constructing an exact section of the vertical thickness of the coal formation has been frequently alluded to by geologists and travellers. It remained for Mr. Logan to demonstrate by a laborious survey, the true thickness of the whole group, in Northern Nova Scotia.* His section is subdivided into eight principal sub-sections or parts, and these again are further divided into the respective members which compose the mass, separately measured by feet and inches. The extent of the labour may be inferred from the fact, that the whole series consists of no less than fifteen hundred and seventy beds or subdivisions, all minutely described, and making up the aggregate thickness of 14,570 feet 11 inches; equivalent to 2¾ miles, 50 feet 11 inches; an amount which far exceeds anything seen in the coal formation in the other parts of the North American continent, to the southward.

One of the most remarkable circumstances which are brought into notice by the section of Mr. Logan, is the extreme thinness of the coal seams in this portion of the Nova Scotia basin. We collect from an examination of the report, the following stratigraphical summary:

	Feet.	Inch.	Coal beds.	Aggregate thickness. Feet.	Inch.
Subdivision No. 3, consisting of	2,134	1	of strata 22 =	5	5
" No. 4, "	2,539	1	" 45 =	37	9½
" No. 6, "	3,240	9	" 9 =		10
	7,913	11	76	44	0½

The average or mean thickness of each of these 76 coal seams, is a fraction less than seven inches. The maximum thickness is shown in No. 7 bed of the fourth sub-section, where there are 3 feet 8 inches of coal in a bed 4 feet 6 inches thick; and in the next thickest, we have No. 29, which has 3 feet 5 inches of coal out of a bed 4 feet thick. At the same time, it may be observed, that the greatest thickness of pure coal in any one seam, is only two feet, and it may be questioned whether any one out of the seventy-six coal beds, in 2¾ miles of strata, will ever be considered as workable.

Thicker coal beds appear to exist to the eastward of the Joggins. On the Macon river, which falls into Cumberland bay, one seam occurs of ten feet in thickness; and the same, if not another, is seen at river Philip.

Several geologists have noticed the presence of erect trunks of trees in the coal strata of the Joggins, particularly in a bed of sand-

* Rapports sur une exploration géologique de la province de Canada. January, 1845.

stone, twelve feet thick. The first notice, probably, being that of Mr. Brown, in Haliburton's "Nova Scotia," in 1829. In 1842, Mr. Lyell saw similar upright trees at more than ten different levels here; all placed at right angles to the planes of stratification. Lithologically, the strata resemble the English coal measures, and those with which the coal and erect trees are associated are more than 2500 feet thick. The grits and shales containing coal plants above these are of prodigious thickness, as we see in Mr. Logan's section just adverted to. Mr. Lyell saw seventeen vertical stumps, varying in height from six to twenty feet, and from fourteen inches to four feet and a half in diameter. The trunks of these trees, which are all broken off abruptly at the top, extend through different strata, but are never seen to penetrate a seam of coal, however thin. They all end, downwards, either in beds of coal or shale; no instance occurring of their termination in sandstone. The exterior coating of these trunks is in the state of coal, while the solid interior usually consists of sandstone or fire-clay.

The exact position of all these beds which contain vertical stems of *Sigillaria*, under these circumstances, are exhibited in Mr. Logan's section, which is illustrated by figures of some of these trunks in the position in which they appear on the cliff.

In the vicinity of the highest coal seams, in the series, viz., between seam No. 43 and 44, a twelve feet stratum of arenaceous schist is penetrated by several erect *calamites*, in one instance one of the plants three inches in diameter, extends its roots, and twenty-one others are visible along the face of the cliff, within the space of twenty yards. Their diameters vary from half an inch to four inches. In this sub-section of 2539 feet thick, Mr. Logan enumerates, among the visible organic remains, fifteen *sigillaires* growing erect, and fifty-six *calamites*, standing apparently in their native beds.* Mr. Lyell states that immediately above the uppermost coal seams and vertical trees, are two strata, probably of fresh-water origin, of black calcareo-bituminous shale; chiefly made up of two species of *modiola* and two kinds of *cypres*.†

The Lower Carboniferous Rocks of Nova Scotia, are described by J. W. Dawson, Esq., a resident of the Pictou coal mines, in the Quarterly Journal Geological Society, London, Feb. 1, 1845.

II. *The Pictou, or Central Coal Basin.*—As we have previously explained, this belt of carboniferous strata, stretches from near Cape St. George on the east coast, to Advocate Harbour at the Bay of Fundy, on the west, and follows the north shore of the Basin of Mines. The area embraces the coal mines of Pictou and Albion, and Dr. A. Gesner states that two seams of coal have been discovered in the forest, ten miles north of Truro, and that outcrops of coal appear in the same belt at Jolly river, at Herbert and Economy rivers, and at Parr's borough.‡ Strictly speaking, this district is not wholly

* Logan's First Geological Report of Canada, Montreal, 1845.
† Lyell, in Proceedings Geol. Soc., 1843, Vol. IV., pp. 178—190.
‡ Edinburgh Cabinet Library, No. XXVI., Vol. II. British America.

separated from the Cumberland region, but is connected for a brief space opposite the east end of the Annan hills, in the vicinity of Pictou. It is, however, most convenient as regards topographical arrangement to treat them as separate districts.

The Pictou region appears to be the richest in coal, yet worked in this province, and it contains the two principal mining establishments of the province.

Mr. Logan's section of this region, made in 1841, is interesting; below are the results of his admeasurements of the carboniferous series; the details we are compelled to omit. Section commencing at the base of the series.

	Feet.
1. Red and drab coloured sandstones, a few coal seams towards the base, the thickness is not stated.	
2. Shales and sandstones with workable beds of coal and ironstone, - - - - - -	5000
3. Limestone, with marine fossils, - - - -	10
4. Coal measures, probably unproductive, - -	1900
5. Limestone, with carbonized vegetable remains, -	10
6. Red and green shales, and red sandstones, -	650
7. Limestone, - - - - - -	20
Total in this part of the coal area of Nova Scotia,*	7590

No. 1, of this section, we presume to be that series which Mr. J. W. Dawson has since investigated and described, under the name of the "*Lower Carboniferous rocks, or Gypsiperous Formation* of Nova Scotia."† This series overlies the Silurian strata, and consists of limestones, gypsum, and soft sandstones; above which are hard reddish sandstones and shales, with limestone; and lastly, red and grey sandstone, shales and conglomerate, with carboniferous plants. Probably these beds pass into the productive coal measures—No. 2, of Mr. Logan's sub-section above.

Mr. Dawson has not been able to ascertain the exact aggregate thickness of the lowest carboniferous rocks; he remarks, however, that in the vicinity of Merigonish, and east of Pictou, the band of carboniferous rocks amounts to 10,000 or 12,000 feet in thickness, all dipping to the north-west at an angle of twenty degrees.

This gentleman discovered a bed of erect calamites in the Pictou coal-field, one mile and a quarter west of Pictou, in a bed of sandstone about ten feet thick. They all terminate, downwards, at the same level where the sandstone rests on subjacent limestone, but their tops are broken off at different heights.‡ This is a repetition of the same phenomena observed, at the distance of one hundred miles, on the shores of Chignecto bay.

Pictou Mine.—At this mine, situated on the West river, there is but one seam, but several miles to the southward, Mr. Logan, in

* Trans. Geol. Soc., London, 1842.

† Quarterly Journal of the Geological Society, February, 1845, p. 26.

‡ Proceedings of the Geo. Soc., Vol. IV., p. 178.

1841, ascertained the existence of more than twelve coal beds, which are thought to correspond with those in the coal-field of Cape Breton Island.

From notes, made in 1833, we were informed that the main coal seam of Pictou was twenty-nine feet thick: but, at that time, only ten feet of the best quality of coal were worked. It has one seam of slate, five inches thick. The shafts were from sixty to two hundred and forty feet deep, of which two hundred and twenty were below the level of the sea. This is a hard, open burning coal, and is worked with powder;* it does not command so high a price in the market as that of Cape Breton. [Sydney and Bridgeport.] The Pictou coal communicates northward from its excellent harbour, with the Gulf of St. Lawrence; while the Cumberland coal passes down the Bay of Fundy, to the southward.

In 1833, the coal was conveyed from the shafts of the Pictou mines, on a railroad of one mile, to a landing place, and from thence about six miles to the shipping. In 1840, a new railroad of six miles, for carrying the coal at once from the pits to the wharves, was put in operation.†

This coal is stated to possess properties which render it well suited to the various branches of the iron manufacture. It is peculiar, according to Mr. Alger, on account of the abundance of mineral charcoal that it contains; and, for domestic purposes, this is thought to give it an advantage over the Sydney and most other bituminous coals, by preventing it from cementing together while consuming.‡

Albion Mines.—Situated on the banks of the East river, in the district of Pictou, and distant about eight and a half miles from the town of that name, a port of safe and easy access from the Gulf of St. Lawrence. A lighthouse was erected on the coast, near the town of Pictou, a few years ago.

The East river is only navigable for the larger craft to within six miles of the Albion mines; so that vessels arriving for coal, formerly, received their cargoes from barges which loaded at the mines, and were towed down to the deep water by the steamers belonging to the association. In 1840, a railroad was completed, and obviated this inconvenience, as well as the breakage which previously took place by the transhipment.

At the Albion mine is a great collection of coal seams all dipping to the north. The number is stated by Judge Haliburton to be ten, and the aggregate thickness to be sixty feet. The only seam worked a few years ago contained twenty-four feet of clean coal, of which about two hundred and forty tons were raised daily. In 1839, the quantity of coal raised per month, was from five thousand to six thousand tons. Above three hundred vessels, of various descriptions, were loaded here during that season.

* Journal of the Senate of Pennsylvania, 1833, p. 570.

† Mining Journal of London, Vol. X., p. 45, and Vol. XII., p. 123.

‡ Alger's edition of Phillips's Mineralogy, p. 591.

There are several shafts at the Albion mine, for raising the coal: one of the engine shafts is four hundred and fifty feet deep.

Quality and Properties.—The Pictou coal is in favourable repute for the use of steamboats. In 1833, the steamer Royal William, of one hundred and eighty horse power and one thousand tons burthen, performed the voyage from Pictou to Cowes with the employment of Albion coal; the trial proving entirely satisfactory. In 1838, the coals for the voyages of the Great Western steamship were supplied from the Nova Scotia mines. They were stated to answer beyond expectation, the quantity consumed being less than the necessary supply of English coal, while the price was lower. The Cunard line of steamers is now supplied with Sydney coal from the depot at Boston.

We have heard less favourable opinions from some persons, yet it has been preferred to the Virginia bituminous coal, which contains more sulphur, and is consequently liable to occasional spontaneous combustion.

The relative value of the Nova Scotia and Cape Breton coals may be inferred from Mr. Johnson's analyses and experiments.*

			Carbon.	
Nova Scotia Pictou Coal,	Cunard's sample,	60.73	100 parts,	The volatile matter being nearly 27 per cent.
	mining association,	56.98	"	
Cape Breton, Sydney coal, mean of two specimens,		67.57	"	

Ranks of coals, according to the several practical characters, out of forty-two varieties.	Nova Scotia or Pictou. Cunard's.	Nova Scotia or Pictou. Mining Association.	Cape Breton or Sydney.	Pennsylvania. Queen's Run.	Virginia. Chesterfield.	Liverpool.
Arrangement in the order of						
Their relative weights, ...	29	15	35	28	40	33
" rapidity of ignition, ...	6	9	15	2	13	7
" completeness of combustion, ...	1	4	7	24	16	19
" evaporative power under equal weights, .	30	33	37	2	20	38
" evaporative power under equal bulks, .	29	23	35	7	30	36
" evaporative power of the combustible matter,	31	30	39	2	23	40
" freedom from waste in burning, . .	33	37	5	19	20	2
" freedom from tendency to form clinker, .	41	40	13	7	28	11
" maximum evaporative power under given bulks,	30	23	30	10	32	33
" maximum rapidity of evaporation, . .	4	25	24	27	1	10
	234	239	246	128	223	229

* Report to the Navy Department of the United States, on American Coals, by W. R. Johnson, 1844.

We select the foregoing results from the "Report on American Coals," 1844, whereby the practical characters of the British American coals will be seen, and compared with those of other bituminous coals.

Out of the forty-two varieties of coal which have been experimented upon, thirty-five are from the United States, and seven from British America and Great Britain. The numbers in the table represent the order in which they take their appropriate rank, from one to forty-two. From the care which we know has been bestowed to obtain these results, we cannot hesitate to receive them, in perfect reliance on their accuracy. By taking the four tables of results of evaporative power, the respective coals in the foregoing synopsis, range themselves in the following order of value:—

1. Pennsylvania coals of Queen Run.
2. Virginia coals.
3. Pictou Mining Association.
4. Pictou—Cunard's.
5. Sydney.
6. Liverpool.

On an average, 80,000 chaldrons of coal and 50,000 cords of wood are shipped annually from Nova Scotia to the United States, which return large quantities of manufactured iron.

At present, the iron imported into Nova Scotia and New Brunswick, amounts in value to £139,000 per annum, while, at the same time, there is not a smelting furnace in any of the British North American provinces, Canada only excepted. This state of things will probably remain until the resources of British America are better known in the mother country; where alone there is capital to improve them. This deficiency of iron works is by no means ascribable to the want of iron ore, which, by the geological statements of Dr. Gesner, is very abundant. There are many varieties of iron ores distributed over this country.

The coal-fields of the Pictou district, and Cumberland county, contain workable strata of the argillaceous oxide and carbonate of iron, known as "clay iron stone." At those places, the ore, the coal for fuel, and the limestone necessary for the flux, are placed side by side. This admirable arrangement, made by Providence, whereby all the materials necessary for the production of iron, are deposited together, is still overlooked in this province, whose metals are imported from foreign countries. All the iron employed for railroads and mining operations, is imported from Great Britain; and having been transported three thousand miles, it is finally thrown into castings, at the very site where thick beds of Nova Scotia ore are seen protruding from the earth; and where a single stratum of coal, thirty-six feet in thickness, is ready to supply the fuel for its smelting and manufacture!*

We may add to the foregoing notice of the prevalence of iron ore, that an enormous deposit of the specular oxide of iron has recently [1847] been discovered at Londonderry Mountain, in Nova Scotia.

* Geology of Nova Scotia, by Dr. A. Gesner. Also Mining Journal, April, 1845.

Dr. Gesner has reported favourably of the ore, and its local position. It is estimated to yield about seventy per cent. of cast metal.

Exportation.—In the year 1836 the quantity of coal exported from Nova Scotia to all parts, was 42,587 tons, the value of which was £38,328 or $185,507, being 18*s*. or $4 48 per ton. In 1839 the quantity exported was 67,632 tons.

The General Mining Association now ships from the Albion Mines to the United States from 40,000 to 50,000 chaldrons of coals annually. Ten seams of coal have been penetrated by the workings at the Albion Mines; the united thickness of these coal beds is upwards of seventy-five feet. The main coal band is no less than thirty-six feet in thickness—of this, the company only work twelve feet, leaving twelve feet of good coal, and twelve feet fit for furnaces and forges.

In 1839, six steam-engines, 100 horses, and 500 men was employed at those mines, and upwards of 48,000 tons of coal were exported to the United States, and to different ports along the coast.

The Province contains about 15,000 square miles. Of that area, there are 2000 to 2500 square miles of coal-field.

"On the coast of Chignecto Bay, the tide rises upwards of fifty feet; and at low water the beds of coal are uncovered by the sea. Upon these beds vessels from New Brunswick and the United States lie aground, and from them receive their cargoes; and as the shore can hardly be said to be inhabited, no notice is taken of such depredations."

"The steamboats that run into Chignecto Bay are propelled by coal *imported from Great Britain;* their keels often pass within a few feet of the coal strata already mentioned, and from which they might be cheaply supplied; but the Mining Association possesses an entire monopoly, which has prevented every kind of mining enterprise in the province; the inhabitants of Nova Scotia have not been permitted to open the earth beyond the depth of the soil, and up to the present hour they are compelled to pay the price fixed by a single company for all the coals they consume. By withholding the coal from the inhabitants of any civilized country, where that mineral is abundantly found, the manufacture of iron and other metals is prevented; manufactories cannot exist; trade will languish, and general industry be greatly retarded. The truth of these remarks is fully proved by the present state of the province—a colony that will never thrive until her resources are liberated from the fetters of unyielding monopolists."* Let these men look at Pennsylvania.

It appears from a memorial or report of a committee of the House of Assembly, in 1846, that twenty of the sixty years of the lease to the association had then expired; and that yet no effort had been made to work a single bed of coal, or other mineral, with the exception of the coal-beds at Pictou and Sydney!

The General Mining Association, in 1847, determined to open a new coal mine in Cumberland county. This establishment will supply

* Mineral wealth of Nova Scotia, by Dr. A. Gesner, July 19th, 1845, Mining Journal.

New Brunswick and the Nova Scotia ports in the Bay of Fundy, and will shorten the distance for the coal vessels from the United States.

Newer Coal formation, on the eastern part of Nova Scotia, in the district of Pictou.

Mr. J. W. Dawson has described in Nova Scotia, a newer coal formation than the usual old coal formation upon which it rests. In a palæontological point of view it possesses considerable interest, as its fossils show the continuance of the coal flora during the deposition of a series of red sandstones of more recent origin; and also of the co-existence of that flora with terrestrial vertebrated animals.

The older coal measures of the Albion Mines on the banks of the East River of Pictou are, according to Mr. Logan, 5000 feet in thickness; and are succeeded in ascending order by a great bed of coarse conglomerate, which, as it marks a violent interruption of the processes which had accumulated the great beds of coal, shale, and ironstone beneath, and, as it is succeeded by rocks of a character very different from that of these older coal measures, forms a well-marked boundary, which Mr. Dawson considers as the commencement of the newer coal formation.

The conglomerate is followed by soft reddish sandstone, above which is a bed of grey limestone, supporting a small bed of coal and a few inches of under clay; and over these are at least 2000 feet thick of reddish and grey sandstones and shales, in which is another seam of coal, eleven inches thick, with an under clay. In the grey sandstones are coniferous lignites, fossilized by carbonate of lime, and *calamites*, *endogenites*, and *lepidodendrons*. Near Pictou, in addition to these, are fossil ferns, *sternbergia*, [Artisia ?] and carbonised fragments of wood, impregnated with iron pyrites and with sulphuret and carbonate of copper. In this series also, and near the town of Pictou, is the bed of sandstone containing erect *calamites*, noticed by Mr. Lyell, in his papers on the fossil trees of the Joggins. In the coast section, westward of the entrance of Pictou harbour, much red sandstone appears, and also a bed of limestone, and a small seam of coal. Some grey sandstones also appear, in which are numerous fragments of carbonized wood, containing sulphuret and carbonate of copper. Proceeding coastwise to Cape John, at the extremity of the Cape is a bed of white granular gypsum, about three feet thick.

Beyond Cape John this newer coal formation skirts the shores of the Gulf of St. Lawrence, to Wallace Harbour.

In the red sandstones, near Tatmagouche, Mr. Dawson had found on a previous examination, a few foot-marks of an unknown animal. They were mere scratches made by the points of the toes or claws, and their arrangement appeared to indicate that the animal was a biped; their form being quite analogous to that of the marks left by our common sandpiper, when running over a firm sandy shore. On a subsequent inspection, a series of foot-marks of another animal was found. In a specimen forwarded to the Geological Society, the tracks were somewhat injured by the rain-marks, which cover the slab. Many other beds in the neighbourhood were observed to be rippled, rain-marked, or covered with worm-tracks; and as such indications of a littoral origin are not unfrequent in other parts of the newer coal formation, it may be anticipated that many interesting relics of terrestrial animals will in future be discovered. Among other fossil remains in the red sandstones of Tatmagouche, Mr. Dawson noticed a fossil plant covered with shells of a species of *spirorbis*, and few small scales of ganoid fishes.

The sections described by the author of the memoir are included in a district extending about fifty miles along the shores of the Gulf of St. Lawrence.

The greater part of the rocks composing the newer coal formation of Pictou, were formerly confounded under the name of new red sandstone. It is conjectured that in other parts of Nova Scotia, this formation will be found to be a well-marked carboniferous group. It is not valuable, however, as a depository of coal; but the existence of such a distinct formation, more than five thousand feet thick, in this country, is as interesting as it was unexpected.* The detection of animal tracks on the coal measures, is the first instance we have had of the probable existence of air-breathing land animals at any period earlier than the new red sandstone.†

Cupriferous Lignites.—In Mr. Logan's published section of the carboniferous strata of the Joggins, lowest sub-section, or No. 8, are noted several beds of mineralized vegetable remains, belonging to the true coal series, which are replaced by gray sulphuret of copper, covered by a thin pellicle of green carbonate of copper. Four seams of these cupriferous lignites occur in this section—within an area or depth of two hundred and six feet.‡ Their aggregate thickness is twenty-one feet. Mr. Dawson describes similar instances in what he denominates the newer coal formation, along the gulf of St. Lawrence, or west coast of Northumberland strait.

At the mouth of French river he observed gray sandstone and shales, containing a few endogenites, calamites, and pieces of lignite, impregnated with copper ore. Beneath these are other sandstones and shales, containing, in a few places, nodules of copper glance. These sandstones are often rippled, and contain branching fucoidal marks. On one of the rippled shales Mr. Dawson found foot prints of animals.

Traces of Animals.—In the ripple marked sandstones of Horton Bluff, Mr. Logan discovered footsteps, which appeared to Mr. Owen to belong to some unknown species of reptile; constituting the first inclinations of the reptilean class known in the carboniferous rocks.

* Proceedings of the Geological Society of London, Vol. I., p. 322, second series.
† Silliman's American Journal, July, 1846.
‡ Rapports sur une Exploration Géologique de la Province de Canada, p. 153.

ISLAND AND COUNTY OF CAPE BRETON.

Topographically the coal area of this island is distinctly separated into three fields or basins, which we shall arrange in the following order, according to the geological map of Dr. Gessner:

I. The Sydney coal-field. II. The Southern coal-field.
III. The Western coal-fields.

I. *The Sydney coal-field* occupies the north-east portion of the Island of Cape Breton. It extends along the coast, and is exhibited in the cliffs, from the north of Sydney harbour to Miray bay; and thence inland to the great entrance of the Bras d'or. This portion is generally estimated to contain 250 miles square of workable coal.

The coal of Cape Breton appears to have been known to the early French settlers prior to the discovery of that mineral in Nova Scotia and Newfoundland. The Abbé Raynal is among the first writers in describing the Cape Breton coal, which he says was worked in horizontal beds in the open cliffs. And he adds that one of these coal seams had been set on fire, and burned with great fury.*

The Sydney coal, as a domestic fuel, is by some asserted to be equal to the Newcastle, and is used by the steamers successfully; but the latter commands two dollars a chaldron more in New York and Boston. The reputation of this coal has so much increased its demand of late years, that the town of Sydney has grown into some importance.

The principal coal seam at Cape Breton is about six feet thick, but the roof not being good, the workmen are obliged to leave a part of the coal—at least this was the case a few years ago, and even then they worked out five feet five inches.

The coal is taken up by shafts, 250 feet deep, by steam power. In 1833 it was hauled one and a half to two miles to the landing, and thence conveyed by lighters to North Sydney, five or six miles up the bay. In quality, this coal may be classed among the soft, close burning, bituminous kinds.

Mr. Richard Brown states that "this coal formation is probably the most recent stratified group in the island; and it is certainly the most important, as it furnishes Newfoundland, Nova Scotia, Prince Edward's Island, and the United States, with an abundant supply of coal, equal in quality to the best of that found in the Newcastle district." The coal-field of Sydney, he continues, situated on the N. E. coast of the island, is the only one that has been sufficiently explored to determine its limits. It extends from Miray bay to Cape Dauphin, averaging about seven miles in width, and occupying an

* History of the Settlement and Commerce of the West Indies, by the Abbé Raynal.

area of 250 square miles. As the general dip of the strata is north-east, or seaward, this great area of coal measures is probably the segment only of an immense basin extending towards the coast of Newfoundland.

Mr. Brown estimates the perpendicular thickness of the exposed coal measures at more than 5400 feet. In this thickness are contained four seams of workable coal, ranging from four to seven feet each, and several small seams of less than two feet.

They contain similar vegetable remains to those of the English coal fields in great abundance; and occasionally trunks of trees, from 1 to 2 feet in diameter. There have also been discovered, recently, fishes' scales, with teeth, fins, bones, and coprolites, in bituminous shale and in a thin seam of impure cannel coal.*

A previous memoir from the same gentlemen contained a section of the general sequence of the coal measures and gypsiferous formations on the north-western end of the Sydney coal field.†

We have, besides, in the same publication, November, 1846, a description of a group of fossil trees in the Sydney coal field of Cape Breton, also by Mr. Brown. One of the most interesting sections of the coal measure, is that afforded by the cliffs on the north-west shore of Sydney harbour, which runs directly at right-angles to the strike of the strata; exposing almost every individual bed, from the old red sandstones, through the overlaying carboniferous limestone, millstone grit, and coal measures.

The author states that the total thickness of the coal measures, calculated from the highest bed of the millstone grit to their abrupt termination on the sea coast, is 1843 feet. These dimensions appear irreconcilable with those previously given, of the same district.‡

The Sydney coal mines are situated on the north-west entrance of Spanish river, or Sydney harbour, which is equal, if not superior, to any in British America, and is accessible in all winds. It is here that the most extensive operations of the company are carried on. The coal is well suited for domestic use, and for steam purposes, being highly bituminous, igniting readily, and leaving but little ash. A railroad which cost forty thousand pounds, conveys the coal, by means of several locomotives, from the pits to a point in the harbour where vessels of any burthen can load with ease, and are well sheltered from the prevailing winds. There are fourteen coal seams above three feet thick each. One of these is eleven feet and another nine feet.

Like the Pictou coal field, this is interrupted by intrusive masses of trap, but it contains a sufficiency of coal to supply the world for ages.

The Sydney mines are included in the lease by the crown, in 1827, to the general mining association.

* Quarterly Journal of the Geological Society of London, May, 1845, p. 210, Vol. I.
† Ibid., February 1st, 1845, p. 25.
‡ Ibid., Vol. II., p. 393.

		Tons.
Production in	1832 - - - -	39,651
	1837 - - - -	70,000
	1844 - - - -	50,000

About half of this went to the United States; the remainder for home consumption. The value of this mine, under active management, has always been held to be considerable. The association, it is said, were offered, previously to 1828, £7000 per annum, which is a smaller sum than they could now obtain, if put unfettered in the market. The coal was first opened above sixty years ago, and has continued from that period to be wrought.

Bridgeport Mines are fifteen miles from Sydney. The coal seam worked here is nine feet thick, of which, formerly but five feet nine inches were worked, the remainder being left for roofing. This coal which resembles that of Sydney, is conveyed by a railroad near two miles to the shipping place, from whence small schooners convey it to the larger vessels, which approach within a mile; or the schooners take it on, at once, to Sydney. The mines are situated on the south side of Indian bay, 1¾ mile from the harbour.

Cost of mining coal—Pictou.—In 1833, this coal was mined at the cost of 1*s*. 9*d*. the cubic yard; and filling 1*s*. 5*d*.; total, 3*s*. 2*d*. the miner finding powder, and the company finding tools.*

Sydney.—In 1833, price of mining, 1*s*. 9*d*. per chaldron of 50 bushels; with 10*s*. a week for rations. Transportation to the landing, 1½ mile, 1*s*. per load, or about 1 cent per bushel.

Table of the production of the Nova Scotia and Cape Breton Coal Mines, and the quantity sent to the United States, in chaldrons of one and a half tons each, but weighing in general, according to the custom of the trade, 3,750 *pounds.*† *Royalty paid in currency,* £3,333.

	1827.	1833.	1835.	1837.	1839.	1844.
Cape Breton, Sydney, ...	8,776	15,302	14,673	35,154	38,199	50,000
Cape Breton, Bridgeport, . .	1,325	9,805	8,265	13,121		
Nova Scotia, Pictou, or Albion, .	4,000	18,698	16,185	36,697	29,433	
	14,101	43,805	39,123	84,972	67,632	

1849.—The quantity of coal exported was 35,527 chaldrons; value £29,528.

Island of Boularderie.—This island, which lies to the north of the port of Sydney, is four miles wide and twenty-six miles long, or one hundred and four square miles, and is wholly composed of the carboniferous formation. It was examined in 1843 by Mr. R. Brown, who invariably found beds of gypsum in the lower part of the coal series, between the coal and the conglomerate. It does not appear

* Senate Journal of Pennsylvania, Vol. II., 1833, p. 570.

† Martin's Statistics of the British Colonial Possessions, p. 230.

that any of the coal seams which are exhibited in the natural sections on every side of this island, have ever been worked.

Geologically speaking this area is commonly included within the limits of the Sydney coal-field.

II. *Southern coal field of Cape Breton.*—This coal district occupies the area between the Grand Lake, the Gut of Canso, and the Atlantic Ocean, on the southern part of the Cape Breton. It appears to be about thirty miles in length and about six in breadth.

III. *Western coal field of Cape Breton.*—On Dr. Gesner's map this region occupies about fifty or sixty miles in length, on the south-west coast of the main island, from St. George's Bay to Salmon River.

Shipments of tons of Coal from Pictou to United States, to August 25th, 1853, *prepared by W. H. Shock, U. S. N.**

Years.	British Ships.	American Ships.	To Brazils in American Ships.
	Tons.	*Tons.*	*Tons.*
1846	28,000	30,303	
1847	49,135	41,343	
1848	43,225	33,333	240
1849	43,350	25,335	
1850	49,450	21,903	
1851	35,174	8,558	
1852	55,277	18,525	1,070
1853 to date.	35,000	9,000	
Total,	338,611	188,300	1,310

COAL TRADE BETWEEN BRITISH AMERICA AND THE UNITED STATES.

During the discussion of the United States tariff bill of 1846, much anxiety was felt and expressed in the United States, but especially in Pennsylvania, as to the effect which the remission of so large an amount of the duty then imposed on the introduction of foreign coals, might have upon her home trade.

It was shown, and may be confirmed by inspection of our own tables, that, while with the 1842 tariff duty of $1.75 per ton, the increase of bituminous coal from the colonies into Boston, its principal market, was, in 1845, sixty-five per cent. over the supply of 1844,—the increase of Pennsylvania anthracite in the same market, and at the same time, was only eighteen and a half per cent. It might, with good reason, therefore, be inferred that, on reducing the duty to about one-third of the sum heretofore paid, the consequence would be a diminished demand for anthracite, and the almost total exclusion of American bituminous coal from the eastern states.

This has not proved to be the result;—for, while the foreign coal

* Mining Magazine, March, 1854.

trade of Boston, for instance, has remained nearly stationary under a low tariff, the home trade in anthracite has trebled.

It seems to us that there is one view, in relation to a reciprocal trade in coal, which has heretofore been overlooked. Thus, Canada, although just now not a very important customer, is a purchaser of American coal to a certain extent. Thus, again, while the provinces of Nova Scotia and New Brunswick obtain a limited number of customers from one or two American ports in their vicinity, the coal proprietors of Pennsylvania, of Ohio, and ultimately of Michigan, will, in their turn, supply the adjacent provinces of Canada with the fuel of which they are in need. The Colonial government imposes no tariff on this importation, although the American duty is thirty per cent. on what is received in the United States; a tax equivalent to sixty-five cents per ton. As there exists no coal formation in all Canada, along a frontier of more than a thousand miles; as the wants of the people increase; as manufactories occasion new demands with an increasing population; as the recent requirements for smelting within the mining regions call for an adequate supply of mineral fuel, it does appear to us that the Canadian provinces are destined to became extensive recipients of American coal; and to an amount, ultimately, that will immeasurably exceed the amount of Nova Scotia coal which may reach the American Atlantic ports.

In consequence of the reduced duty on coal imported into the United States, an additional impulse was given, towards the close of 1846, to the trade in coal from the British colonies. Several barques of from three hundred to four hundred tons burden each, were, on the passing of the act of Congress of July, 1846, at once chartered in London for this trade. The deep waters of the north-eastern coast allow the largest class of vessels to take in and deliver cargoes of Nova Scotia and Sydney coal, and hence they could bring it at a lower rate than the small vessels which convey the Pennsylvania and Virginia coals; independently of avoiding the heavy charges on the American coal, by railroads and inland navigation.

For four years the admission of Nova Scotia coal had been increasing in the eastern ports, for the iron and other manufactures, for the supply of the Cunard steamers, and for various uses, in the face of a protective duty of $2.25 per chaldron. With a diminished duty, therefore, it is probable a considerable demand for this description of coal will take place in those ports.

1848. The expectation suggested in the last paragraph, has not been exactly realized. That there has been no larger demand for the provincial coal, we ascribe only to the simple fact, that no bituminous coal will hereafter be able to supplant the use of anthracite for general purposes, and especially for domestic use.

COAL TRADE OF BRITISH NORTH AMERICA.

The principal exportation of bituminous coal from Nova Scotia and Cape Breton, is to the ports of Boston and New York.

This coal is sold to the American merchants by the nominal chaldron of one ton and a half, weighing three thousand three hundred and sixty pounds, or forty-two bushels; but it is understood that the large measurement brings up the chaldron to forty-eight bushels; even measure, three thousand seven hundred and fifty pounds. By the tariff of 1842, the duty was levied on the chaldron of thirty-six bushels of eighty pounds, which is, generally, two thousand eight hundred and fifty-two pounds weight. The retailer at Boston sells a chaldron of two thousand five hundred pounds, and sometimes two thousand seven hundred pounds, the nominal price being influenced by the weight. The Nova Scotia ton is considered equivalent to thirty-six bushels, even measure.

The agitation of the tariff subject on foreign imported coal in the United States, brought forward a great many facts in relation to the trade with the British colonies, some of which details it may be useful to preserve; as the prevailing customs of the trade were but little understood out of the immediate market.

Cape Breton, or Sydney and Bridgeport coals, command a higher price, at all times, than the Nova Scotia or Pictou coals.

Price of Sydney coals in 1846, \$3.20 to \$3.30 per chaldron, delivered on board the vessel. If five hundred chaldrons be taken by one person, and paid for, forty cents on the chaldron, mine measure, is refunded to the purchaser. Sydney coals overrun the Boston measure about eighteen to twenty per cent. No other allowance is made.

Nova Scotia or Pictou coal, in 1846, cost from \$3.00 to \$3.20 per chaldron. If one thousand chaldrons be purchased and paid for by one person, thirty cents on the chaldron, mine measure, is deducted. The above prices are for ninety days credit. The Pictou coal overruns the measurement about twenty-five per cent.

A large portion of the Sydney and Pictou coals are carried to the United States in British vessels. The American vessels in this trade are generally chartered to proceed to Sydney and Pictou, and back to Boston or New York. In such case, they generally go in ballast. Some vessels are occasionally loaded with American produce, which goes to Newfoundland and St. Peters, and then these vessels come back to the mines and load with coal for Boston, &c. Corn and flour are sometimes carried to Sydney and Pictou for sale, in small quantities, say one hundred barrels of flour and five hundred bushels of corn; though in general the American vessels carry nothing.

The Cape Breton [Sydney and Bridgeport] coal mines are about two miles from the shipping places, by rail-road. There is no extra charge for putting coals on board. The Pictou coal [Nova Scotia] is conveyed from the mines by a rail-road of about six miles, to the shipping place.

American Tariff on foreign imported coals, by act of Congress, passed in August, 1842 = \$2.25 per chaldron of 36 bushels of 80 lbs. each, \$1.75 per ton of 28 bushels of 80lbs. each, or 2240 lbs.

British Tariff, commencing July 5, 1843, by the colonial legislature. Coals—*free*, both import and export.*

Rate of Toll, in 1846, for American mineral coal, passing the Welland canal, 2*s.* 6*d.* per ton, for the whole distance. Sea coal—free.

Freight from the mines to Boston. Boston measure;—average, in 1846, about $2.75 per chaldron; in 1847, $2.50 per chaldron. To Providence, $2.87½.

BOSTON COAL TRADE WITH SYDNEY AND PICTOU.

Imports of Cape Breton or Pictou, and Nova Scotia or Sydney and Bridgeport coal into Boston, U. S., in the following years. The purchases are made in Pictou by the large chaldron of 1½ tons, but the Custom-house returns are on the chaldron of 1¼ tons or 2,800 pounds, [36 bushels,] [48 bushels, even measure.]

Years.	Chaldrons of 1¼ tons, or 2,800 pounds.	Tons of 2,240 pounds.	Years.	Chaldrons of 1¼ tons.	Tons of 2,240 pounds.
1835		17,650	1843	20,184	25,230
1836		30,453	1844	20,334	25,417
1837	29,691	37,114	1845	33,628	42,035
1838	26,610	33,762	1849	34,531	
1839	37,986	47,482	1850	32,486	
1841		37,536	1851	30,330	
1842	21,899	27,374	1852	40,754	

The following statement exhibits the comparative cost of Pictou or Sydney coal, delivered at Boston, by the chaldron of 36 bushels, [2,812 pounds,] and the ton of 2,240 pounds, under the American tariff duties of 1842 and 1846, respectively.

Details.	Under the tariff of 1842, 6¼ c. per bushel.		Under the tariff of 1846, 30 c. per bushel.	
	Chaldron of 2,812 lbs. [36 bushels.]	Ton of 2,240 lbs. [28.7 bushels.]	Chaldron.	Ton.
	Dolls.	*Dolls.*	*Dolls.*	*Dolls.*
Mean price of coal per chaldron of 48 bus. weighing 3,750 pounds, [nominally 1½ tons,] from $3 to $3 20, say $3 10,	2 32	1 85	2 32	1 85
Freight to Boston, per chaldron of 36 bushels, $2 75,	2 75	2 19	2 75	2 19
Duty, ...	2 25	1 75	70	55
Cost, exclusive of any discount or allowances, . .	7 32	5 79	5 77	4 59
Retail prices in Boston, in 1846, from $8 to $9, per chaldron,				
Cost of a ton of Pennsylvania anthracite, in Boston, .		5 75	to	6 00
Current retail prices, in 1847–8, ...		6 75	to	7 00

* Pope's Journal of Trade, 1844, p. 481.

Cost of Pennsylvania anthracite, in Boston, as compared with Nova Scotia coal, in 1846.

Price on shipboard at Philadelphia,	$4.00 p. ton, to		$4.37½, red ash.
Freight to Boston,	$1.50 to 1.75	"	1.62½, "
Anthracite,	$5.75		$6.00

1853. Pennsylvania anthracite, freight charges from Philadelphia to Boston, $3 to $3.10. In former seasons the rates were $1.75 or $2.00.

The cargo prices of coals of all descriptions are generally about $1 per ton below the retail prices.

To these charges may be added, as the case may be, insurance 2 per cent., and commission 2½ per cent.

A preference will always be given in the eastern ports, where bituminous coal is required, in favour of the Nova Scotia coal, over that of Richmond, in Virginia, on account of the large amount of sulphur in the latter.

	Per ton. 2,240 pounds.		Colonial currency.
New York.—Average cost of Sydney coal, exclusive of duty in 1836, at Sydney and Pictou,	$3.60	=	18*s*. 0*d*.
In 1837, " "	2.90	=	14*s*. 6*d*.
In 1846, " "	3.02	=	15*s*. 0*d*.

			Prices at N. York.
In 1844—By the statement of a New York coal importer, it appears that the mean price of Sydney coal imported by him, in that city, from the spring of 1842 to the fall of 1844, was, per New York chaldron of 36 bushels,	$3.88 or	3.01	
Add the duty of 1842,	2.25	1.75	
Average cost in the New York market,	$6.13	4.76	
Wholesale price same time, from $6.00 to $7.00, say - - - - - - -			$6.50
Comparative price of Pennsylvania anthracite, during the same time, in New York, per ton.			
Cost at Pottsville, - -		2.00	
Freight and toll to N. Y. -		2.60	
		4.60	
Wholesale price, from $5.00 to $6.00, - -			$6.00
1846—Cost of Nova Scotia coal, with an ad valorem of 30 per cent., - - -		2.69	
Freight to New York, - - - -		2.00	
			$6.50
		4.69	

		Per chaldron.
Price of Pictou & Sydney coal, in N. Y.	1832,	$9.50 = 7.39
	1833,	8.50 = 6.70
By the cargo,	1842 to 1844,	8.75 = 6.90
$6.50 + 2.25 duty = 8.75	1846,	8.75 = 6.90
Current retail price of anthracite,	1847-8,	= 6.00
" " Pictou and Sydney,	1847-8,	= 6.50

Providence.—The following statement is the result of an actual purchase and sale of a cargo of Pictou coal received at this port, 8th August, 1846.

Cost on board at Pictou, $3.30, from which a discount of 30 cents is allowed, making the cost price of $3.00 per Pictou chaldron.

Quantity, 162 Pictou chaldrons, measuring at the Custom house, 7776 bushels = 48 bushels to the chaldron, or 216 chaldrons of 36 bushels each.

Weight of the same, 271 tons, 18 cwt. 2 qrs. 8 lbs. = 607,500 lbs.

Therefore the actual weight of the Pictou chaldron of 48 bushels, was 3750 lbs.

That of the Boston chaldron of 36 bushels, 2812 lbs.

The proportionate weight of 28 bushels is 2191 lbs., therefore there are 28.7 bushels to 2240 lbs.

Cost of the cargo of 162 Pictou chaldrons at $3.00,	$4.86
Freight to Providence, $2.87½, on 216 chaldrons of 36 bushels, - - - - - -	6.21
	$11.07

Amount of duty under the tariff of 1842, - -	$4.75
Amount of duty under the tariff of 1846, - -	1.46

	Per Chal. of 2,812 lbs.	Per Ton of 2,240 lbs.
Cost price of the coal at Providence, under the old tariff, - - - - - -	$7 23	$5 82
Cost price of the coal at Providence, under the new tariff, - - - - - -	5 80	4 60

Philadelphia.—Aggregate of nine years importation of foreign coals into Philadelphia, viz., from 1833 to 1841:—

	Bushels.	Tons of 28 bushels.
Nova Scotia coal, - - -	1,204,732	
English coal, - - - -	188,532	
	1,393,264 =	43,026
Annual average, - -	154,607 =	4,780

1841—Importation of Nova Scotia and Cape Breton coal into Philadelphia:

From Cape Breton, { Sydney,	-	38,808	
From Cape Breton, { Bridgeport,	-	129,401	
Nova Scotia, Pictou, - -	-	45,221	
		213,430 =	7,622
English bituminous coal, -	-	31,158 =	1,113
		244,588 =	8,735

1841—Price of Pictou coal on board, in the Schuylkill:

25 cents per bushel, - -	$7 00
Richmond coal, Virginia, 28½ cents,	8 04
Alleghany coal, 28 cents, - -	7 80

1846—Estimated cost at Philadelphia:	With the Tariff of 1842, per ton.	With the Tariff of 1846, per ton.
Price charged on board the vessel, at the place of export, $3 37 per chaldron of 48 bushels, or 1½ ton, -	$2 25	$2 25
Fair average freight from Pictou to Philadelphia, - - - - -	2 50	2 50
Duty, - - - - - -	1 75	67
	$6 50	$5 42

1847—Wholesale or cargo price of bituminous Alleghany coal at Philadelphia, from 18 cents to 20 cents per bushel = $5 00 to $5 60 per ton, - -	5 60
1848—Alleghany bituminous coal, Jan. 1st, 25 cents per bushel, - - - - - - -	7 00
Virginia or Richmond coal, Jan. 1st, 22 cents per bushel, - - - - - - -	6 16
1848—Wholesale cargo price current of anthracite, Jan. 1st, in Philadelphia, - - - -	4 00
1853—Price of Pictou coal per miner's chaldron, for the Gas Works, Philadelphia, exclusive of duty, -	3 50
Average freight and duty per ton of 2000 lbs., according to freight, make it - - $5 25 to	7 00
Sidney coal at mines, - - - - - -	3 60

By the late reciprocity treaty between the United States and the British provinces, the importation of coal from the British provinces to the United States will be allowed free of duty. This treaty has been signed by both the contracting parties, and a publication to that effect having been made on the 11th of September, it is now in full and complete operation. The treaty endures for 10 years. Sept. 16th, 1854.

Magdalene Islands.—These, according to the Report of Sir J. E. Alexander, Jan. 5th, 1846, are parts of a great coal formation.

Prince Edward's Island, formerly St. John's.—In respect of its geology, this island is, apparently, a continuation of the great Nova Scotia and New Brunswick coal-field, but no mines have been worked here.

According to an official report of Sir J. E. Alexander, dated Montreal, Jan. 5, 1846, Prince Edward's Island is, in fact, one continuous coal-field.

Gypsum, Grindstones, &c., exported from Nova Scotia.—The Nova Scotia and New Brunswick blue grindstone bed, which crops out from beneath the coal, is forty-four feet thick, and is probably similar to the millstone grit of England. Eighteen hundred tons of grindstones were annually exported from hence to the United States, some years ago, and probably more of late. The price in the United States is from fourteen to eighteen dollars a ton.

Number of grindstones exported in 1832, 19,240, valued at 30*s.* each = £28,860 = $139,682.

Of *Gypsum*, one hundred thousand tons were exported annually from hence to the United States, the value in 1830 being $119,234.

By returns to a circular, addressed by the Secretary of the Treasury of the United States, in 1845, it appears that there are at this time about 200,600 tons of foreign plaster annually imported into the United States.

This plaster is admitted free of duty, but the amount is exaggerated, for the returns for 1832 give quantity of gypsum exported from Nova Scotia and Cape Breton, as 46,136 tons, valued at 10*s.* per ton, or £23,270 5*s.* = $112,627. In 1844–5, the official value was reduced to $77,990.*

From an official statement of the commerce of Nova Scotia for 1852, the total value of exports of the Province was £970,780. Imports, £1,194,173.—Total, £2,164,953. Of the exports, £62,675 were to Great Britain; to the United States, £257,849. Of the imports, £429,532 were from Great Britain; £347,843 from the United States.

* Official United States Report, December 3d, 1845.

NEWFOUNDLAND.

THE entire western side of this great island, along a space of three hundred and fifty miles, and from forty to sixty miles in breadth, is occupied, according to Mr. Jukes and Sir R. H. Bonnycastle, by secondary and carboniferous rocks. This country has been very imperfectly explored, and the interior is almost entirely unknown. Of the extent or absolute area of coal in the carboniferous region, we are very imperfectly informed. That which is best known is the south-west part of the island; and it has been traced, at intervals, along a space of a hundred and fifty to two hundred miles to the north-east. Some of the points, where coal seams are intersected by the rivers, are known only through the reports of the Red Indians.

The southern part of the coal basin, best known to Mr. Jukes, he states to be about twenty-five miles wide, by ten in length.

Hitherto it does not appear that coal has formed any part of the exports of Newfoundland.*

In regard to the area which is occupied by coal formations we have no information. It is probably not less than five thousand square miles, nor more than 10,000.

Peat.—Large quantities of this fuel exist on the island. Vast tracts of peat-bog were noticed by Mr. Cormack, in 1823, who states also, that beneath its surface occur the trunks and roots of trees, much larger than any which are now growing on the island.

Emigration from Great Britain to British America and the United States.—Before closing our statistics of North America, it may not be wholly out of place to insert a statement of the annual number of persons who have emigrated hither during a few years past, from the several ports of the United Kingdom.

Years.	Destination.		Years.	Destination.	
	British America.	United States.		British America.†	United States.†
	Persons.	*Persons.*		*Persons.*	*Persons.*
1833	28,808	29,109	1846	35,617	
1834	40,060	33,074	1847	100,000	
1835	15,573	26,720	1849	41,367	219,450
1840	32,293	40,642	1850	32,061	212,706
1841	38,164	45,169	1851		239,255
1842	54,123	64,215	1852		300,992
1843	23,518	28,351	1853		284,945
1844	22,923	43,661			

* Martin's Statistics Brit. Emp., p. 269.
† Hunt's Magazine.

Emigration from Great Britain during the twenty years from 1825 to 1844 inclusive.

The British American colonies,	- -	551,386 persons.
To the United States, *via* the colonies,	-	569,633 "

1846—The number of emigrants who landed at Quebec and Montreal in 1846, was 32,753; of these there were Irish, 21,000. Emigrants arriving in West Canada, through the United States, 2,864. Emigrants and passengers arriving in New York from Europe in 1846, being upwards of 300 per day, 115,230. Total emigrants to the United States, arrived in 1846, 158,648. 1853, 284,945.

The number of emigrants arrived at the ports of Quebec and New York respectively for four years.

Years.	Quebec.	New York.
1849	38,494	220,603
1850	32,292	212,796
1851	41,076	239,601
1852	39,176	300,992
1853	36,176	209,765

1847—Of the 100,000 persons that emigrated to Canada in this year, full 25,000 died of the "Ship Fever," either on the voyage, or immediately after their arrival.—"Report of the Montreal Immigrant Committee for 1847."

The following is a statement of the arrivals, tonnage, and passengers at the port of Quebec from 1841 to 1851, inclusive:

Years.	Vessels.	Tons.	Passengers chiefly emigrants.
1841	1,246	446,642	28,086
1842	864	307,687	43,811
1844	1,214	458,981	19,698
1846	1,439	573,208	32,903
1847	1,444	542,505	
1848	1,350	494,247	
1849	1,328	502,613	
1850	1,328	502,613	
1851	1,469	573,397	

Custom receipts at Quebec and Montreal, on imports at sea, and of receipts at inland ports, 1849—£443,337, gross receipts.

	Tons.
Tonnage cleared at Quebec and Montreal in 1846,	592,577
Cleared for the Lower Provinces, - - -	6,558
Total, - - - - -	599,135

Population of Quebec, - - - 42,052
" Toronto, - - - 30,763

Kingston, Upper Canada.—The number of steamers and propellers belonging to this port in 1846, was 115.

HUDSON'S BAY TERRITORY.

ORIGINALLY STYLED RUPERT'S LAND.

Royal Charter and Powers of the Hudson's Bay Company—granted by Charles the II., 2d of May, 22d year of his reign. A. D. 1670.—The extraordinary magnitude of the powers, privileges, and resources of this company being but little known or understood, we have made an abstract of the Royal Charter for the purpose of exhibiting them.

The title of the company,—"The Governor and Company of Adventurers of England, trading into Hudson's Bay."

The grant comprises the sale, trade and commerce of all the seas, bays, straits, lakes, rivers, creeks and sounds, in whatsoever latitude they shall be, that lie within the entrance of Hudson's straits; and all the lands and territories upon the countries, coasts, and confines of the same seas, bays, &c., that were not already in possession of, or granted to, our other subjects, or the subjects of any other Christian Prince or State; together with rights of fishing therein; and the Royalty of the sea upon the coast within the limits aforesaid; and all mines royal, of gold, silver, gems and precious stones, within the said limits. "The said land shall henceforth be reckoned and reputed as one of our plantations or colonies in America, called *Rupert's Land*—Prince Rupert being the first Governor thereof; to be held as of our Royal Manor of Greenwich, in the county of Kent, in free and common socage; yielding and paying yearly to us, our heirs and successors, for the same, *Two Elks and two Black Beavers*, whensoever and as often as we, our heirs and successors, shall happen to enter into the said countries, territories and regions hereby granted."

It shall be lawful for the Governor and company to make and ordain such reasonable laws, constitutions, orders, and ordinances, as shall appear necessary, and at their pleasure to revoke and alter the same; and they may lawfully impose and ordain such pains, penalties and punishments, upon all offenders against such laws and ordinances, as the Governor and Company shall deem necessary or convenient; and the same fines and amerciaments shall and may be made to the use of the said company, without any account to us, our heirs or successors. They shall have not only the entire and only trade to and from the territories specified, but also the whole and entire trade to and from all havens, bays, creeks, rivers, lakes and seas, into which they shall find entrance by water or land, out of the territories aforesaid, and with all the natives and people inhabiting the same, and with all nations adjacent.

No part of the said territories, nor the islands, havens, ports, cities, towns or places thereof shall be frequented or haunted by any other of our subjects contrary to the true meaning of this grant; and all such persons are prohibited from visiting, trading or trafficking in the said territories, upon penalty of the forfeiture and loss of the goods and other things which shall be seized, as also the ships wherein such goods shall be found; and such offenders for their said contempt shall become bound unto the said Governor in the sum of one thousand pounds at the least, at no time thereafter to trade in any of the said places or territories.

And we further grant that all lands, islands, territories, plantations, forts, fortifications, factories or colonies, within the scope of this grant, shall be from henceforth under the power and command of the said Governor and company, saving the faith and allegiance due to the Crown; and they shall have power to judge all persons, in all causes, civil and criminal, according to the laws of England, and to execute justice accordingly. And free liberty and license is granted to the said Governor and company to send ships of war, men, and ammunition, unto any of their plantations, forts or factories, for the security and defence of the same, and to grant commissions to the commanders and officers, and to give them power and authority to make peace or war with any prince or people whatsoever, that are not Christians, in any place where the company shall have factories, forts or plantations, or adjacent thereto. And it shall be lawful for the company to build such castles, forts, fortifications, garrisons, colonies or plantations, towns or villages, in any places within the limits granted, and to send out from England all kinds of clothing, ammunition and implements, necessary for such purpose; and to transport over such number of men, being willing thereunto, as they shall think fit, and also to govern them in such legal and reasonable manner, as the company shall think best; and to inflict punishment for misdemeanors, fines or breach of orders.

They shall have power to seize upon all English which shall sail into Hudson's bay, or shall inhabit any of the countries hereby granted to the company, without their leave and license first obtained, or that shall contemn or disobey their orders, and shall send them prisoners to England, there to receive such condign punishment as the cause shall require.

The company shall have power to examine upon oath all factors, masters, pursers, supercargoes, commanders of castles, forts, &c., touching any matter not repugnant to the laws of the realm. And all admirals, vice-admirals, justices, mayors, sheriffs, constables, bailiffs, and all other officers, ministers, liegemen, and subjects whatsoever, are commanded to aid and assist the said Governor and company, as well on land as on sea, whenever they shall be required.*

This charter is still in operation.

The boundaries of this vast territory, as may be perceived, are not very satisfactorily defined by this Charter. This point was of very little consequence at that time; but it afterwards proved the cause of very serious and long-continued disputes, between the company and a rival association, called the "North-west Company," which was established in 1783. The union, formed in 1821, between that company and the Hudson's bay company, has greatly enlarged its territorial limits, so that it now claims a kind of proprietorship over the whole of British America, with the exception of the settled provinces or governments.

We have previously cited the Charters of the "General Mining Association," and that of "The New Brunswick and Nova Scotia Land Company."

Area of the British possessions in North America.

The Provinces of the Canadas, New Brunswick, Nova Scotia, &c., is 425,062 square miles.† Territories owned by Great Britain, including the Hudson's Bay Company's possessions, deducting bays, lakes, &c., 2,574,938. British Honduras in Central America, 62,740. Total, 3,062,740.

The entire area of the United States, including Texas, Oregon, and western territories, also California, in 1853, according to Mr. Kennedy, is 3,230,572 square miles. Of Russian America, 900,000.‡

* Colonial Statistics of the British Empire, Martin, Appendix III.
† McCulloch's Gazetteer.
‡ Commerce and Resources of British America, Hunt's Magazine, Vol. X., 1844.

ARCTIC OCEAN.

Greenland or Gröenland—partly colonized by Denmark, but formerly considered part of North America. A regular coal formation on the east coast of this peninsula was first discovered, we believe, by Captain Scoresby, the limits of whose survey extended from N. Lat. 69° to 72° 30′. Northward of this point, the exploration was continued by Captain Clavering to N. Lat. 76°.

The coal formation is described by Captain Scoresby as corresponding with that which prevails around Edinburgh, and with all the coal-fields of England and Scotland. The fossil vegetation appears to be analogous to that of the European coal measures. The examination of the Greenland coal beds was not carried on beyond Lat. 71°.

Captain Scoresby remarked the prevalence of masses of secondary trap intruding among the coal strata.

Hasen Island, Greenland.—Brown or Bovey coal, in which amber is interspersed, prevails here. In these lignite beds occur the mineral resin called retinasphalt. Beds of peat and turf are also encumbent upon granite.

West coast of Greenland.—Disco Island.—In both these situations, secondary and tertiary formations prevail, although primary rocks are by no means absent. Limestone, containing fossil fishes, and beds of shale and slate, with brown coal and amber, abound. The island of Disco is mainly composed of trap rocks, and the tertiary formations including the lignites referred to. Mineral charcoal is announced as occurring in the island.

Byam Martin's Island.—Secondary or true coal series.—A portion of this island consists of rocks of the primitive class: but there are also secondary formations, among which is a coal-field. Captain Parry has reported that the greater part of the superficial area of this newly discovered island consists of secondary red sandstone, in close proximity to which is the coal. A carboniferous sandstone, for so it appears to be considered, not only in this island but in Melville island, and in various positions which were subsequently discovered further to the south and south-west, belongs to the true coal formation, as in Europe, and other parts of the world. The brick red sandstone, which is described as occurring here, and also seen at many parts of the adjacent American continent by later explorers, and horizontally disposed along the cliffs of Melville and Byam Martin's islands, probably represent the old or the new red sandstone, or portions of both; but the relative positions of these formations do not appear to have been ascertained.

A fossil dicotyledonous tree was found on the shore of Byam Martin's Island.

Melville Island.—True coal formation.—This is the most westerly point hitherto attained by any exploring expedition from the Atlantic side. It lies in N. Lat. 74° 26′, and in W. Long. 113° 46′; a position where the summer lasts but a few weeks.

Here, an extensive coal formation prevails in secondary sandstone, overlying the carboniferous or mountain limestone. This sandstone contains remains of arborescent ferns, and casts and impressions of the usual coal vegetation. In the specimens collected here that were best preserved from the influence of the atmosphere, the coal possessed a slaty structure; colour, brownish black: after burning, leaves grayish white ashes, and emits no unpleasant smell under the process. Another species of coal was also brought from Melville Island. According to the Wernerian nomenclature, this would be denominated *transition glance coal* or anthracite, in contradistinction from the other variety, which held the name of the *first or oldest secondary coal formation.** This so called "transition coal," is associated with a sandstone or micaceous quartz rocks containing trilobites,[?] and traversed by whin dykes or trap veins. We much doubt the existence of coal formations of separate ages, as is here indicated, and the narrative of the expedition by no means countenances such a view of the case.

The secondary coal would, of course, be deprived of its bitumen and all volatile matters, in the vicinity of the intrusive masses of trap, as is commonly the case: it would, consequently, assume the appearance of glance coal or anthracite. This bituminous coal of Melville Island belongs, according to Mr. Lyell, to the true carboniferous series.

It seems, therefore, to be fully settled, that nearly the whole of this island is composed of horizontal coal sandstone and red sandstone, except at Table Hills, where the carboniferous limestone made its appearance. This latter rock, it has since been ascertained, is extended over a very large space in these northern regions, occupying nearly an equal area to that of the primary rocks. According to the excellent authority of Professor Lindley, the Melville island coal vegetation is decidedly that of the true coal formation, and consists of the usual sigillariæ, stigmariæ, calamites, ferns, &c.

Prince Regent's Inlet.—On the west side of this inlet is the country named by Captain Parry, North Somerset; and on the east side is that called by him Prince William's Land; both composed of a magnesian limestone, which is supposed to correspond with the mountain limestone of Europe, and with the metalliferous limestone of the United States. In association with this rock, are other formations which we are led to infer are all of a later origin; also fibrous brown iron ore and a species of brown coal. Above the limestone reposed thick beds of gypsum and a newer slaty limestone. The coal spoken of is probably not so modern as the tertiary.

To the southward of the Inlet, primary rocks occupy the largest areas apparently.

* Professor Jameson's Arctic Geology.

NORTH-WEST TERRITORY.

BROWN COAL FORMATION. SUPER-CRETACEOUS STRATA.

Shores of the Arctic Sea.—Dr. Richardson, who accompanied Captain Franklin's expedition of discovery, in the capacity of naturalist, describes much bituminous shale which formed precipitous banks. In many places these cliffs were observed to be on fire; attributable to the great admixture of sulphur in the shale.

Brown coal is more subject to spontaneous combustion than the true coals. The super-cretaceous coal beds, which extend many hundreds of miles in breadth along the upper Missouri valley, were observed by Lewis and Clarke, in 1804, and by subsequent travellers ever since, to be on fire in numerous places on the borders of the great rivers. The same phenomena prevail in Australia in coal of the like age.

Tertiary and other Coal formations east of the Rocky Mountains.—In a preceding part of this work, we traced the southern portion of this great area of tertiary coal through the upper Missouri valley, within the United States limits. Commencing at the boundary line of N. lat. 49, where this formation is full four hundred miles wide, we proceed to trace it in its progress northward.

It follows the general range of the Rocky Mountains in a zone which gradually contracts in breadth to the north. It is intersected by all the great streams which descend eastward from the Rocky Mountains, and the coal seams thus exposed are from one to eight feet in thickness. In numerous places these lignite beds have, from the period of their earliest discovery, been on fire, and in one locality it has continued on fire for more than forty years.

Near Edmonton, on the north branch of the Saskatchewan, Mr. Drummond found beds of a beautiful bituminous coal, which Dr. Buckland, from its peculiar fracture, considered to be tertiary. Captain Franklin saw beds of lignite and tertiary pitch-coal at Garry's island, off the mouth of the Mackenzie river. There occurs an extensive deposit of it near the Babbage river, on the coast of the Arctic sea, opposite to the termination of the Richardson chain of the Rocky Mountains. There were also seen beds of tertiary pitch-coal opposite Herschel island.

On the west side of Great Bear lake, Dr. Richardson discovered strata of brown coal, earthy coal, and bituminous shale and clay, overlying a vast region of magnesian limestone and dolomite, [Iowa and Wisconsin limestone.] He also describes the lignite formation on Mackenzie's river, as lying in horizontal strata, in four seams. It is bituminous, and, when recently detached, is pretty compact, but soon splits into rhomboidal pieces. It burns with a fetid smell, and was found by the blacksmith to be unfitted for welding iron when used alone, but it answered when mixed with charcoal, although the stench it created was a great annoyance. Different beds, and even different parts of the same seam, presented specimens of the fibrous brown coal, earth coal, conchoidal brown coal, and the trapezoidal brown coal of Jameson. These beds in some places were on fire in 1789, when visited by Mackenzie, and were still burning in 1827.

Beds of lignite were seen by Captain Franklin at the junction of the Great Bear river and the Mackenzie.

Not far from the base of the Rocky Mountains, and ranging parallel between them and the western boundary of the great limestone formation of the north, the scientific explorers traced, at numerous points, coal deposits which varied much in quality, from the brown wood-coal to an excellent pitch-coal, the fractured surface of which is marked with very peculiar concentric semicircular depressions. It is interesting to know that this coal, which would be excellent fuel for a steam vessel, occurs on the coast of the Polar sea, near the Mackenzie, in considerable quantity. It was also traced from the 49th to the 69th degree of north latitude.*

We believe that we have collected and examined all the published details which throw any light upon the coal formations of the extreme north, more especially those which establish the continuity of the immense deposits of the tertiary age. But it would seem probable that coal deposits older than these tertiary lignites, do also appear in these northern regions. Without adverting to the true coal-field of Melville Island, and the accompanying sandstones which extend from thence to the south-west as far in that direction as Great Bear lake, it appears, on the authority of Dr. Richardson and Captain Franklin, that a formation of the oolitic period exists in one part of the Mackenzie valley, near the junction of Great Bear lake with the Mackenzie; the sandstone strata contain ammonites. These ammonites were referred by Mr. Sowerby to a part of the oolite series, near the *Oxford clay*. With these fossils occur, likewise, carbonized impressions of ferns and coal plants, lepidodendrons, &c. The splitting or separation of the lignite into series of rhomboids as mentioned by Dr. Richardson, we have often observed in the semi-bituminized wood of the Oxford clay, in Europe.

It deserves inquiry, therefore, whether at this place we may not have the equivalent of the carboniferous strata, which form a conspicuous portion of the oolite series in Yorkshire, and at Brora, in Scotland.

Again, the fossils collected at the point called the Ramparts, on Mackenzie river, were all referred by Mr. Sowerby to the *cornbrash*, another member of the oolite group. At Great Bear lake, certain strata were observed which had a remarkable resemblance to the numerous thin beds of *lias* or alum shale of Whitby. On the border

* Captain Franklin's Narrative.

of the Arctic ocean, east of the Mackenzie towards Cape Bathurst, the cliffs offered a singular resemblance to those of the *alum shale* in Yorkshire, upon which the *inferior oolite* rests. Thus, there seems some probability that a part of the oolite series really exists in these latitudes, and that some of the coal seen may be as old as that of Yorkshire; thus forming an intermediate deposit, between the true coal of Melville Island, on the one side, and the tertiary coal range on the other.

In the vicinity of the Hudson Bay Company's Fort called Edmonton, in about north latitude 53°, and west longitude 112°, a seam of coal, of about ten feet in thickness, can be traced for a very considerable distance, along both sides of the river Saskatchewan. Sir George Simpson thus describes this coal. "It resembles slate in appearance; and though it requires a stronger draft than that of an ordinary chimney, yet it is found to answer tolerably well for the blacksmith's forge. Fossil remains are also found here in abundance; and at the fort there was a pure stone, which had once been a log of wood, of about six feet in length, and four or five in girth."*

Peel River, Rat River, and northern termination of the Rocky Mountains.—This important stream [Peel river,] falls into the Mackenzie from the south-west, in north latitude 67° 42′. The geology of its vicinity and of that of the Rat river, has been sketched by Mr. Isbister. There is little difference in these districts.

Peel river.—Below the alluvium are thick beds of aluminous shale, alternating with which are seen thin strata or brown coal; a formation which seems to be extensively distributed over all the country north of Slave lake. A loose red sandstone prevails in the district west of the Peel river, and is apparently the general underlying rock to these carboniferous deposits. To this red sandstone, succeeds below, a limestone formation, which is not particularised by the author, but is doubtless the same as exists throughout a vast extent of the northern part of America, and perhaps an extension of that in Illinois and Michigan.

The ranges of the Rocky Mountains opposite the newly established post of Fort McPherson, north latitude 67½°, are here dwindled down to a comparatively insignificant elevation: few of the peaks rise above six or seven hundred feet in height. Viewed from the west, they present soft undulating outlines, rising in a series of terraces. The inferior or western ridges consist, generally, of sandstone, while the higher are capped by limestone. As we trace these mountains towards the south, the transition and primitive rocks appear, and they increase in ruggedness and altitude. There are, at one part of the chain, ten of these parallel ranges, which occupy a breadth of from fifty to ninety miles.

North of Rat river, and opposite the mouth of Peel river, the continuity of the main range dies away, and exhibits only irregular ridges and solitary peaks, stretching towards the Arctic sea. It is

* Overland Journey round the World, by Sir George Simpson, 1847, p. 69.

observable here, that the succession of formations, rising from the secondary to primary, is from the west to the east; the eastern aspect being the most abrupt and precipitous.*

Extent of the Tertiary Lignite Formation.—Beds of brown coal have been observed to the east and west of the mouth of the Mackenzie, along the borders of the Arctic sea. Whether it be continuous with that observed on each side of Icy Cape and as far as Behring's Strait, we have no direct or conclusive evidence. But there seems now no doubt but there is a continuous belt of this formation from the Frozen sea, to near the sources of the Platte, the Arkansas, and the Canadian rivers in the United States territory; nearly as low down as north latitude 35, which, following the oblique direction of the range, from point to point, is little short of 2,500 miles. How far to the southward this tertiary formation extends is still doubtful: but there is a formation of coal, of some kind, as low as Sierra Verde in New Mexico, in about latitude 32°, and M. Humboldt states that lignite occurs in many parts of New Spain.

The breadth of this belt is but ill defined at the present day. For several hundred miles it probably averages one hundred miles wide, increasing towards the south; but subsequently diminished in that direction; and at the boundary between the British and American territories, is four hundred miles broad. A vast breadth of country between the Upper Missouri and the Platte rivers is overspread by this formation, which partially covers the cretaceous beds of Nicollet.

It is impossible to arrive at any certainty in relation to the superficial area, but we cannot estimate it at less than 250,000 square miles.

Until the final settlement of this matter be effected by more geological evidence than we at present possess, we fear we must leave the question undetermined. Unfortunately, the position is too remote, and the difficulties in the way of investigation are of such a formidable character, that it may be long before this interesting and important question in geology is satisfactorily decided.

Even far to the southward, on ground much more frequently trod, the geology is very partially and obscurely developed. Colonel Long states, that the sandstone which flanks the east side of the Rocky Mountains acquires considerable height and breadth, near the sources of the Missouri, the Platte, and the Arkansas; forming a belt from two to many miles in width, and containing fossils. Dr. James says, that this sandstone "contains organic remains similar to those in the sandstones of the coal formations." This rock is described as arising above the plain, abruptly, like a vast rampart; often highly inclined or vertical; while the strata of the plains [containing the tertiary coal] are horizontal.

Thus, at various and remote points, along the range east of the Rocky Mountains, we have references, more or less obscure, and frequent, of an older coal formation than the mere lignite range which stretches along the plains. Whatever doubt may attach to the pre-

* Journal of the Royal Geograph. Soc., Vol. XV., 1845.

sence of the former, there can be little or none as relates to the prodigious extent of the tertiary coal deposit.

Western Territory beyond the Rocky Mountains.—In the neighbourhood of Fort McLaughlin, in Millbank Sound, lat. 52° 10′ north, coal has been found "of excellent quality, running in extensive fields, and even in dumpy mounds; and most easily worked, all along that part of the country."*

Vancouver's Island.—According to the narrative of Captain Wilkes, United States Navy, coal of good quality is found here, and specimens were collected by the exploring expedition. He remarks, that the Hudson's Bay Company had made trial of it; but owing to its being taken from near the surface, its quality was not very highly thought of.†

The Port of Camosack, Vancouver's Island, with its excellent harbour, promises to become of great importance, especially on account of the coal-field of the north-eastern district. The coal is worked so near the surface, that the Cormorant, steam-sloop, was supplied by the natives with sixty-two tons within three days. It is not much inferior to the coal of South Wales, although it yields considerable ash.‡

Vancouver's Island is very nearly in the same latitude as the British Islands, and enjoys a climate very similar, but milder and more equable. It possesses a fertile soil, magnificent forests, and immense fields of good coal, reaching to the water's edge on various parts of the coast.

This coal is peculiarly valuable, and an American company is about to run a line of steamers from the Columbia river to Panama, and another company proposes to establish steam communication between the west coast of America and China.

From these and other inducements, an export trade in coal from Vancouver's Island appears to be under consideration, if the support and encouragement of the British government can be secured.

1853. Accounts from California state that a number of vessels were engaged in the coal trade between Vancouver's Island and San Francisco, distant 900 miles from San Francisco; the coal in this island being as abundant and much more easy of access than that of Oregon.

Vancouver's Island Coal.—The circular of Bonnard, Johnson & Co., dated San Francisco, Nov. 15th, contains the following: "Coal—Importations, 4,543 tons. During the last fortnight, two small cargoes have been received here, taken from a mine opened some time since in Vancouver's Island, the first importation from that quarter. The experiments that have been made with this coal have proved, we are informed very satisfactory; and the opinion is expressed that it will answer well for steamer's use. One cargo sold at $15, and the other at $17."

Queen Charlotte's Island.—Coal is also found here, according to Prof. J. D. Dana.

* Dunn's History of the Oregon Territory, 1844.
† Report to the Secretary of the Navy, by Lieutenant Wilkes.
‡ Athenæum, February 3d, 1849.

RUSSIAN AMERICA.

AREA 900,000 SQUARE MILES.*

NORTH of Behring's Strait, Cape Beaufort, is described by Captain Beechy as composed of carboniferous sandstone containing petrified wood and vegetable impressions, and traversed by narrow seams of coal, ranging in an east north-east direction. This coal deposit is doubtless continuous to the northward, as the same navigator traced it at Icy Cape; and lumps of coal were also dredged up, off the coast.

Beyond the Icy Cape, and Point Barrow, an abundance of coal was observed upon the beach. Still further north, at Point Franklin, the surface of the beach was covered with a fine sand; but by digging a few inches down, it was found to be mixed with coal.

The trade of the Russian American colonies appears to be, in great measure, absorbed by China, who gives her teas in exchange for the American peltries, besides other things, to the amount of more than a million of francs, annually forwarded to Moscow. Measures have been lately taken by the Russian American Company, to facilitate the communications between the coasts of Siberia and the Russian colonies of North America. The government proposes to make examinations in the Bay of Aiane, upon the shores of the sea of Okhostk, in the hope of finding a more safe port and of more easy access, than that of Okhostk, for centralizing her commercial operations.†

Respecting the area of the country claimed by Russia on the North American continent, we have seen no estimate.

The extent of the colony of Russian America was estimated by Hassel at 24,000 square miles. M. Kœppen, of the Academy of Sciences of Petersburg, calculates it at 17,500 square miles only. This appears to embrace only what was considered as belonging to the settled part of the territory; but if we take the boundary assigned to the entire Russian claim, that is, all above 54° 40′ of north latitude, and west of 140° west longitude, extending to the Artic ocean, the actual area belonging to Russia is about 900,000 square miles. Between lat. 54° 40′ and 60°, the Russian American Company possesses on the mainland only a strip, which nowhere exceeds thirty miles in depth. The rights of hunting and trading over the greater part of this last mentioned area have been lately leased to the Hudson's Bay Company.‡

* Compte rendu du Commerce Russe, en 1843.
† Sir George Simpson's Overland Journey, p. 124.
‡ Commerce and resources of B. America. Hunt's Mag., Vol. X., 1844.

SOUTH AMERICA,

COMPRISING

1. REPUBLIC OF NEW GRANADA.
2. REPUBLIC OF ECUADOR.
3. REPUBLIC OF VENEZUELA.
4. REPUBLIC OF PERU.
5. REPUBLIC OF CHILI.
6. PATAGONIA.
7. REPUBLIC OF LA PLATA.
8. EMPIRE OF BRAZIL.
9. BRITISH GUIANA.
10. FALKLAND ISLANDS.

SOUTH AMERICA.

CARBONIFEROUS FORMATION.

So far as we have any knowledge, the South American continent, even more than that of Africa, is singularly deficient in coal of the carboniferous age. It was long doubted whether, on either of these southern continents, any coal formation existed of an earlier date than the tertiary epoch. In Africa, however, it is now ascertained that true coal exists in more than one position.

In South America, if any exist there, it is probably within the empire of Brazil. Brown coal, of the tertiary age, has been traced through a vast space on both sides of the Andes, but especially next to the Pacific, at intervals, from Patagonia up to Panama. There is reason to conceive that this great chain of tertiary coal deposits, is of the same geological age as that which we have described as existing along a range of between two and three thousand miles of the North American continent. The interval between the tenth and thirty-fifth degrees of north latitude remains almost unexplored; and with the exception of two or three known points within that interval where tertiary coal appears, we remain without any data wherewith to fill up the vacant space in reference to coal.

In relation to South American geology, more especially on the Pacific border, we have perhaps received more information from M. D'Orbigny* than from any other naturalist. There are extensive exhibitions of the silurian, devonian, and even of the carboniferous rocks. Carboniferous limestone occurs at Lake Titaca, in Peru. The base of the Moro of Arica is stated to be of the same rock, and the same formation acquires an elevation of thirteen thousand feet to the east and west, of the great Bolivian system. In the Chiquitian system, it forms summits five thousand feet high. But in none of these, nor at a number of other points where similar formations occur, has this author ascertained the presence of regular coal beds of the ancient series.

Mr. Darwin, who devoted four years, from 1832 to 1835 inclusive, to the investigation of the natural history and geology of South America, judging from the position of the tertiary deposits which

* M. D'Orbigny on South American Geology. Jameson's Edinburgh Journal, 1843–4.

exist on both sides of the southern Andes, entertained the opinion that the primary chain must have had a great elevation anterior to the tertiary period. In Chili, the Cordilleras are divided into two chains. That on the west consists of stratified sedimentary rocks, resting upon granite. The eastern chain is composed of sandstones and conglomerates, which are more recent than the rocks of the western chain, being partly made up of their debris. Mr. Darwin conceives that these eastern formations are of the same age as the tertiary deposits of Patagonia, Chiloe, and Conception, and that like them they contain brown coal or lignite and fossil wood. He noticed at one escarpement of the Andes, a wood of petrified trees, in a vertical position. Some of these were perfectly silicified and were dicotyledonous; in others the wood was replaced by carbonate of lime. Close to this clump of silicified trees, a gold mine has been worked. The latter details are exemplified in a transverse geological section from Valparaiso to Mendoza, and Mr. Darwin expresses his conviction that the granite [now rising into central peaks, fourteen thousand feet in elevation,] must have been in a fluid state since the tertiary group was deposited.*

REPUBLIC OF NEW GRANADA.

Province of Veragua.—West of the province of Panama, coal beds have been discovered, but of their nature and extent little is known. We are assured that this mineral, which appears to be brown coal, is here in great abundance, and in ample quantity for the supply of a large extent of country around. The existence, also, of further deposits in the mountains of the two Provinces of Panama and Veragua, is spoken of with some confidence.†

Isthmus of Panama, Island of Muerta, &c.—Brown Coal Formation.—This region was explored in 1841, by Mr. Wheelwright, for the purpose of searching for coal, for the use of the steamers in the service of the Pacific Steam Navigation Company. After giving a cursory examination to the island of Boca Brava, in which there was observed abundant evidence of the existence of coal, a more specific and practical exploration was entered upon in the island of Muerto, [Death] and some outcrops of coal beds were discovered upon the beach *dipping due west*, at the foot of a small cliff twenty feet high. The place selected for mining operations offered the greatest facilities, as the steamers could approach within a hundred

* Martin's Statistics of the British Colonies, p. 144. Also Proceedings of the Geological Society, London, Vol. II., pp. 367, 212.

† On the union of the Atlantic and Pacific Oceans, at or near the Isthmus of Panama. J. A. Bryan, 1846.

yards of the shaft. The further presence of coal in other parts of the province of Panama was, at the same time, ascertained, while this investigation was proceeding.

Muerto is one of the extensive Archipelago of islands, which border this coast and is situated in north latitude 8° 20′, and in 82° 8′ west from Greenwich. It is without inhabitants and is covered with a dense forest. No other works are undertaken here beyond the ascertainment of the coal. Respecting that which was experimented on, in the steamer, by Captain Peacock, he reported that it burned freely, leaving a white residuum. He considered its practical value, as compared with English coal, in the proportion of thirteen to eighteen, and stated that it bore a strong resemblance to the Talcuhuano coal, in Chili, and probably might, when mined from a greater depth than that penetrated by this trial, be sufficiently available for steam purposes.

The town of St. David de Cherokee is distant fifteen miles from the opening of which we speak, and is about forty miles due south from the fine harbour of Boca del Toro, in the Atlantic. At the latter place coal of precisely similar character to that of the Muerto was known, prior to these explorations. At St. David de Cherokee, and at various intermediate points, this coal also prevails. Thus, from the best information attainable, Mr. Wheelwright was led to the conviction that a coal area of undetermined dimensions, stretches entirely across the Isthmus of Panama, in this parallel at least, and intersects it in the 82d degree of longitude.* It is generally admitted that this coal, like that of Talcahuano, is not older than the tertiary period but the parties immediately concerned in these investigations did not pretend to any geological skill. If Mr. Wheelwright's views are correct, that a desirable route might be found for a canal, or even for a good road, from Boca del Toro, on the Atlantic, to Cherokee on the Pacific, the distance being only forty miles, and the harbours at either end being excellent, it would present, among other singular features, the remarkable circumstance of passing from ocean to ocean through a continuous coal formation.

1853. We learn that various companies, English and American, are working the coal mines in New Grenada.

Island of Santa Clara.—In a correspondence of the governor of Guayaquil it is announced that coal of good quality can be obtained.

Santa Fé de Bogota.—Coal occurs abundantly on the south side of the city, and even within the limits of the city itself. This fuel is reputed to burn extremely well, and to give out a great heat. We have received this information from a resident of Bogota, familar with the use of this combustible. A specimen of this coal has been presented to the cabinet of the Geological Society of London. From the character of the fossils which accompany the formation in which the coal is imbedded, it evidently belongs to the cretaceous period, and probably is of the age of the Gault of England. These fossils

* Wheelwright's Report on the Coal Mines on the Isthmus of Panama.

have been figured and described by Professor Forbes;* and, at an earlier period, similar fossils from the same locality, by Von Buch. They appear to have a strong agreement with the cretaceous fossils first brought by Lewis and Clarke; subsequently by Mr. Nuttall,† and yet more recently by Mr. Nicollet, from the cretaceous beds of the Upper Missouri Valley.‡ It seems, therefore, not improbable that the formation of Bogota, containing wood coal, is about the same geological age as the formation, containing cretaceous fossils, with thin seams of coal, and petroleum in Upper Missouri.§

Province of Choco.—Near the shores of the Pacific, fossil wood abounds, mixed with rolled fragments of basalt and greenstone. This deposit is celebrated for containing gold and platina.‖

The bitumen of Murindo, near Choco, is of a brownish black colour: soft; and has an earthy fracture. It has an acrid taste; burns freely with a smell of vanilla, and is said to contain a large quantity of benzoic acid. This arises, apparently, from the decomposition of trees which contained benzoin.¶

In this province, coal, so called, is found at an elevation of seven thousand, six hundred and eighty feet, which is about the same level as the coal of New Mexico, of Upper California, and of eastern Oregon, in the northern continent.

REPUBLIC OF ECUADOR.

In the environs of the city of Guayaquil, a considerable deposit is said to occur of a new species of resinous mineral, which Mr. Johnson, to whom the specimens were submitted, gave the name of Guayaquillite. Two varieties have been examined, and they have been declared to be of organic origin.**

* Quarterly Journal Geol. Soc. of London, May 1st, 1844, p. 174.

† Recognised and described by Dr. S. G. Morton, in 1829, and in Silliman's Journal, 1830.

‡ Described by Dr. Morton in Proceedings Acad. Nat. Science, October, 1841.

§ Harris, in Proceedings of the Academy, May, 1845.

‖ Humboldt's Personal Narrative.

¶ Dr. Ure's Dictionary of Arts, Mining, &c.

** Philosophical Magazine, November, 1837.

REPUBLIC OF VENEZUELA.

Island of Margarita.—A vast abundance of mineral pitch flows out at various points.

Gulf of Cariaco.—At the Punta d'Araya, at Cape Cirial, and near Cape de la Brea, M. de Humboldt observed a stream of Naphtha, issuing from mica slate, containing garnets and cyanite. A continuation of the same phenomena is repeated in all the large West India Islands, from Trinidad to Cuba, where the bitumen appears chiefly to exude from magnesian and modified rocks. M. de Humboldt considered it a singular circumstance that this spring, the produce of which covers the sea to a great extent, should issue from mica slate: as all others, he observes, belong to the secondary class.*

Porto Cabello.—About fourteen miles south of this place, and seventy miles from Caracas, a body of what was termed excellent coal was discovered a few years ago. Whether this be a deposit of brown coal, or a bed of solid bitumen or asphaltum, like the chapapote of Cuba, we have no information. We should rather infer the latter, reasoning from the vast amount of bituminous matter that prevails along the northern border of the South American continent, and the great abundance of the same substance in nearly all the West India Islands.

Maracaybo.—Compact mineral pitch, like that of Cuba, and copious streams of petroleum, occur opposite the city and on the borders of the lake. The petroleum is employed here, as at Havana, for paying the sides and bottoms of vessels. Towards the north-east margin of this lake, which is two hundred and fifty miles in circumference, is a remarkable mine of asphaltum, [pix montana,] "the bituminous vapors of which are so inflammable, that, during the night, phosphoric fires are continually seen; which in their effect, resemble lightning. They are more frequent during times of great heat than in cool weather, and go by the local name of 'the Lantern of Maracaybo,' because they serve both for lighthouse and compass to the Spaniards and Indians, who, without the assistance of either, navigate the lake."†

Magdalena River.—According to M. Bousingault, bitumen prevails along the margin of this valley.

This naturalist, who has published a dissertation on bitumen, shows that the immense reservoirs of mineral pitch, which exist on the northern shores of New Granada, on the banks of the river Magdalena, at Payti, in Colombia, and upon the shores of Peru and Venezuela, have a geological position precisely similar to that in which we find bituminous impregnated sands in Europe: that is to say, in formations which we must refer to the super-cretaceous group.‡

* Travels and Researches of Alexander Von Humboldt, 1799.
† McCulloch—Geographical Gazetteer.
‡ Philosophical Magazine, 1837.

REPUBLIC OF PERU.

Cerro Pasco.—The Director general of the mines belonging to the Republic of Peru, has drawn up a memoir on the coal-fields of his district.

Near to Cerro, he informs us, from four to ten leagues, there are numerous beds of fossil charcoal, of which the chief deposits, near Raucas, are of very good quality, and are several leagues in extent. He has found in these coal-fields a considerable quantity of yellow amber, but could not discover any impression of the remains of plants or animals. This coal is used in heating steam engines, &c.* We infer from this concise yet decisive statement that the deposit is of tertiary age.

In the immediate vicinity of the celebrated silver mines of Cerro Pasco, at an elevation of 14,278 feet above the sea, coal, "of all descriptions," is found in abundance. This convenient supply of fuel is of particular importance to the extensive population of the city. There are here, also, numerous beds of fossil charcoal, of a quality that may be used for heating steam engines, and for the like purposes.†

Of the geological circumstances which attend the position and characterize the age of the coal alluded to, we are uninformed. It does not appear to have been examined by the scientific gentlemen who were attached to the United States Exploring Expedition. From the private narrative of Dr. Pickering, with the perusal of which he has kindly favoured us, we perceive that the plateau of the Cordilleras, including the silver mines and the highest peaks, consist of sedimentary rocks. He describes these as consisting of the series which ascend from the lias up to the cretaceous period, inclusive; and he properly suggests whether the coal be not a lignite, rather than a true carboniferous formation? In this respect he is no doubt correct. We have ourselves seen casts of what appeared to be tertiary fossils, from hence, and indeed from the height of 14,000 feet above the ocean.

There are a few specimens among the collection of the U. S. Exploring Expedition, in the Washington collection, of slaty coal slate, and some thin flakes of impure coal, from Peru; but the locality is not stated.

On the western slope of the Andes, opposite to Truxillo, Lieut. Maw observed what he considered a seam of coal. This might perhaps be a continuation of the Cerro beds.‡

* Annales des Sciences Nat., 1829.

† Smyth and Lowe's Narrative of their Journey from Lima to Paris, 1834.

‡ Lieut. Maw. Descent of the Amazon river.

Coal is said to be prevalent in various parts of the country, at the distance of from two to seven leagues around Pasco. The price is one real for an araba, which might be much reduced if the business were properly attended to.*

Asphaltum, of Coxitambo in Peru. This substance, which may be considered the type of the species, has been carefully submitted to analysis by M. Bousingault. It has a fracture which is eminently conchoidal, and possesses a high degree of lustre. Specific gravity, 1.080. Carbon, 75.0. Hydrogen, 9.5—Oxygen, 15.5, per cent. The residuum, after burning before the blow pipe, was found to be 0.16 only.

REPUBLIC OF CHILI, OR CHILÉ.

Regulations, made in 1842, *as to money, weights, and measures in Chili, and their corresponding values and denominations in French and English standards.*

	Denominations.		
	Chilian.	French.	English.
Money, ...	1 Piastre = 8 rials = 100 cts.	5 fr. 40 cents,	
	Silver rial, 12½ cents,	0 67.5	
	Cent,	0 05.4	Metre.
Linear measure, .	National vare,	0 metre, 836	Yard, 0 914; Foot, 0 304; Inch, 0 025
Superficial measure,	Square vare,	0 metre, sq. 6,987	Sq. inch, 0 0006; Sq. foot, 0 0929
Liquid measure, .	Arrobe = 9 gallons,	34 lit. 0 65	Lit.
	Bottle ordinary,	0 841	1 gallon, 3,785
	1 Gallon, 1-9th of an arrobe,		1 pint, 0 473
Measures of weight,	1 Quintal = 4 arrobes.	46 Kilogrammes,	100 pounds.
	1 Arrobe,	11 Kilogrammes,	25 pounds.
	1 Pound = 2 marcs,	0 K. 460	16 ounces.
	1 Ton = 2,000 pounds,	920 Kilogrammes,	2,000 pounds.

Tariff of 1844, *fixing the official values of imported articles in Chili. Stone coal is chargeable as follows.*†

	Values.	According to value.	Specific duties.
	Piastres. Cents.		*Piastres.*
Chilian Unities, per the Quintal,	0 40	20 per °/o	0 38 cents.
		Duties ad valorem.	Specific duties.
French Unities, per 100 Kilogrammes,	4 fr. 70 cents.	20 per cent.	0 fr. 94 cents.
English Unities, per ton,	£2 0 0 =	20 per cent.	£0 8 0

* Macgregor's Progress of America, Vol. I., p. 951.

† Documens sur le commerce exterieur, Dec. 1844, Paris. Chili is supposed to be the

An exploration of the coal beds that exist so abundantly in Chili, and the other republics of South America, has been made by some of the most scientific engineers, miners, and geologists of that continent. The account so far given by them, is most satisfactory. It appears certain that these countries will be enabled to supply themselves with fuel of a superior quality to the wood fuel. Arrangements are making for working some of these mines, and for constructing a railroad from the great commercial port of Valparaiso to the capital of Santiago, a distance of about 135 miles. We await the reports of these scientific investigators.

Tertiary or Wood Coal Formation of the Chilian Coast.—Talcahuano, Arauca, Chiloe, &c.—In making arrangements for the introduction of the South Pacific steam navigation, it was a primary—indeed, the most essential—point, to ascertain the existence on the west coast of South America, of a combustible suited for the purposes of steam. That a certain description of coal prevailed in Talcahuano, 36½° south latitude had been ascertained several years previously, although its properties and amount remained uninvestigated.

In 1834, Mr. Wheelwright, subsequently superintendent of the company's affairs in that quarter, made a voyage to the port of Talcahuano and obtained samples of the coal, which, on experiment, seemed adequate to the object required.

In January, 1841, the coast between Valparaiso and Talcahuano was hastily examined, and satisfactory evidence was at once obtained as to the presence of a vast continuous strata of this coal. Previously to this, Mr. Wheelwright had been furnished with samples of a description of anthracite, from the Cordillera of the Andes; probably from the metamorphic secondary strata there; but it was in too remote a position to be made available. He also received, as coal, a mineral pitch or asphaltum, from the province of Piura.

On landing at Talcahuano, Captain Peacock and Mr. W. proceeded to the Moro, a range of hills in the vicinity of the town, and found seams of coal, visible in the broken cliffs. Heretofore this fuel had been simply taken from the surface, and no subterraneous mining had been attempted. This work was now commenced. On examining the eastern and northern sides of the bay, extensive coal strata appeared; not differing, it was judged, from that which had been experimented upon. The result of these researches appeared to demonstrate the prevalence of continuous coal beds along that entire section of Chilian coast. About forty labourers were set to work, and forty tons of coal were sent to Valparaiso on trial.

In order to ascertain the most favourable positions for a mining establishment, it was determined to explore the coast of Arauca, and to proceed as far south as the island of Chiloe, which extends to south latitude 43° 58′.

Coal mines had for some time been opened near Conception, and

only American State, formerly subject to Spain, whose commerce has increased, since the separation from the mother country.

had already become a considerable article of trade and consumption at Valparaiso.

Captain Fitzroy, R. N., found it in great abundance at the mouth of the Laraquita, where it was also subsequently examined by Mr. W. Its appearance was similar to that of Talcahuano; the formation being decidedly the same. Seams of similar coal were traceable, even from the vessel, in the cliffs of the coast along which the steamer passed, on the voyage; and no doubt remained of the continuity of these deposits, to a very great extent. Passing Arauca, the party proceeded on to Valdivia, and up the river, about eighteen miles, to the town. Here were obtained samples of the same kind of coal; but the place was considered too distant to suit the required purposes. On arriving at Chiloe, researches were commenced. No seam of coal was, at first, observed in situ; although large pieces were picked up; indicating its existence in the neighbourhood of San Carlos.

An experiment made on this fuel, during this exploring excursion of the steamer Peru, showed a comparative consumption of thirteen tons of English coal to sixteen tons of the South American; a result which was considered fully satisfactory. It was further determined, that the influence of the latter upon the fire bars and boilers was favourable: that it made no clinkers, and that the residuum lay lightly upon the bars, without adhering in the slightest degree. On her second voyage, the Peru steamed fifteen hundred miles with this fuel; which fact seems calculated to set at rest all doubts and fears, as to its practical purposes. The seam from whence this supply was derived, has a floor, composed as usual, of shale or indurated clay—fire clay—and a roof of carboniferous sandstone. About five thousand tons were mined, at an expense of about fifteen shillings = $3.65 per ton: a cost which, as may readily be supposed, would be materially diminished during subsequent operations, and by later improvements and experience in the manner of working.

Already this Talcahuano coal has been worked to the depth of more than a hundred feet, and, at the last report, a shaft was being sunk to reach a lower seam which was thought to be of a more firm and compact quality. Machinery, shops, railroad, mole, and breakwater have been constructed, and the ships of the company were employed in transporting the coal. Although these explorers made no pretension to geological knowledge, they express a passing opinion of "the evidently modern formation" of this coal.*

In 1825, Captain Beechy, R. N., made some trial of this fuel, or rather of that which was supplied to Conception; and, as we were prepared to learn, pronounced it to be of inferior quality, and fit only for the forge. He states that the beds occur in a *red sandstone* formation, and that the coal, at that time, sold for nine dollars a ton.† The correspondent of a Boston paper, evidently of very slender scientific attainments, describes the Talcahuano coal as much resembling the English cannel. Recently its cost, including the putting it on

* Report on the Mines and Coal of Chilé, by W. Wheelwright, 1843.
† Voyage to the Pacific, 1825.

board the steamers, has been only $2 50 per ton; a great saving over the price of English bituminous coal, which used to be brought out to these ports at $10.00 per ton. At Penco, near Valparaiso, an inexhaustible supply of similar coal is now attainable.*

In a communication in Silliman's Journal from Mr. Wheelwright, prior to the report to which we have adverted above, he merely adds that the coal of which he had mined several thousand tons, was of excellent quality—a phrase of universal application—and that, "in fact the whole southern country is nothing but a mine of coal."†

Volume I. of the Proceedings of the Academy of Natural Sciences of Philadelphia, contains a description of a specimen of the Arauca coal, by W. R. Johnson. He observes that "in external appearance it is nearly related to many of the richest bituminous coals of America and Europe." His analysis appears to confirm this view; for we know of no lignite which contains such an amount of carbon as this; being no less than 67.62 per cent. The greater part of the mass is represented as "of a dull or pitchy black colour. Its locality is said to be in the province of Arauca, thirty miles south of the Rio Bio river."‡

From a recent Journal we see the following statement of a new and extensive mine of bituminous coal having been discovered at Lota a commodious harbour on the south coast of Chili, about 30 miles south of the river Rio Bio, which enters the Pacific ocean in latitude 36° 55′ south, The coal is said to sell at $6 a ton.

In some statements made in 1845 by practical operators at Valparaiso, we observe that they complain that the coal of this country is not adapted for copper smelting, "in as much as it contains too much sulphur and iron;"§ and coal for that purpose has been brought out from England, at enormous expense. The tertiary deposits of Chiloe and Conception were examined by Mr. Darwin, and are described as composed of beds of sandstone and carbonaceous shale without shells, but containing many silicified trunks of dicotyledonous trees, and alternating with beds of lava.‖ In 1844 there were upwards of twenty coal mines open in the neighbourhood of Conception.¶ In 1845 a railroad was projected from Valparaiso to Santiago. The plan is recommended on account of the scarcity and extreme dearness of carbonic fuel, arising from the insufficient inducements to work the extensive coal beds of the interior.**

The reports, in 1846, of the progress of Tulcuhuano coal mining, are not equal to the anticipated results; but we are at the same time informed that the South Pacific Mining Company, having exhausted their first mines, have struck another richer seam, which promises to

* Boston paper, 1841.
† Silliman's Journal, July, 1842.
‡ Proceedings Acad. Nat. Science, Philadelphia, May 18th, 1841.
§ Memorial of Copper Smelters, 1845.
‖ Proceedings Geol. Soc. London, Vol. II., p. 211, 1835.
¶ Niles's Register, 1845.
** Mining Journal, November, 1845.

produce more coal than the steamers can require. It is destined perhaps to similar results.

Importation of English Coals into Chili.—In the year 1845, 15,149, in 1846, 8,864 tons.

The indiginous coal in the vicinity of Conception, still continues an object of research. From information which has reached us from Valparaiso, towards the close of 1847, it appears that new mines are occasionally opened in that country. At Talcahuano, a new seam of four and a half feet was proved. In the tide-way of Penco they are working a bed, at some two hundred yards from the beach, and have cut in vertical depth, ten feet without passing through the coal. A third seam has been opened at Perales, on the road to Conception. Altogether, they speak of five new mines, and commend the quality, of course in the usual manner.

PATAGONIA.

A GREAT southern tertiary formation has been described by Mr. C. Darwin, forming extensive groups on both sides of the chain of the Andes. These appear to be the prolongation of the series which is so largely displayed in Chili. Mr. Darwin thinks that the tertiary deposits of Patagonia may be separated into distinct periods, as they have already been done in Europe, and subsequently in North America. In S. Lat. 50°, and elsewhere, he found fossil shells of this period, with bones of the mastodon, and megatherium, and five or six other quadrupeds. Little is said by the author respecting beds of lignite, which are so abundant in higher latitudes.* We are not informed whether the coal range on the eastern flank of the Andes corresponds in geological age with that on the western side.

Tertiary Lignite range of South America.—From the evidence, incomplete as it is, which has been adduced in the foregoing pages, it will be seen that a vast belt of tertiary deposits, which contain brown coal and lignite, occupies the larger portion of the countries bordering upon the Pacific Ocean, from N. Lat. 10° to at least as low down as S. Lat. 50°. The intervals to which our information does not extend, or remain as matters of inference, are the south-western portion of Colombia, the southern part of Peru, and the northern part of Chili; but if the tertiary strata which are described as flanking the Cordilleras are co-extensive with those regions, as is generally supposed, the whole length of the tertiary range is scarcely short of that of the entire continent. At any rate we think we do

* Proceedings Geol. Soc. of London, Vol. II., 211.

not exceed probability in suggesting two thousand five hundred miles as the aggregate length of the tertiary formation, in the greater part of which, we are informed, lignites abound. Looking to the northern continent, where a similar zone has been traced for nearly the same distance, and following the same range, we cannot but be struck with the contemplation of this extraordinary development of a single member of the geological series.

REPUBLIC OF LA PLATA, OR ARGENTINE REPUBLIC.

UNTIL of late years the existence of mineral coal in La Plata had had not been suspected, nor can we, even now, speak with certainty as to that fact. Along the Cordillera, bituminous shale and indications of coal are affirmed to be abundant; and it is also said that there are extensive beds of coal in the extreme south-west angle of the country.* These are probably not older than the tertiary period, and form part of the great zone of that formation which we have already indicated. Mr. Darwin made some geological examination of this part of the Andes, and along the Rio Negro, between the years 1832 and 1835. He also crossed from the Rio Negro to Buenos Ayres, by Sierra de la Ventana, a chain almost unknown to travellers. The tertiary formation occupies a wide area in the south-western part of the country.†

* McCulloch, Geographical Dictionary.
† Proc. Geol. Soc. of London, Vol. II., pp. 211, 367.

EMPIRE OF BRAZIL.

It is still doubtful as to the presence of coal here. Specimens of coal were exhibited in 1845, which were the production of the Isle of Santa Catherina, S. Lat. 27°, and of the continental part of the province of that name. The examination to which these coals have been submitted, appears to leave for Brazil, at least for the present, small hope of drawing from her own soil the essential combustibles for her steam navigation and industrial purposes.* Much, however, remains to be effected ere we can arrive at a knowledge of the vast regions in the interior of this country. This may yet be accomplished in the course of a few more years. An English company has been formed to establish steam navigation up the Amazon river and its tributaries, to form settlements and to commence mining operations. The Amazon river alone can be navigated over two thousand miles; and it is proposed to effect a junction between this river navigation and a railroad to Arica, in Peru. This object is patronized by the governments of Brazil, Bolivia, Escuador, and Peru.†

M. Karsten has furnished two analyses of coals, said to be the production of Brazil. We give the results, and think that they, or one of them, may have been derived from the extreme western limits of the empire, and may probably belong to the brown coal series.

Specimen 1.		Specimen 2.	
Carbon,	57.90	Carbon,	38.10
Volatile matter,	40.50	Volatile matter,	33.50
Ashes,	1.60	Ashes,	28.40
	100.00		100.00
Specific gravity,	1.289	Specific gravity,	1.483

A small quantity of coal is annually imported into this country from England, as shown in the following table: the increase of late years is considerable.

Years.	Tons.	Declared value.	Current prices per ton at Pernambuco.			
			Years.	French.	English.	Par value of £1 sterling in London, 7 milreis, 117 reis of Rio Janeiro.
				fr. *cts.*	£ *s.* *d.*	
1831	840					
1833	1,863					
1840		£9,718				
1842		17,552				
1844	20,601	9,507	1843	27 03	1 1 0	
1845	30,038	17,732	1845	21 62	0 17 6	

* Documens sur le Commerce exterieur, 1844–5.

† Mining Journal of London, August 2d, 1845.

Value of coal imported from Europe into Brazil, and entered for consumption in the financial year 1842–3, paying an import duty of five per cent.

From Great Britain, 708,722 rials: France, 5,037: Portugal, 804: Hanseatic towns, 134,653: United States, 6,881. Total, 856,097 rials.

Lignite.—In the neighbourhood of Crato, a town about three hundred miles due west from Pernambuco, within the limits of the cretaceous formation, a bed of lignite about two feet thick has been described by Mr. Gardner. An enormous area of rocks of the chalk period, according to this traveller, exists in this country. "Between the cretaceous series and the primary stratified rocks, there are no traces either of the carboniferous or of the oolite formations; nor in any part of Brazil through which I afterwards travelled, did I meet with any signs of them." In a note to the foregoing paragraph, the author observes, "Dr. Parigot appears to have found coal abundantly in the island of Santa Catherina, in the south part of Brazil."*

This latter gentleman was employed by the government to make geological surveys in the province of Santa Catherina, with especial reference to coal. In a report which he published in 1841, he mentions a bed of coal of about three feet in thickness, and of considerable superficial extent. Dr. Parigot, also, has reported upon the existence of a carboniferous stratum, which is from twenty to thirty miles in breadth, and about three hundred miles in length; running from south to north through the province. The best vein of coal which he opened he designated as "*half bituminous:*" it occurs between thick strata of the hydrous oxide of iron and bituminous schist.†

The coal which Spix and Martins informs us exists near Bahia, Dr. Parigot found to consist of beds of lignite, and Mr. Gardner thinks they may be equivalent to those which he found at Crato.

* Gardner's Travels in the Interior of Brazil, London, 1846, p. 208.

† Macgregor's Progress of America, 1847; Vol. I., p. 1455.

BRITISH GUIANA.

Post-tertiary lignites.—An alluvial belt, thirty or forty miles wide, borders the coast and occupies the deltas of the principal rivers. In a thick mass of variously coloured clays beneath the diluvium, are two deposits of fallen trees, decayed wood, and other vegetable matter, in a semi-carbonized state. The first is at twelve feet below the surface, the second is fifty feet below the surface, and is twelve feet thick. This clay has been penetrated to the depth of one hundred and forty-five feet. The trees are recognized to be of the same species as are now growing in the vicinity, and called couridas, and indicate two or three distinct epochs and levels of surface on which they have grown.*

Coal has not, we believe, been discovered in this vast, but little explored country; yet it seems not altogether improbable that the coal formation may yet be found in the interior. Sir R. Schomburg states that Maravacca, near the Orinoco, rises to eleven thousand feet; and Roraima, the culminating point of the Pacaraima mountains, is eight thousand feet above the sea. They are composed of the older red sandstone, and exhibit mural cliffs one thousand and sixteen hundred feet high.†

FALKLAND ISLANDS.

S. Lat. 51° to 51° 30′. Peat. These islands, destitute of coal or timber, are in some degree compensated by their extensive fields of peat, which vary in depth from two to four feet.‡ In the absence of all other descriptions of fuel, this species of combustible may, at some future period, be of great service to the inhabitants. In fact, even now, we are told that the want of wood is abundantly supplied by the peat, which is found in every part of this group of islands, and is collected with very little labour.§

The geological features of these islands have been described by Mr. Darwin.‖

* Byam Martin's Statistics of the Colonies of the British Empire, p. 120.
† Report of the British Association, for 1845.
‡ Hunt's Merchants' Magazine, February, 1842.
§ Martin's Statistics of the British Colonies, p. 144.
‖ Proceedings Geol. Soc. London, March 25th, 1846.

WEST INDIA ISLANDS,

COMPRISING

1. CUBA.
2. JAMAICA.
3. PORTO RICO.
4. BARBADOES.
5. GUADALOUPE.
6. ANTIGUA.
7. TRINIDAD.
8. GRENADA.
9. ST. LUCIE.
10. MADEIRA.

THE WEST INDIES.

ISLAND OF CUBA.—[Belonging to Spain.]

Vicinity of Havana.—Bituminous substance called Chapapote, Asphaltum or Solid Bitumen.—Of this inflammable mineral substance there are many varieties, to which we shall refer in the progress of this book. They differ in consistency, from a thin fluid to a solid compact mass, with conchoidal fracture, externally resembling coal, and in the West Indies, not unfrequently mistaken for that combustible.

German,	Erdpech, Judenpech,	*Spanish,*	Asfalto, Chapapote,
Dutch,	Jodenlym,	*Portuguese,*	Asphalto,
Latin,	Asphaltum, Bitumen Judaicum,	*Russian,*	Asfalt,
		French,	Bitume.
Italian,	Asfalto,		

These comprehend the several species named after their respective qualities, Naphtha; Petroleum; Maltha, or Sea-wax; Elastic Bitumen, or Mineral Caoutchouc; Compact Bitumen, or Asphaltum; Mineral Pitch; Bitumen Candidum; Mineral Oil, the Seneca or Genesee Oil of the United States.

The chapapote of Cuba, *commonly called coal,* is worked or mined much in the same manner as the latter mineral, and appears in several positions in the vicinity of Havana and Matanzas. We are enabled to speak of this substance from personal acquaintance with the localities. In the Transactions of the American Philosophical Society, Vol. VI. p. 191, and in the London and Edinburgh Philosophical Magazine of March, 1837, are notices of a vein of so called "bituminous coal," near Havana, by Richard C. Taylor. We refer to our original notes on which those communications were based.

Casualidad Mine.—Situated six miles from the city of Guanabacoa, three leagues from Havana, and two miles from the sea or place of embarkation. In a region of metamorphic and magnesian rocks, of which the most prevalent are serpentines, diorites, and euphotides, accompanied by veins of quartz, of chalcedony, and often of copper, occurs the substance denominated chapapote. Instead of a coal seam in the formation appropriate to that mineral, which we had been invited to inspect, we saw in the midst of these stratified rocks, true wedge-formed veins there, where they appear at the surface, but enlarging downwards to the breadth or thickness of several feet. The strike of the Casualidad vein is nearly north and south, conforming to the local range of stratification, although the general range is nearly east and west, following the direction of the island. At the point excavated by the negro workmen, the vein was laid bare, to the width and depth of near forty feet, each way; its character being, for that space, fully developed, or sufficiently so to enable a plan and section to be constructed. At the outcrop the vein is scarcely a foot thick, but at the depth of thirty feet it is enlarged to nine feet, descending nearly vertically. Thus, at the rate at which it continued to increase, in the short depth proved, it was anticipated the mass beneath must acquire enormous magnitude. Several lateral branches pass upwards from the main vein, both in its vertical and longitudinal direction, all apparently ramifying from a voluminous mass below. Strictly speaking, the solid bitumen was in no case enclosed between the walls, but seemed rather to occupy fissures in the ancient rocks, and cavities larger than we could venture to speculate upon. The outcrop was easily traced about two hundred to three hundred yards, but beyond this no effort had been made to prove the vein. Miserably inadequate as was the system adopted for the extraction of this coal, we could not but infer that an enormous amount of this substance might very cheaply be obtained. Under the management then going on, all the water, as well as the materials, was hoisted up by hand, in small vessels, and conveyed to a distance by a gang of negroes; economy in labour being in no respect consulted, and no kind of machinery, not even a windlass or wheelbarrow, was employed in the so called mine.

In regard to the arrangement of the matter of the vein itself, we noted, that the asphaltum was disposed in horizontal laminæ, whatever might be the inclination of the veins or branches; thus essentially differing from the usual character of coal seams, whose lamination is always parallel to the direction of the strata.

An analysis was made by Mr. T. G. Clemson; the result is as follows:—

Carbon, - - - - - -	34.97
Volatile matter, - - - - -	63.00
Ashes or cinder, - - - - -	2.03
	100.000

Specific gravity, in three different specimens, 1.142—1.189—1.197. Streak—dark, bistre brown.

Externally it is of a deep jet-black; having the horizontal surfaces of the laminæ covered with curious conchoidal markings, like the impress of a seal upon black wax. These impressions are marked with concentric, or rather with excentric rings, not unlike the lines of growth on the flat valves or upper shells of some bivalves. They vary greatly, in diameter, from only half an inch, to a foot.

A considerable quantity of this coal or asphalte we found excavated and stored; some of which had been employed by the smiths and workers of iron in Havana. From various causes, we understand that the mine has been prosecuted very feebly, and latterly has not been in operation; nor do the proprietors appear to have a ready market for the material.

Near Havana.—We see it announced that a combustible similar to that we have described, has been tried by the Spanish steam frigates, and had been pronounced on very favourably. The analysis is as follows:—

Carbon, - - - - -	71.84
Oxygen and hydrogen, - - -	14.66
Ashes and cinder, - - -	13.50
	100.00

From the great amount of impurities in this specimen, we presume that it was derived from some other source than that of Casualidad mine.

Six leagues from the mine of Casualidad, towards Mantanzas, a body of chapapote exists, from whence a few tons have been forwarded to Philadelphia, Liverpool, and London. The geology of the vicinity is of a corresponding character to that we have described.* The chapapote is, however, far more compact and solid than that of the Casualidad vein. It emits, when rubbed, an agreeable odour, resembling that of amber. It is very pure, free from all extraneous matter; its specific gravity is greater than that near Havana, and the mineral is more resinous and less friable.

There can be no doubt but this is an admirable combustible, where much flame is a desideratum, for such purposes as evaporation, and for heating surfaces; and in this respect it must be superior to many descriptions of fuel whose proportion of volatile matter is less. For the generation of steam, for boiling or concentrating the juice of the sugar cane, or for the manufacture of gas, this flaming coal appears to be singularly well adapted. In other respects it cannot, of course, compare with the intense, enduring, and concentrated heat of anthracite.

As it contains no sulphuret of iron, the gas would be wholly free from any deleterious admixture. The chapapote might also be profitably employed in the manufacturing of lamp-black. For domestic purposes it gives out far too much smoke, in burning, to form a desirable fuel.

Chapapote near Havana.—Other positions, in the neighbourhood of the principal mine of this substance, shows its prevalence in the

* Silliman's American Journal of Science, 1842.

country. We have ourselves examined and reported upon some excavations, at two leagues from Havana; but they were not of so promising a character as at the Casualidad mine.*

Mine Prosperidad.—Partido de San Miguel.—Asphaltum.—Six miles from Havana, on the road to Taposte.

An article of M. Castáles, in the *Diaria de la Habana* of 1842, has appeared in the scientific journals of the United States, accompanied by the analysis last quoted, but which was derived from another source.

The substance here denominated bituminous coal, is of the two varieties to which we have alluded, and is developed to a surprising extent. Two shafts have been sunk here, forty-five yards apart. In the principal one of these, the coal or chapapote was reached at the depth of seven yards, and continued therein to the depth of forty yards—the bottom of the shaft. From the four sides of this shaft four straight exploratory galleries have been conducted, in opposite directions, thirty yards in length; in all which space the mass of bitumen continues horizontally, and without any interruption. At the bottom of the shaft, or of the forty yards above mentioned, instead of driving further in the chapapote, the miners proceeded to bore perpendicularly down, about fifteen yards more;—always in coal. One of the galleries communicates with the other shaft, forty-five yards distant, still continuing entirely in coal. At four hundred yards from the principal shaft, a third pit has been sunk, which reaches the coal at the depth of fourteen yards.

The results of the explorations are these. In the small space indicated, a body of coal, asphaltum, or solid bitumen, is thus far proved to be forty-eight yards [one hundred and forty-four feet] perpendicular, and more than one hundred and eighty feet in surface or horizontal extent: that is to say, and it is to be understood, so far only as had been bored without reaching the bottom. The mass is spoken of as almost horizontal; but its true form cannot satisfactorily be ascertained from the foregoing data, and, moreover, the position of the stratified rocks is stated to be almost vertical.

According to the report of an English engineer, this is one of the most extraordinary mines in the world. By his account, which however is not particularly intelligible, the upper part was highly charged with bitumen, and was convertible into good coke. The lower portion consisted of an improved quality, being, as he thinks, less bituminous and much more compact. A railroad, we understand, has, of late, been constructed from the mine to the port.†

Punta Icacas.—The existence of solid bitumen in rocks, near the north coast of Cuba, not far from Mantanzas, was known to the celebrated Von Humboldt. This mass, he observes, reminds us of the asphaltum of Valorbe, in the Jura Limestone.‡

* Philosophical Magazine, R. C. T., March, 1837, and Transactions American Philosophical Society.

† Mining Review, October, 1840, p. 76. Also Silliman's Journal of Science, for 1842.

‡ Essai Politique sur l'Isle de Cuba.

Something of the same kind also occurs at Puy de la Lège in France.

Asphaltum, in various degrees of density, occurs among the serpentine and magnesian rocks at other points on the Island of Cuba than those we have indicated. From the direct observations that we have been enabled to make, it seems very probable that all the bituminous matters, whether known under the names of Chapapote; Asphalte; Mineral Pitch; Petroleum; mastic bitumen; liquid or fixed bitumen, and other terms,—simple varieties of the same mineral substance, appear at the surface, at the points of fracture in the disturbed and metamorphic regions. In other words, "in the centres of dislocation of the beds."*

Petroleum.—Springs are abundant near Havana, rising from fissures in the serpentine rocks at Guanabacoa, and have been known for two centuries at least. In fact, the whole country is impregnated with bituminous matter, to a surprising degree. Even the solid quartz, the serpentine rocks, and the veins of Chalcedony, have cells and cavities filled with liquid pitch; and the air is scented with it, when these rocks are broken by the blows of a hammer. In this respect it resembles the mineral pitch found filling the cavities of Chalcedony and calc-spar, in Russia.†

Even in the bay of Havana, the shore, at low water, abounds with asphalte and bituminous shales, in sufficient quantity for the paying of vessels, as a substitute for tar. It is stated that, in buccaneering times, signals used to be made, by firing masses of this chapapote, whose dense columns of smoke could be recognized at great distances, and served as signals to vessels at sea.

It is matter of history that Havana was originally named, by the early visitors and settlers, CARINE;—"for there we careened our ships, and we pitched them with the natural tar which we found lying in abundance upon the shores of this beautiful bay."‡

Petroleum leaks out in some, indeed in numberless, places, in this delightful island, from amidst the fissures of the serpentine, and perhaps has deeply seated sources. We are acquainted with abundant springs of petroleum between Holguin and Mayari, in the eastern part of the island, and also possess notices of others in the direction of Santiago de Cuba.§

In fact, the entire chain of the West India and Windward islands present similar phenomena of petroleum springs, beds or veins of asphaltum, and accumulations of mineral pitch, and traces of metamorphic and volcanic rocks, in great abundance.

M. Bousingault, in a dissertation on the bitumens of France, remarks that the only contradictory fact opposed to his conclusion that the geological position of mineral pitch is in formations referable to the supercretaceous group, is that given by M. de Humboldt, who

* Office de Publicite.

† Allan's Manual of Mineralogy, p. 291.

‡ Early history of Cuba.

§ Essai Politique sur l'Isle de Cuba.

saw at Punta d'Araya, on the coast of Carracas, petroleum issuing from mica slate. To these exceptions might be added many more; for we have seen in the greater part of the larger islands of the West India chain, that petroleum, mineral pitch, and asphaltum, in various degrees of solidity, appear between the fissures of ancient rocks, particularly of the magnesian class, serpentine, euphotide, &c., and in regions where no supercretaceous rocks occur, in their neighbourhood. So also, in Europe, bitumen occurs in older formations, from the coal measures down to granite.

The chemical results of these inquiries, however, are these:—That glutinous bitumens are mixtures of two substances, which we can isolate. One of these principles is solid and fixed, and in its nature approaches to asphalt. The other is liquid, oily and volatile, and resembles petroleum in some of its properties.*

Vegetable remains in Tufa.—The recent calcareous tufa deposits, so common in the north-eastern portions of Cuba, contain vast quantities of vegetable casts, impressions of stems and leaves, and seeds of plants, such as abound in the vicinity at the present time.† We have collected abundant specimens of these, at the base of the metamorphic limestone range of mountains on each side of Gibara.

SHIPMENTS OF COAL FROM GREAT BRITAIN TO THE WEST INDIES.

The trade in coals from Great Britain to the West Indies is limited. They are partly required for furnaces, but the principal quantity consists of a particular description of coal for steam purposes, under contract with the British government, and is a trade of comparatively recent origin.

The government stations are Jamaica, Antigua, and Barbadoes, and some coals go to St. Thomas's. The average price of the coals there is about 45*s*. to 47*s*. per ton, [= $10.90 to $11.40,] according to the demand. They have been freighted from London, costing 20*s*. per ton there. The freight from Newcastle to the West Indies is 27*s*. 6*d*. to 30*s*.‡

English Bituminous coal imported into the British West Indies.§—In 1831, 48,536 tons; 1832, 43,980; 1840, 82,564; 1841, 71,311; 1844, 77,338; 1845, 102,339.

British coals imported into the Foreign West Indies.—In 1844, 26,592 tons; 1845, 22,154.

In the West Indies the price of coal varies from 45*s*. to 47*s*. per ton, for government contracts; it has been occasionally much higher.

Spanish tariff on coal imported—valuation $3.75 per ton. Duty, 32½ per cwt.

The importations of copper ore from Santiago, aud other ports of Cuba, constitute a very considerable portion of the trade of Swansea.

* Philosophical Magazine, 1837.
† Trans. Amer. Phil. Society, Vol. IX., p. 210.
‡ Evidence on the Coal Trade of London, in 1838, p. 104.
§ Official Tables of Revenue, Commerce and Population.

The ships employed in this trade are from 300 to 500 tons burden. The chief back freight for these ships is Welsh coal. It was feared by the shippers of this Welsh coal that the discovery of a supposed bituminous coal, of high value, at more than one point, within a few miles of a shipping port of the island of Cuba itself, would materially diminish, if not entirely cut off, the market for the supply of the free-burning coals of South Wales.* Owing, however, to other circumstances, rather than to any deficiency in the quality of the Cuba asphaltum, there has not, at present, been experienced any change in the importation of foreign coals; but the demand in a tropical climate can never, we think, for obvious reasons, be very extensive.

II. JAMAICA.—[Great Britain.]

Three or four thin seams of *true coal*, embedded in shale, were described in 1825, by Sir Henry De la Beche, near the north-eastern extremity of the island.† None of these beds were in sufficient thickness to constitute a profitable or workable coal stratum. It appears, also, that bituminous coal exists on the other, or south side of the island, within ten miles of Kingston. It burns with a clear, bright flame, and is said to be good; but the thickness of the seam there is not mentioned.‡

Island Tariff.—On coals, [except those for the Royal Mail Company,] 6*d*. per ton; duties under the British act, 1842, 4 per cent.

III. PORTO RICO.—[Spain.]

In the returns of exports from this Spanish island, coals are mentioned. We possess no further information.

* Remonstrative to Sir Robert Peel, March, 1842.
† Trans. Geol. Society of London, Vol. II., second series, p. 143.
‡ Mining Journal, London, April, 1837.

IV. BARBADOES.—[Great Britain.]

Compact bitumen or asphalte abounds here, as in several other West India Islands. In this, it supplies, in a great measure, the place of coal. Dr. Wilkins has published some "observations on the green mineral *naphtha* of Barbadoes."

The calcareous rocks here are frequently impregnated with bitumen. There is a *petroleum* or burning spring, at St. Andrew's parish. It goes by the name of green tar, and often supplies the want of pitch and lamp oil.*

Tariff on Coals.†—Colonial duty on coal imported, 2*s*. per ton. Crown foreign duty, 20 per cent.

May, 1845.—It is announced that some "very superior coal" has been discovered on Grove Plantation estate, "and various parts of the island, which, for plantation purposes, is considered fully equal to the imported English coal." This substance may, perhaps, be the solid bitumen or asphalte above alluded to.

We have recently seen an analysis of this bituminous coal by Mr. Herapath, as follows. We place by its side in another column, the analysis of the Cuba chapapote, whereby the analogy of the two is satisfactorily shown :—

	Barbadoes.	Cuba.
Bitumen, resolvable by heat into tar and gas,	61.60	63.00
Coke, or Carbon, - - - - -	36.90	34.97
Ashes, (no sulphur,) - - - - -	1.50	2.03
Total, - - - - -	100.00	100.00

Mr. H. observes, that "the large proportion of bitumen, in proportion to the carbon, will prevent this coal from being used as common fuel, unless it be mixed with some substance more fixed in the fire. Hard charcoal, more refractory coal, and even perhaps earthy substances, would be beneficial. It could be employed in the production of gas, of which it would furnish a large quantity, and of a very rich quality, even exceeding that of cannel coal—the best for that purpose hitherto known."

The Chairman of the Barbadoes Railway has announced that the geological formation of the Scotland district of the island, which he had opportunities to inspect, leaves little doubt that it contains coal measures to a great extent.

The Barbadoes Standard confirms the above, and states that the

* Dr. Skey in Geol. Trans., Vol. III., 1816.

† Mining Journal, January 24th, 1846.

result of a scientific examination of the parishes of St. Andrew and St. Joseph, leads to the confident belief of the existence of useful coal. This combustible, it is stated, is different from that bituminous substance so long in use in Barbadoes, of which the analysis of Mr. Herapath is furnished in a preceding paragraph.*

GUADALOUPE.—[France.]

Contains a volcano, rising five thousand five hundred feet above the sea. It has no regular crater, but smoke issues out of three or four different spots. Not far from the shore, south-west of the volcano, is a place in the sea which sends up boiling hot water.

ANTIGUA.—[Great Britain.]

Although somewhat celebrated for the abundance of petrified tertiary wood that it contains, specimens of which, when polished, are exceedingly beautiful, it does not appear that coal has yet been discovered within the limits of the island.†

The monocotyledonous structure of the stems of palms is beautifully preserved in these lignites, and no examples surpass them in beauty and interest.‡

Four distinct species of fossil palm occur in the pliocene tertiary of the island of Antigua.§

* Mining Journal, February 14th, 1846.
† Dr. Nugent.
‡ Mantell's Medals of Creation, Vol. I., p. 70.
§ Professor Unger.

TRINIDAD.—[New Granada.]

The *pitch lake*, lagoon, basin, or plain, [for it has been called by all these epithets,] is sufficiently remarkable to require notice in our lists of deposits or accumulations of bituminous substances. It is described as three miles in circumference; but its depth is unknown, being incapable of admeasurement. This substance is used for paying the bottoms of ships, and probably differs little, except in density, from the chapapote of Cuba. We may not greatly err, if we ascribe them to a common origin.*

The pitch lake of Trinidad is, perhaps the most remarkable locality of asphaltum in the world. It occupies the highest land in the island, and emits a strong smell, *sensible at ten miles distance.* Its first appearance is that of a lake of water; but when viewed more nearly it seems to be a surface of glass. In hot weather, its surface liquifies to the depth of an inch, and it cannot then be walked upon. The geological data in the vicinity exhibit traces of geological agency: and not only in the lake itself, but in the neighbourhood, are seen holes and fissures, sometimes containing liquid bitumen.† Fissures of great length, from four to six feet wide, traverse the surface of this lake in every direction, and are generally filled with water. The consistence and general appearance of the asphalte is that of pit-coal; only the colour is rather grayer. It is very brittle, and breaks into small cellular, glassy fragments. Some of the more elevated parts of the surface are covered with thin brittle scoria.

We know not if any practical employment of a mineral substance, here so astonishingly abundant, has yet been suggested or undertaken, on an extensive scale. It surely was not placed there in vain. Beside the purpose above mentioned, that of paying ships, and thereby protecting them from that pest of the West Indian seas, the *teredo*, or borer, it is capable of being used as an ordinary varnish, and in a variety of minor matters.‡

It has been attempted to apply the asphaltum, brought from this lagoon, to the same objects as pitch and tar; but it is found to require so large an admixture of oil that it becomes too expensive. If it could be economically applied, Trinidad might furnish abundant supplies for the whole world.§

Petroleum.—South of Cape de la Brea is a submarine volcano, which occasionally boils up and discharges a quantity of petroleum.

* Allan's Manual of Mineralogy, art. Bitumen.

† Dr. Ure's Dictionary of Arts, &c.

‡ Essay on Bitumen; its uses in remote ages, and its revival in modern times, and applicability to various purposes, 1839.

§ Trinidad Almanac for 1840. App. c. 4.

Another occurs on the east side of the island, which throws up on the shore masses of bitumen, black and brilliant as jet.

Coal.—Schistose plumbago has been discovered in Trinidad, and near it is a mine of coal, about five miles from the sea-shore.* We have seen no details.

Lignite.—Mr. Link has made microscopical observations on some lignites from Trinidad, and has recognized therein the structure of the wood of the palm.†

GRENADA.—[Great Britain.]

We have not heard of the discovery of coal beds, but it is reported by Dr. Simpson that the red secondary sandstone of this island contains vegetable fossils; such as the leaves and stalks of plants.‡

ST. LUCIE.—[Great Britain.]

In corroboration of the geological evidence, so frequently, and we may even say so universally presented, that the entire group or range of the Antilles, from Trinidad to Cuba, has been, from time to time, subjected to volcanic influence, which is occasionally felt even at the present day, we add here, that this island yet contains an active volcano. Its summit is more than four thousand feet above the sea level, and within its crater are several depressions, filed with boiling water and mud. From one of these rises, at invervals, a column of smoke. The last eruption of this volcano, of which we have any information, took place in 1812.§

* Martin's Statistics of the British Colonies, p. 25.
† Annales des Mines, tome XVII., p. 575.
‡ Martin's Statistics of the British Colonies, p. 43.
§ Geography of America and the West Indies, p. 26.

ISLAND OF MADEIRA.—[Portugal.]

Brown coal or lignite occurs on the north side of the island, on the banks of one of the tributaries of the St. George. Professor Johnstone considers it to be the dried relict of an ancient peat bog, and that its lustre, compactness, and rhombodial fracture, are due to the action of the basalt which overlyes it. An analysis gave

Carbon, - - - - -	60.70
Hydrogen, - - - - -	5.82
Oxygen and nitrogen, - - -	33.48
	100.00

and 20.05 of ash. This is the organic constitution of true peat; but no peat exists at present in Madeira, nor has been noticed so near the equator. It is suggested, therefore, that this deposit may indicate a former colder climate in that latitude.

SUMMARY

OF THE

COAL-FIELDS OF THE WORLD.

A LATE English paper estimates the annual produce of coal in Europe as follows:—

	Tons.
British Islands, - - - -	37,000,000
Belgium, - - - - -	5,000,000
France, - - - - -	4,200,000
Prussia, - - - - -	3,500,000
Austria, - - - - -	700,000
Spain, - - - - -	550,000
	50,950,000

The produce of coal in British America, amount not ascertained.

United States, 1853—
Pennsylvania, anthracite and bituminous, 7,000,000
Maryland, - - - - - - 533,980
Other States not ascertained.

For the area of the coal-fields in Europe and America, see Introduction, p. 28.

In Russia, on the northern shores of the Black sea, bituminous coal (brown) has been found in abundance. The richest Russian coal-field is on the shores of the sea of Azof, between the Dnieper and Donetz rivers; it is said to be equal in quality to the best English, and may be delivered at a port on the Dnieper or Don rivers, for about 4*s.* to 5*s.* per ton.

A section of the works at Lissitchia and at Balkia, on the Donetz, shows that, in a depth of 900 feet, there are 12 seams of coal, the united thickness of which amounts to 30 feet.*

* Bulletin de la Société Géologique de France, 138, p. 237, from 1st edition of Taylor's Statistics on Coal.

The capital invested in the British coal trade is estimated at £10,000,000, and the value at the mouth of the pit, £10,000,000. At the points of consumption, including expenses of transportation and other charges, the cost is said to be £20,000,000. For the supply of London alone, 3,637,878 tons of coal are required for manufacturing and domestic purposes; the coasting vessels conveyed, in 1850, upwards of 9,360,000 tons to various ports in the United Kingdom, and 3,350,000 tons were exported to foreign countries and the British possessions. The number of persons employed in mining the coal was 120,000.*

Coal is reported to have been found on the Nile, near Assuan. The first vein, six feet thick, is 100 feet below the surface. The second is deeper, 3 feet thick, and of excellent quality.† It is also said to be found in various parts of Africa and Asia.

In China, there is evidence that there exists several varieties of coal—tertiary, or brown coal, bituminous of various kinds, cannel coal and anthracite; all of which have been in use for ages for domestic purposes, and in the manufacture of iron and other metals in this remarkable country.

There is also abundance of coal in Australia and New Zealand, besides various other localities.

We have taken the liberty of quoting from a work just published,‡ the following Table of the amount of Iron produced at the present time, 1854, in Europe and United States.

	Tons.
Russian Empire,	200,000
Sweden and Norway,	155,000
Great Britain,	3,000,000
Belgium,	300,000
Prussia,	150,000
Saxony,	7,000
Austrian Empire,	225,000
Rest of Germany,	100,000
Switzerland,	15,000
France,	600,000
Spain,	25,000
United States of America,	1,000,000

* English Paper, 1853.

† Athenæum, February 10th, 1849.

‡ Whitney's Metallic Wealth of the United States.

MONEY, WEIGHTS AND MEASURES.

GREAT BRITAIN.

Money, Weights, Measures.

Gold.—The standard gold sovereign containing 1-12th alloy, weighs 123.274 grains.

1 lb. troy of this standard gold, computed at £3. 17. 10½ per oz., is coined into 46.74 sovereigns = £46. 14. 6. No duty is charged on its coinage.

Value of the sovereigns in the United States currency = $4 83c. 8m.

Silver.—The standard silver contains 18-240ths alloy.

1 lb. of this silver is coined into 66 shillings, of which 4 are taken as seignorage, or mint duty; nearly 6 per cent. Value per lb. standard £3. 6. 0. Each shilling contains 87.27 grains standard.

1 lb. troy of the same silver is coined into 32 Company's Rupees, of which 2 per cent. are taken as mint duty.

Copper is valued at £224 per ton = 24 pence to the pound avoirdupois.

Comparative Currencies.

	English.			French.		United States.
	s.	*d.*		*franc*	*ct.*	*dollar c.*
Spanish Piastre,	4	4½		5	43	1.05
Prussian Rix dollar,	3	0				0.72
United States dollar, 4*s.* 16*dec.*,	4	1½		5	18	1.00
Hindostan, Sicca Rupee,	2	0				0.48
French, Franc,	0	9	69	1	00	0.19.3
English, Shilling,	1			1	16	0.21

Weights.

1 lb. avoirdupois, = 453.544 Grammes of France = 0.97 lb. of Berlin.

112 lbs. = 1 cwt. of which 20 cwt. makes 1 ton = 2240 lbs. English = 1 quintal Engl.

1 English ton = 10.1465 metrical quintals, or 1,015 Kilogrammes of France.

1 Kilogramme usuel of France is *2lb. 3oz.* 4½*dwt.* avoirdupois.

1 metrical quintal of France is 220 *lbs.* English = 100 Kilogrammes.

Weight of 1 Newcastle chaldron of coals, 2,675 Fr. Kilog. = 2 tons and 13 cwt. Engl.

Weight of 1 Last about 3⅓ tons, = 99.54 cubic feet.

26½ cwt. was the legal weight, by act of Parlt. 1831, of 1 London chaldron of coals.

53 cwt. 1 Newcastle chaldron.

1 barrel of meat = 200 lbs., 1 tierce of do. = 304 lbs.

1 sack of flour = 280 lbs.

Measures of Capacity and Solidity.

1 English Winchester bushel contains 2150.42 inches = 35.236 French Litres.

1 English Imperial bushel contains 2218 inches, weighing 80 lbs. of water.

2.84 English Winchester bushels = 3½ cubic feet = 22 imperial gallons = 1 hectolitre.

26⅔ English bushels to 1 ton, when dry.

1 English bushel weighs 84 or 85 lbs.

1 ton = 252 imperial gallons.

On the old system of selling coals by measure.

1 sack of coals - - - - -	= 3 bushels.
1 vat - - - - - -	= 3 sacks.
1 heaped London chaldron - -	11 sacks. 36 bushels. 4 vats.
21 chaldrons - - - - -	= 1 score.
1 Imperial bushel even, - -	2218 cubic inches.
1 Imperial bushel heaped, - -	2815 do.
1 Boll = 36 Winchester gallons, -	= 9676 do.
1 Fother, - - - - -	= 77,414 do.
1 Last, [now used in the Hanse Towns.]	= 99,540 cubic feet.
7½ bolls, - - - - - -	= 1 cub. y'd coal.
6 bolls, - - - - - -	= 1 chaldron.

In Ireland, 1 barrel or kish represents 6½ cwt. or about 3 barrels to 1 ton.

In Scotland, 36 cubic yards of coal are equivalent to 32 tons weight.

In England, 1 cord of wood is 4 feet high, 8 feet long, and 4 feet deep = 128 cubic feet.

Measures of Length and Area.

1 French mètre is = 3*ft.* 3*in.* 37*dec.* = 3*ft.* 28*dec.* = 39*in.* 371*dec.*

1 English imperial acre = 4840 sq. yards, { 0.4046 Hectars of France. 1.561 Morgen of Prussia. 0.7025 Joch of Austria.

The imperial acre, English, is to the Scotch acre as 1 to 1.261.
do. do. Irish acre as 1 to 1.62.

1 Irish acre contains 1 acre 2 roods, 19 poles English.

30¼ Irish acres are equal to 49 imperial acres.

The English commercial and financial year closes on the 5th of January, annually.

KINGDOM OF FRANCE.

Money, Weights, and Measures, and their equivalents.

Gold.—1 Kilogramme of standard gold is coined into 155 Napoleons or 20 franc pieces.

1 Gold Napoleon or new Louis = 16*s.* 2*d.* Eng. = $3.86 U. S. = 20 francs.

1 Livre = 1 sovereign = £1 sterling = 24 francs 76c. = $4.84 U. S.

Par value in London of £1 sterling = 25 francs 57 cen.

Silver.—1 Kilogramme [one-tenth alloy] is coined into 200 francs.

250 Francs of the Paris mint are equal to 100 Sicca Rupees.

1 Franc = 9*d.* 69*dec.* Eng., generally estimated at 10*d.* = $0.19⅓ U. S.

The franc is divided into 10 decimes = 0*d.* .969 each. 100 centimes = 0*d.* .0969 each.

5 Franc piece = 4*s.* 04*dec.* Eng. currency = 4*s.* 8*d.* Canada currency = $0.97 U. S.

1 English sovereign = 24 francs 76 cen.

1 English shilling = 1 franc 16 cen.

5 Shilling-piece = 1 crown = 5 francs 80 cen.

1 United States dollar = $1.00 = 5.18 francs = 4.16 Eng. shilling.

1 Million of francs = £40,000 = $193.500.

Weights.

1 Kilogramme = 1000 grammes = 2 lbs. 3 oz. 4½ dr. Eng. = 2 lbs. 204 dec. avoir.

1 Gramme is the 1000th part of a kiligramme.

1 Livre usuelle or French pound = 500 grammes = 1 lb. 1 oz. 10½ dr. = 1 lb. 1 dec. = 7717 English grains.

1 Metrical quintal = 0 ton 0985 = 220 lbs. = 486 dec. English.

10.1465 Metr. quin. [of 100 kilog.] = 1 ton English = 2240 lbs.

1014. 65 Kilogrammes = 1 ton Eng. = 10.146 metr. quintals.

1 Metrical tonne = 1000 kilogrammes, for coal, 10 metr. quintals = 20 German centners or quintals.

1 Metrical tonne = 2207 lbs. English.

1 Ton English = 10 quin. 15,940.

1 Metr. quin Eng. = 0.09843.

Measures of Length.

1 { Pied usuel = foot 1 inch 122 dec. = 1-6 of 1 toise.
French foot = 3.048 decimetres = 12.78 Eng. inches. }

1 French metre = 3 feet 3 inches 37 dec. = 39 in. 371 dec. = 3 feet 28 dec. Eng.

1 Centimetre = 0.39371 inches.

1 Decimetre = 3.9371 inches.

1 Decametre = 32.809 feet.

1 Hectomètre = 109 yards 1 foot 1 inch.

1 Kilomètre = 10 93 yards 633 dec.

161.024 Kilomètres = 100 English miles.

1 Myriamètre = 10,936 yards, 330 dec. = 6 miles 1 furlong, 28 poles.

1 English mile = 1.6 kilometres; 3 English miles = 4.8 kilometres; 5 English miles = 8.0 kilometres; 10 English miles = 16 kilometres; 20 English miles = 32 kilometres; 25 English miles = 40½ kilometres; 30 English miles = 48 kilometres; 40 English miles = 64 kilometres; 50 English miles = 85½ kilometres; 60 English miles = 96 kilometres; 70 English miles = 113 kilometres; 80 English miles = 129 kilometres; 90 English miles = 145 kilometres; 100 English miles = 161 kilometres.

1 English mile = 1640 metres.

1 Toise usuelle = 6 feet 6 inches 735 dec. = 6 metres 57 dec.

1 Aune usuelle = 1 yard 3 dec. Eng. = 3 feet 11¾ inches Eng.

1 League = 2000 toises = 2 miles 743 yards = 4263 yards English = 3898 kilom.; 36,214 leagues = 100 English miles.

Measures of Capacity.

1 Litre is a cube of which the side is 1-10th of 1 metre = 1.763 pint = 0.264 gallon.

1 Decalitre = 2 gallon 1 pint 60 dec.

1 Hectolitre = 90 kilogrammes = 2.838 bushels = 3½ cubic feet = 22 imperial gallons.

11.26 Hectolitres = 1 ton English.

1 Wheel-barrow load = ¾ of 1 hectolitre = 150 lbs., nearly.

1 French Boisseau usuel = 0.354 English bu.

1 London chaldron = 25½ quintals or cwt. = 1295 kilog. = 17.26 hectolitres.

1 Newcastle chaldron = 53 quintals or cwt. = 2991 kilog. = 36 hectolitres.

1 Muid = 1.124. hhds.

Measures of Solidity.

1 Stere = 1 cubic metre = 35.3174 cubic feet English, for stone, peat, timber, &c.

1 Deci-stere = 3.53 cubic feet.

1 Deca-stere = 353.174 cubic feet.

1 Corde of wood = 2¾ steres = 97 cubic feet; but with local variations.

1 Banne of charcoal = 50 cubic feet.

1 Cubic metre = 35.31 English feet cube.

A stere of wood, produces on an average, 68 kilogr. 40 of charcoal.

Measures of Area.

1 Arpent of land = 1.043 acre.

1 Are of land = 0 rood 098 dec. = about 1-40th of 1 acre = 100 square metres.

1 Hectare = 2.471 acres = 11,960 square yards = 10,000 square metres.

4.046 Hectares = 10 acres English = 100 acres French.

1 Square kilometre = 0.386 English square miles = 1,196,044 square yards.

1 Square mile English = 3,097,600 square yards English.

Rules.

To reduce French kilogrammes to English pounds, multiply by 2.205.
" " quintal metriques to Eng. pounds, multiply by 221.
" " quintals to English pounds, multiply by 113.37.
" " metres to English pounds, multiply by 39.375.
" " cubic metres to Eng. cubic feet, multiply by 35,32.
" " grammes to English grains, multiply by 15,4340.
" " loths to Eng. pounds, multiply by 0.032,208,435.

RUSSIAN EMPIRE.

Weights, Measures, and Money.

1 Pood, Pud or Poud = 36 lbs. 11 drachms English. Sometimes 40 Russian pounds to 1 pood.

110 lbs. Russia is equivalent to 100 lbs. English.

43 Russian lbs. = 1 poud or pood = 16 kilog. 38 = 36 lbs. 1 oz. 11 dr.

56 Pouds = 1 ton Eng. There is also another of 63 poods to the ton; and another of 50 poods to the ton.

The Russian pound is rather heavier than the avoirdupois pound.

1 Chetivert = 12,800 Eng. cubic inches = 5.952 imperial Eng. bushels = 209.740 Fr. litres.

1 Last = 13,8 Quarters.

The Moscow foot = 13.17 Eng. inches = 3.343 Fr. decimetres.

1 Russian Archine, or Arsheen, or Ell = 28 Eng. inches.

1 Russian Verst = 1500 archines = 3500 Eng. feet = 1167 yards = 1,066 Fr. kilom. Therefore a verst is two-thirds of an English mile = 5 furlongs 76 yards = 53½ chains.

150,814 Russian versts = 100 Eng. miles.

Produce of the Gold Mines of Russia, [Ouraland Siberia,] from 1830 to 1842, belonging to the Crown and to individuals:

	Russian Pouds.	Eng. lbs.	Tons. lbs.
Mines belonging to the Crown, 1830 to 1842, - - - - -	2,052 =	73,872 =	32 00
Mines belonging to individuals, -	4,119 =	148,284 =	66 444
	6,171 =	222,156 =	98, 444

Money.

Gold Currency.—Ducat of 1796,	$2.29c.	U. S.	not in circulation.
Gold Ruble of 1799,	0.73.1	"	not in circulation.
Imperial of 1801,	7.83.6	"	not in circulation.
Demi Imperial,	3.86.6	"	in circulation.

Platina Coins.—£1 sterling.

1 Silver copec = 0fr. 4c. = $\frac{1}{100}$ part of a silver rouble = 3*s.* 2*d.*

1 Silver Rouble = 4 Fr. francs = $0.77 United States.

Consequently 1 silver copec = 77c. U. S.

100 Copecs = 1 rouble. But the silver rouble varies in value from 3*s.* 2*d.* to nearly 4*s.* English, according to the distance from the capital, = $0.75 to $1.00

1 Paper Rouble is worth about 10*d.* and is usually considered

equivalent to 1 fr. = 19.33*c*. U. S., and is also in mercantile transactions divided into 100 copecs, [represented in copper coin.]

1 Copper Copec is therefore = 1 centime of a franc = *c*.0.193 U. S.

Par value for £1 sterling in London, 6 roubles, 40 cop. of St. Petersburg, at 3*s*. 1½*d*. per each rouble.

KINGDOM OF PRUSSIA.

PRUSSIAN SYSTEM OF CURRENCY, WEIGHTS AND MEASURES.

Prussian Currency.

1 Frederick single gold ducat, = $3.97½, United States.

1 Thaler or Prussian Rix-dollar, 3.711 francs, = 3*s*. English—current value, which is divided into 30 silbergros, and each silbergros into 12 pfennigs.

Par value of £1 English in London, 6 dolls. 27 s. gr., and of 1 Prussian dollar = 2*s* 9*d*½.

£1 English = 5 thalers, 15 gros.

10 Florins = 16*s*. 8*d*.

1 Florin = 20 pence, or 1*s*. 8*d*. English, = $0.40 United States.

1 Pfennig, = 1-12th of 1 silbergros, = 0.0103 franc.

1 Silbergros, [silver groschen,] = 1.198*d*. English.

Weights.

The Prussian tonne of coal, &c., = 4 quintals, or centners, or scheffels, of about 110 lbs. = 113.38 lbs. avoir., therefore there are nearly 5 Prussian tonnes to 1 English ton; 4 scheffels of 6⅖ bushels each.

1 Lain of coal, = 1000 lbs.

1 Berlin scheffel, or quintal of coal weighs generally 110 lbs. = 113.3 lbs. English, = 51.58 kilog., = 54.94 litres.

1 German tonne, = 1000 kilog., 10 metrical quintals, = 20 centimes.

1 Prussian quintal, = 55.44 kilog., = 121.98 lbs. English. 18.2 quintals 1 ton English.

The Prussian livre, = 0.47 kilog., is divided into 32 laths.

1 Foudre, = 30 centners of 110 lbs., = 3300 lbs., 1½ tons, nearly.

100 lbs. Cologne, = 103 lbs. avoir. English.

116 lbs. Cologne, = 1 quintal. The quintal of the Rhine, = 50 kilog.

1 centner, or quintal, of 110 lbs., Cologne weight, = 51.6 kilog., = 113.38 lbs. avoir.

Measures of Capacity.

The Prussian coal measure called a scheffel, or boisseau, is a fraction less than 1½ Imperial English bushel: about 20 of these will weigh one ton.

4 Scheffels, = 1 Prussian tonne, = 54.943 litres.

1 Berlin scheffel, = 3,180 English cubic inches, = 1,479 English bushels, = 52.107 Fr. litres.

1 English Imperial bushel, = 2,150 English cubic inches.

1 Last of wood, = 75 cubic feet Prussian, = 2.32 cubic French metres.

1 Corde of wood, = 3.34 steers, = 1 klafièr.

Measures of Area.

1 Prussian morgen, = 3053 square yards English, = 1.52 English Imperial acres.

15,853 Morgens = 10 English acres.

Measures of Length.

1 Rhenish foot, legal measure in Prussia, = 0.314 metres French.

1 Berlin foot, = 12.19 English inches, = 3.097 decimeters.

1 Prussian lactre or lachter, = 1,884 metres, = 6.17 feet English.

6 Rhenish feet, = 2.09 metres.

1 Square lachter, 4,378 metres.

1 German mile, = 7.4089 kilometers, = 4.6 miles English.

Saxony Money.

1 Rix-dollar, = 23 groschen, = 228 pfennigs, = 3*s.* sterling.

£1 Sterling, = 6 rix-dollars and 13¾ gros.

1 Convention dollar, = 32 groschen, = 4*s.* 1½*d.* English.

Weights and Measures.

Coal in Saxony is computed commonly by the bushel or scheffel.

The Saxon boisseau, or scheffel, = 1.743 hectolitre, = 3.045 bushels English, which is about 9.21 bushels to 1 ton.

100 Dresden scheffels are equivalent to 195 of Berlin.

1 Saxon scheffel, therefore, is = 2.884 English bushels, and

10 scheffels are equal to 1 ton English.

1 Last of coals is about 6000 lbs. weight, but is seldom used; =2⅔ tons English.

SPAIN.

SPANISH SYSTEM OF CURRENCY, WEIGHTS AND MEASURES.

Currency.

THE gold doubloon, or quadruple pistole of 1801, = $15.53½ U. States.

The gold pistole of 1801, 16*s*. 9*d*. English, = $3.88.4 U. States.

1 Piaster = 20 rials = 5f. 43ct. = *s*.4.384 = 4*s*. 4½*d*. Engl. = $1.05 U. S.

1 Rial, or Réal de Veillon = 0 fr. 27 c. = *d*. 2.616 = 2½*d*. Eng. = $0.5 U. S. = 34 maravédes, = 8½ cuartos.

19 Rials are considered = 5 francs of France.

90 Rials, = £1 English.

Par value of £1 English in London, = 6 dolls. 2¾ rials.

1 Libra Catalan, = 2*s*. 4*d*. English, nearly.

1 Réal of Barcelona, = 4*d*. English.

Weights.

1 Quintal of Asturias, by which coal and iron are sold, = 69 kilogrammes, = 155 lbs. avoir., = 14.5 quintals to 1 ton English.

1 Quintal of Catalonia, = 91 lbs.

1 Quintal of Castille, = 46 kilog. 10 grammes, = 101 lbs. Eng. = 100 lbs. Spanish. It is divided into 4 arobes of 11.50 kilog. each, = 25.3 lbs. Eng.

1 Quintal macho, = 77½ kilogr. = 170½ lbs. Eng. = 13 to 1 ton.

Another quintal of 50 kilogrammes.

4 Arobes, = 1 quintal, = 102 lbs. English.

1 Livre or Spanish pound = 0.46 kilogrammes. The livre is divided into two marcs.

100 lbs. of Barcelona, = 82,215 lbs. avoirdupois.

Measures of Length.

1 Vara or ell, [Vare Castellane,] = 0.8359 metre, = 2.742 ft. = 33.38 inches.

1 Burgos standard foot, in general use, 0.2786 metre, = 0,914 ft.

1 Madrid foot very little used, = 0.2826 metre, = 0.927 ft.

1 Spanish foot, = 2.826 Fr. decimetres, = 11.12 English inches.

107.913 Spanish feet, = 100 English feet.

100 Spanish varas, = 92½ English yards.

1 Spanish league =7,416 English yards, =6.781 Fr. kilometres, nearly 4¼ English miles.

100 English miles= 27.732 Spanish leagues.

Measures of Area.

1 Square league, = 17.75 English square miles.
1 Spanish fanegada, = 5,500 English square yards.
10 English acres, = 8.800 Spanish fanegadas.

Measures of Capacity.

1 Farrega of corn, = 3439 cubic inches, = 1,599 bushels, =56.351 Fr. litres.
1 Farrega of corn, = 7.79 Imperial bushels.
1 Farrega of salt, = 25 lbs Spanish.

Import duty on coal in foreign vessels, 3 reals de Veillon, = 0 fr. 75 cts. per quintal, = 12*s*. 8*d*. English; = $3.06 U. S. per ton.

Import duty on coal in Spanish bottoms, 9*s*. 8*d*. English, =$2.34 U. S.

Export duty, in 1844, 2*s*. English, = $0.48 U. S.

KINGDOM OF BELGIUM.

System of Weights, Measure and Currency.

Belgium has adopted the weights and measures of the French metrical system; the fundamental principle of which is the measure of length. Its unity, the *mètre*, is the ten millioneth part of a quadrant of the meridional circle of the earth. The length of the metre is nearly an inch less than the English yard and half a quarter;—that is, 3 fr. 28 dec.

The unit of superficial measure, the *are*, is a square, of which the side is ten mètres.

The unit of the measure of capacity, the *litre*, is a cube, of which the side is the tenth part of a mètre, = 61.028 cubic inches.

The *stère* is a cubic metre, = 35,317 cubic feet.

The unit of the measure of weight is a *centimètre* cube of distilled water; that is a cube of which the side is a hundreth part of a metre.

The itinerary measures are the *decamètre*, = 10 mètres; the *kilomètre*, = 1000 mètres; and the *myriamètre*, = 10,000 mètres.

Land is measured by the *hectare*, containing 10,000 square *mètres;* the *decare* of 1000 square mètres, or 1196 square miles; the *are*, containing 100 square miles; and the *centiare*, which is one square mile.

For solid measure, are used the *stère* and the *decistère;* that is, a cubic mètre and its tenth part.

For the measure of weight are used the *gramme*, the *decagramme*, or 10 grammes; the *kilogramme*, or 1000 grammes; and the *quintal*, or 100 kilogrammes.*

Table of Corresponding Measures, English, and Belgian or French.

Belgian.	English.	Belgian.	English.
Metre,	3.28 feet=39.37 inches.	Hectare,	2.471 acres=11,960 sq. yds.
Millimetre,	0.039 inch.	Litre,	1.760 pint=61.03 cubic in.
Centimetre,	0.393 inch.	Decalitre,	2.201 gall.=610.28 cubic in.
Decimetre,	3.937 inches.	Hectalitre,	22.009 gall.=284 W. bush.
Myriametre,	6.213 miles=10,936 yds.	Gramme,	15.434 gr. troy.
Metre Carre,	1.196 square yards.	Kilogramme,	2.680 lbs. troy=2 lb. 8 oz. 3 dwt.
Are,	0.098 sq. rood=119.6 sq. yds.		2.605 lbs. avoir.=2 lb. 3 oz. 4 dwt.
Decare,	1196.0 square yards.	Millier or Bar,	9 tons, 16 cwt. 3 qrs. 12 lb.

The Belgian kintal, = 103 lbs. English, = 47 French kilometres.
21 Belgian kintals, = 75 lbs. = 1 ton English, = 2240 lbs.
10,1465 metrical quintals = 1 do do.

The hectolitre, used in coal measure { 90 kilog., = 284 Winches. bu. / $3\frac{1}{2}$ cubic ft., = 22 impe. gals.

11.26 hectolitres, = 1 ton English.

1014.65 kilogrammes, =1 ton, in ordinary calculations, 1000 kilogrammes are held as 1 ton.

1 hectare of land, = 2.471 acres English.

1 vierkantebunder, = 119.6 English yards square, = 1 French acre.

1 metrical mile, = 1093 English yards, = 1 French kilomètre.

1 mudde, = 6102 cubic inches English, = 2,837 bushels, = 100 French litres.

Currency.

The franc is the monetary unit of Belgium, and its divisions are made according to the decimal system.

1 franc, = 9.69 pence English, = $0.19\frac{1}{3}$ U. S. currency.

20 francs, = 1 Napoleon, = 1 new louis, = 16*s.* 2*d.* English, = \$3.86 U. S.

1 English sovereign in Belgic money, = 25 francs, 20 centimes.

1 English shilling, = 1 franc, 16 cent.

* Chiefly derived from McCulloch's Gazetteer, and Loudon's Tables.

AUSTRIAN EMPIRE.

Weights, Measures, and Money.

The Austrians, Hungarians, and Galicians, use the same standards, in most respects.

Measures of Length.

The German mile = 8101 English yards, = 7,407 French kilometres, = 4 miles 1061 yards English, 21,725 German miles to 100 English miles.

The Vienna foot = 12.45 English inches, = 3.161 French decimetres.

Weight.

The centner or quintal, or cwt of Austria or Vienna = 123½ lbs. English; therefore there are 18.2 cwt. to 1 ton English. (100 livres) = 56 kil. 00.12.

The scheffel of Prussia is 110 lbs., being about 20 scheffels or quintals to 1 ton English.

The Vienna lb. = 1.235 lb. avoirdupois.

The Livre de Vienne = 0 kil. 56.

The commercial Last, = 2905 kil. 6000 lbs. weight, =2⅔ tons English.

16 German Loths = 1 Marc = 7oz. 2dwts. 4grs. Troy.

Measures of Capacity.

The Vienne Metzen, = 3753 cub. Eng. inch. = 1.745 bush. = 61.496 Fr. litres.

The Metz of Gressburg, dry measure, = 1.745 imperial bushel English.

The Koretz of Galicia = 123 litres = 3.3 imperial bushels.

The Last of Hamburg for grain, &c., contains 11.2 Winchester quarters, = 99,540 cubic feet, = 88 Winchester bushels.

The Joch, or Austrian acre, = 1.46 English acre.

1 Oke = 43.3 oz. avoirdupois.

1 Centner or Quintal = 44 Okes.

1.785 quintals of wood = 1 French tonne.

Currency.

Gold Coinage.—The Souverain, 3 dollars 38 cents 7 mills United States. Double Ducat, 4 dollars 59 cents 3 mills United States. Hungarian Ducat, 2 dollars 29 cents 7 mills United States. Gold Ducat of Kremnitz = 4½ Florins = 6*s.* 6*d.* English = \$1.57.

Silver.—The Austrian, Hungarian, Frankfort, Nassau, and Rhenish florin is = 60 kreutzers = 15 batzen = 2*s.* English = 2 francs, 61c. = \$0.48 United States.

9 florins and 35 kreutz. = £1 English at par = \$4.84, or 4*s.* 8*d.*

1 Livre di Milano was introduced by the French = 1 franc = 22.8 kreutz.

1 Austrian Lira = 100 centimes = 0 fr. 87 cts.

Par value in London of £1 sterling, 9 florins, 50 kreutz. of Vienna.

The Convention florin of Hesse Darmstadt, Bavaria, Baden, and Holland, is = 1*s.* 8*d.* English, divided into 60 kreutzers.

The Vienna florin = 2 francs 59c. = 2*s.* 1*d.*

1 franc of 1809 = 9¾*d.* English.

5 francs = 4*s.*

Copper.—	1 Grosh,	\$0 3cts.	14	American currency.
	1 Gould,	51	85	
Austrian,	1 Rexdollaar,	77	77	
	The Kreutzer,			0 francs 87 cts. = \$0.008 U S.

TABLES OF ANALYSIS

OF ABOUT ELEVEN HUNDRED SPECIES OF MINERAL COMBUSTIBLES, DISPOSED IN GEOGRAPHICAL ORDER, IN CONFORMITY WITH THE ARRANGEMENT OF THIS WORK.

AMERICA.

Semi-bituminous or dry Coal.

State and county.	Locality.	Designation of coal beds.	By whom analysed.	Specific gra'y.	Analysis.		
					Carbon.	Volatile matter.	Ashes.
TENNESSEE.							
	Cumberl'd mountains,	Kimbrow's Vein,	Dr. Troost,	1.450	71.00	17.00	12.00
	"	Gillenwaters,	"	1.450	69.00	14.00	17.00
KENTUCKY.							
	Hawsville,	Splint or cannel coal,	Dr. Jackson,	1.250	48.40	48.80	2.80
	Caseyville,	Bitumin's coal,	Johnson,	1.392	44.49	31.82	23.69

Fat Bituminous Coals in Western Virginia.—State Reports.

County.	Locality.	Designation of coal beds.	Analysis.		
			Carbon.	Volatile matter.	Ashes.
[Upper coal series.]	Clarksburg,	Main seams,	56.74	41.66	1.60
	"	"	49.21	45.43	5.36
	Pruntytown,	"	57.60	39.00	3.40
	Morgantown,	"	60.54	37.30	2.14
Lower coals—Valley of the Ohio. Kanawha,	1. Coal creek,	Judge Summers's bank,	55.55	41.85	2.60
"	2. Grand creek,	"	52.75	43.20	4.05
Logan,	3. Wolf creek, Big Sandy river,	Burning spring,	47.15	48.00	4.85
Kanawha,	4. Big Coal River,	(Lewis')	50.20	47.10	2.70
"	5. Three mile creek,	Cartrell's,	45.95	50.30	3.75
"	6. Elk river,	Friend's mines,	55.90	39.90	5.20
Logan,	7. Logan Court-house,	Lawson's,	58.35	39.50	2.15
"	8. Guyandotte,	Traa fork,	56.50	42.00	1.50
"	9. Big Sandy river,	Pigeon creek,	55.00	41.00	4.00

Moderately Bituminous Coals in Western Virginia.

County.	Locality.	Designation of coal beds.	By whom analysed.	Analysis. Carbon.	Volatile matter.	Ashes.
Formation No. XI., Rogers.						
		Little Sewell Mountain,	Wm. B. Rogers,	80.24	17.48	2.28
		"	"	77.64	17.36	5.00
	Big Sewell Mn. E. side, W. flank,	Rogers's seam,	"	75.88	22.32	1.80
		Tyree's bed,	"	67.84	30.08	2.08
	"	Deem's bed,	"	71.73	27.13	1.14
Fayette,	Mill creek,	Paris's bank,	"	71.88	26.20	1.92
	Scrabble creek,		"	63.36	29.04	7.60
	Bell creek,		"		32.16	
Lower Coal series in the Valley of the Kanawha.						
	Keller's creek,	Hansford's,	W. B. Rogers's State report,	60.92	37.08	2.00
	Second seam,	Storkton's mine,	"	74.55	21.13	4.32
	Campbell's creek,	Ruffner's 2d sm.	"	55.76	32.44	11.80
	"	Noyes's seam,	"	64.16	32.24	3.60
	"	"	"	65.64	31.28	3.08
	Cox's creek,	3d seam,	"	51.41	42.55	6.04
	Faure's bank,	Upper seam,	"	53.20	35.04	11.76
	L. Ruffner's b'k.	"	"	49.84	44.28	5.88
	Bream's bank,	3d seam,	"	57.76	33.68	8.56
	Smither's bank,		"	54.52	29.76	15.76
	Hughes's bank,		"	62.32	32.88	4.80
	D. Ruffner's b'k.	Upper seam,	"	57.28	35.08	7.64
	Warth's bank,		"	54.00	39.76	6.24
Semi-Bituminous or Dry.						
Montgomery,	Thom's creek,	Strouble's run,	Wm. B. Rogers.	80.20	13.60	6.20
	Lewisburg,		"	78.84	14.16	7.00
Botetourt,	Catawba,		"	78.50	16.50	5.00
Hampshire and Hardy counties,—Basins containing the Lower Coal Series.						
Hampshire,	Brantzburg, N. br. Potomac,	2 m. above m. of Savage,		72.40	19.72	7.88
"	Olwer's tract,	12 ft. seam,		79.08	16.28	4.64
Maryland,	Nr. Western-port,	Sigler's mine,		82.60	15.76	2.64
"	Lonaconing,	12 ft. seam,		77.43	19.37	3.20
	Abraham's cr.	Macdonald's 3d seam,		74.00	18.60	7.40
	1 mile from top of Alleghany,	Nr. Turnpike,		77.12	19.60	3.28
	Vandover's,	N. W. Turn-pike,		61.44	14.28	24.28
Hardy,	Kitzmiller's,			79.76	15.48	4.76
"	Falls of Stony river,	Lower seam,		79.16	15.52	5.32
	Abraham's cr.	Michael's,		72.40	15.20	12.40
	Stony river,	N. of Turnpike,		83.36	13.28	3.36
	Michael's,	Upper part,		45.24	14.96	39.80

County.	No.	Locality.	Designation of coal beds.	By whom analysed.	Analysis. Carbon.	Volatile matter.	Ashes.
WESTERN VIRGINIA. Preston and Mongalia Counties,—Basins containing the Lower Coal Series.							
Preston,		Kingswood,	Fairfax's,	State reports,	53.77	31.75	14.48
"		"	Middle seam,		65.32	27.77	6.91
"		"	Forman's basin,		73.68	21.00	5.32
"		Deck Hollow, c.	Martin's,		65.42	23.42	11.16
"		Buffalo Lech run,	Beatty's,		62.56	29.60	7.84
"		Big Sandy,	N. Brandon's,		67.60	22.40	10.00
"		N. Brandonville,	Morton's,		65.28	30.80	3.92
"		Cheat river, n. Kingswood,	Price's,		60.36	25.00	14.64
"		Big Sandy, W. side,	Seaport's,		66.64	27.12	6.24
"		Kingswood,	Hagan's,		68.32	26.48	5.20
"		"			67.28	29.68	3.04
"		Big Sandy basin,	W. side Cheat,		60.04	26.88	13.08
"		Kingswood,	Cresaps,		64.24	30.24	5.32
Bituminous Coals in EASTERN VIRGINIA, in the Chesterfield, Powhatan, Goochland, and Henrico Basins.							
South side James river,	1	Stonehenge,	Chesterfield,		58.70	36.50	4.80
Chesterfield,	2	Maidenhead,	Engine shaft,		63.97	32.83	3.20
"	3	Heth's pit,	"		62.35	37.65	2.80
"	4	Mill's & Reid's,	Creek pit,		57.80	38.60	3.60
"	5	Wills's pit,			62.90	32.50	4.60
"	6		Green hole sh't,		67.83	30.17	2.00
"	7	Heth's deep shaft,	Bottom seam,		53.36	35.82	10.82
"		"	Middle seam,		66.50	28.40	5.10
"		"	Top seam,		61.68	28.80	9.52
"	8	Powhatan pits,	Finney,		59.87	32.33	7.80
"	9	Winterpock cr'k,	Cox's mine,		65.52	29.12	5.36
		Cloverhill, Appomattox river,	Slate coal,	G. W. Andrews, M. D.,	55.00	38.50	6.50
		"	Mean of 4 spec.	Johnson,	54.83	33.04	10.13
		Richmond coal,		Andrews,	59.25	32.00	8.75
		Mid Lothian,	Wooldridge's p.	Johnson,	61.08	28.45	10.47
		"	Mean result, av. size coal,	"	53.01	33.25	14.74
		Creek Coal Co.,	Mean of 6 trials,	"	60.30	31.13	8.57
		Black Heath pits,	Mean of 4 spec.	"	58.79	32.57	8.64
		Tippecanoe pits,	"	"	54.62	36.01	9.37
North side of James R.	10		Randolph's,	W. B. Rogers, State report,	66.15	30.50	3.35
	11	Coalbrook dale,	Second seam,	"	66.48	29.00	4.52
	12	Anderson's pit,	First seam,	"	66.78	28.30	4.92
	13	Barr's pits,	"	"	70.80	24.00	5.20
	14	"	Second seam,	"	54.97	22.83	22.20
	15	"	Third seam,	"	65.50	24.70	9.80
	16	"	Fourth seam,	"	56.07	21.33	22.60
	17	Crouch's Lower shaft,	Upper seam, 110 ft. from surface,	"	64.60	30.00	5.40
		"	Mean of 4 spec.	Johnson, State	67.32	23.96	8.72
	18	Scott's pit,		report,	60.86	33.70	5.44
	19	Waterloo shaft,		"	55.20	26.80	18.00
	20	Deep Run pits,		"	69.84	25.16	5.00
		"	Mean of 40 sp.	"	67.96	21.57	10.47
		Wills's pit,	Upper vein,	T. G. Clemson,	66.60	28.80	4.60
		Anderson's pit,	Bottom seam,	R. C. Taylor,	64.20	26.00	9.80
		Chesterfield,	Called natural coke,	W. B. Rogers, State report,	80.30	9.98	9.72
		"	"	"	70.00	16.00	14.00
		"	"	Prof. Bailey,	68.00	17.00	15.00
		"	Mineral charc'l,	T. G. Clemson,	83.30	10.70	6.00

Semi-bituminous or dry Coals in the State of Maryland.—The Cumberland or Frostburg Coal region, occupying a small part of Pennsylvania.

State and County.	Locality.	Designation of Coal Beds.	By whom analysed.	Specific grv'y.	Analysis.		
					Carbon.	Volatile matter.	Ashes.
PENNSYLVANIA. Somerset,		1. Hoyman's new 8 foot bed,	W. R. Johnson,	1.343	69.90	22.00	8.10
"		2. Uhl's up. vein,	"	1.319	75.75	20.20	4.05
"		3. Korn's,	"	1.386	68.46	20.10	11.44
"		4. Schaeffer's,	"	1.370	70.70	18.80	10.50
"		5. Hoyman's 8 ft. as above,	"	1.363	71.50	18.30	10.20
"		6. Hoyman's 6 ft.	"	1.362	68.54	19.80	11.66
"		7. Uhl's 7 ft vein,	"	1.388	68.44	19.50	12.06
"		8. Weller's 4 ft.	"	1.321	69.10	19.99	11.00
"		9. Church land vein,	"	1.480	68.56	18.70	12.74
"		10. Hardin's v'n,	"	1.491	66.36	17.60	16.04
		Mean results of the ten veins,		1.382	69.73	19.59	10.68
MARYLAND. Alleghany,	Maryland company,	Hoffman's mine, on main seam,	Silliman and Shepard,	1.380	82.01	15.00	2.99
"	Cumb'd coal,	"	W. Hayes, (Boston,)		77.86	15.60	6.54
"	Savage river,	"	Dr. Jones, (Washington,)		78.00	19.00	3.00
"	"	"	D. Jackson, (Boston,)	1.321	77.09	16.05	7.06
"	Maryl. comp.	"	Dr. T. P. Jones, (Washington,)	1.291	72.50	22.50	5.00
"	"	"	"	1.333	81.00	15.00	4.00
"	"	Frost's mine,	Dr. Ducatel,		70.00	20.50	9.50
"	Dan's mount.	Av. of 40 spec.	Johnson,	1.311	73.59	16.04	10.37
"	Cumb'd coal,		Prof. Daniel,		66.30	19.40	14.30
"	"	Maryland com'y,	Johnson,	1.431	67.26	14.42	18.32
"	"	Frostburg Neff's,	"	1.332	74.53	15.13	10.34
"	"	Howell's estate,	Silliman,		76.77	14.66	8.57
	"	"	Prof. Renwick,		81.00	13.00	6.00
"	"	Easby's,	Johnson,	1.305	77.25	16.23	6.52
"	George's creek.	Main vein, Lonaconing,	Dr. Ducatel,	1.386	79.25		
"	"	Third coal,	"	1.552	80.08		
"	"	Fourth coal,	"	1.584	85.00		
"	Lonaconing company,	George's creek, thick bed,	Johnson,	1.346	70.75	16.03	13.22
"	Maryland company,	Eckert mine on main seam,	"	1.437	68.56	15.62	15.82
"	Frostburg.		Chilton,		77.00	12.00	11.00
"	"	Mean of 2 analy.	Dr. J. Percy,		78.80	9.47	11.73
"	Big vein,	" 5 "	Dr. Higgins,	1.320	88.05	8.54	3.41
"	6 ft. vein,	" 5 "	"	1.340	86.01	8.68	5.31
"	44 in. vein,	" 5 "	"	1.390	74.24	7.13	18.63
"	Oakland,	" 5 "	"	1.290	73.34	12.54	5.12

Fat, Bituminous Coals in the State of Ohio.

County.	Locality.	Designation of Coal Beds.	By whom analysed.	Specific gr'y.	Analysis.		
					Carbon.	Volatile matter.	Ashes.
Portland,	Talmadge,	Upson's mine,	W. W. Mather,	1.264	53.404	44.298	2.288
Jackson,	Lick Town'p,		"	1.283	49.882	47.327	2.221
"	Madison T'p,		J. L. Cassels,	1.560	39.950	44.800	14.620
"	"	Cannel Coal,	"	1.410			
	Carr's Run,		R. C. T.	1.270			
	Pomeroy,		Dr. J. Percy,		76.70	18.70	4.60

Fat, Bituminous Coals in Pennsylvania.

County.	Locality.	Names of Coal Seams.	By whom analysed.	Specific gr'y.	Analysis.		
					Carbon.	Volatile matter.	Ashes.
Venango,	Shippensv'le,	Sandy Ridge,	H. D. Rogers's State report,		49.80	43.20	7.00
"	6 M. F. of Franklin,		"		29.54	52.78	17.68
Beaver,	Greersburg,		"		30.12	36.00	33.88
Crawford,	Conneaut l'e,		"		59.45	38.75	1.80
Mercer,	Greensville,		"		57.80	40.50	1.70
	"		R. C. T.	1.275			
	Orangeville,		State report,		53.45	43.75	2.80

Moderately bituminous, dry, and close burning Coals in Pennsylvania.

County or District.	Locality.	Designation of Coal Beds.	By whom analysed.	Specific gr'y.	Analysis.		
					Carbon.	Volatile matter.	Ashes.
Tioga or Blossburg Coal-field.	Blossburg,	Coal Run, upper vein,	Taylor and Clemson,	1.371	75.40	16.40	8.20
	" Bear cr'k,	Clement's coal,	"	1.398	73.74	15.00	11.26
	"	Bloss's coal,	"	1.405	73.00	15.60	11.40
	"	"	State report,		62.80	32.80	5.20
	" Johnson's Run,	Splint coal,	Taylor and Clemson,	1.493	69.30	14.60	16.10
	" Coal Run,	Slaty variety called Cannel,	"	1.750	33.40	8.40	58.20
	" "	Pitch coal,	"	1.500	54.26	18.50	27.24
	Head of Tioga,	New Hope vein,	R. C. T.	1.429			
	Arbon company,	Coal run, mean of 4 specim.	Johnson,	1.323	73.11	16.12	10.77
Ralston and Lycoming Creek District.	Ralston,	Big vein,	State report,		74.50	20.50	5.00
	"	"	Johnson,	1.387	71.54	14.50	13.96
	Queen's Run,	Av. of 40 spec.	"	1.331	73.44	18.81	7.75
	"		State report,		73.68	21.50	4.60
Bradford or Towanda Coal-field.	Schrœder, branch of Towanda cr'k,	Lower bed in three parts.	W. R. Johnson,	1.515	62.60	15.00	22.40
				1.448	70.00	17.40	12.60
				1.465	63.90	19.10	17.00
	"	Miller's old coal drift,	"	1.377	68.10	20.50	11.40
				1.378	65.50	19.20	15.30
				1.349	74.97	19.30	5.73
	"	Mason's coal, upper part,	"	1.388			
		lower part,		1.400			
Centre county,	Snow-shoe,		State report,		76.73	21.20	2.07
"	Farrands-ville,	Select port'n of Diamond vein,	Bache and Rogers,	1.339			5.50
"	Lick Run,		State report,		66.21	20.72	13.07
Clearfield co.,	Karthaus,		W. R. Johnson,	1.263	68.15	26.80	5.05
"	"	Salt Lick,	"	1.292	80.49	12.83	6.68
				1.275	76.64	22.27	5.09
"	"	Upper seam,	State report,		78.20	13.00	8.80
"	"	Lower seam,	"		70.50	24.80	4.70
"	Curwensville,	Reed's vein,	"		67.70	27.00	5.30
"	Caledonia,		"		54.50	37.00	8.50
"	"		"		54.60	38.20	2.70

Moderately bituminous, dry, and close burning Coals in Pennsylvania.—(Continued.)

	County or District.	Locality.	Designation of Coal Beds.	By whom analysed.	Specific gravity.	Analysis. Carbon.	Volatile matter.	Ashes.
Moshannon district.		Karthaus,	Main or 6 ft. s'm,	W. R. Johnson,		68.15	26.80	5.05
		Philipsburg,	Best coal,	W. R. Johnson and R. C. T.,	1.308	70.00	22.30	7.70
		"	"	State report,		64.40	29.50	6.10
		"	"	Dr. Goddard,	1.360	70.00	20.00	10.00
		near "	Showalter's 3 ft. vein,	R. C. T.	1.358			
		" "	Goss's 6 ft. vein,	"	1.357			
		" "	Coal hill,	"	1.500			
		16 miles,	Steed's mine,	State report,		68.40	20.40	11.20
		17½	Leech's mine,	"		67.93	20.32	11.75
	Cambria,	Blair's Gap,	Portage Rail R.	T. G. Clemson,		77.00	15.00	8.00
	"	"	Mineral charc'l,	"		66.40	6.60	27.00
	"	"	Large bed,	State report,		65.00	31.00	4.00
	"	Summit,	Portage Rail R.	Johnson,	1.406	69.59	21.36	9.05
Semi-bituminous Coals in Pennsylvania.	Dauphin,	Short Mn.	South drift,	Dr. Ellet, R. C. T.	1.330	75.50	15.30	9.20
	"	"	Big Flat's vein,	J. C. Booth,			16.90	
				M. C. Lea,		71.20	17.32	13.46
				R. C. T.	1.395			
				Rogers's reports,		76.94	15.06	8.00
	"	Rattling Run Gap,	Perseverance vein,	J. C. Booth,			15.80	
				M. C. Lea,	1.391	78.80	13.20	8.00
				"		76.10	16.90	7.00
				Rogers's reports,		74.55	13.75	11.70
				Johnson, mean of 6 exp's,	1.531	72.22	14.29	11.49
	Lebanon,	Yellow Springs,	Back bone vein,	H. C. Lea,	1.389	74.70	14.80	10.50
	"	"	Kugler vein,	J. C. Booth,			8.10	
				M. C. Lea,	1.391	80.33	8.86	10.80
				"		81.20	9.80	9.00
				"	1.395	77.50	11.00	11.50
				Rogers's reports,	1.410	79.55	10.95	9.50
	"	Cold Spring,	Six feet vein,	H. Lea,	1.403	76.30	11.10	12.60
	Bedford,	Broad Top M.	Riddle's bank,	Clemson,	1.700	70.10	16.70	13.20
	"	"	Hopewell,	Rogers's reports,		84.80	11.20	4.00
	"	"		W. R. Johnson,		77.60	16.00	6.40
Soft, free burning Red Ash Anthracite in Pennsylvania. Coal seams of the South Mn.	Dauphin,	Lyken's Valley,	Bear Gap mines,	R. C. Taylor,	1.318			
	"	"	1st sample,	W. R. Johnson,	1.391	87.95	7.60	4.45
	"	"	2d,	"	1.404	89.30	5.95	4.75
	"	"	3d,	"	1.416	85.70	10.00	4.30
	"	"	4th,	"	1.374	88.70	4.60	6.70
	"	"	5th,	"	1.376	87.75	8.35	3.90
	"	"	6th,	"	1.395	88.65	8.30	3.05
	"	"	7th,	"	1.382	87.20	7.84	4.15
	"	"	8th,	"	1.398	83.99	11.85	4.15
	"	"	9th,	"	1.378	87.00	7.30	5.70
	"	"	mean of 9 s'pls,	"	1.390	87.36	8.06	4.57
	"	"	Third Bed,	H. D. Rogers's report,		88.25	8.85	2.90
	Schuylkill,	Lower Mohantongo,	Klinger's or Rausch Gap,	J. R. Chilton, M. D.,		89.71	4.48	5.81

Anthracite of Pennsylvania.

	Description and Localities of Anthracite Coal Beds.	By whom examined or analysed.	Specific gravity.	Analysis of 100 parts of Anthracite.		
	Hard, White Ash Coal.			Carbon.	Water, hydro. and volatile matter.	Ashes, silica, earthy matter, iron, &c.
Schuylkill Eastern Region.	Mauch Chunk,	Olmsted,	1.550	90.10	6.60	3.30
	"	Vanuxem,	1.494			
	" Summit Mines,	W. R. Johnson,		92.30	6.42	1.28
	" "	Karsten,		86.00	8.00	6.00
	" "	M. C. Lea,		87.00	7.30	5.70
	" 14 feet vein,	Rogers's reports,		88.50	7.50	4.00
	" Hardest variety,	"		87.70	6.60	5.70
	" mean of 2 results,	Dr. J. Percy,		92.60	5.15	2.25
	Nesquehoning,	Taylor,	1.558			
	" 10 feet vein,	Rogers's reports,		86.60	6.40	7.00
	Tamaqua vein, D. east,	M. C. Lea,		91.00	5.50	3.50
	" D. "	Rogers's reports,	1.570	92.07	5.03	2.90
	" E. "	"	1.600	89.20	4.54	6.26
	" R. "	"	1.550	87.45	7.55	5.10
	Tuscarora,	"		88.20	7.50	4.30
	Forest Improvement, av. of 4 specim.	Johnson,	1.477	92.12	4.83	3.05
Middle Coal-field.	Hazleton,	Taylor,	1.550			
	Sugar Loaf mountain,	W. R. Johnson, 3 samples,	1.591	88.18	6.99	4.83
			1.574	85.91	5.36	8.73
			1.550	90.70	7.06	2.24
	Beaver Meadow,	Taylor,	1.600			
	"	W. R. Johnson,	1.630	85.34	9.60	5.06
	"	"	1.560	91.64	6.89	1.47
	"	"		92.30	6.42	1.28
	Girardville,	Taylor,	1.600			
	Broad Mountain, W. W. Branch,	"	1.700			
Pine Grove District.	Big vein, Lorberry creek,	"	1.472			
	" "	M. C. Lea,		85.90	7.20	6.90
	Locust M't Coal and Iron Company,	Blake, (Boston),		96.77		3.33
Red Ash c'l,	Sharp Mountain,	Rogers's reports,	1.540	80.57	7.15	3.28
Red Ash.—Free burning coal, with slight flame. *Stony Creek Estate, Lebanon Co. First Coal-field.*	Black Spring Gap, 4 feet vein,	Taylor,	1.528			
	" "	Rogers's reports,	1.440	82.47	9.53	8.00
	" Peacock vein,	Taylor & M. C. Lea,		88.60	7.10	4.30
	" Grey vein,	M. C. Lea & Taylor,	1.379 1.395	86.00	4.50	9.50
	" The black compact part,	Rogers's reports,	1.440	81.02	9.78	9.20
	" The Grey central part,	"	1.330	81.40	11.40	7.20
	" Fishback vein,	M. C. Lea,		84.00	6.50	9.50
	" Lea vein,	Rogers's reports,	1.350	85.84	8.96	5.20
	Gold Mine Gap, Peacock vein,	M. C. Lea,		83.0	9.0	8.0
	" "	Rogers's reports,	1.410	83.15	10.95	6.90
	" Heister vein,	"	1.410	81.47	10.43	8.10
	Rausch Gap,	M. C. Lea,		78.90	11.00	10.10
	" Pitch vein,	R. C. Taylor,	1.387			
		"	1.454			
	" Heister vein,	M. C. Lea,		77.10	10.90	12.00
		Rogers's reports,	1.450	77.23	10.57	12.30
White Ash Coal. *South and Mid. Coalfield.*	Broad Mountain,	Dr. C. T. Jackson,	1.593			7.80
	Lehigh or Summit Company, 1st,	W. R. Johnson,	1.613	87.48	7.51	5 01
	" " 2d,	"	1.594	91.69	4.31	4.00
	" " 5th,	"	1.612	86.06	9.23	3.71
	Mauch Chunk,	Dr. J. Percy,		84.98	4.82	10.20
	Buck Mountain,	W. R. Johnson,	1.559	91.02	5.90	3.08
	Shamokin, (Snyder's)	Rogers's reports,		89.90	6.10	4.00
	West Mahanoy,	Taylor,	1.371			
White Ash Coal. *North or Wyoming Coal-field.*	Wilkesbarre, Blacksmith's coal,	"	1.472			
	" Warden's vein,	Rogers's reports,	1.403	88.90	7.68	3.49
	Wyoming,	J. F. Frazer,		91.20	4.50	4.30
	Lackawanna,	Dr. C. T. Jackson,	1.609	79.20	9.20	11.60
	" mean result,	Johnson,	1.421	88.98	6.36	4.66
	Scranton anthracite,	Prof. H. D. Rogers,		87.88		
	Carbondale,	Rogers's reports,	1.404	90.23	7.07	2.70
Red Ash coal. *Pottsville District. First or South Coal-field.*	Peach Mountain, Delaware co.,	Taylor,	1.446			
	" mean of 40 specimens,	Johnson,	1.464	86.09	6.96	6.95
	" N. American Co.,	Dr. C. T. Jackson,	1.569			7.20
	Peach Orchard,	"	1.532			6.60
	Salem vein,	Taylor,	1.574			
	Plumbago vein, Sharp Mount,	"	1.412			
	Black Mine vein,	H. Lea,		88.40	6.80	4.80
	Gate vein,	Dr. C. T. Jackson,	1.609			6.60
	Shenoweth vein,	Rogers's reports,	1.500	94.10	1.40	4.50
	Nealey's Tunnel, 3d vein,	"	1.550	89.20	5.40	5.40

Anthracites of the United States.

Description of Coal Beds.		Details.		Analysis of 100 parts of anthracite.		
State.	Locality of Mines.	By whom examined or analysed.	Specific gravity.	Carbon.	Water, hydrogen and volatile matter.	Ashes, silica, earthy matter, iron, &c.
Rhode Island,	Portsmouth mines,	Dr. C. T. Jackson,	1.850	85.84	10.50	3.66
	" "	"		87 50	7.00	5.50
	" "	"		77.50	13.00	9.50
	" "	"	1.770	84.50	10.00	5 50
	" "	L. Vanuxem,		90.03	4.90	5.07
	" "	"		77.70	6.70	15.60
	Cumberland, "	Dr. C. T. Jackson,		77.00	7.60	15.40
	" "	"		39.70	7.80	52.10
	Providence, "	"		72 00		28.00
	Portsmouth old mine,	"		74.00	10.00	16.00
	Case's mine,	"		97.00		3.00
Massachusetts,	Mansfield mine,	"	1 690	87.40	6.20	6.40
	" "	"	1.780	92.00	6.00	2.00
	" "	"	1.710	92.00	6.00	2.00
	Worcester plumbaginous anthrac'e,	Dr. J. Percy,		28.35	3.08	68.57
North Carolina,	Near Leakesville, middle secondary rocks,	W. B. Rogers,		83 12	7.76	9.12
*England,	Borrowdale, } Plumbago,	L. Vanuxem,		83.37	1.23	10.40
*Pennsylvania,	Bustletown, } Plumbago,	"		94.40	.60	5.00
*England,	Cornwall, } Plumbago,	Saussure,		96 00		4.00

Bituminous Coals.

State and County.	Locality.	Designation of Coal Beds.	By whom analysed.	Spec. gr'y.	Analysis.		
					Carbon.	Volatile matter.	Ashes.
INDIANA,							
Parke county,	Sugar creek,	Foundry,	D. D. Owen,	1.219	75 00	21.00	4.00
Vermilion,	Brouillet's creek,		"	1.270	52.00	39.00	9.00
Vigo,	Honey creek,		"	1.240	70.00	27.50	2.50
Sullivan,	Busseron,	Lick Fork,	"	1.240	70.00	28.00	2.00
Fountain,	Wabash,	Coal creek mouth,	"	1.260	60.00	25.00	15.00
Spencer,	Anderson creek,		"	1.270	45.00		
	White river,		"	1.270	56.40		
	Terre Haute,		"	1.210	50.80		
	Cannelton,	Cannel coal,	W. R. Johnson,	1 272	59.47	36.59	3.94
ILLINOIS,	Rock river,	Coal,	Dr. D. D. Owen,	1.340	45.50	44 50	10.00
	Vermilion,	Danville,	A. Morfit,		48.50	47.20	4.30
	Western port,		Johnson,	1.290		32.80	
	Ottowa,		J. F. Frazer,		62.60	35.50	1.90
	Rockwell,		C. U. Shepard,	1.273	46 50	47.50	6.00
IOWA,	Duck creek,	W. bank of the Mississippi river	Dr. D. D. Owen,	1.270	48 50	44.00	7.50
MISSOURI,	Cote-sans dessein, Callaway co.,	Mastodon vein, forty-six feet thick,	Booth and Boye,		46.83	40.05	13.12
			J. R. Chilton, M. D.	1.252	50.81	34.06	15.13
		Mammoth vein, twenty-four feet,	"	1.250	50.78	34 20	15.02
	Osage river,		W. R. Johnson,	1.200	51.16	43.50	5.34
ARKANSAS,	Johnson county,	Spaldre's bluff,	J. F. Frazer,	1.396	62.60	28.90	8.50
MAINE,		Peat,	Dr. Jackson,		21.00	72.00	7.00
Miscellaneous Analysis,							
Isle of Cuba,	Near Havana,	Asphalt,	T. G. Clemson,	1.190	34.97	63.00	2.03
	Near Matanzas,	Asphaltum,	"				13.50
South America,	Peru,	Coxitambo,	M. Bousingault,				
	Chili,	Arauco,	W. R. Johnson,	1.324	67.62	30.00	2.38
	Brazil,		Karsten,	1.289	57.90	40.50	1.60
	"		"	1.483	38.10	33.50	28.40
Madeira Island,	Brown coal,	Or lignite,	Johnstone,				20.05
British America Bitum's Coal.							
Nova Scotia,	Pictou,	Cunard's sample,	Johnson,	1.325	60.73	26.76	12.51
		Mining Associa'n,	"	1 318	56.98	29.63	13.39
Cape Breton,	Sydney,	Mean of 2 species,	"	1.338	67.57	26.93	5.50

* Mr. Vanuxem adds, by way of comparison, the analysis of 3 varieties of plumbago or graphite.

EUROPE.

Fat, bituminous Coals of England and Scotland.

Description and Locality of Veins.	By whom analysed and described.	Specific gravity.	Analysis of 100 parts of Coal.		
			Carbon.	Bitumen, volatile matter and water.	Ashes and cinders.
Alfreton, furnace coal,	D. Mushet,	1.235	52.46	45.50	2.04
Butterley, " Derbyshire,	"	1.264	52.88	42.83	4.29
Derbyshire, cannel c'l, Alfreton,	"	1.278	48.36	47.00	4.64
Wigan, cannel coal,	Dr. Thomson,		52.60	44.00	3.40
"	Kerwan,	1.272	75.20	21.68	3.10
"	R. C. T.	1.275	61.73		
Lancashire, cannel coal,	Dunn,		56.40	41.00	2.60
Woodhall, near Glasgow, cannel coal,	Dr. Ure,	1.228			
Liverpool coal,	Johnson,	1.260	54.90	40.48	4.62

Fat, bituminous adhesive Coal—coked previously to using in the Furnace.

Description and Locality of Veins.	By whom analysed and described.	Specific gravity.	Analysis of 100 parts of Coal.		
			Carbon.	Bitumen, volatile matter and water.	Ashes and cinders.
Newcastle-upon-Tyne Birtley Works,	Dufrenoy and Berthier,	1.270	60.50	35.50	4.00
" "	Thomson,		65.90	32.60	1.50
" "	Karsten,	1.256	67.65	31.50	0.85
Northumberland, Tyne Works,	Dufrenoy and Berthier,		67.50	30.00	2.50
Staffordshire, Apdale Works,	"		62.40	34.10	3.50
Redesale, Newc'le, N. of Tyne,	M. Dunn,		49.95	51.00	3.05
Wylam,	"		48.49	37.60	13.91
Garesfield and Auckland,	"		72.71	25.90	1.39
Newcastle coal, (mean)	W. R. Johnson,	1.257	57.00	37.60	5.40

Bituminous Coal—used crude in the Hot Air Furnace.

Description and Locality of Veins.	By whom analysed and described.	Analysis of 100 parts of Coal.		
		Carbon.	Bitumen, volatile matter and water.	Ashes and cinders.
Staffordshire, Tipton, Wednesbury coal works,	Berthier,	67.50	30.00	2.50
Derbyshire, Butterley, cherry coal,	"	57.00	40.00	3.00
Derbyshire, Dodnor Park, soft coal,	"	51.50	45.50	3.00
Soft or mixed English,	D. Mushet,	53.00		

Bituminous and Semi-bituminous Coals of South Wales, on the eastern side of the Coal Basin.

Description and Locality of Coal Beds.		Thickness of each bed. ft. in.	An. of 100 parts co'l by Mushet. Carbon.	Bit'n, vo'e matter, &c.	Ashes or cinders.	The Coal described and its uses.
Abersychan,	Meadow vein,	8 6	65.98	29.40	4.62	thin laminæ, furnaces.
	Old coal,	2 6	71.10	27.40	2.50	laminæ, "
Golynos Iron works,	Three quarter,	5 6	71.88	25.50	2.62	"
	Rock vein,	7 0	69.60	27.40	3.00	thin laminæ, run out fires.
	Meadow vein,	7 0	68.00	27.50	4.53	dense, furnaces.
	Old coal,	2 4	73.40	20.60	6.00	laminæ, "
Verteg Iron works. Furnace coal,	Red vein,	4 0	69.45	26.30	4.25	" "
	Big vein,	4 0	66.05	30.70	3.25	irreg'ly laminated, "
	Droydeg or rock vein,	4 0	64.45	32.30	3.25	oblong, forge and mill.
	Three quarter,	5 6	67.90	29.60	2.50	rather friable, furnace.
	Meadow vein,	7 0	69.25	30.50	9.25	thin laminæ, "
Blaenafon Iron works,	Three quarter,	5 6	65.63	31.25	3.12	" "
	Droydeg or rock vein,	5 6	65.55	28.95	5.50	cubical, refineries.
	Meadow vein,	2 10	72.00	26.00	2.00	furnace.
	Old coal,	6 0	75.21	22.29	2.50	laminated, "
Clydach or Llanelly Iron works,	Red vein,	3 0	76.58	20.80	1.62	forge and mill.
	Big vein,	5 0	73.42	24.58	2.00	thin layers, "
	Three quarter,	2 9	72.70	25.30	2.00	broad, "
	Droydeg or rock vein,	7 0	72.13	21.87	6.00	cubical, "
	Tach coal,	3 0	70.05	25.57	4.38	cones, household.
	Yard vein,	2 9	78.68	19.32	2.00	hard, furnaces.
	Old coal,	5 6	77.55	18.95	3.50	alternating laminæ, "
Nant-y-glo Iron works,	Elled coal,	3 6	81.87	17.13	1.00	lamellar, "
	Three quarter,		82.65	15.10	2.25	coking for the refineries.
	Droydeg vein,		77.14	17.86	5.00	twisted, forges and mills.
	Old coal,	5 6	78.75	18.75	2.50	imperfect, furnaces,
Sirhowey Iron works,	Ell coal,		82.04	16.71	1.25	sectional, "
	Three quarter,		83.50	12.00	4.30	with numerous partings.
	Big vein,	8 0	81.52	15.10	3.38	rhomboidal, furnace.
	Mudelog v'n or droydeg,		80.50	11.87	7.63	partially granulated.
	Yard coal,		82.24	15.88	1.88	twisted, with clod.
	Engine coal,	4 0	82.46	15.41	2.13	irreg'ly granular, furnaces.
	Three quarter engine,		75.78	13.22	11.00	hard.
	Old coal,	4 0	78.50	13.00	8.50	conchoidal partings.
Ebbw Vale Iron works,	Yard coal,	3 2	81.04	15.83	3.13	friable, forges and mills.
	Three quarter,	3 0	80.25	17.00	2.75	cubical, blast furnaces.
	Big vein,	4 8	81.37	17.00	1.63	cubical, " and forges.
	Bydelog or droydeg,	2 10	72.88	16.87	10.25	coarse, " "
	Ell coal,	3 2	82.32	16.30	1.38	granular, " "
	Old coal, top,	5 6	79.28	18.22	2.50	cubical, forge and mill.
	" bottom,		81.77	15.73	2.50	less granular, blast furn'ce.
Beaufort,	Old coal,	4 6	76.82	20.80	2.38	cubical, fur'ce and refin'ry.
Pont-y-mister,	Forge coal,		75.06	22.22	2.72	cubical, furnace and forge.
Blaina,	Big vein.	5 6	72.14	25.86	2.00	granular, "
	Three quarter,	5 6	77.38	21.12	1.50	cubical, "
Tredegar Iron works,	Big vein, top.	3 0	80.45	16.55	3.00	compact, blast furnace.
	" lower part,	3 0	81.56	14.94	3.50	very compact, "
	Red vein, upper part,	1 0	79.44	14.06	6.50	laminated.
	" under part,	3 0	80.26	18.49	1.25	reedy laminæ, blast fur'ce.
	Bydelog, (droydeg)	3 0	78.90	17.97	3.13	fine laminæ, forge.
	Yard vein,		82.26	15.36	2.38	reedy laminæ, blast fur'ce.
	Penmark vein,		80.11	14.06	5.83	reedy, forges and mills.
	Three quarter,		80.20	13.80	6.00	laminæ with clod.
	Engine coal or old coal,	5 6	77.30	15.20	7.50	reedy.
Rhymney Iron works,	Big vein, upper part,		81.26	16.24	2.50	partially reedy, blast fur'e.
	" lower part,		82.33	13.17	4.50	strong, "
	Raslas vein,	7 0	82.79	12.96	4.25	reedy, "
	Brassy vein,	1 3	83.38	14.37	2.25	twisted, workmen's fires.
	Red coal,	3 6	84.25	12.75	3.00	granular, furnaces.
	Yard coal,		80.92	16.20	2.88	irregular, "
	Four feet coal,		80.15	15.10	4.75	irreg. l'ina, forges & mills.
	Fire clay coal,	1 0	80.60	17.40	2.00	reedy.
	Black vein,	2 6	82.51	14.99	2.50	" forge.

Semi-bituminous or Steam Coals of South Wales.

	Description and Locality of Coal Beds.		Thickness of each vein. ft. in.	An's of 100 pt's of c'l, by D. Mushet. Carbon.	Bit. vol. m'r, water, &c.	Ashes or cinders.	The Coal described and its uses.
On the northern side of the Coal Basin.	Bute Iron works,	Red vein,	4 0	83.04	14.58	2.38	thin laminæ.
		Big vein,	9 0	83.53	13.74	2.73	"
		Raslas vein,	9 0	84.06	12.44	3.50	"
		Four feet vein,	4 0	78.30	16.45	5.25	in compact, forges.
	Dowlais Iron works,	Big v. upper part,	11 0	80.88	15.62	3.50	reedy, blast furnaces.
		Middle part,		85.00	11.87	3.13	thin layers, brittle.
		Bottom part,		82.81	13.44	3.75	irregular, shining.
		Raslas vein,					
		Upper part.	9 0	84.08	13.02	2.90	imperfect cleavage.
		Lower part,		85.02	13.23	1.75	strong, laminæ thin.
		Upper four feet,	4 0	85.75	12.75	1.50	prisms, furnace coal.
		Lower four feet,	4 0	85.35	12.40	2.25	shining laminæ.
		Cwmcenol,	2 9	88.63	9.74	1.63	in layers, furnace coal.
		Four feet forge,	5 0	88.78	7.97	3.25	part reedy, forge coal.
		Little vein,	3 0	86.90	11.72	1.38	thin laminæ, forges.
	Pen-y-darren Iron works,	Foes-y-frane,	7 0	85.04	11.46	3.50	compact, blast furnaces.
		Raslas,	7 0	87.69	10.31	2.00	incompact, "
		Three feet coal,	3 0	85.07	12.18	2.75	broad, "
		Four feet coal,	4 0	88.50	10.00	1.50	hard, "
		Upper vein,	3 6	86.08	10.42	3.50	" forges and mills.
		Roof-pin vein,	2 0	83.30	11.20	5.50	" mixed quality.
		Big vein,		86.01	12.24	1.75	crystallized, with clod.
		Bottom coal,	4 0	87.20	11.30	1.50	compact, forges.
	Plymouth and Duffryn Iron works,	Upper or yard coal,	3 9	79.06	15.34	5.60	mixed, "
		Four feet coal,	4 0	84.98	11.77	3.25	" less shining.
		Clynmil coal, top part,	3 6	86.62	12.00	1.38	conical, best furnace coal.
		" bottom,	4 6	85.48	12.39	2.13	granular, blast furnaces.
		Raslas, top part,	4 6	82.05	13.95	4.00	bright, "
		" bottom part,	3 6	83.84	13.33	2.83	both reedy and granular.
		Upper Dingle coal,	3 6	77.00	20.00	3.00	broad, forges and mills.
		Lower Dingle,	2 6	80.34	16.66	3.00	less bright and shining.
	Cyfarthfa and Ynnis Vach. Merthyr Iron works,	Cyfarthfa big vein,	9 0	90.28	7.97	1.75	slightly reedy.
		Cwm-dhu pit,	6 0	88.78	9.22	2.00	regularly laminated.
		Cwm-mynedd,	5 6	88.87	9.00	2.13	slaty, forges and mills.
		Cwm-y-glo,	4 0	89.29	6.58	4.13	incompact, blast furnaces.
		Upper yard vein,	3 0	86.80	11.20	2.00	regular, "
		Gelly-deg,	3 0	91.86	6.14	2.00	specular, forges.
		Mountain vein,	3 6	90.02	8.48	1.50	crystallized, furnaces.
		" ironstone,	1 6	92.11	6.14	1.75	reedy and granular.
	Aberdare Iron works,	Four feet coal,	4 0	87.00	11.50	1.50	compact, blast furnaces.
		Raslas vein,	10 0	88.89	9.11	2.00	" forges.
		Two feet coal,	2 0	88.12	10.00	1.88	mixed reedy and granu'r.
		Small vein,	5 6	80.42	10.83	8.75	compact, refineries.
		Foes-y-frane,	4 0	84.67	8.33	7.00	" furnaces.
		Two feet 9 inch coal,	2 9	88.51	9.99	1.50	bright reedy coal.
		Yard vein,		82.99	14.26	2.75	reedy, furnaces.
		Black mine coal,		91.18	6.82	2.00	bright reeded coal.
		Upper vein,		82.12	12.13	5.75	granular, with clod.
	Hirwain Iron works,	Big vein,	9 0	88.94	7.18	3.88	crystallized, forges.
		Four feet coal, or Glo-wynn,	4 0	90.26	7.86	1.88	slightly reedy, furnaces.
		Six feet coal,	6 0	89.96	8.04	2.00	regular reeded, forges.
		Pit vein,	3 0	87.15	8.85	4.00	reedy and granular.
South side the Coal Basin.	Argoed in Cwm Buchan,	Upper vein,	3 0	76.54	22.50	0.96	friable, weak.
		Lower vein,	3 0	74.30	23.40	2.30	harder, more compact.
	Oakwood, Glamorgansh. second coal. series,	Wern Dhu,	2 3	78.02	20.18	1.80	used for smelting iron.
		Wern Pistill,	2 6	80.06	17.46	2.48	hard, compact.
		Rider vein,		80.67	18.52	0.81	second coal series.
		Tor Mynydd,	1 8	81.18	18.00	0.82	"
	Cwm Buchan, third coal series,	Four feet, upper vein,	4 0	91.67	7.70	0.63	strong, heavy coal.
		" lower part,	4 0	88.65	10.00	1.35	reedy, furnaces.
		Nine feet coal,	9 0	83.74	15.20	1.06	large and lumpy.
		Yard coal,		78.90	20.00	1.10	clean and bright.
	Pyle, lower series,	Vein fawr, 1,	9 0	72.82	24.68	2.50	very bright, reedy coal.
		" 2,	9 0	72.58	24.42	3.00	cannel and bitu's mixed.

40

Bituminous and Semi-bituminous or Steam Coals of South Wales, on the South-eastern side of the Coal Basin.

By the term "reedy coal," is locally understood a coal in which the vegetable impressions are conspicuous and abundant in its substance,—"Clod coal," having a soft, laminated, vegetable texture,—"Splint coal," which does not lose its form in coking, angular,—"Semi-bituminous coal coke," where the angles of the cokes are rounded, and having considerable adhesion,—"Partially bituminous," where the coked masses have rounded edges, and slightly adhere together,—"Bituminous cokes," those which dissolve and enter into fusion, forming a compact mass.

Description, Locality, and names of Coal Seams.		Thickness of each vein. ft.	in.	Anal. of 100 p'ts of coal, by D. Mushet. Carbon.	Bitumen, vol. matter, &c.	Ashes or cinders.	The Coal described, and its uses.
The white ash or furnace Coals. Mynyddys-lwyn vein, sale coals for household purposes, *Upper or red ash coals,*	Cyfarthfa furnace,			88.07	8.50	3.43	irregular crystallized.
	Powell's,	4	0	66.58	27.92	5.50	cubical, compact, reedy.
	Morrison's,	4	0	68.58	36.92	4.50	" "
	Penner vein,	4	0	60.25	33.00	6.75	slightly granular, reedy.
	Cwm Dows, (Morrison,)	3	4	68.86	27.14	4.00	cubical.
	Prothero's,	4	0	64.95	33.30	1.75	" compact.
	Rosser Williams,	4	0	68.50	30.00	1.50	no sulphur, clean.
	Crossfields,			69.34	24.16	6.50	cubical, compact.
Beddws vein, *Lower red ash coal,*	Cwm Tillery,	2	4	64.45	24.80	10.75	shining bright.
	Phelps's,	2	4	68.00	30.00	2.00	compact, hard, strong.
	Abercarne,	2	9	66.88	28.37	4.75	" cubical, reedy.
	Cwm Carne,	2	9	62.63	31.10	6.25	" " "
Risca veins,	Upper Rock vein,	3	6	66.11	31.14	2.75	for steam-packets.
	Lower Rock vein,	3	6	61.78	34.28	3.94	oblong masses, reedy.
	Big vein,	12	0	66.02	29.15	2.83	irregular, no sulphur.
	Red vein,			61.25	33.80	4.95	pyrites, strong.
	Sun vein, [or Meadow vein,]	3	0	67.28	31.34	1.38	compact, reedy.
Cwm Brane coals,	Yard vein,	2	0	63.03	32.60	4.37	with layers of splint.
	Rock vein,	6	0	62.22	34.78	3.00	cubical.
	Red vein,	3	6	60.65	31.35	8.00	oblong reedy.
	Meadow vein,	5	8	66.34	28.16	5.50	strong, bright, shining.
	Old coal,	3	0	68.30	27.70	4.00	cubical, tarnished.
Blaen-dare furnace coals,	Rock vein,	10	0	68.86	28.64	2.50	used for blast furnaces.
	Meadow vein,	6	0	67.84	29.16	3.00	" "
Pen Twyn furnace coals,	Big vein,	4	6	71.88	25.50	2.62	laminæ regular, reedy.
	Droydeg, or Rock vein,	3	6	68.20	24.80	7.00	cubical.
	Meadow vein,	7	0	63.65	32.60	3.75	cross or sectional, pure.
	Old coal,	2	6	68.50	27.50	4.00	incompact, friable.
Abersychan British Iron Company.	Red vein,	4	0	72.95	25.30	1.75	used raw in blast furna's.
	Big vein,	7	0	67.05	25.70	7.25	" "
	Rock vein,	8	0	69.30	25.70	5.00	cubical splint.
Bit's c'ls of S. Wales, s'th side the c'l b'sn. Park, south veins of the S. Wales coal basin, between Pyle and Llantrissant.	Cribbwr Vach,	4	6	72.36	26.14	1.50	bright, furnaces.
	Bedws vein,	10	0	70.68	25.82	3.50	" in thin laminæ.
	Maesteg issa,	5	0	79.69	19.26	1.25	" blast furnaces.
	Llangonydd,	5	0	60.40	38.60	1.00	"
	" No. 2,	2	6	69.64	27.86	2.50	imperfectly rhomboidal.
	" 20 inch,	1	8	70.22	28.28	1.50	broad, reedy coal.
	Hirwain common,	4	6	69.34	29.16	1.50	smooth fracture.
	" 2 y'd coal,	4	0	73.75	22.50	3.75	slaty, partially reedy.
	Llanharry,	6	0	65.75	33.00	1.25	strong, reedy coal.
	Collenna 3 feet,	3	0	75.06	23.44	1.50	broad, reedy structure.
	" cannel coal,	3	0	63.25	34.12	2.63	laminated, oolitic.
Mellin Criffin and Pentyrch, near Cardiff,	Little vein,			70.66	27.34	2.00	workmen's fires.
	Brassey vein,	2	4	61.00	30.00	9.00	heavy, forges and mills.
	Pentyrch hard vein,	4	0	71.25	23.75	5.00	crystallized, blast furna's.
	" forked vein,	5	6	64.63	31.87	4.50	reedy, tin plate works.
	" wing coal, pure anthracite,	5	0	95.69	2.81	1.50	" furnace and forge.
Very dry Coals, with excess of carbon in South Wales.	Dowlais iron works,			79.50	17.50	3.00	lamellar, does not cake.
	Cyfarthva "			78.40	18.80	2.80	" cakes.
	Pen-y-daran, "			76.80	20.00	3.20	" "

Anthracite of South Wales, towards the western extension of the Basin.

Description and Locality of Coal Beds.		Thickness of each vein. ft.	in.	Analysis of 100 parts of coal, by D. Mushet. Carbon.	Vol'e matter, &c.	Ashes or cinders.	The Coal described.
Pwll-feron, in Neath valley, Neath Abbey furnaces,	Pwll feron vein, 1st bed,	18	0	89.34	6.66	4.00	mixed with coke.
	2d "			86.56	6.94	6.50	very hard and reedy.
	3d "			86.24	12.00	1.76	anthracite.
	4th "			90.59	8.50	0.91	"
	5th "			91.08	8.00	0.92	more brilliant.
	6th "			81.00	9.00	10.00	more reedy.
Yyiscydwn Iron works,	Big vein,			88.70	7.80	3.50	
	Brass vein,			88.70	7.80	3.50	
	Cwm Phil vein,			89.60	6.66	3.74	bright and shining.
Swansea,	Swansea peacock coal,			89.00	7.50	3.50	surface smooth.
Ystal-y-ferra, Iron works, Swansea valley,	Big vein, 1st bed,	6	0	91.42	7.08	1.50	blast furnaces.
	" 2d "			92.89	5.61	1.50	flat, boarded coal.
	" 3d "			91.99	6.51	1.50	reedy and granular.
	Cefn v. upper bed,			91.26	7.24	1.50	pitchy, bright, shining.
	" lower part,			91.89	6.61	1.50	partially granular.
	Brass v. upper part,			92.46	6.04	1.50	bright, laminæ irregular.
	" lower part,			91.52	6.98	1.50	irregular, reedy.
	Black vein,			93.14	5.36	1.50	rough, crystallized.
	Little vein,			90.64	7.86	1.50	regular but twisted.
	Pentyrch wynn,	5	0	95.69	2.81	1.50	reedy, forge and furnace.
Cwm Neath,	Three feet vein,	3	0	94.10	4.90	.93	analysis by D. Schafhaeutl.
	Eighteen feet vein,	18	0	91.43	6.24	2.28	
	Nine feet vein,	9	0	93.12	5.22	1.59	
Bituminous coals of the forest of Dean, Gloucestershire.	Cinderford furnace, or Lower High coal Delf,	3	0	62.00	36.00	2.00	strongly reeded, bright.
	Park-end coal,	4	0				free from splint.
	Coleford High Delf, top part,	4	6	63.72	32.03	4.25	thin, bright laminæ.
	Middle part,	to		63.61	34.89	1.50	reedy, bright, pyritous.
	Bottom part,	6	0	60.96	37.29	1.75	smooth fractured.
	Church-way coal, top,	5	0	60.33	35.67	4.00	irregular fracture.
	Church-way coal, bottom,			64.13	34.74	1.23	smooth, straight, reedy.
	Rocky vein,	2	0	61.73	36.14	2.13	strong, partially reeded.
	Starkey coal,	2	6	61.53	36.72	1.75	partially reedy.
	Park-end, Little Delf,	1	8	58.15	36.35	5.50	compact, bright, reedy.
	" Smith end,	2	0	63.36	34.89	1.75	heavy, compact, reedy.
Bituminous coals of Staffordshire. Corbyn's Hall,	Tow coal, part of the 10 yard coal,			51.90	40.60	7.50	strong, reedy coal.
	" "			56.00	42.50	1.50	irregularly laminated.
	Heathing coal,			54.17	43.33	2.50	strong, bright, reedy.
	Brooch coal,			50.49	47.76	1.75	bright, pitchy fracture.
South Staffordshire,	New mine top coal,			52.77	45.10	2.13	reedy, without splint.
	Fire clay coal,			51.40	46.35	2.25	weak, friable, reedy.
	New m. bottom coal,			53.98	44.27	1.75	reedy, mixed with clod.
Bentley Estate, South Staffordshire,	Ten yard coal,			54.05	42.70	3.25	pitchy, bright coal.
	" bottom p't,			63.57	34.18	2.25	bright and thin splint.
	Four feet coal,	4	0	53.18	44.82	2.00	bright, shining, smooth.
	Three feet coal,	3	0	54.82	43.12	2.00	bright, pitchy, reedy.
	Fire clay coal,	5	0	54.84	42.91	2.25	strong, reedy, uniform.
	Bottom vein,	7	0	62.87	32.00	5.12	hard, splint coal.
	Five ft. splint c'l,	5	0	49.42	45.83	4.75	laminæ minute.
	Bottom coal,			79.78	10.72	9.50	imperfectly crystallized.
Lane end, North Staffordshire,	Bassey mine coal,			58.30	38.70	3.00	reedy, dull surfaced.
	Little mine coal,	4	0	62.30	35.20	2.50	compact.
	Great row coal,	9	0	57.38	39.74	2.88	smooth, thin laminæ.
	Best furnace coal,	10	0	65.20	32.30	2.50	bituminous looking coal.
	Ashes coal,			61.32	37.18	1.50	bright, shining, cubical.

Bituminous Coals of Staffordshire, Shropshire, North Wales, Derbyshire, and Yorkshire.

District	Description and Locality of Coal Veins.	Thickness of each seam. ft.	in.	An. of 100 parts co'l by Mushet. Carbon.	Bit'n, vol-matter, &c.	Ashes or cinders.	The Coal described and its uses.
North Staffordshire. Kidsgrove.	Little Row coal,	4	0	63.08	34.67	2.25	bright, with clod partings.
	Seven feet coal,	7	0	67.90	30.47	1.63	thin layers, furnace.
	Stony vein,	8	0	65.17	33.33	1.50	compact.
	Banbury or Harecas,			63.84	35.16	1.00	bituminous.
	Knowles's coal, Delph Lane,	10	0	59.64	37.86	2.50	bright, free burning.
	Peacock coal, Fenton Park,	9	0	60.42	37.08	2.50	cubical, furnaces.
Golden H.	Spendcroft vein,	4	0	58.67	39.58	1.75	broad, potteries.
	Ten feet coal,	7	0	58.89	39.11	2.00	uniformly reedy, potteries.
	Great Row coal,	8	0	60.80	37.70	1.75	cubical, pott's & salt w'ks.
	Little Row coal,	4	0	62.47	34.53	3.00	hard, " "
Shropshire.	Shropshire stone coal,			58.17	39.20	2.63	bright, clod partings.
	Sulphur coal,			55.72	42.03	2.25	broad, for inf. house purp.
	Clod coal,			63.79	35.58	1.63	reedy, furnaces.
	Randle coal,			64.19	32.81	3.00	" "
	Flint coal,			60.63	36.87	2.50	hard, sale or smith's use.
	Top coal,			64.10	34.77	1.13	regular.
	Best fungous coal,			63.33	35.67	1.00	minutely lamin. no pyrites.
	Double coal,			57.87	41.38	.75	hard, blast furnace.
North Wales. Brymbo Coals.	Three yard coal, part not coked,			61.31	35.80	2.89	fine, "
	" part coked,	3	0	62.70	35.70	1.60	strong, "
	Two yard coal coked,			69.98	28.60	1.42	bright, "
	Brassy vein coked,	5	6	64.58	34.10	1.32	cubic, "
	Crank coal,	2	6	73.56	25.70	.74	mixed, furnace and sale.
	Drowsall vein,	5	0	62.69	36.70	.60	not firm, free.
	Powell vein,	5	0	63.41	34.80	1.79	shining.
Dee Bank.	Five yard vein, top part,			61.89	36.20	1.91	laminated, free.
	" middle,			62.72	36.00	1.28	more compact, iron mak'g.
	" bottom,			63.79	32.85	3.36	thin laminæ, with clod.
	Three yard coal,			62.88	36.00	1.12	compact, free.
	Two yard coal,			60.61	38.47	.92	free, shining fracture.
	Bone coal,			55.20	40.00	4.80	hard, lead works.
North Wales.	Pankey Iron works, stone vein,	2	0	61.95	35.67	2.38	broad, partly crystallized.
	Pant Iron works, blast furnace c'l,	1	7	67.25	31.25	1.50	granular, for blast furnace.
	Coed Talon, "	9	0	58.50	40.00	1.50	surface, "
	Sweeny colliery, brassy vein,	3	0	49.94	34.56	15.50	alternate layers, "
	Cefn colliery, near Rhuabon w'ks,	7	0	57.49	36.56	6.25	"
	" brassy coal,	3	0	66.37	32.13	1.50	laminated.
	Black Park coal, 2 yard vein,	6	0	57.50	40.00	2.50	firm, with splint.
	" 1½ yard vein,	4	6	59.88	38.12	2.00	strong, with clod.
	Llwyn-y-onnion, ½ yard coal,			62.85	34.40	2.75	smooth, reedy, blast fur'es.
	Chirke bank colliery, strangers c'l,	5	6	57.00	40.00	3.00	hard.
	Delf colliery, y'd v'n n'r Rhuabon,			64.89	34.11	1.00	compact.
Derbyshire.	Kirby, upper hard or main coal,	6	0	64.15	33.85	2.00	mixed, shining parting.
	Dunshill near Swanwick,			55.77	40.73	3.50	strong, breaking oblong.
	Swanwick, main coal,			60.27	38.23	1.50	" twisted laminæ.
	Main, upper hard, Duckmanton,			64.47	32.03	3.50	" blasted furnaces.
	Normanlon Com. Codnor Park c'l,			56.21	41.66	2.13	free, forge and mills.
	Main soft coal,			56.49	37.76	5.75	" clod, spar.
	Alfreton works, lower hard coal,	4	0	62.60	35.15	2.25	strong, blast furnaces.
	Butterly p'k col'y, " "			61.14	34.11	4.75	splint, "
	Morely p'k w'ks, " "			55.89	37.86	6.25	" "
	Chesterfield, " "			61.65	35.10	3.25	mixed, good cleavage.
	Double or Minge coal,			60.66	37.34	2.00	bright, furnaces.
	Clod coal,			61.21	37.29	1.50	smooth fracture.
	Buckland Hollow or Kilburn coal,			58.62	40.00	1.38	broad, smooth fracture.
	Moreley Park, cannel coal,			45.00	45.05	9.95	conchoidal.
	Cannel coal near Alfreton works,			55.27	40.73	4.00	beautiful, specular.
Yorkshire.	Lower Moor, better bed,	2	4	67.06	32.19	.75	dense, furnace and forge.
	" black bed,	1	8	71.42	27.08	1.50	friable, domestic use.
	Bowling, better bed,	1	10	64.25	32.55	2.00	bituminous, fur, and forge.
	" crow coal,	1	8	66.15	33.85	1.00	distinct, with clod, sale.
	Parkgate, main coal,	7	0	67.14	30.73	2.13	4 feet, blast furnaces.
	Old Parkgate vein,	4	6	65.09	33.28	1.63	hard, in laminæ, "

Fat, bituminous Coals of Yorkshire and Scotland.

Description and Locality of Coal Veins.	By whom analysed and described.	Thickness of each seam. ft. in.	Analysis of 100 parts of coal. Carbon.	Bitu'n, vola'e matter, water, &c..	Ashes or cinders.
YORKSHIRE.					
Parkgate, top c'l, upper part of the 7 ft. c'l,	D. Mushet,	1 4	62.51	36.49	1.00
" bottom part,	"	1 8	66.94	31.56	1.50
Birkenshaw coal,	"		64.96	32.54	2.50
Worsboro' furnace coal,	"	9 0	60.32	38.18	1.50
" another specimen,	"		56.45	40.85	2.50
Milton, main coal, splint part,	"	9 0	69.40	27.60	3.00
" roof or soft part,	"		62.71	36.04	1.25
Thorncliffe, thin furnace coal,	"	[illegible] 6	[illegible]3.98	35.52	.50
Smithy, wood coal,	"	2 6	54.60	44.27	1.13
Easley Park,	"	1 7	69.12	30.00	.88
Yorkshire Kent, coal,	"	5 0	66.40	32.72	.88
Strafford, main coal, 5 feet bottom part,	"	3 0	62.08	35.67	2.25
" " top part,	"	2 0	68.12	30.20	1.68
Silkstone, main coal,	"		65.08	32.29	2.63
" soft or clod coal,	"	5 0	63.10	35.15	1.75
SCOTLAND.					
Clyde, upper vein, top,	"		37.00	41.50	21.50
" " bottom,	"		53.45	44.80	1.75
" " second vein,	"		42.10	48.34	9.56
" third or furnace,	"		51.20	45.50	3.30
" fifth splint coal,	"		53.40	42.40	4.20
Calder, furnace coal, top,	"		49.98	43.82	6.20
" " splint part,	"		50.67	47.48	1.85
" " main coal, top,	"		49.60	49.39	1.01
" " middle,	"		52.30	39.95	7.75
" " bottom,	"		51.60	44.51	3.89
Glen Buck, furnace coal,	"		53.20	45.20	1.60
" inferior coal,	"		48.80	44.20	7.00
Cleugh, furnace coal,	"		47.08	42.25	10.67
Omoa, splint "	"			46.62	
" bright "	"			47.29	
Marystone Pyat, shaw coal, top,	"		49.60	49.31	1.09
" pine splint,	"		51.82	46.57	1.61
" heavy splint,	"		54.67	39.25	6.08
Fat, bituminous Coals.					
Govan coal, first vein, top part,	"		49.55	44.65	5.80
" lower part,	"		49.41	48.92	1.67
" second vein,	"		42.20	48.34	9.46
" fifth vein, splint,	"		48.84	49.79	1.37
" sixth vein, lower main,					
1. Craw coal,	"		51.58	44.60	3.82
2. Head coal,	"		48.08	49.38	2.54
3. Ground coal,	"		45.57	51.00	3.43
4. Foot coal.	"		52.27	44.15	3.58
Lismahago, cannel coal,	"		39.43	56.57	4.00
Dry Coals, not very adhesive.					
Clyde, splint coal,	"		59.00	36.80	4.20
" "	Thomson,		55.23	35.27	9.50
" clod coal,	D. Mushet,		70.00	26.50	4.50
" soft coal,	Thomson,		42.25	47.75	10.00
" near Glasgow,	Dufrenoy & Berthier,		64.40	31.00	4.60
Calder, "	"		51.00	45.00	4.00
Monkland, "	"		56.20	42.40	1.40
Middlerig,	Dr. Fyfe,		50.50	42.00	7.50
Scotch coal,	W. R. Johnson,		48.81	41.85	9.34
Scotch cannel,	"		60.34	36.95	2.71
"	Dr. Ure,		39.40	53.60	4.00

Anthracites of Europe.

Localities.	By whom examined and analysed.	Specific gravity.	Analysis 100 parts. Carbon.	Hydrog'n, water and volatile matter.	Silica and ashes.
SOUTH WALES.					
Anthracites.					
Welsh anthracite, Cwm Neath,	Dr. Schafhaeutl,		92.42 94.10	5.97 4.90	1.91 .93
Ynis Cedwin, Crane,	Jno. F. Frazer,		86.60	7.60	5.80
"	W. R. Johnson,	1.336 1.372	87.60	9.18	4.32
Welsh stone coal,	D. Mushet,	1.368	89.70	8.00	2.30
Welsh slaty stone coal,	"	1.409	84.17	9.10	6.73
Mean of several varieties of Welsh coal,	Dr. Ffye's Exper.	1.354	71.40	17.80	10.80
EUROPEAN CONTINENT.					
Anthracitous Coals.					
The Alps, Isere, Canton of Launure,	M. Robin,				4.0
Canton of Lanton, near Grenoble,	"	1.072			
Westphalia,	M. Karsten,	1.358			
"	"	1.336			
Mean analysis of twelve varieties of anthracite,	Berthier,		79.15	7.37	13.25
Dry or slightly Bituminous Coals.					
IRELAND.					
Kilkenny, Leinster,	D. Mushet,	1.602	92.88	4.25	2.87
" slaty or cannel,	"	1.445	80.47	13.00	6.53
Boolavoonein, stone coal,	"	1.436	82.96	13.80	3.24
Corgee, "	"	1.403	87.49	9.10	3.40
Queen's county, Leinster,	"	1.403	86.56	10.30	3.14
Kilkenny, cannel,	Karsten,		74.47	25.01	.50
Kilkenny,	Dr. C. T. Jackson,		79.60	12.00	8.40
SCOTLAND.					
Coal, under Basalt, Renfrewshire,	D. Mushet,		69.74	16.66	13.60
FRANCE.					
Mean of twelve specimens,	M. Berthier,		79.15	7.37	13.25
Côte d'or, Sincey,	De Nerville,		82.60	8.60	8.80
Mais Salze,	M. Varin,		83.00	7.50	9.50

Bituminous Coals of France.

Departments, Coal Basins, and varieties of Coal.	Locality.	Concessions.	By whom analysed.	Specific gravity.	Analysis. Carbon.	Volatile matter.	Ashes.
Coal of the Bourbonnais, department of Allier. Basin of Fins,	Montet,		M. Baudin,	1.38	75.23	24.77	
	Gabeliers,		“	1.34	74.92	25.08	
	Deux Chaises, Chapelle,		“	1.48	74.22	25.78	
	Fins,		“		68.28	31.72	
	Noyant,		“	1.30	62.49	37.51	
Basin of Bussiere-la-grue,			“	1.34	54.76	45.24	
Basin of Bert,			“	1.36	58.47	41.53	
					64.20	35.80	
Basin of Commentry,	Commentry,		M. Regnault,		63.20	36.60	1.20
	Chambled,	Marais,	“	1.38	87.85	12.15	
	Commentry,	Amenat,	“	1.27	60.00	40.00	
	Néris,	Great bed,	“	1.35	58.87	41.13	
	Id.,	Ferrieres,	“	1.31	56.76	43.24	
Basin of Doyet,	Doyet,	La Souche,	“	1.30	61.23	38.77	
	Monticq,	Bourdignat,	“	1.28	59.58	40.42	
	Id.,	Chauvais,	“	1.30	58.61	41.39	
	Bezenet,	Grande masse,	“	1.32	56.84	43.16	
Champagnac in the coal basin of Haute-Dordogne—Cantal,	Mines de Lempret,	New bed, 1	M. Baudin,	1.26	66.60	30.19	4.60
		2 metres, 2		1.33	56.30	29.80	13.90
		3		1.36	53.20	30.30	16.50
		Upper bed, 4	“	1.28	65.70	30.10	4.20
		5		1.27	62.40	30.60	7.00
	Mine de Madie,	First or lowest bed, 6	“	1.28	60.70	32.90	6.40
		2d bed, 7	“	1.28	64.00	31.60	4.40
	Mauriac,	Madie,	M. Berthier,		61.00	34.50	7.50
Coal basin of la Haute-Dordogne Cantal,	1. Messeix,	Clydance,	“	1.39	85.49	14.51	
	2. Singies,	Morilleux,	“	1.38	71.27	28.73	
	3. Lempret,	New bed,	“	1.38	69.10	30.90	
	4. Madie,	2d bed,	“	1.27	68.97	31.03	
	5. Prodelles,	3d bed,	“	1.40	67.28	32.72	
	6. Vendes,	Champlaix,	“	1.38	66.08	33.92	
	7. Madie,	1st bed,	“	1.31	66.05	33.95	
	8. Singles,	Ginguette,	“	1.32	65.14	34.86	
	9. Lempret,	C. de l'air,	“	1.35	63.08	36.92	
	Mandailles,	Lignite,	“	1.29	41.45	58.55	
	Chambeuil,	Lignite,	“	1.28	40.88	59.12	
Auverge, Central France, department of Puy-de-Dôme, *Coal Basin of Brassac.*	1. Charbonnier,	Great bed,	M. Baudin,	1.43	86.41	13.59	
	2. La Combelle,	“	“	1.38	80.31	19.69	
	3. “	La Ronziere,	“	1.36	78.82	21.18	
	4. Armots,	Fontaine-du-Chien,	“	1.38	77.48	22.52	
	5. “	Chamas,	“	1.41	76.79	23.21	
	6. Gras Ménil,	Great bed,	“	1.35	75.31	24.69	
	7. Fondary,	Les Vignes,	“	1.30	75.15	24.85	
	8. La Taupe,	Arrest,	“	1.32	73.89	26.11	
	9. Mégécoste,	6th bed of 4 ft.	“	1.34	73.01	26.99	
	10, La Taupe,	Great mass,	“	1.33	71.80	28.20	
	11. Les Barthes,	Batard of 3 ft.	“	1.36	74.40	28.60	
	12. “	8 feet bed,	“	1.39	70.14	29.86	
	13. Mégécoste,	7th bed of 8 ft.	“	1.34	70.09	29.91	
	14. Les Barthes,	3 feet,	“	1.45	70.07	29.93	
	15. Mégécoste,	6 feet bed,	“	1.34	69.86	30.14	
	16. Les Barthes,	3 feet,	“	1.35	68.64	31.34	
	17. Mégécoste,	7 feet,	“	1.49	67.08	32.92	
	18. Les Barthes,	Le Feu,	“	1.33	66.78	32.22	
	19. Brioude,	Preissat,	“	1.32	59.43	40.57	

Bituminous Coals of France, Department of Puy-de-Dôme.

Departments, Coal basins and varieties of Coal.	Locality.	Concessions.	By whom analysed.	Specific grav'y.	Analysis. Carbon.	Analysis. Volatile matter.	Analysis. Ashes.
Basin of Brassac, mines of La Taupe and Arrest,	Near Guinguette,	Cingles,	M. Baudin,		63.90	32.00	4.10
		Agassiz bed,	"	1.340	65.00	26.40	8.60
		La Louise bed,	"	1.310	68.50	26.80	4.70
		Four feet bed,	"	1.320	66.20	27.00	6.80
		La Felicite bed,	"	1.300	67.50	26.00	6.50
		La Trouelle,	"	1.330	66.10	26.30	6.60
Basin of St. Eloy, or Montaigne,	La Roche,		"	1.300	59.80	40.20	5.20
	La Vernarde,		"	1.300	60.40	39.60	8.70
Basin of Bourg Lastic,	Messeix,		"	1.390	86.24	13.76	5.20
	Singles,		"	1.380	75.00	25.00	13.00
	"		"	1.320	68.00	32.00	8.20
	Puy S. Gulmier,	De Chiex,	"	1.340	84.45	15.55	
	La Besette,	Vazazène,	"	1.280	66.75	33.25	
Saone and Loire. Basin of Creusot and Blanzy,	St. Bérain,	Molliere,	M. Gruner,		74.00		28.52
		Jumeaux,	"		68.75		22.63
		" 1st classe,	"		68.75		24.00
		Vignes,	"		73.52		26.50
		Quatre Bras,	"		66.80		19.80
		St. Charles,	"		69.25		22.10
	Maillot,		"		59.50		12.20
	Communautés,		"		67.75		5.45
	Montchanin,		"		61.25		9.00
	Longue Pendue,		"		60.00		8.00
	Ragny,		"		63.20		8.55
	Blanzy,	Montceaux,	"		58.00		5.00
	Lucy,	Lucy,	"		65.00		14.00
Saone et Loire,	Basin of Epinac,		Regnault,		61.10	36.40	2.50
Provence. Basse Alpes,	Volx,	Lignites,	M. Diday,		51.70	42.50	5.80
	Dauphin,				49.20	46.30	4.50
	Volx,				39.20	45.80	15.00
	Dauphin,				34.70	53.10	12.20
	Sigonce,				40.60	52.40	7.00
	Manosque,				31.20	50.50	18.30
	Villemus,				38.90	51.50	9.60
	Pierre-vert,				28.00	45.80	26.20
Provence. Var.	Cadiere,				48.10	44.80	7.10
	St. Zacharie,				32.40	61.60	6.00
	Beausset,				40.60	39.40	20.00
Vaucluse,	Méthamis,				40.90	50.00	9.10
	Piolenc,				26.60	51.10	22.30
	"				41.50	52.30	6.20
	Montragon,				36.80	48.20	15.00
Fat Coals, very rich in carbon, first class. Basin of Saint Etienne,	Du Soliel,	Montcel,	M. Gruner,		77.59	19.60	2.81
	St. Marie,	Chaney, 1	"		74.81	21.67	3.52
	St. Claude,	Méons, 1	"		74.31	24.17	1.52
	"	" 2	"		73.80	23.13	3.07
	St. Marie,	Chaney, 2	"		73.78	24.33	1.89
	Reveux,	Reveux,	"		72.73	22.83	4.44
	St. Claude,	Méons, 3	"		72.13	24.47	3.40
	Caraude,	Cote-Thiolière,	"		69.13	25.67	5.20
		Grande couche du cros.	"		69.27	24.50	6.23

Bituminous Coals of France, Department of Puy-de-Dôme.

	Departments, Coal basins, and varieties of Coal.	Locality.	Concessions.	By whom analysed.	Analysis. Carbon.	Volatile matter.	Ashes.
Saint-Etienne.	2d class. Ordinary Coals of Saint-Etienne,	Chené,	5. De la Roche,	M. Gruner,	67.96	28.47	3.57
		Vincent,	5. Bérard,	"	66.66	29.20	4.14
		Deville,	3. De la Roche,	"	65.72	31.90	2.38
		"	3. "	"	61.05	32.54	6.41
		"	7. "	"	66.35	28.27	5.38
		St. André,	Méons,	"	65.68	27.83	6.49
		Pompe,	7. Du. Treuil,	"	62.89	29.73	7.38
		"	7. "	"	58.81	26.03	15.16
		Vincent,	7. Bérard,	"	63.71	25.27	11.02
		"	6. "	"	63.94	27.13	8.93
	3d class. Fat Coals, longue flamme,	Montrambert,	Great bed, 1st quality,	"	57.82	34.10	8.08
		"	2d "	"	54.56	35.43	10.01
		Littes,	Beraudière,	"	58.79	35.57	5.64
		Great bed,	1st quality,	"	59.29	35.20	5.51
Lignite Beds. Peat. Lignite Beds.	Bouches du Rhone,	Lignites of Greasque,	Mène du haut, 1	M. Diday,	48.20	49.00	2.80
			Bleu, 2	"	48.50	48.00	3.50
			Menette, 3	"	45.60	49.60	4.80
			Maitre Jean, 4	"	41.90	51.60	6.50
			La Fortune, 5	"	44.70	52.30	3.00
			La Saoude, 6	"	43.70	47.80	8.50
			La Ravette, 7	"	39.10	53.90	7.00
	Ardennes,	Séchwal,	Peat,	M. Sauvage,	22.00	69.70	8.30
	Basses-Alpes,	Lauzanier,	"	M. Diday,	9.00	58.00	33.00
	Basses-Pyrénées,	Bayonne,	Lignite,	M. Gruner,	48.20	38.10	13.70
	Lozère, and	Rosiers.	"		50.70	47.60	1.70
	Aveyron,	Peyre lau	"	M. Cochon,	49.10	46.20	4.70
	Gard,	Alais,	St. Christol, lig.	M. Varin,	34.00	46.00	20.00
	Bouches-du-Rhone,	Rocher Bleu,	Great lig'te bed,	M. Diday,	50.20	46.30	3.50
		Belcodène,	Smaller beds of lignite,	"	45.20	52.40	2.40
					26.50	53.50	20.00
	Var,	Collobrieres,	Coal,	"	63.00	28.00	9.00
Bituminous Coals in France.	Coals in Arrondissement of Alais. Department of Gard,	La Grande Combe,	1. Fournier,	M. Varin,	76.00	16.00	8.00
			2. Plomb,	"	70.00	17.00	13.00
			3. L. Barraque,	"	80.50	14.50	5.00
			4. Abilon,	"	81.00	14.00	5.00
			5. Velours,		75.00	14.00	11.00
			6. Bosquet,	"	74.00	20.00	6.00
			7. Rotschild,		73.50	16.50	10.00
		Prescol,	8. Levade,		77.00	18.00	5.00
			9. Trois-Machoires,	"	77.50	18.00	4.50
			10. Cing-pans,	"	78.00	19.50	2.50
		Partes,	11. Taranière,		65.00	18.50	16.50
			12. Rowière,	"	78.50	14.00	7.50
		Bessége,	13. Great bed,	"	67.50	20.50	12.00
		Champelauzon,	14. Champelauzon,		79.50	13.00	7.50
		St. Jean-de-Valorisle,	15. Remise,	"	72.00	9.00	19.00
		St. Paulet,	16. Lignite,	"	36.50	51.00	12.50
		Connaut,	17. "	"	35.00	51.00	14.00

Bituminous Coals of France, Department of Puy-de-Dôme.

Departments, Coal basins, and varieties of Coal.	Locality.	Concessions.	By whom analysed.	Specific gravity.	Analysis.		
					Carbon.	Volatile matter.	Ashes.
Coal basin of Alais, department of Gard,	1. Bességé,	Coal,	M. Varin,		68.50	25.50	6.00
	2. Saint Cristol,	Lignite,	"		34.00	46.00	20.00
	3. Grand Combe,	Coal,	"		80.50	17.00	2.50
	4. "	Pin bed,	"		74.00	18.50	7.50
	5. Bességé,	Coal,	"		63.00	24.50	12.50
	6. Grand Combe,	Plomb,	"		77.50	19.00	3.50
	7. Cessous,	Mean of 3 exper.	"		78.30	17.70	4.00
	8. "	Masse bed, 3 exper.	"		58.50	26.50	15.00
	9. "	Salze bed, 2 exper.	"		83.00	7.50	9.50
	Rochebelle,		M. Berthier,		68.00	21.60	11.40
Hérault,	Saint-Gervais basin,	Concessions of Bousquetd'orbe and Graissesac, 2	M. Garella,		56.50	17.00	26.50
		3	"		77.50	19.00	3.50
		4	"		70.50	16.00	13.50
		5	"		69.70	15.00	15.40
		6	"		65.20	18.50	16.30
		Graissesac,	M. Gruner,		68.80	31.20	15.30
		St. Gervais,			85.16	14.84	14.05
					78.30	16.40	5.30
Aveyron,	Basin of Aubin, or Decazeville,	Paleyret, No. 3,	M. Senez,		67.50	26.60	5.90
		" 4,	"		61.00	32.80	6.20
		Bourran,	"		71.50	24.60	3.90
		Fontanges,	"		63.00	27.20	4.20
		Fareiret,	"		53.00	38.30	8.70
		Bouquiès,	"		69.80	25.10	5.10
		Cransac,	"		53.20	39.30	7.50
		Lavergne,	"		50.00	42.00	8.00
		Le Poux,	"		55.00	38.00	7.00
		Lagrange,	M. Regnault,		61.20	34.20	4.60
Tarn,	Basin of Carmeaux,	Grand-Vein,			72.60	23.60	3.80
		Castillan,			74.50	20.90	4.60
Aude,	Basin of Ségure,		M. Berthier,		56.00	24.00	20.00
			M. Bois,		60.00	22.00	18.00
			M. Leplay,		71.60	24.00	4.40
	Basin of Durban,		M. Berthier,		49.00	33.50	17.50
Nord,	Basin of Valenciennes,	Anzin, bitumens,	Chevalier,	1.284	74.25	25.00	0.75
		Fresnes anthrac'e,	"	1.360	89.30	7.20	3.50
		Anzin,	Berthier,	1.284	71.50	25.00	3.50
Haute-Saône,	Gémonval,	Corcelles & Lure,	M. Drouot,	1.440	48.90	36.60	14.50
Vosges,	Norroy,		M. Regnault,	1.410			
Rhône,	Rive-de-Gier,	Grand-Croix,		1.298	67.20	31.00	1.80
				1.302	68.80	29.80	1,40
		Cimitière,		1.288	68.40	28.00	3.60
				1.294	67.00	30.00	3.00
		Couzon,		1.298	62.80	34.50	2.70
				1.311	62.10	32.60	5.30
		Corbeyre,		1.315	74.00	23.00	7.00
		Couzon,	M. Gruner,		63.55	30.93	5.52
		Grézieux,	"		62.54	25.10	12.36
		Couzon,	"		62.57	30.07	7.34
Doubs,		Morteau,	M. Boyé,		29.50	53.50	17.00
		Flangebouche,	"		30.00	62.00	8.00
Jura,		Orbagna,	"		30.50	57.50	12.00

Analysis of Combustibles, Europe.

Departments, and varieties of Coal.	Locality.	Designation of Coal Beds.	By whom analysed.	Analysis. Carbon.	Volatile matter.	Ashes.
Deux Sévres,	Chantonnay,	Main coal,	M Boyé,	62.70	20.00	17.30
Vendee,	Basin of Vouvant,	Faymoreau,	M. Berthier,	61.15	29.50	6.15
				65.10	27.50	7.40
Loire Inférieure,	Ancenis,	Guignardiere,	M. Sentis,	58.50	31.00	10.50
Finistèrre,	Plogoff,	Cap. Sziain,	"	63.00	25.00	12.00
FRANCE. Coals of Maine et Loire.	Layon et Loire,	Pits of St. Barbe,	M. Sentis and M. Lechatelier,	80.21	17.00	3.79
				77.59	18.00	4.41
		Du Bocage,		82.39	13.20	4.41
		Des Barres,		67.03	16.60	16.37
		The West,	"	69.92	18.00	12.08
	Mont-jean,	St. Nicholas,	"	65.88	23.40	10.72
				73.76	22.34	3.90
		Beau-soliel,	"	74.02	19.20	6.78
	St. Georges, Sur Loire,	The Arch,	"	63.40	27.87	8.73
	Chaudefonds,	St. Barbe,	"	73.57	10.00	16.43
		St. Nicolas,	"	71.78	11.60	16.62
	St. Georges, Chatelaison,	Conception,	"	72.71	15.60	11.69
				65.00	23.80	11.20
		Adele,	"	80.99	9.40	9.61
		Du Pavé,	"	78.10	18.40	3.50
	Doué,	De Minières,	"	65.28	27.80	6.92
Haute-Loire,	Basin of Langeac,		M. Baudin,	74.00	26.00	7.10
Feroe Islands,	Suderoë I.	Brown coal,	Durocher,	24.50	37.50	38.00
Hesse Cassel, Coals.	Meissmer,	Anthracite,	M. Kuhnert,	70.19	14.34	15.47
	"	Pechkohle,	"	56.60	40.97	2.43
	Hirschberg,	"	"	60.83	38.36	0.81
	Habichtwald,	"	"	57.26	35.41	7.33
	Hirschberg,	Dry coal,	"	66.11	31.13	2.76
Lignite.	Habichtwald,	Lignite, passing to coal,	"	54.18	42.49	3.33
	"	Inferior lignite,	"	52.98	42.10	4.92
	"	Middle "	"	54.96	41.84	3.20
	Rigenkuhl,	Woody "	"	51.70	47.01	1.29
	Stillberg;	Lignite,	"	50.78	42.27	6.95
		Peat or turf,	D. Mushet,	25.20	72.60	2.20
			Marcher,	65.00	22.00	13.00
			"	37.00	48.00	15.00
Southern Russia,	Cossack, Don,	Country of the Don best anthracite,	Voskressensky,	94.23		
	Tiflis,	Inferior,		63.64		
ITALY. Principality of Monaco,	Menton,	Earthy,	M. Diday,	49.20	29.90	20.90
SPAIN. Coals of Asturias.	Cueva,	One yard coal,	I. T. Cooper,	66.00	31.80	2.20
	Emanuela,	Three yard coal,	"	67.90	30.90	2.10
	Viena Alta,	Four yard coal,	"	63.60	33.90	2.50
		Mine of Clausel,	M. Berthier,	35.00	53.00	12.00
		Del Regueron,	"	43.00	44.00	13.00
		Mean of 5 other mines,	"	53.00	40.00	7.00
	Tudela,		M. Paillette,	65.80	32.27	1.93
	Mieres,		"	57.60	39.40	3.00
	Lama,		"	56.69	41.51	1.80
	Oloniego,		"	60.40	36.55	3.05
	Arnao,		"	45.69	45.11	9.20
	Ferrones,		"	46.98	46.91	6.11

Bituminous Coals in Belgium.

Countries, Provinces, and varieties of Coal.	Locality.	Designation of Coal Beds.	By whom analysed.	Specific gravity.	Analysis. Carbon.	Oxygen and hyd'n.	Ashes.
Province of Hainault. Near Mons,	Dour,		Berthier,		71.50	23.30	5.50
	"		"	1.307	88.00		2.50
	"		"		85.00	12.70	2.30
			M. Canchy,	1.276	84.67	13.23	2.10
			"	1.292	83.87	12.47	3.68
	Basin of Mons,	Plate seam,	M. Chevalier,	1.273			1.98
	"		"	1.263			1.27
	"		Karsten,	1.307	85.50	12.00	2.50
	Canton of Dour,		Berthier,	1.270	71.50	23.30	5.20
	Near Mons,	Bouleau,	"	1.270	65.30	33.00	1.70
		Grand Gaillet,	"		58.50	38.50	3.00
		Gade vein,	"		51.00	44.00	5.00
Province of Liege.	Liege,	St. Margarite,	C. Davreax,		78.30	17.80	3.90
			"		76.00	19.60	4.40
	"	Olisson.	"		69.90	23.40	6.70
			"		72.60	24.20	3.20
	"	Cerisier,	"		68.50	21.20	10.30
Dry Coals.	Harion,	L'Harbe S. Michael,	M. Delvaux,	1.365	81.90	9.00	9.10
	Chokier,	Petite Hareng,	"	1.286	71.68	16.36	11.96
	Bonnier,	Moset seam,	"	1.318	91.38	8.00	6.12
GERMANY. *Bituminous Coals.*							
Upper and Lower Silesia,	Waldenberg,	Glanz coal,	Richter,		57.20	36.40	6.40
	Sabrze,	"	"		63.20	32.93	3.90
	Bielschowitz,	"	"		58.17	37.89	8.93
	Leopoldinen-grube,	"	Gay Lussac,		61.50	35.62	2.88
	Frederich zu Zawada,		Karsten,	1.263	57.90	42.00	2.10
	Gustaw Grube,		"	1.270	68.00	30.10	1.90
Saxon States,	Sälzer,	Newark,	"	1.288	81.60	17.70	0.70
"	"	"	Gay Lussac,		80.10	18.90	1.00
Prussian Saxony,	Circle of the Saale,	Wettin or Wittenberg,	Karsten,	1.466	56.70	18.90	24.40
Germany,	Brown coal,	Shraplau,			20.25	62.25	17.50
"	Eschweiler,	Flotz Gyr,	Gay Lussac,		82.40	16.42	1.18
"		"	Karsten,	1.300	80.23	18.60	1.17
Saxony,	Pottschapel,	Gate Schicht,	"	1.454	41.00	31.30	27.70
"	Planitz,	Pitch coal,	"	1.860	63.40	35.50	1.10
Bohemia,	Elbogen,	Brown coal,	M. Balling,		37.18	56.16	6.66
"	Schlakenwerth,	Carbonized peat,	M. Debette,		67.00	30.00	3.00
Wurtemburg,	Kœnigsbrunn,	Raw peat,	M. Berthier,		24.40	70.60	5.00

ASIA.

Country.	Locality.	Designation of coal beds.	By whom analysed.	Specific gravity.	Analysis. Carbon.	Volatile matter.	Ashes.
HINDOSTAN.							
Pres. Bengal,	Nerbudda,	Fatephur,			22.00	14.00	64.00
	Chirra Poonje,	Slaty,		1.447	41.00	36.00	23.00
Assam,	Cossyah or Kosya hills,	Few ashes,		1.275	60.70	38.50	.80
Pres. Bengal,	Prov. Delhi,	Hurdwar,		1.368	50.00	35.40	14.60
Birmese coast,	Aracan,			1.308	33.00	66.40	0.60
Turkey in Asia,	Anatolia,	Heraclea,	Prof. Hitchcock,		62.40	31.80	5.80
Syria,	Mt. Lebanon,	Asphaltum,	"		24.40	68.00	7.60
"	Mt. Hermon,	Anti-Libanus,	"		14.00	72.60	13.40

ANALYSIS

OF

OTHER COMBUSTIBLE MINERALS REFERRED TO IN THIS WORK.

	Authority.	Carbon.	Hydrogen.	Oxygen.	Nitrogen.
Naphtha, bitume napthe, ...	Thomson,	82.2	14.8		
	Saussure,	87.60	12.78		
Petroleum, bitume petrole, ... }	similar,				
Seneca oil of New York, ... }					
Earthy bitumen, earthy mineral pitch,	similar,				
Elastic bitumen, bitumen flexible, of England,	Henry,	52.250	7.496	40.100	0.154
Of France, ...	"	58.260	4.890	36.746	0.104
Compact bitumen, asphalt, Matanzas, . .		71.84		14.66	
" Peru, Coxitambo, . . .	Bousingalt,	75.00	0.50	15.50	
Amber, Succin, mineral resin, ...	Drappier,	80.59	7.31	6.73	
"	Ure,	70.68	11.62	7.77	
Halchettine, mountain tallow, ...	Johnston,	76.437	12.479		
Shererite, in lignite, ...	Macaire,	73.00	24.00		
Ozokerite, used as fuel in Moldavia, . . .	Glocker,	85.204	13.787		
Mellite, Honeystone, Thuringia, ...	Klaproth,				
Retinasphalt, bitumen fragrans, ...	Hatchett,				
Fossil copal, Highgate resin, ...					

INDEX.

ADDITIONS AND CORRECTIONS.

Page 45 line 25, make of iron in England, for "2,250,000," read about 3,000,000.*
" 45 " 26, for "mining," read "mining coal"; this paragraph belongs to p. 37, last line of the note.
" 46 " 31, Table, for "2,250,000," read about 3,000,000.
" 131 " 41, for "Lepedofloyos," read Lepedophloeus.
" 178 " 4, for "bar iron," read puddled iron with 242 kilog. peat.
" 178 " 5, for "116 kilog. pig iron," read puddled iron.
" 225, 366, 494, for "State of Oregon," read Territory.
" 256 " 32, 33, value of exports of iron from the United States, for "875,621," read 1,875,621

2,255,698

" 366, " 33, The price of labour at Pottsville.—Wages have varied very little since 1847 up to this year, when they have increased 30 per cent. Miners received $7 per week; inside labourers, $6, and outside labourers, $5.50 per week, up to 1854. Now miners receive $11 and $12 per week; inside labourers $8, outside $7 per week. Miners now work by the ton, which pays them about $50 per month.—*Pottsville, Sept.* 1854.

* By a London correspondent of the New York Commercial Advertiser, dated July 10th, 1854, we find it stated that the aggregate exports of iron are 1,950,000 tons per annum, (pig iron) which may he taken as two-thirds of the entire production.